OXFORD PAPERBACK REFERENCE

A Dictionary of Chemical Engineering

D1484212

Professor Carl Schaschke is a chemical engineer having worked first at BNFL at Sellafield in Cumbria. After then completing a PhD, his academic career began at Napier University. He is currently professor at the University of Strathclyde. He has had several secondments including to the Fawley oil refinery and BBC TV's *Tomorrow's World*. His teaching and research interests include chemical engineering applications under extreme conditions. He is a Fellow of IChemE and UK representative of the EFCE Working Party on High Pressure Technology. He is married with two daughters, Emily and Rebecca.

( ) SEE WEB LINKS

Many entries in this dictionary have recommended web links. When you see the above symbol at the end of an entry go to the dictionary's web page at www.oup.com/uk/reference/resources/chemeng, click on **Web links** in the Resources section and locate the entry in the alphabetical list, then click straight through to the relevant websites.

Oxford Paperback Reference

The most authoritative and up-to-date reference books for both students and the general reader.

Many of these titles are also available online at www.Oxfordreference.com

A Dictionary of
Chemical
Engineering

CARL SCHASCHKE

OXFORD
UNIVERSITY PRESS

OXFORD
UNIVERSITY PRESS

Great Clarendon Street, Oxford, OX2 6DP,
United Kingdom

Oxford University Press is a department of the University of Oxford.
It furthers the University's objective of excellence in research, scholarship,
and education by publishing worldwide. Oxford is a registered trade mark of
Oxford University Press in the UK and in certain other countries

Published in the United States of America by Oxford University Press
198 Madison Avenue, New York, NY 10016, United States of America

British Library Cataloguing in Publication Data

Data available

ISBN 978–0–19–965145–0

Printed in Great Britain by
Clays Ltd, St Ives plc

Contents

Preface

The purpose of this dictionary is to provide a quick, useful, and comprehensive reference to commonly used and, in some case, less commonly used terms from the field of chemical engineering. As with any dictionary, it is intended to provide definitions to words; it is not merely a brief glossary of terms, nor is it intended to be encyclopedic, with lengthy and overly long explanations. It is aimed at students at school and undergraduate students who will encounter, perhaps for the first time, unfamiliar technical terms. It is also aimed at postgraduates engaged in chemical engineering research as well as practitioners of chemical engineering in industry who may require clarification regarding terms. This dictionary is also aimed at the general reader who in the course of their work or daily lives may encounter unfamiliar terms.

The focus of the dictionary is scientific and engineering terms. It includes core and fundamental terms commonly encountered across all degree programmes of chemical engineering worldwide. It includes many scientific and engineering concepts, laws, theories, and hypotheses. It includes significant organizations, international legislation, and biographical notes of influential scientists and engineers who have contributed to the development of the discipline. There are definitions of many types of specialist process equipment encountered in chemical engineering. This dictionary should therefore enable the reader to distinguish between a lute and a dead leg or a Hortonsphere and a holley-mott. Being a diverse discipline, there is an emphasis on established processes across a wide range of industries spanning nuclear, mineral, oil and gas, food, and pharmaceutical processing. Some older or former processes are also included where their usage was pioneering at the time or influenced later processes. Products, raw materials, and feedstocks are included, though to a far lesser extent; only those upon which major industries are based, such as crude oil, natural gas, minerals, and ores, are included. The full details of chemicals and their properties are included in the sister dictionaries such as the *Oxford Dictionary of Chemistry*.

As a branch of engineering in its own right, the roots of chemical engineering extend back to the nineteenth century. While many of the original and familiar terms are still in use today (such as *unit operations* attributed to Arthur D. Little), chemical engineering in the twenty-first century has expanded considerably and diversified into many new technological fields such as renewable energies, nanotechnology, and biomolecular engineering. Many students and professional engineers alike encounter new terms almost daily with which they may not be familiar or entirely clear. This dictionary therefore aims to provide up-to-date, clear, concise terms and definitions, and other useful and valuable information that can be used as a quick reference source.

The dictionary features over 3,000 of the most commonly encountered terms, although the number actually used by chemical engineers is far greater! There are many cases where words are used uniquely within a particular industry, or within a single industrial organization, and are not be found anywhere else. These have not been included. In providing a definition of each of the included words, the aim has been to be inclusive of all aspects of chemical engineering without being too general. If one starts with the very name *chemical engineering*, there are no doubt as many definitions as chemical engineers! Founding member of the Institution of Chemical Engineers Norman Swindin once described chemical engineering as *engineering without wheels*. An amusing definition but it falls a long way short of being helpful or informative.

The SI system of units has been used throughout although it is recognized that British Imperial and American customary units are still widely used in many industries. Reference has been made to commonly encountered units and conversions presented where appropriate.

In the preparation of this dictionary, I am indebted to many people who have assisted in suggesting words, their comments and corrections. Any errors, omissions, misprints, or obscurities are entirely my own. My thanks to the editorial staff of Oxford University Press and in particular Judith Wilson, Jamie Crowther, and Clare Jones, as well as thanks to the copy-editor, Marilyn Inglis, and the proofreader, Sarah Chatwin, for their attentive and invaluable work. Finally, this book could not have been produced without the support of my wife Melodie and my daughters Emily and Rebecca.

Carl Schaschke

ABE fermentation Another name for the *Weizmann process used for the production of acetone, butanol, and ethanol using the acid-resistant bacterium *Clostridium acetobutylicum.*

ablation The removal of material by *erosion, *evaporation, or *chemical reaction. For short-term protection against high temperatures as a form of fire protection or fireproofing of process equipment, sacrificial materials are used such that during a fire there is resistance and protection to the equipment beneath for a sufficient period of time.

ablimaton *See* SUBLIMATION.

abscissa The horizontal or x-coordinate in a two-dimensional Cartesian coordinate system such as a chart or graph. The *ordinate is the vertical or y-coordinate.

absolute Denoting a number or a measurement that does not rely on a standard reference value.

absolute density The mass per unit volume of a substance. It is the density of the actual substance and does not include any free space that may be between particles. The SI units are kg m^{-3}.

absolute error The difference between a measured value and its true value.

absolute filter A type of filter used to remove all particles that may be present in the flow of gas into or out of a process. Absolute filters are used for ensuring the sterile flow of air or oxygen to biological reactors as well as for clean rooms and sterile cabinets used for analytical work. Unlike an *air filter, the pore sizes are smaller than the expected particle size. With a typical uniform pore size of 0.2 μm, the pressure drop is greater than that of air filters made from fibrous materials.

absolute humidity The amount of water in air expressed as the mass of water vapour per unit mass of dry air for a particular temperature and pressure condition. The SI units are $kg_{water} \, kg_{air}^{-1}$.

absolute pressure The measurement of gas or air pressure relative to the pressure in a total vacuum. In comparison, the *gauge pressure is measured above atmospheric pressure, which is variable.

absolute roughness (Symbol ε) The roughness of a solid surface expressed as the average height of undulations and imperfections. It is measured using an instrument that draws a stylus over the surface. The roughness of the inner surface of a pipe wall used to transport fluids with turbulent flow has the effect of increasing frictional pressure drop. Expressed as a ratio with internal pipe diameter, it is used in determining the friction factor of fluids flowing in pipes with turbulent flow. *See* RELATIVE ROUGHNESS.

absolute temperature *See* KELVIN.

absolute viscosity *See* VISCOSITY.

absolute zero The lowest possible thermal energy state of a material. This corresponds to 0 K.

absorbed dose *See* DOSE.

absorber 1. A material that is capable of stopping ionizing radiation. *Alpha particles can be readily stopped by a sheet of paper whereas beta radiation can be resisted by a centimetre of aluminium. Gamma radiation is absorbed by materials with a high density, such as steel and concrete. Neutron absorbers include boron, hafnium, and cadmium and are used in the control rods in nuclear reactors. **2.** A shortened name for an *absorption tower or column.

absorption A mass transfer process in which one or more gases in a gaseous mixture is transferred into a liquid solvent or a solid. It is the most common form of separation of low molecular weight materials. Absorption is often used to remove gases from gas streams that may be harmful downstream or when released from the process. The **absorption factor** is used to determine the ease with which a component will absorb into the liquid phase and is based on liquid and vapour flow rates as well as the vapour liquid equilibrium for the component. For example, ammonia can be absorbed from a gas stream using water as the scrubbing liquid. *Compare* ADSORPTION.

absorption tower A tall vertical column containing a packing material in which a gas is absorbed by intimate contact with a liquid flowing downwards under the influence of gravity. The gas can be admitted either countercurrent or cocurrent to the flow of liquid in which one or more of the gaseous components are absorbed into the liquid. The minimum flow rate of scrubbing liquid required to achieve an absorption duty requires an infinite height of packing. In practice, a higher liquid rate is used to achieve a compromise between capital cost (i.e. height of column) and the operating cost (i.e. liquid flow rate). It is also known as a *scrubber.

absorptivity The portion of radiant thermal energy falling on a surface which is converted to heat with the remainder being either reflected or transmitted. The absorptivity is dependent on the wavelength of the energy and the properties of the surface including colour. *Compare* REFLECTIVITY; TRANSMISSIVITY.

accelerant A substance used to initiate and develop a fire. Flammable liquids are the most common form of accelerants.

acceleration (Symbol a) The rate of change of speed or velocity with respect to time. If the acceleration is constant then the final velocity, v, of a body that is initially moving with a velocity u after time t, is $v = u + at$. If the acceleration is not constant, then the acceleration can be found from:

$$a = \frac{dv}{dt} = \frac{d^2 s}{dt^2}$$

where s is the distance moved by the body. In the case of motion in a circle, the acceleration is v^2 / r and directed to the centre of the circle of radius r.

acceleration due to gravity (Symbol g) The acceleration experienced by a body due to the Earth's gravitational field. The acceleration is normally taken as $9.806\ 65\ m\ s^{-2}$ although it does vary by small amounts over the Earth's surface and with altitude.

acceleration phase The rapid growth of the culture of microorganisms in a bioreactor prior to the *log phase. After the medium within a bioreactor has been inoculated with a small population of microorganisms, there is an initial *lag phase of no growth in which they adjust to their new environment. Cell division then occurs at an increasing rate until the maximum growth rate is reached. The log or exponential phase corresponds to the rapid cell division such that the logarithm of the population increase with time is constant. As the substrate eventually becomes exhausted, this is then followed by a deceleration phase prior to the *stationary phase.

accelerator A substance that alters the rate of a chemical reaction such as a *catalyst.

accumulator A device used to smooth the rate of flow from a reciprocating pump and prevent the destructive effects of *water hammer from occurring. It consists of a vessel located on the pipe close to the pump with a *non-return valve preventing return flow back to the pump. The vessel contains a gas or a bladder bag although some use springs. As the pump discharges, some of the fluid enters the accumulator compressing the gas or spring. At the point of valve closure, the gas or spring expands allowing the accumulated volume to discharge through the pipe.

accuracy A measure of the closeness or agreement of a numerical value to a true value. It is expressed as either *significant figures or decimal places depending on whether proportional or absolute accuracy is important. For example, a number written as 5.425 normally assumes that the four figures are meaningful. It would be incorrect to write the number to a precision of five significant figures unless the *error in the estimate is indicated such as 5.4250 ± 0.0005. *Compare* PRECISION.

acentric factor A parameter used in *equations of state to estimate physical and thermodynamic properties. It is used to characterize the acentricity of molecules in reduced-state correlations along with reduced pressure and reduced temperature.

acetate process A process for the production of cellulose fibres used for textiles. There are two methods: **1.** The cellulose is obtained from wood pulp and dissolved in carbon disulphide and sodium hydroxide. The thick brown liquid that contains cellulose xanthate is forced through orifices into acid. The xanthate decomposes to leave a cellulose fibre known as viscose rayon. **2.** The cellulose obtained from wood pulp and cellulose acetate is formed by dissolving in acetone. The solution is forced through orifices and the solvent is allowed to evaporate leaving a cellulose fibre of acetate rayon.

ACHEMA (Ausstellungstagung für chemisches Apparatewesen) A triennial trade fair for chemical technology and biotechnology held in Frankfurt, Germany.

• Official website of ACHEMA.

Acheson process A process used for the production of graphite. It involves heating coke mixed with clay to a very high temperature. At a temperature in excess of 4,000°C, silicon carbide is formed leaving graphite. It is named after the American inventor Edward Goodrich Acheson (1856–1931) who patented the process in 1896.

acid A chemical compound or material containing hydrogen that has the tendency to lose protons and form hydrogen ions in solution. Solutions of acids have *pH values less than 7.

acid egg An egg-shaped vessel used to transport highly corrosive acids. The container has inlet and outlet pipes and is filled with a charge of liquid to be transported. Another pipe is used to admit compressed air or another gas. The pressure of the gas on the liquid surface forces the liquid through the discharge pipe that extends down into the liquid. The acid egg is not very efficient as the compressed air or gas is usually blown off when the operation is completed. *See* MONTEJUS.

acid gas Natural gas, which consists mainly of methane, but also contains significant amounts of carbon dioxide, hydrogen sulphide, and other acidic contaminants. Natural gas from offshore reservoirs that contain these corrosive and toxic contaminants are required to be removed or reduced at the platform before export using an *amine gas treating process. *Compare* SOUR GAS.

acid number A measure of the acidity of oils such as crude oil, mineral oils, and biodiesels. It is expressed as the mass in milligrams of potassium hydroxide titrated in one gram of the oil required to neutralize it.

acid rain A precipitation of rain that has a pH below that of typical rain, which is around pH 5.6. Rainwater is naturally acidic due to the absorption of carbon dioxide from the air to form carbonic acid. However, rainwater will also absorb other gases such as sulphur dioxide and various oxides of nitrogen that have been released into the atmosphere as pollutant gases through processes such as the combustion of fossil fuels and from car exhausts. The dissolved gases form sulphuric and nitric acids with pH values of less than 5.0 and have an adverse effect on trees and plants. Acid rain causes damage to leaves and increases the acidity of the soil preventing further growth. The water run-off into rivers and lakes also prevents freshwater fish from thriving, leaving the water sterile, and has a major impact on the ecosystem.

activated carbon A compound of powdered or granular amorphous carbon mainly made from coconut shells. It has a very high specific surface area used to adsorb vapours and gases. With a surface area typically of around 1,000 m^2 per gram, it is widely used to adsorb vapours and gases. The amount of substance that can be adsorbed is proportional to the absolute temperature and pressure. The activated carbon can be reactivated for reuse using steam to strip the adsorbents and recover the carbon. Activated carbon is used in water and air purification, and used in gas masks for the removal of harmful gases. It is also known as **activated charcoal** and **active carbon**.

activated sludge process A process used in the treatment of sewage and wastewater. *Sludge is formed when air is bubbled through the sewage resulting in the aggregation of flocs. These contain denitrifying bacteria that are capable of decomposing organic substances. Aeration ensures a high level of dissolved oxygen and helps to reduce the *biological oxygen demand. Stirring of the sludge can also aid the process.

activation energy (Symbol E_a) The minimum energy required to activate one mole of a substance to cause a *chemical reaction to take place. For a chemical reaction to proceed,

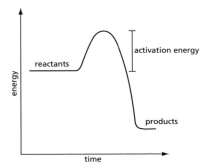

Fig. 1

the reactants are converted to products in which the energy increases to a maximum and falls to the energy of the products (see Fig. 1). The activation energy is the difference between the maximum energy and the energy of the reactants and is therefore the energy than needs to be overcome. *See* ARRHENIUS EQUATION.

active site 1. An available location on the surface of a *catalyst available for reactants to bind and result in a *chemical reaction taking place. The active site can be blocked by a chemical agent or *poison thereby reducing the effectiveness of the catalyst. **2.** On the surface of an enzyme, the active site is the location to which a substrate binds. The binding of the substrate to the enzyme is dependent on the conformation or 3-D shape of the protein. Inhibition prevents the binding from taking place by altering the conformation thereby preventing the substrate from binding, or by blocking the site.

activity 1. The change in one condition to another expressed as a ratio. Examples include *water activity, and chemical activity. The activity of a chemical reaction is used in place of concentration in equilibrium constants for reactions involving non-ideal gases and solutions. **2.** A quantitative term used to characterize the number of atomic nuclei that disintegrate in a radioactive substance per unit time. It is measured in *becquerel (Bq) where one Bq is equal to one disintegration per second. The unit of activity replaces the former unit of the curie (Ci) where one Ci is equal to 37×10^{10} Bq. The specific activity is the activity per unit mass of a pure radioisotope. **3.** (Symbol U) The amount of an enzyme present in a biologically catalyzed reaction. It is usually expressed in terms of units of activity based on the rate of the reaction that the enzyme catalyzes. The international unit of activity is the amount of enzyme that will convert one μmol of substrate to a product in one minute under defined conditions. These are usually 25°C and the optimum pH. **4.** A thermodynamic parameter that measures the so-called active concentration of a substance, a, in a chemical system, and is in contrast to the molecular concentration, c. It is related $a = fc$ where f is a dimensionless parameter and approaches unity in dilute solutions.

activity coefficient (Symbol γ) A correction factor that allows for the deviation from ideal behaviour of a gas or solution.

ADC *See* ANALOGUE-TO-DIGITAL CONVERTER.

additive A substance added in a small amount to another or mixture to improve the performance or properties in some way. Additives are added to polymers to enhance their stabilizing properties. Additives are added to foods as preservatives, and to enhance colour and flavour. Additives can also provide corrosion resistance, alter surface tension, and viscosity etc.

adiabatic A thermodynamic process that takes place without heat transfer to or from an external source. When a fluid is compressed adiabatically, there is an increase in temperature of the fluid. Likewise, **adiabatic cooling** occurs when the pressure of the fluid is reduced without any heat exchange to the surrounding. **Adiabatic expansion** of a fluid occurs without any heat transfer with the surroundings. **Adiabatic compression** is the compression of a gas without any transfer of heat to the surroundings. It results in an increase in the temperature of the gas undergoing compression.

adiabatic efficiency 1. The ratio of the work required for adiabatic compression to the real work input. **2.** The ratio of kinetic energy of a fluid through a valve to the kinetic energy obtained through the process of *adiabatic expansion.

adiabatic flame temperature The theoretical temperature of a flame during the combustion of a fuel in oxygen considered when there is no loss of energy. The temperature is dependent on whether the combustion process occurs at either constant pressure or constant volume. At constant pressure, the adiabatic flame temperature is due to the complete combustion of the fuel with no heat transfer or changes in kinetic or potential energy. Constant volume combustion results in a lower flame temperature since some of the energy is otherwise used as work to change the pressure.

adiabatic flash Another name for *flash evaporation, which involves the rapid isenthalpic evaporation of a saturated liquid into a liquid and vapour by the reduction in pressure.

adiabatic process A physical or chemical process without the loss or gain of heat. The **adiabatic equation** $pV^\gamma = k$ describes the relationship between the pressure of an *ideal gas and its volume where γ is the ratio of the specific heat capacities of the gas and k is a constant.

adiabatic saturation temperature The equilibrium temperature attained when a liquid and gas are brought into contact with no work or heat transfer done.

adjutage A tube inserted into a vessel to obtain a measure of its pressure or to allow the discharge of its contents.

adsorbate A substance that is adsorbed from a gas or liquid onto a solid surface or **adsorbent** during an *adsorption process.

adsorption A process in which components in gases, liquids, or dissolved substances are selectively held on the surface of a solid. It is used to remove components that may otherwise be harmful if released into the environment or may cause process difficulties further downstream such as causing the poisoning of a catalyst. Adsorption usually takes place in *fixed beds.

adsorption isotherm The relationship between the mass of *adsorbate taken up per unit mass of adsorbent at constant pressure, if a gas, or at constant temperature, if in a solution. The *BET, *Langmuir, *Freundlich, and *Temkin adsorption isotherm equations are empirical equations used to describe the surface available for adsorption at constant pressure for gases and constant temperature for solutions.

advection The natural movement of a fluid such as air resulting in horizontal motion caused by local pressure differences. It differs from *convection since it does not include the effects of diffusion.

aeration The introduction and movement of air or oxygen at a low flow rate through a liquid medium such as a *bioreactor or *activated sludge process. Aeration is used to provide oxygen to microorganisms that are responsible for biologically catalyzed reactions. The oxygen is usually introduced through a *sparger as small bubbles that have a high surface area. Aeration is used to promote effective mass transfer of the oxygen to the liquid medium and therefore microorganisms.

aeration number A dimensionless number, N_a, used in the aeration-mixing of bioreactors and relates the gas flow rate, G, to the impeller speed, N, and diameter, D, as:

$$N_a = \frac{G}{ND^3}$$

aerobic process A biochemical process involving microorganisms that require the presence of oxygen, usually in the form of air. Many organisms require the presence of oxygen to survive and grow, such as plants, animals, and many microorganisms. They are dependent on oxygen for the breakdown of sugars into carbon dioxide and water, and for the release of energy through aerobic respiration. In comparison, anaerobic respiration releases energy in the absence of oxygen.

aerogel A highly porous material based on metal oxides or silica. It has a very low density below 10 kg m^{-3} and has excellent heat and electrical resistance as well as acoustic properties. Aerogels can be formed using the process of *supercritical drying using carbon dioxide to remove a solvent such as ethanol used in their formation. Being supercritical and without a gas–liquid interface, it avoids the crushing effects of capillary forces on the porous structure during a conventional drying process.

aerosol A dispersion of fine droplets of liquid or particles of solid within a gas such as air. The particles are often very small and colloidal in size. An aerosol spray can contains propellants that are liquefied under pressure and used to create an aerosol when released into the air.

agglomeration The process of bringing a suspension of small or fine particles together to form larger and more coarse particles or aggregates.

aggregated fluidization *See* FLUIDIZATION.

aggregation The formation of large groups of molecules or particles. With particles, aggregation consists of both *flocculation and *coagulation.

agitated vessel A vessel in which the contents are stirred by mechanical means through the use of an agitator, paddle, or stirrer. Impellers and propellers are commonly used to provide good mixing characteristics. It is also known as a **stirred tank**.

agitation intensity A measure of the power consumption of the shaft of an agitator used to mix a liquid in a stirred tank or *agitated vessel. Agitation intensities are expressed as the power supplied per unit volume of liquid. The SI units are W m^{-3}. The magnitude of the agitation intensity is dependent on the nature of the liquid being stirred. Biological solutions containing flocculating materials are significantly affected by the level of agitation.

agitator A simple stirring device used to provide turbulence and mixing of the contents of a vessel containing a liquid. It is typically used to provide homogeneity, provide good oxygen transfer in fermentation vessels, and in the prevention of particles settling.

An agitator consists of blades attached to a rotating shaft. Impellers have flat blades and provide radial flow patterns whereas propellers provide axial flow movement. Paddle agitators consist of tilted flat blades providing a combination of radial and axial flow movement. Selection of the appropriate agitator depends on the processing requirements, the fluid properties, and the materials of construction.

AIChE *See* American Institute of Chemical Engineers.

air An odourless and colourless mixture of gases and vapours that surround the Earth. At sea level, the composition of dry air is mainly nitrogen (78.09 %) and oxygen (20.95 %), with an average relative molecular weight of 29. Other gases include argon (0.93 %), carbon dioxide (0.03 %), neon (1.8×10^{-3} %), helium (5.2×10^{-4} %), and lesser amounts of methane, krypton, hydrogen, nitrous oxide, xenon, and radon in decreasing amounts, respectively. Air is a common source of oxygen used in many processes such as *combustion.

air conditioning The process of controlling the environmental air conditions in buildings through control of the temperature and level of relative humidity, as well as through filtration of particles to provide human comfort. The movement and cleanliness of the air are also involved.

air filter A type of filter used to remove particles such as dust, soot, and microorganisms from the flow of air. They are often used for ensuring a sterile flow of air or oxygen to bioreactors as well as for clean rooms and sterile cabinets used for analytical work. The pore sizes of the filter are larger than the particle size to be removed such that the filter relies on the depth of the filter to entrap the particles within a fibrous mesh structure. Fibrous filters are relatively cheap and robust, and have a low pressure drop in comparison with *absolute filters.

air-lift A pumping device used to raise a liquid from a depth such as a well. It consists of a vertical pipe extending down into the well into which compressed air is injected at the bottom. As the air bubbles rise, the reduced hydrostatic pressure results in a flow of liquid up the leg. The air or gas is disengaged from the liquid at the top of the leg. It is used for raising oil from wells.

air-lift reactor A type of bubble column reactor into which air is sparged at the bottom as bubbles to promote oxygen transfer and cause circulation of the liquid. The reactor is cylindrical and mounted on its axis. It has an inner tube up which the air or oxygen rises. An external-loop air-lift-type reactor consists of a U-tube within which the sparging takes place promoting oxygen transfer and liquid circulation.

air lock 1. Trapped air or some other gas or vapour within a pipe that prevents the intentional flow of a liquid. **2.** The intentional seal in a process that relies on a differential pressure to prevent the undesirable loss of material.

air pollution The release of particles, vapours, and gases into the environment that are harmful to human health and to the environment such as plants, forests, and animals. Carbon dioxide is a product from the combustion of fossil fuels in power stations, vehicles, aeroplanes, and numerous industrial processes, and is a greenhouse gas responsible for contributing to the warming of the Earth's atmosphere. Methane is another greenhouse gas as are chlorofluorocarbons (CFCs), which were once widely used as refrigerants and as aerosol propellants but are now banned due to their known damaging effect on the Earth's ozone layer. Sulphur dioxide is another product of the combustion of fossil fuels and is known as the cause of *acid rain.

In the UK, an Act of Parliament was introduced in 1956 to reduce the level of air pollution. It was a landmark in environmental protection and was responsible for reducing the level of smoke pollution as well as sulphur dioxide emitted into the environment.

In the US, the Clean Air Act introduced in 1963, together with its subsequent amendments as a federal law, has been responsible for controlling air pollution. Other governments have also taken measures to control air pollution and limit the emission of carbon dioxide and other greenhouse gases. The Kyoto Protocol is an international agreement between countries to reduce the emissions of carbon dioxide emissions and restrict or ban the emission of certain chemicals such as CFCs. One way of restricting carbon dioxide emissions is to raise the level of taxation on fuels so that people and industrial companies have greater incentives to conserve energy and pollute less.

(⊕) SEE WEB LINKS
• Official website of Environmental Protection UK.

air separator A device used to separate solid or liquid particles from air in which centrifugal force is used. The device has a cylindrical body with a conical base. The particle-containing air enters tangentially and the particles leave from the bottom while particle-free air leaves from the top. It is also known as a *cyclone separator.

air-to-close A type of pneumatically operated control valve that automatically opens in the event of a loss of instrument air pressure. An **air-to-open** valve is a pneumatically operated control valve that automatically closes in the event of a loss of instrument air pressure. For example, the fuel supply to a furnace should automatically shut on air failure.

ALARA An abbreviation for **as low as r**easonably **a**chievable, it is a management tool used in the controlling of risks. For example, it is used to manage the exposure to chemicals and ionizing radiation doses in humans working in the nuclear industry. *Compare* ALARP.

alarm An indicator used to alert operators and personnel that there has been a significant deviation from an expected measured *process variable or process condition. The alarm may be audible in the form of a siren, bell, or other noise, or may be a flashing or continuous light signal. Alarms are a feature of control panels where the process is displayed on screens with associated alarms. **Alarm flooding** is a condition in which alarms appear on control panels in *control rooms at a rate which exceeds that which an operator can comprehend or respond to quickly or effectively. It therefore prevents the operator from identifying the cause of the process upset and consequently limits the scope for an effective response.

ALARP An abbreviation for **as low as r**easonably **p**racticable, it is a management tool used to determine the level to which risks are to be assessed and controlled. It involves a rigorous and systematic assessment of the minimization of risk and the costs in terms of time, money, and effort to achieve it. As a form of good practice requiring judgement between risk and societal benefit, it was developed through the UK parliamentary Health and Safety at Work Act (1974). Outside the UK, similar forms of engineering practice are used and this includes *ALARA (as low as reasonably achievable) in the US for radiation protection.

(⊕) SEE WEB LINKS
• Official website of the Health and Safety Executive UK offering risk assessment advice.

algorithm 10

algorithm A mathematical method or operation that follows a scheme of calculations or steps designed to be repeated such that the result from one calculation forms the basis of the next. The stage-by-stage computation of the liquid and vapour flows and compositions in a distillation process is based on a defined algorithm.

aliquot A portion of a total amount of something. For example, a prepared solution of reactants may be fed to a process in aliquots.

alkali A metal hydroxide that produces hydroxyl (OH-) ions in solution.

alkane A saturated aliphatic hydrocarbon that has the general formula C_nH_{2n+2}. Forming a homologous series, the smallest is methane (CH_4), followed by ethane (C_2H_6), propane (C_3H_8), butane (C_4H_{10}), etc. The smaller alkanes are gases at ambient temperature. Methane is found in oil and gas reservoirs, and in lesser amounts in coal seams. Methane is also the product of the decay of organic material by bacteria and produced from *anaerobic digestion processes. Mixtures of short-chain alkane gases can be separated by distillation, either by condensing the liquids using low temperature or by pressurization, or a combination of the two. They were formerly known as **paraffins** although this name is still used in certain industries such as petroleum refining.

alkene An unsaturated aliphatic hydrocarbon that has the general formula C_nH_{2n}. They comprise one or more carbon–carbon double bonds. They were formerly known as **olefins**. The series starts with ethene (ethylene) with the formula C_2H_4, followed by propene (propylene) C_3H_6, butene (butylene) C_4H_8, etc. Isomerism occurs with the higher alkenes beginning with butane for which there are two isomers: but-1-ene and but-2-ene that differ by the position of the double bond. Alkenes can undergo *polymerization to form thermoplastics such as polyethene (polyethylene).

alkylation A process in which an alkyl group is added to another organic molecule such as by removing a hydrogen atom from an *alkane and adding a methyl group. In the refining of *crude oil, it is used to upgrade petroleum through the alkylation of isobutane with *alkenes (olefins) such as propene, in the presence of either sulphuric or hydrofluoric acid as a *catalyst. The reaction takes place as a two-phase reaction at ambient temperature. The reaction products are a mixture of branched hydrocarbons with a high *octane rating. The octane number of the mixture depends mainly on the kind of alkenes used. Iso-octane has an octane rating of 100 and is the result of reacting isobutane with butene (butylene).

alkyne An unsaturated aliphatic hydrocarbon that has the general formula C_nH_{2n-2}. These are characterized by a triple carbon–carbon bond. Alkynes that feature a single triple bond form a homologous series beginning with ethyne (acetylene) C_2H_2, followed by propyne (propylene) C_3H_5, butyne (butylene) C_4H_6, etc. They were formerly known as acetylenes.

allotropy The existence of different forms of the same element in the same phase, known as **allotropes**. Carbon has the allotropes of graphite, diamond, graphene, and fullerenes. Many other elements exhibit allotropy.

alloy A material that consists of two or more metals, or a metal and a non-metal such as carbon. Pewter is made from tin with lesser amounts of lead along with small amounts of antimony and copper. Steel is made from iron alloys and contains a small amount of carbon.

alpha particle A positively charged particle emitted by various radioactive materials such as uranium during radioactive decay. The particle consists of two neutrons and two protons, and is therefore identical to the nucleus of a helium atom. The result of this *radioactive decay is that the original element is gradually converted into another element with a decreased atomic number and mass. Alpha particle emissions, or **alpha decay** may occur at the same time as *beta decay. It can be stopped by a sheet of paper and is harmful to humans only if the substance emitting the alpha particles is ingested, inhaled, or enters the body through wounds.

alternator Another name for an electromagnetic *generator used to produce alternating current in a power station.

Amagat's law A law that states that for an ideal gas, the total volume occupied by a gaseous mixture is equal to the sum of the pure component volumes:

$$V = V_A + V_B + V_C + \ldots$$

It is named after French physicist Emile Hilaire Amagat (1841–1915).

amalgam 1. An alloy of mercury with another metal, such as silver, used in dentistry. Most metals form an amalgam with mercury with the exception of iron and platinum. **2.** A white mineral consisting of mercury and silver that occurs in deposits of silver and cinnabar, which is a bright-red mineral form of mercuric chloride found near areas of volcanic activity and hot springs.

ambient temperature The temperature of the surrounding atmospheric air. Ambient air temperature can affect the operation of process equipment, instruments, and control. It is sometimes referred to as room temperature.

American Institute of Chemical Engineers (AIChE) A professional society based in the US with a membership of over 43,000 chemical engineers in a hundred countries. Founded in 1908, AIChE was established to provide its members with a focal point to share ideas and grow the discipline. Today it provides its members with technical resources and organizes major conferences, as well as setting accreditation standards for chemical engineering education, and setting guidelines for government agencies.

((⊕)) SEE WEB LINKS
• Official website of the American Institute of Chemical Engineers.

American National Standards Institute (ANSI) An American not-for-profit organization responsible for the accreditation of organizations that write industrial standards. It was founded in 1918.

((⊕)) SEE WEB LINKS
• Official website of the American National Standards Institute.

American Petroleum Institute (API) An American professional trade organization that represents all aspects of the US oil and natural gas industry. It was formed after the First World War (1914–18) as a consortium of oil and gas companies to help the recovery from the war by working together. It was formally established in 1919 as a means of co-operation with the government in all matters of national concern and to develop the wider interests of the petroleum industry. It sets standards and recommends practices, covering all aspects of the industry, and promoting the use of safe, proven, and sound engineering practices.

((⊕)) SEE WEB LINKS
• Official website of the American Petroleum Institute.

American Society for Testing Materials (ASTM International) An international standards organization that develops and publishes voluntary technical standards for materials, products, systems, and services. Unlike ANSI, it is not a national standards body. It has been responsible for developing and maintaining more than 12,000 standards and the *Annual Book of ASTM Standards* consists of 77 volumes.

((⊕)) SEE WEB LINKS
• Official website of ASTM International.

American Society of Mechanical Engineers (ASME) A professional organization based in the US that provides its members with technical resources focusing on technical, educational, and research matters. It also produces standards such as ASME VIII, which is an accepted code for the design of pressure vessels and heat exchangers covering design, material selection, fabrication, inspection, and testing.

((⊕)) SEE WEB LINKS
• Official website of the American Society of Mechanical Engineers.

AMIChemE Post-nominal letters used after a person's name to indicate that they are an Associate Member of the *Institution of Chemical Engineers.

amine gas treating process A post-combustion process used to remove *acid gases such as carbon dioxide, hydrogen sulphide, and mercaptans from natural gas using an amine chemical solvent to react and form reversible compounds. Carbon dioxide is required to be removed since it reduces the calorific value of natural gas and forms carbonic acid in water which is corrosive as well as having a *global warming potential. The process involves the reversible reaction of the gas with an amine to form an amine salt. Various amines are used including monoethanolamine. The amine solution is sprayed into a large tower and absorbs the hydrogen sulphide as well as carbon dioxide from upflowing gases. A regenerator operating at a higher temperature is used to strip the amine solution of the gases for reuse. *See* GAS SWEETENING.

ammonia-soda process *See* SOLVAY PROCESS.

amorphous A non-crystalline solid form of matter in which the atoms or molecules are arranged at random within a three-dimensional structure. Glass is an example of an amorphous solid. *Compare* CRYSTAL.

amount of substance (Symbol n) A measure of the number of entities present in a substance, such as atoms, molecules, ions, and electrons, etc., expressed in moles. For example, the amount of an element is proportional to the number of atoms present where one mole of that element is equal to $6.022\,1367 \times 10^{23}$ atoms, which is *Avogadro's constant. It is given by:

$$n = \frac{N}{N_A}$$

where N is the number of atoms and N_A is Avogadro's constant. The SI unit is the mole. It is also known as **chemical amount**.

ampere (Symbol A) The SI unit of electric current, it is the constant flow of current that is maintained between two parallel conductors of infinite length and of negligible cross section that produces a force of 2×10^{-7} newtons per metre (Nm^{-1}) between them. It is named after the French physicist and mathematician André Ampère (1775–1836), who made significant discoveries in electricity and magnetism.

ampere-hour A practical unit of electric charge as the quantity that flows in one hour through a conductor carrying a current of one ampere. It is equivalent to 3,600 coulombs.

amplitude The maximum value of varying quantity from its mean or base value. For example, in simple harmonic motion the amplitude of a wave is half the maximum peak-to-peak value.

a.m.u. *See* ATOMIC MASS UNIT.

anaerobic digester A type of bioreactor used for the *anaerobic digestion of organic waste liquids from domestic and industrial sources. The biological process involves the use of bacteria in the near absence of oxygen to produce a mixture of methane and carbon dioxide, known as *biogas. *Continuous stirred-tank reactors are used for the treatment of industrial waste with a continuous inflow and outflow. *Batch processes are used for smaller domestic, community, or farm-scale processes.

anaerobic digestion A biochemical process in which bacteria break down organic matter in the absence of oxygen into a mixture of carbon dioxide and methane known as *biogas. The main stages involve hydrolysis, acidogenesis, acetogenesis, and methanogenesis. An *anaerobic digester can be operated at a steady-state condition through control of temperature for psycrophilic, mesophilic, and thermophilic bacteria, pH, the carbon-to-nitrogen ratio, organic dry matter content, hydraulic retention time, degree of mixing, the availability of nutrients and trace elements, and rate of biogas removal.

analar reagent A high-purity chemical reagent used for chemical analyses with a defined level of purity.

analogue signal An electrical signal used in the control of processes as a current or a voltage representing temperature, pressure, level, etc. The commonly used electrical current signal has a range of 4–20 mA. The voltage range commonly used is 0–5 volts DC.

analogue-to-digital converter (ADC) Electronic hardware used in the control of processes that converts analogue signals such as electrical voltage, current, temperature, and pressure into digital data that a computer can process.

analogy A form of general agreement or similarity between problems, reasoning, methods, or logic. It is used to compare the results from one particular problem to those of another from a known similarity between them.

analysis The detailed examination of something such as a mathematical problem using the theories of calculus, a chemical substance into its constituent parts, the study of a physical process and its function or operation, the economics of a chemical process or business, etc.

analysis of variance (ANOVA) One of a number of statistical techniques used to resolve and observe the variance between sets of statistical data into components. These techniques are used to determine whether the difference between samples is explicable as random sampling variation from within the same statistical populations. ANOVA techniques are used in *quality control.

analyte A substance that is being determined in an analytical procedure.

analytical reagent A chemical compound of a known and high purity used in a chemical *analysis.

ancillary equipment Mechanical equipment used to support or assist a primary item of equipment in meeting its functional duties. Pumps, blowers, and heating equipment are all ancillary items of equipment used to support main process plant items.

Andrews, Thomas (1813–85) An Irish scientist noted for his work on gases. He studied chemistry at the University of Glasgow before undertaking further studies in Paris. He then attended Trinity College, Dublin before completing his medical studies in Edinburgh and then returning to Belfast to set up practice as a physician. When Queen's College opened in 1845, he was appointed professor of chemistry, and also the first vice president of the college. During this time, he carried out his most important studies on gases. His three main areas of work concerned thermochemistry, the nature of ozone, and the continuity of liquid and gaseous states of matter. He was offered a knighthood but declined on the grounds of ill health.

Andrussov, Leonid (1896–1988) A chemical engineer born in Riga who is noted for developing a process for the production of hydrogen cyanide based on the oxidation of ammonia and methane over a platinum catalyst.

Andrussov process A catalytic process used for the production of hydrogen cyanide by the reaction of ammonia, methane, and air at a temperature of around 1,000°C using a platinum catalyst:

$$2CH_4 + 3O_2 + 2NH_3 \rightarrow 2HCN + 6H_2O$$

The ammonia in the product gases is removed by gas absorption with sulphuric acid and the hydrogen cyanide is absorbed in water. The hydrogen cyanide is used as the preliminary product for the synthesis of polyamide 66 or nylon, and for polymethyl methacrylate. It is also called **Andrussov oxidation** after the inventor who patented the process in 1930s.

anemometer An instrument used to measure the speed of a gas such as air. It comprises cups or vanes that rotate freely and are linked to a tachometer. **Hot-wire anemometers** feature a heated wire over which the gas or air passes. Since the electrical resistance of certain metals such as tungsten is dependent on temperature, the cooling effect of the gas over the wire changes its resistance from which the velocity is inferred.

aneroid An instrument used to measure barometric or atmospheric pressure. It has metal bellows as a sensing device.

angel's share An amount of Scotch whisky lost by evaporation during the process of maturation in wooden casks. Scotch *whisky is stored for a minimum of three years over which time the level of whisky can drop by as much 2 per cent per year.

angström (Symbol Å) A unit of length equal to 10^{-10} m. It is used to measure the wavelengths of electromagnetic radiations and was formerly used for the measure of intermolecular distances. It has now been replaced by the nanometre (10^{-9} m). It is named after the Swedish astronomer and professor of physics Anders Jonas Angström (1814–74).

angular momentum (moment of momentum) A measure of the momentum of a body caused by its circular motion around an axis of rotation. It is the vector product of the position vector and the tangential component of velocity of an object moving about a centre of rotation. The angular momentum of a mass m of fluid is $mv_\theta r$ where v_θ is the tangential velocity.

angular velocity (Symbol ω) The rate of change of angular displacement with time:

$$\omega = \frac{d\theta}{dt}$$

The rotational speed of shafts for mixers, centrifugal separators, and centrifugal pump impellers are sometimes expressed in radians per second.

annealing A heat treatment process used to relieve internal stresses in ferrous and non-ferrous metals. It involves heating the metal to a specified temperature over a specified period of time to soften it. It is then allowed to cool slowly. The annealed metal is less brittle with reduced internal stress and is therefore easier to work or machine. A similar process is applied to glass.

annular flow A two-phase flow regime of a gas and a liquid in a vertical pipe or tube characterized by a continuous gas core with a wall film of liquid. The flow regime occurs at high gas velocities compared with the liquid. There is often a simultaneous flow of the liquid phase entrained in the gas as a fine dispersion of droplets. In horizontal pipes, the effect of gravity causes the film to become thicker on the bottom of the pipe. As the gas velocity is increased, the film becomes more uniform around the circumference. *See* MULTIPHASE.

annulus The region between two concentric circles. The area of an annulus is equal to $\pi(d_1^2 - d_2^2)/4$ where d_1 and d_2 are the outer and inner radii. An **annular gap** is the clearance between two concentric pipes or tubes. The use of concentric pipes or tubes is found in the nuclear industry as a form of double containment. The central pipe is used to carry a radioactive liquid such as plutonium nitrate and the gap in the annular gap is kept under reduced pressure. In the event of leaks, the radioactive liquid is retained within the annular gap and recovered without release into the environment.

anode A positive electrode in an electrolytic cell. In the process of electrolysis in which electricity is passed through an electrolyte, the electrode attracts electrons from an external circuit. *Compare* CATHODE.

anodize An electrolytic process used to provide a hard, smooth, and corrosion-resistant surface to aluminium and some other metals. The piece for coating is connected to the anode of a DC circuit and is immersed in an acid solution. The flow of current liberates oxygen at the surface which reacts with the aluminium to form aluminium oxide. Chromic, oxalic, and sulphuric acids are commonly used. The anodized surface may typically have a thickness of between 0.005 mm and 0.018 mm.

ANOVA *See* ANALYSIS OF VARIANCE.

anoxic reactor A type of anaerobic bioreactor in which oxygen is excluded from the cultured bacteria. **Anoxia** is the absence of molecular oxygen in living tissue cells used to indicate the reduction of the oxygen content of the blood below physiological levels.

antilogarithm (antilog) The inverse function of a *logarithm. That is, a number whose logarithm to a given base is a given number. For example, the antilogarithm of 2 to the base 10 is 100. In natural logarithms, the antilogarithm of x is e^x.

antithixotropic fluids Shear thickening fluids that thicken with time. The viscosity of such fluids increases when a shear stress is applied, as in stirring, and is also dependent on the time that the shear stress has been applied. *Compare* THIXOTROPIC FLUIDS. *See* RHEOPEXY.

Antoine equation An empirical equation used to determine the vapour pressure of a substance as a function of temperature:

$$\log_{10} p = A - \frac{B}{T+C}$$

where p is the vapour pressure, T is the temperature and A, B, and C are empirically determined constants. The pressure is given in mmHg. It is named after C. Antoine who published the equation in 1888.

Antonov's rule An empirical equation used to describe the surface tension between two liquids in equilibrium being equal to the difference between the surface tension of the two liquids when exposed to air.

APCChE (Asian Pacific Confederation of Chemical Engineering) Founded in 1975, it is a not-for-profit organization that brings together various societies, associations, and institutions of chemical engineering in the Asia Pacific region. This covers the thirteen countries of China, Korea, Japan, New Zealand, Thailand, India, Philippines, Indonesia, Singapore, Australia, Malaysia, Taiwan, and Hong Kong. The American Institute of Chemical Engineers and the Institution of Chemical Engineers are corresponding members.

API gravity A measure of the density of petroleum oils used in the US and related to *specific gravity:

$$^{\circ}API = \frac{141.5}{SG} - 131.5$$

The specific gravity and API gravity refer to the weight per unit volume at 15.6°C (60°F). Most crude oils range between 20 and 45°API.

apparent density The mass per unit volume of a material that includes voids. It is a measure of the bulk of the material. *Compare* SPECIFIC DENSITY.

apparent viscosity (Symbol η) The viscosity of a fluid as a measure of the ratio of the shear stress to shear rate and used for non-Newtonian fluids such as drilling muds.

approximation A mathematical process used to describe roughly the value of a quantity of something that is not exact but is sufficiently close to a known or correct value within acceptable boundaries of error.

aqueous Used to denote solutions in which water is the solvent.

Archimedes of Syracuse (287–212 BC) A Greek mathematician and philosopher credited with the principles of levers, the **Archimedean screw** as a pump, and a method of successive approximations which allowed him to determine the value of π to a good approximation. King Hiero is said to have asked Archimedes to check if a crown was pure gold throughout or contained a cheap alloy. While in a public bath and pondering on how to do this without damage to the crown, Archimedes is supposed to have suddenly thought of the possibility of immersing it in water and checking its density by way of displacement, and to have been so excited that he ran naked through the streets shouting 'Eureka! Eureka! I have found it! I have found it!' He was killed by a soldier in the Roman siege of Syracuse.

Archimedes' principle A principle that states when a body floats it displaces a weight of liquid equal to its own weight. The principle was not stated by Archimedes but is connected to his discoveries in hydrostatics. When a body is partially or totally immersed in a liquid, there is an upthrust on the body equal to the weight of the liquid displaced by the body.

area The extent of a plane figure or surface. The area of a rectangle is the product of the length and base. The area of a circle of diameter d is $\pi d^2 / 4$. The SI unit is m^2.

Argand diagram A graphical way of representing complex numbers in the form $z = x + jy$ in which real and imaginary parts of the complex number are the x and y axes, respectively (see Fig. 2). The modulus is the distance z and the angle of z is the argument. It is named after Swiss mathematician Jean-Robert Argand (1768–1822) and is useful in understanding the stability of controlled processes.

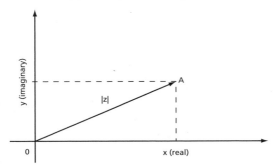

Fig. 2

Arrhenius, Svante August (1859–1927) A Swedish physicist and chemist who did fundamental work on physical chemistry. He worked with *van't Hoff in Amsterdam and proposed a theory of activated molecules and established a connection between rate of reaction and absolute temperature. He also developed a theory for electrolytic dissociation based on van't Hoff's results and stated that any acid, base, or salt dissolved in water is partly split up into positively and negatively charged ions, and that they move in opposite directions on electrolysis. He was awarded a Nobel Prize for Chemistry in 1903.

(⊕) SEE WEB LINKS
- Official website of the Nobel Prize organization, with a transcript of Arrhenius' lecture of 1903.

Arrhenius equation An equation that represents the effect of temperature on the velocity of a chemical reaction expressed as:

$$\frac{d \ln k}{dT} = \frac{E_a}{RT^2} \text{ or } k = Ae^{\frac{-E_a}{RT}}$$

where k is the *rate constant for the reaction, E_a is the *activation energy, R is the gas constant, T is the absolute temperature and A is frequency factor. An Arrhenius plot of $\ln k$ against $1/T$ gives a straight line of slope $-E/R$ and is valid for a large number of chemical reactions. It is named after Swedish chemist and physicist Svante August *Arrhenius (1859–1927).

aseptic A condition in which all contaminating microorganisms are eliminated, not present, or allowed to reproduce. Substances that provide aseptic conditions are known as antiseptics. **Aseptic processing** involves ensuring sterility and therefore a process that is free from microbial contamination. It is used in the packaging of foods, pharmaceuticals, and medical products. Sterility is achieved using a flash-heating process and the product packaged into aseptic containers. The container is required to be robust and provide a tight seal against outside contamination sources. The container and its contents have the benefit of not requiring refrigeration. *Compare* STERILITY.

ash The non-volatile products and residues that remain after a combustion process. *Electrostatic precipitators are used to remove ash particles from flue gas streams.

ASME *See* AMERICAN SOCIETY OF MECHANICAL ENGINEERS.

aspect ratio The ratio of the height to width or diameter of an item of process plant equipment such as a column or storage tank.

asphyxia A state of unconsciousness as the result of anoxia or *hypoxia and increased carbon dioxide in blood and tissue. *See* SUFFOCATION.

assay An analytical procedure used in the laboratory for assessing a sample of something either qualitatively or quantitatively in terms of an amount of a substance, its composition, or some other entity under investigation. Biochemical assays of samples taken from bioreactors that often feature complex mixtures involve procedures for determining cell content, substrate utilization, and product formation. The complexity of the samples that require various prescribed steps to be followed does not permit conventional forms of chemical analysis such as titration.

association The grouping together of atoms or molecules often in the vapour state or in solution to form conglomerates of high molecular weight. *Compare* DISSOCIATION.

assumption A statement that is used in order to simplify a problem, in reaching a solution, or where the full understanding of the problem is not actually known. Assumptions arise in all forms of chemical engineering. For example, in distillation, liquid and vapour are often assumed as being at equilibrium on a theoretical stage; a reaction mechanism may assume no side reactions; the flow from a vessel may assume no vortex formation or a constant discharge coefficient.

ASTM *See* AMERICAN SOCIETY FOR TESTING MATERIALS.

asymptote A straight line that is closely approached by a curve so that the perpendicular distance between them decreases to zero at an infinite distance from the origin.

ATEX An EU directive that describes the work that may be safely carried out in an explosive atmosphere. The areas or zones in a process plant are classified according to the type of hazards, the location, and size, and the likelihood of an explosion. It is applied to mining operations, offshore processing, petrochemical plants, and flour mills, where potentially explosive atmospheres may exist. The name is derived from the French title for the EU directive *Appareils et systèmes de protection pour les atmosphères explosibles.*

((⊕)) SEE WEB LINKS
• Official website of the Health and Safety Executive, UK, outlining information on ATEX and explosive atmospheres.

atmolysis The separation of a mixture of gases by diffusion through a porous membrane such as hollow fibres. Each gas in the mixture has a different rate of diffusion, allowing them to be separated.

atmosphere A layer of gases of largely oxygen (21 per cent) and nitrogen (79 per cent) surrounding the Earth's surface which comprises the troposphere, stratosphere, and ionosphere and traceable to an altitude of around 800 km. The barometric pressure varies with altitude with *standard atmospheric pressure at sea level being taken to be 101,325 Pa or 1,013 mbar.

atom The smallest particle of an element that can exist and which can take part in a chemical reaction and cannot be chemically divided any further into smaller parts. It is identifiable as that element by its nucleus. The nucleus contains neutrons and protons and is surrounded by a cloud of orbiting electrons. The number of electrons equals the number of protons such that the overall charge is zero.

atom balance A material balance based on the number of atoms of specified elements.

atomic bomb A nuclear weapon whose explosive force is due to the energy released through the process of nuclear fission. It involves bringing together a mass of fissile material sufficient to result in a chain reaction that proceeds explosively. Uranium-235 and plutonium-239 are examples of fissile material used in nuclear weapons. The explosive force of nuclear weapons is quoted in kilotonnes or megatonnes of *TNT equivalents. The atomic bombs that were dropped on Hiroshima (uranium-235 bomb) and Nagasaki (plutonium-239 bomb) had the explosive energy equivalent to 13 and 22 kilotonnes of TNT, respectively.

atomic energy *See* NUCLEAR ENERGY.

atomicity The state of being made up of atoms and is the number of atoms in molecules. For example, carbon dioxide (CO_2) has an atomicity of 3; hexane (C_6H_{14}) has an atomicity of 20, etc.

atomic mass The mass of an isotope of an element expressed in *atomic mass units. It is short for *relative atomic mass.

atomic mass unit (a.m.u.) A unit of mass used to express atomic and molecular weights. It is equal to one twelfth of the mass of an atom of carbon-12 and is equivalent to 1.66×10^{-27} kg.

atomic nucleus *See* NUCLEUS.

atomic number The number of protons in an atomic nucleus. The classification of elements is based on the increasing order of atomic number.

atomic pile An early name for a *nuclear reactor that used graphite as the *moderator. *See* WINDSCALE NUCLEAR ACCIDENT.

atomic power An alternative name for *nuclear power.

atomic volume The *relative atomic mass of an element divided by its density.

atomic weight *See* RELATIVE ATOMIC MASS.

atomization The creation of very small droplets of a liquid within a gas. The droplets may range in size from 10 micrometres to 1 millimetre and consequently have a very high surface area, thereby permitting rapid chemical reaction, drying, heat, and mass transfer. Atomization is particularly useful for fuels in combustion processes and for drying or dehydration of liquid products in spray dryers using an *atomizer.

atomizer A device used in the process of atomization to produce very small droplets of a liquid within a gas. Such small droplets can be produced by forcing a liquid through a very small aperture under high pressure or by contacting the liquid with a high-speed rotating plate or disc.

auriferous A rock or ore containing gold.

austenitic stainless steel An alloy of iron that contains at least 8 per cent nickel and 18 per cent chromium. It is noted for its very good corrosion resistance, heat resistance, and creep resistance, and is also non-magnetic. It is used extensively for process pipes and vessels.

autocatalysis A catalyzed chemical reaction in which one of the products is the catalyst for the reaction. The chemical reaction starts slowly as the catalyst is formed and continues rapidly until the point when the reactants are depleted.

autoclave A sealed and heated thick-walled pressure vessel used for the thermal sterilization of biological agents and tinned food products using steam. It is also used for carrying out chemical reactions at elevated temperatures.

autoignition temperature The temperature at which a material ignites in air or some other oxidant at a specified pressure without the aid of a spark or flame. The minimum autoignition temperature is determined by an *ASTM test method. It is also known as the **autonomous ignition temperature**.

automatic control *See* FEEDBACK CONTROL.

autoradiolysis The dissociation of molecules contained within a substance or mixture through *ionizing radiation arising from radioactive materials such as in highly active *nuclear waste.

autothermal A system, process, or reaction that is completely self-sufficient in terms of its energy requirements. Some *exothermic reactions are autothermal. Some *anaerobic digesters are operated in this way in which the methane liberated is used to fuel the process.

average velocity Also known as the *mean velocity, it is the total volumetric flow rate of a fluid per unit flow area. It is a useful parameter particularly where there may be local variations in velocity and hence flow across a flow area due to the effects of turbulence or obstructions in a pipeline, duct, or stack. The SI units are m s^{-1}.

aviation gasoline A hydrocarbon fuel produced in petrochemical refineries. It is used by aircraft with piston engines. It has a high *octane rating and more closely resembles motor gasoline or petrol than diesel fuel. *See* JET FUEL.

Avogadro, Amedeo (1776–1856) An Italian chemist and physicist who provided Avogadro's law as a way of calculating molecular weights from vapour densities. He was educated and graduated in ecclesiastical law; however, he had a keen interest in the natural sciences and received private tuition in physics and mathematics. He published his hypothesis, known now as *Avogadro's law, while working as a schoolteacher. He was appointed to the first chair in mathematical physics at Turin University in 1820. The importance of his work was first recognized by the Italian chemist Stanislao Cannizzarro (1826–1910) in 1858, shortly after Avogadro's death.

Avogadro's constant (Symbol N_A) The number of atoms in one mole of a substance. It has the value of 6.022 1367(36) × 10^{23} and was formerly known as **Avogadro's number**.

Avogadro's law A law that states that equal volumes of gases at the same temperature and pressure contain the same number of molecules. This was first stated as a hypothesis by the Italian chemist and physicist Amedeo *Avogadro (1776–1856) in 1811. However, this law was not generally accepted until after his death when the Italian chemist Stanislao Cannizzaro was able to explain why there were some exceptions to the hypothesis.

axenic culture A microbial culture in a biological process that involves only one species of microorganism.

axial In the direction of the axis of a pipe, tube, cylinder, or a rotating shaft. A propeller provides a flow of fluid in the direction of the shaft whereas an impeller provides *radial flow of fluid in the direction of the radius.

axial compressor A mechanical device used to move air or a gas at high pressure. The gas to be compressed is drawn through alternate rows of radially mounted rotating and fixed aerofoil blades in which the kinetic energy is converted to pressure energy.

axial dispersion model A mathematic model used in the design of tubular *plug flow reactors. The model is based on the *axial mass transport of material corresponding to an effective or apparent longitudinal diffusivity but with a constant *radial concentration.

axial-flow fan A power-driven mechanical device used to move air or a gas. It consists of a rotating shaft with blades or a propeller in which the flow of air or gas is parallel to the axis of the shaft. It operates with low static pressure and high air flow. *Compare* RADIAL-FLOW FAN.

axis 1. A fixed reference point or line about which a graph or figure is plotted. **2.** A line about which a body rotates such as an impeller in a centrifugal pump.

azeotrope A mixture of two liquids that boils at a constant composition. That is, the composition of the vapour is the same as the composition of the liquid. It is therefore not possible to separate an azeotropic mixture by conventional distillation. Azeotropes occur due to deviations in *Raoult's law leading to either a maximum or minimum in the boiling point-composition diagram. The composition of the azeotrope is dependent on pressure.

azeotropic distillation A method of separating azeotropic mixtures by distillation in which conventional distillation is often not suitable or possible. It is used for mixtures that have a *relative volatility near unity or which form *azeotropes and would otherwise require large numbers of theoretical plates and high reflux ratios *See* SUPERFRACTIONA-TION. It is therefore necessary to increase the relative volatility, which entails an increase in the non-ideality of the mixture. An *entrainer is therefore added which is fairly volatile and forms an azeotrope, with one or more of the original components, and leaves overhead, allowing one component to leave in a fairly pure state at the bottom. This new azeotrope must be either heterogeneous or readily separable by some other means such as by *liquid–liquid extraction.

The typical layout with heterogeneity (usually on cooling, see Fig. 3) consists of an over-head heterogeneous azeotrope in which there is an entrainer-rich layer that is returned to the column and an A-rich layer that is sent to an entrainer recovery column. The latter produces more azeotrope overhead, which is sent to the common condenser-cooler, and A of the desired purity leaves at the bottom. The A-rich layer is not necessarily the upper layer in the decanter, and an addition of entrainer to the column may have to be made to make up for losses. An example of this system is the use of butyl acetate (entrainer) to remove water (A) from acetic acid (B). Without the entrainer the relative volatility is very low.

Another example is the use of cyclohexane to separate isopropyl alcohol (IPA) and water in which crude IPA is pumped to the first tower or 'dryer' and cyclohexane and water leaves the top of the tower, is condensed, and separates into two layers. Cyclohexane in the top layer is sent as reflux to the tower and the lower water aqueous IPA layer pumped to another tower for part water removal. The amount of cyclohexane in the system is regulated by the level in the reflux drum.

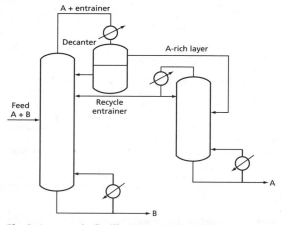

Fig. 3 Azeotropic distillation with heterogeneity

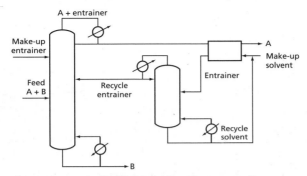

Fig. 4 Azeotropic distillation without heterogeneity

If heterogeneity does not occur, some other means of splitting the A-entrainer azeotrope is required such as liquid–liquid extraction (see Fig. 4).

azeotropic drying The process of removing water from a liquid by the addition of another liquid that forms an *azeotrope with the water. It therefore allows the removal of water at a temperature below the normal boiling point of 100°C.

Babbitt metal One of a number of *alloys originally based on tin, copper, and antimony and now includes lead. It is used in bearings for motors and was invented by American inventor Isaac Babbitt (1799–1862).

Babo's law A law that states that the vapour pressure of a solution is reduced in proportion to the amount of solute that is added. It was discovered in 1847 by German chemist Lambert von Babo (1818–99).

backflushing A cleaning process used to dislodge particulate material in a pipe, column, or filter, etc. It involves reversing the flow of fluids to the normal direction of flow.

background radiation A measurable low-intensity *ionizing radiation that is present all around due to the presence of radioisotopes in rocks such as granite, in the soil, and the atmosphere. The radioisotopes in the atmosphere are naturally forming, the result of nuclear fallout, and emissions for nuclear reprocessing, or emanating as waste gases from power stations. The level of background radiation must be taken into consideration when measuring the radiation from a source.

back-mixing The propensity of reacted materials to become mixed with unreacted materials that are fed to stirred vessels or chemical reactors. The design of *continuous stirred-tank reactors (CSTRs) is based on the assumption of instantaneous homogeneity. In reality, the flow of materials short-circuit and leave the vessel or reactor before the expected time, while some reside for longer periods. Back-mixing is a concern in the design of *plug flow reactors, which lead to a departure in ideality.

back-mix reactor *See* CONTINUOUS STIRRED-TANK REACTOR.

back pressure The resistance to a moving fluid to its direction of flow caused by an obstruction, bend, or friction in a pipe or vessel. It is often used to describe the discharge pressure from a pump or compressor. The term often refers to a pressure greater than atmospheric.

backwashing A method used to clean a fixed bed reactor for reuse. Under normal operation, the flow of a fluid is down through the fixed bed that may act as a support but over time may have become fouled or blocked. Backwashing therefore deliberately reverses the flow of the fluid up through the bed causing *fluidization, detaching, suspending, and washing out of undesirable particles. It is often used in *sand filtration and *ion exchange resin beds.

baffle A plate used in an item of equipment to influence the rate or direction of a flow of material. The plate may be flat or curved and used in vessels such as tanks or heat exchangers to increase turbulence or prevent the formation of a vortex. An impingement baffle is

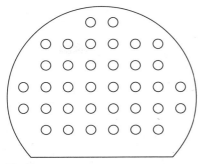

Fig. 5

used to minimize the disturbance of a liquid–liquid interface in a decanter device and can also be used in *shell and tube heat exchangers to reduce the effects of erosion by forcing a vapour stream around the baffle causing any liquid drops to collide with the baffle.

Baffles are often used in shell and tube heat exchangers to direct the fluid stream across the tubes to increase the fluid velocity and increase the rate of heat transfer. The baffle is a circular plate that is similar in size to the tube plate with many holes through which the tubes fit. A segmental baffle allows the shell-side fluid to move under the baffle or over the baffle through the segmental space (see Fig. 5). The **baffle cut** is the segment height removed to form the baffle, expressed as a percentage of the baffle disc diameter. The **baffle spacing** is the distance between the baffles.

Bakelite One of a class of thermosetting resins that are used for making plastic ware and electric insulators. It is named after Belgian-born US inventor L. H. Baekeland (1863–1944).

Baker plot A widely used dimensionless plot representing two-phase flow in pipes (see Fig. 6). Published by O. Baker in 1954 using experimental data for water, various flow

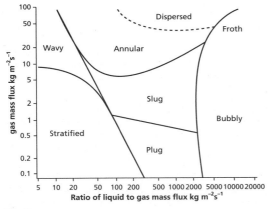

Fig. 6

regimes can be distinguished such as annular, bubbly, froth, plug, slug, stratified, and wavy flow. Corrections are made for two-phase flow mixtures with other densities and viscosities. Improved plots have subsequently been developed by other researchers.

ball mill A mechanical device used to reduce ores to small particles and powder. It consists of a rotating cylindrical chamber mounted on its side containing steel balls and the material to be ground. The slow rotating action of the chamber allows the balls to tumble and cascade over one another thereby crushing the material. They can be operated as a batch or as a continuous operation.

ball valve A type of valve used to control the flow of a fluid in a pipe. It consists of a sphere with a hole through it. When the ball is rotated by 90°, the hole is perpendicular to the pipe axis and the valve is shut preventing flow. With a comparatively low pressure drop, its action is fast and is suitable for dirty and viscous fluids. It is also capable of tight shut-off.

bar A c.g.s. unit of pressure equal to 10^5 newtons per square metre ($N\ m^{-2}$) or pascals (Pa). The millibar (mbar) is commonly used by meteorologists (one thousandth of a bar). *Standard atmospheric pressure is 1,013 mbar.

barn (Symbol b) The unit of area defined as $10^{-28}\ m^2$. It is used to express the effective nuclear cross section of atoms or nuclei in the scattering or absorption of particles.

barometer An instrument used to measure atmospheric pressure. Early barometers consist of a tall vertical glass column containing mercury. The top of the column is sealed, while the bottom is open in a reservoir of mercury. The pressure of air on the reservoir balances the vertical column of mercury, from which the pressure can be read directly. *Standard atmospheric pressure (101.325 kPa) corresponds to a height of 760 mmHg. The space above the mercury in the top of the leg is called a *Torricellian vacuum. It is not a true vacuum but is the vapour pressure of the mercury.

barometric leg A tube immersed in liquid that is used to provide a liquid seal for equipment under vacuum. The tube is of sufficient length such that the static pressure exceeds the atmospheric pressure.

barrel A unit of volume used largely in the oil industry. In the US, one barrel is equal to 42 (US) gallons and equivalent to $0.158\ 98\ m^3$. One barrel (British) is equal to 36 Imperial gallons and equivalent to $0.163\ 659\ m^3$. The abbreviation is *bbl. An oil refinery production and throughput is often quoted in terms of **barrels per calendar day** (BPCD) or **barrels per stream day** (BPSD), which are the average rates based on a 365-day year, or on the number of days per year the refinery was actually operating, respectively.

barrel of oil equivalent (BOE) A measure of the amount of a combustible material which when burnt releases the same amount of energy as the combustion of one *barrel of crude oil. Multiplications of this unit include kilo, million, and billion barrels of oil equivalent (kBOE, MBOE, BBOE). It is also expressed as a rate such as million barrels oil equivalent per day (*MBOED). It is used for financial purposes to combine both oil and gas into a single measure.

Barton, Derek Harold Richard FRS (1918–98) A British chemist noted for his contribution to the pyrolysis of chlorinated hydrocarbons and many other areas of organic chemistry. After gaining his doctorate from Imperial College, London, he held many academic positions and visiting professorships including Regius professor of chemistry at the University of Glasgow and distinguished professor at Texas A&M University. He was elected Fellow of the Royal Society in 1954 and was awarded the Nobel Prize in Chemistry in 1969. He received many other awards in recognition of his contributions to chemistry, which formed the basis of many industrial processes.

barye A unit of pressure used in the c.g.s. system. It is equal to one dyne per square centimetre and is equivalent to 0.1 pascal.

base 1. A substance that has the tendency to gain protons and form hydroxyl ions in solution. A base is neutralized by an acid to form a salt and water. **2.** Used in mathematics, the base is the number of different symbols used in a numbering system. In the binary system the base is 2, while in the decimal system the base is 10.

BA set An abbreviation for **b**reathing **a**pparatus set and is a form of *personal protective equipment that is worn to supply uncontaminated air through a face mask or mouthpiece. It is used by personnel working in hazardous environments.

base unit A unit that is defined arbitrarily and not related in combination with other units. In the SI system, there are seven base units: kilogram (kg) for mass, metre (m) for length, second (s) for time, kelvin (K) for temperature, mole (mol) for the amount of substance, ampere (A) for electrical current, and candela (cd) for luminous intensity. *Derived units are defined as combinations of base units such as the newton (N) which is a unit of force where 1 N is equal to 1 kg m s^{-2}.

basic-oxygen process A process used for the production of high-grade steels that involves a charge of molten pig iron and scrap in a tilting furnace being blown with high-pressure oxygen on the surface through a water-cooled lance. The process has largely replaced the earlier *Bessemer process and *open hearth process.

basis of calculation A statement used at the beginning of a calculation for the quantity of material used entering or leaving a process. The choice of quantity is appropriate for the process and may be expressed in mass or moles. For continuous processes, mass or molar flow rates are used such as kg s^{-1} or kmol h^{-1}, respectively. The total quantities may be used or calculations based on a limiting reactant for a specified product.

basis of design (BOD) A document that is prepared prior to the design and development of a process. It includes the rationale for the design, and includes assumptions and decisions on the options identified for the design as well the codes, standards, and regulations required in the design. The document is used as the basis for the design, development and construction of the process.

basket centrifuge A type of centrifuge used to separate solid particles from liquids. It consists of a perforated bowl or basket that allows the passage of the filtrate and is lined with some form of filter cloth used to retain the solids. It therefore operates as a form of centrifuge filter and operates at relatively low rotational speeds of 1,000 rpm.

BAT An abbreviation for **b**est **a**vailable **t**echniques. It is usually applied to reduce the level of pollution emanating from a process and is considered to be more rigorous than *ALARP.

batch culture A biochemical or biotechnological process such as fermentation in which all the nutrients apart from oxygen are placed in the bioreactor at the start of the operation. The fermentation is inoculated with microorgansims such as yeast or bacteria and the biological reaction is allowed to proceed until completion whereby a *limiting substrate or nutrient is depleted. It is at this point that the product is usually harvested.

batch distillation A type of *distillation process in which the liquids to be separated are placed in a *still. During operation, there is no further addition of liquid for separation. Scotch *whisky is distilled by this technique. Initially, alcohol-rich vapour known as *foreshots is produced and collected before the heart is collected. **Feints**, which are lean in alcohol, are finally recovered. Both the foreshots and feints are returned to the distillation process for the next batch. The process operates at unsteady state and can be described by the *Rayleigh equation.

batch process A process using a fixed quantity of material that is placed within the process equipment and the operation carried out to completion. Described as a *closed system, no material is transferred to or from the system during the time of interest, and can be described using differential material and energy balances. The process may involve several sequential steps but in each the material is kept and processed together. Commonly used for speciality chemicals, pharmaceuticals, and food processing, batch processes include mixing, reaction, and separation. Reactor volumes are generally correspondingly larger than those of continuous processes. The operating costs, including labour costs, are also higher due to materials handling and greater charging and down-times costs. Scotch whisky production is an example of a batch process in which malted barley grain is batch fermented and batch distilled.

batch reactor A vessel within which a controlled chemical or biochemical reaction takes place. They require a charge of reactants, which may be liquid or solid including the use of catalysts. The chemical reaction is allowed to proceed until a required concentration of product has been achieved. The reaction may require heating or cooling depending on the thermodynamics of the reaction. This can be achieved by the circulation of a heat transfer medium through a surrounding jacket or through internal coils and tubes. Once the reaction is complete, the reaction materials are recovered. This type of reactor is considered to be versatile since it can be used for many types of processes.

battery An electric cell used to produce electrical energy. Archaeologists claim that there is evidence from *c.*200 BC Bagdad of batteries being used in early electroplating. Italian physicist Count Alessandro Volta (1745–1827) developed a battery in 1796 that was the first primary source of electrical energy in the form of direct current (DC). His 'voltaic pile' consisted of stacked silver and zinc plates in an acid electrolyte, and could deliver a useful amount of power over a period of several minutes. However, it was not possible to recharge it. French physicist Georges Leclanché (1839–82) developed a carbon-zinc battery in 1866, which was later developed to produce a dry cell battery that is still used today. Rechargeable batteries were first constructed in 1866 by French physicist R. L. Gaston Planté (1834–89) and offered a more convenient electrical source by allowing the electrochemical reaction to be reused and the electrical energy to be replenished. The most common rechargeable battery is the lead-acid battery used in conventional cars.

battery limits The geographical perimeter that surrounds a processing area and includes process equipment, piping, and associated buildings and structures of a process plant. It excludes utilities and process services such as boiler houses and laboratories.

Bayer process A method of making alumina (Al_2O_3) from aluminium ore or bauxite. It involves crushing the bauxite and separating it from the oxides of iron, silica, and titanium that are also contained in the ore. It is then mixed with caustic soda and heated under pressure. The alumina dissolves in the caustic soda forming a solution of sodium aluminate. After *filtration, crystals of aluminium hydroxide are added to the solution. The alumina precipitates out as crystals and are collected by filtration. The crystals are then dried by heating to around 1,200°C to leave a fine white powder. It was invented in 1887 by Austrian chemist Karl Josef Bayer (1847–1904).

bbl An abbreviation for (US) barrel and is a volumetric unit for crude oil and petroleum products where one barrel is equivalent to 158.978 litres. Also used is **BO** representing **b**arrels of **o**il and **BOPD** as a volumetric measure of flow as **b**arrels of **o**il **p**er **d**ay. The unit of volume originates from the volume of spent whisky barrels that were once used to hold oil in the nineteenth century.

bead mill A mechanical device used for rupturing the cell walls of microorganisms to release their intracellular protein products. Laboratory-scale bead mills consist of a chamber filled with small glass beads and a suspension of the cells requiring rupture. The cell wall and membrane are disrupted by collisions between shear force layers generated by high-speed agitation and grinding action of the beads in the chamber. The vigorous agitation can lead to denaturation of released proteins due to the shear forces and local heating effects.

Becher process A process used for producing synthetic rutile (titanium dioxide) as a titanium concentrate from ilmenite ore. The ore contains largely rutile and the process removes iron oxide impurity by roasting with sulphur at 1,200°C in a rotary kiln. The iron is reduced to metal and removed by magnetic separation along with coal and ash. The reduced ilmenite is washed in water containing ammonium chloride as a catalyst and air blown through to convert metallic iron to a flocculent precipitate of iron oxides. These are then removed. The process was invented by Robert Gordon Becher in the 1960s.

becquerel (Symbol Bq) The SI unit of radioactivity corresponding to one disintegration per second. It is named after Antoine Henri *Becquerel (1852–1908) who discovered radiation.

Becquerel, Antoine Henri (1852–1908) A French physicist who accidently discovered the existence of radioactivity in 1896. By chance, he put away in a drawer some unexposed photographic plates wrapped in black paper. In the drawer there was also a specimen of uranium salt. Later, the plates were found to be fogged and led to the conclusion that the uranium had emitted radiation that was sufficiently powerful to penetrate the wrapping. In 1903 he was awarded the Nobel Prize for physics with Marie and Pierre *Curie. His father (Antoine César) and grandfather (Alexandre Edmond) were also eminent physicists and the three held, one after the other, the position of professor of physics at the Musée d'Histoire Naturelle from 1837 to 1908.

bed volume The total volume that is occupied by the support material in a *fluidized bed or *packed bed, such as by a catalyst.

bell metal An alloy of copper and tin used in casting bells. It often contains some zinc and lead.

Benedict–Webb–Rubin (BWR) equation of state An extended *virial equation of state used to describe the vapour–liquid equilibrium data of various substances in terms of molar density. The equation features eight constants for which published data is available for numerous hydrocarbons. The *Lee–Kesler equation of state is an extended form of the BWR equation of state.

beneficiation One of a number of processes involving the separation of ores into valuable components and waste materials called *gangue. They include crushing, grinding, magnetic separation, froth flotation, etc. The separated or dressed ore is then refined. It is also known as **ore dressing.**

Benfield process A process used for the removal of carbon dioxide, hydrogen sulphide, and other gas components from *acid gas, *syngas, and *natural gas streams by scrubbing with a hot aqueous solution of potassium carbonate.

$$K_2CO_3 + H_2S \rightarrow KHS + KHCO_3$$

The high temperature operation prevents condensation of hydrocarbons. It is named after its inventor H. E. Benfield who developed it in 1952.

Bergius, Friedrich Karl Rudolph (1884–1949) A German chemist noted for the invention of the Bergius process used for the production of synthetic fuels from coal. He gained his PhD at the University of Leipzig and worked for a time with Fritz *Haber and Carl Bosch at the University of Karlsruhe before becoming professor at the University of Hanover. His work on high-temperature, high-pressure processes resulted in the *Bergius process in 1913 for production of synthetic fuels by the hydrogenation of lignite. He was awarded the Nobel Prize for Chemistry together with Carl Bosch in 1931 for his work on high-pressure processes.

Bergius process 1. A process used to make hydrocarbon fuels from coal. It was developed by German chemist Friedrich Karl Rudolph Bergius (1884–1949) as a motor fuel in the First World War. It involves heating coal mixed with tar in the presence of a catalyst at a temperature of 450°C in hydrogen at a pressure of 200 atmospheres. There have subsequently been a number of process improvements particularly with more effective catalysts. **2.** The hydrolysis of cellulose with concentrated hydrochloric acid production of sugar from wood.

Berl saddle A saddle-shaped packing material made from an inert, non-reactive material with good mechanical properties such as ceramic (see Fig. 7). As a packing material used in

Fig. 7

adsorption towers, it has a good surface-to-volume ratio while allowing gases and liquids to move past easily offering a low pressure drop across a fixed bed of the material.

Bernoulli, Daniel (1700–82) A Swiss mathematician and physicist noted for his work on fluids. He was one of eleven eminent mathematicians in his family spanning four generations. Born in Groningen in the Netherlands, he was educated at Basel, Switzerland, where his father had been appointed professor of mathematics following the death of Bernoulli's uncle who had previously held the post. Bernoulli gained his master's degree at the age of 16 and his doctorate at 21. He held the position of professor of mathematics at St Petersburg Academy in Russia from the age of 25, but returned to Basel in 1732 to become professor of anatomy and botany at the University of Basel before becoming professor of natural philosophy in 1750. His important work on hydrodynamics demonstrated that pressure in a fluid decreases as the velocity of fluid flow increases, which he published in his book *Hydrodynamica* in 1738. He also made a first statement on the *kinetic theory of gases.

Bernoulli theorem A theorem in which the sum of the pressure-volume, potential, and kinetic energies of an incompressible and non-viscous fluid flowing in a pipe with steady flow with no work or heat transfer is the same anywhere within a system. When expressed in head form, the total head is the sum of the pressure, velocity, and static head. It is applicable only for incompressible and non-viscous fluids. That is:

$$\frac{p_1}{\rho g}+\frac{v_1^2}{2g}+z_1=\frac{p_2}{\rho g}+\frac{v_2^2}{2g}+z_2$$

where p is pressure, ρ is density, g is gravitational acceleration, v is velocity, and z is elevation. It is effectively a statement of the law of the conservation of energy. It was formulated and published in 1738 by Daniel *Bernoulli (1700–82).

Berzelius, Baron Jöns Jacob (1779–1848) A Swedish chemist who discovered several elements including cerium, selenium, lithium, thorium, and vanadium. He invented the present notation of chemical elements and electrochemistry, and worked on the atomic weights of many substances.

Bessel function A type of function denoted by the letter J to represent the solution to a type of differential equation that has independent solutions which can be expressed as infinite series. An example is the harmonic equation:

$$\frac{d^2h}{dr^2}+\omega^2h=0$$

The general solution is $h(\omega r)=AC(\omega r)+BS(\omega r)$ where A and B are constants of integration. The solutions form an infinite series and are listed in tables and plotted. They are used in the study of fluid mechanics and heat transfer, and named after German astronomer Friedrich Wilhelm Bessel (1784–1846).

Bessemer process A process first used for the mass production of steel from pig iron. Named after British engineer Sir Henry Bessemer (1813–98) who in 1856 took a patent on the process; it operates at around 1,250°C in a tilting vessel known as a **Bessemer converter**. The process involves the removal of impurities from iron by oxidation with air blown in at the base through the molten iron. The product is then converted into steel by the addition of spiegeleisen, which is iron containing a high proportion of manganese and carbon. The oxidation process is exothermic and helps to keep the mass molten in which oxides of silicon and manganese are skimmed off while other oxides are either in the form of gas or solid *slag. The furnace is tilted and the molten steel poured off.

best efficiency point The most efficient operation of a centrifugal pump in terms of flow rate and delivery pressure or head. It is usually represented as an identifiable point on the *characteristic curve for a centrifugal pump. The efficiency of the pump is the ratio of the power output (the product of pressure and flow rate) to the power input (electrical energy).

beta decay The spontaneous emission of negatively charged electrons or beta particles by a heavy radioactive element. The radioactive decay results in the original element being gradually converted into another element. For example, the radioactive isotope of lead-210 loses a beta particle to give a radioactive isotope of bismuth-210. Beta decay may occur at the same time as decay of *alpha particles.

beta particle An electron or positron emitted by a radioactive element during radioactive *beta decay or nuclear fission. **Beta radiation** is a stream of beta particles.

BET equation An equation used in the theory of multilayer adsorption of atoms onto a surface. It is based on the assumption that the forces that produce condensation of moisture on a surface are also responsible for the binding energy of multilayer adsorption. It is named after Brunauer, Emmett, and Teller.

Betts process An electrolytic process used for the refining and purification of lead with the associated recovery of silver and gold. The process involves a solution of lead fluorosilicate and hydrofluorosilicic acid as the electrolyte and lead electrodes. Invented by Anson G. Betts in 1901, the process is dependent on cheap forms of electricity. *See* PARKES PROCESS.

billet A small, part-finished piece of metal that is rectangular, circular, or square in shape. Metals such as iron, steel, and plutonium are made into billets for further processing.

billion One thousand million (10^9). In the UK this was formerly one million million (10^{12}) and was changed when the British Treasury started using the term billion in the American sense in the 1960s.

bimetallic strip A metallic strip consisting of two metals that have different thermal expansion coefficients and are firmly joined together. As the strip heats or cools, the strip bends. It is used as a thermal on/off switch in electric circuits.

bimolecular reaction A chemical reaction involving two molecules. The reaction of nitrogen dioxide and carbon monoxide is an example of a bimolecular reaction.

$$NO_2 + CO \rightarrow NO + CO_2$$

binary 1. A system involving two components. A binary separation involves the separation of two components such as ethanol and water by distillation. **2.** A compound or alloy formed from two elements. **3.** A mathematical numbering system involving two symbols (0 and 1) used in digital computers. The numbers represent units, twos, fours, eights, etc. For example, the number 6 in the decimal numbering system is 110 in the binary system (i.e. 4+2+0=6).

binary distillation The separation of two components by the process of distillation. The ease of separation is based on the difference between the boiling points of the two components. Simple graphical techniques such as *McCabe–Thiele and *Ponchon–Savarit can be used to illustrate the separation for continuous binary distillation processes.

binding site A specific location on the surface of a molecule that can combine with another molecule. An example is the binding site of an enzyme with a substrate. *See* ACTIVE SITE.

Bingham plastic A *non-Newtonian fluid that exhibits a yield stress that must be exceeded for flow to occur. Once the fluid has begun to flow, the rate of shear versus shear stress curve is linear. The applied shear stress is expressed as:

$$\tau = k\dot{\gamma} + \tau_o$$

where τ_o is the *yield stress, k is a constant, and $\dot{\gamma}$ is the *shear rate. Examples include toothpaste, mayonnaise, and drilling muds. It is named after American chemist Eugene Cook Bingham (1878–1945).

biocatalyst A catalyst that is, or derived from, a living organism or cell. Enzymes are biocatalysts and are used to catalyze biochemical reactions.

biochemical engineering The principles of chemical engineering applied to biological systems and includes the synthesis of compounds from biological origins, decomposition of organic material, and the development of technical solutions for biomedical purposes.

biochemical oxygen demand (BOD) The amount of oxygen that is required for the complete oxidation of microbiologically degradable material in water. It is measured by the direct oxygen consumption of materials such as sewage and contaminated water, and represents a measure of the amount of organic pollutants in water. As a form of index used for the degree of organic pollution in water, it is expressed as the amount of oxygen used for biochemical oxidation by a unit volume of water at a given temperature and for a given time. It involves incubating a sample at 25°C for a fixed time (often five days), and the amount of oxygen removed determined by chemical analysis. The result is expressed in milligrams of oxygen per dm^3 of water. The higher the value, the greater the level of pollution. It is also known as the **biological oxygen demand**.

biochemistry A branch of chemistry that is concerned with the chemistry associated with living organisms.

biodiesel A liquid fuel that can be used in the place of, or mixed with, mineral diesel fuel in modified engines. It is made from a variety of plant oil feedstocks such as rapeseed, soya bean oils, and waste cooking oils. It is converted to fatty acid methyl esters (FAME) by a transesterification process involving the chemical reaction with an alcohol such as methanol in the presence of a strong alkali or acid as the catalyst. The FAME product is separated from the glycerol.

bioelement An element that is essential in the molecules of all living organisms. The three main elements are oxygen, carbon, and hydrogen. Sulphur and phosphorus are dominant is most organisms, and calcium in humans. Other elements in lesser amounts include sodium, potassium, magnesium, and copper.

bioengineering A branch of engineering concerned with the application of techniques in the development, design, and manufacture of equipment and devices for use in biological systems. It includes products to replace lost bodily functions or supplement defective functioning of body parts or organs such as orthopaedic prostheses, artificial limbs, heart valves and pacemakers, artificial livers, hip joints, hearing aids, etc.

bioethanol The most widely used liquid *biofuel. It is produced by the *fermentation of starch or sugars including maize (corn), sugar beet, and sugar cane. Bioethanol is often blended with petrol for use in some specially converted internal combustion engines, or can be used as a pure fuel.

biofilm A layer of growing microorganisms that adheres to a surface. The adhesion by chemical means from the microorganisms involves polysaccharides and glycoproteins. Biofilms are used in sewage filter beds and some forms of cell-immobilization processes. The formation of biofilms causes fouling in industrial water processes and can lead to process blockage.

biofuel A gaseous, liquid, or solid fuel that contains carbon derived from a biological or organic source. There is an increasing interest in the use of biofuels in the quest for alternative sources and forms of energy. Examples include *biogas, *biodiesel derived from plants, and *bioethanol from crops such as maize and sugar beet.

biogas A *biofuel that consists of a mixture of methane and carbon dioxide formed as a result of the decomposition of organic matter by anaerobic digestion. The process uses methanogenic bacteria to produce methane, at a concentration of between 50 per cent and 70 per cent. Biogas can be used directly or mixed with other gases, or by removing the carbon dioxide for use in power engines. Biogas is a major source of energy in underdeveloped countries.

biological agent A microorganism, cell culture, or human endoparasite, whether or not genetically modified, that can cause infection, allergy, toxicity, or otherwise create a hazard to human health.

biological oxygen demand (BOD) *See* BIOCHEMICAL OXYGEN DEMAND.

biomass Any vegetation or biological waste whose energy can be harnessed in a fuel. It includes waste from agricultural and food processing and can be converted to high-energy fuel or other useful chemicals by biotechnological processes.

biomass to liquids (BTL) A combustion reaction used to convert *biomass to liquid fuels. Involving two main steps, the first is the conversion of biomass such as wood into a *synthesis gas consisting of hydrogen and carbon monoxide. The catalytic *Fischer–Tropsch process is used to convert them to synthetic fuels.

biomaterial A biologically naturally occurring, or synthetically produced, substance that is suitable for use in implanted medical devices or as an implant for the replacement of an organ without rejection.

biomolecular engineering The use of chemical engineering principles and practices in the field of biology, and in particular the manipulation of *biomolecules that are used in the pharmaceutical, medicine, food, environment, bioenergy, and agricultural industries. It involves the structural and functional transformation of proteins, carbohydrates, lipids, and nucleic acids along with the detailed knowledge and understanding of biomolecular mechanisms, thermodynamics, and biochemical reactions kinetics. *See* RECOMBINANT DNA TECHNOLOGY.

biomolecule A molecule that is involved in the maintenance and metabolic processes of living organisms. This includes carbohydrates, proteins, lipids, nucleic acids, and water molecules.

biopolymer A polymer that occurs naturally such as a protein, polysaccharide, and nucleic acid. *See* POLYMER.

bioprocess engineering A specialist branch of (chemical) engineering that involves the design and operation of processes used for the production of biological products such as foods, pharmaceuticals, and biopolymers.

bioreactor A vessel used for biological processing containing the growth of living cells or tissues, either as the product themselves or as biocatalysts in the production of other products. There are many designs of bioreactors. The most common are cylindrical and range in capacity from a few litres to many cubic metres. Small bioreactors are made of glass while large bioreactors are fabricated from stainless steel.

The biological processes contained within the bioreactor may be in the form of a suspension of cells or immobilized, and depending on the living organism, operated aerobically or anaerobically. The mode of operation is batch, fed batch, or continuous. An example of a continuously operated bioreactor is the *chemostat.

Constant agitation within a bioreactor can be maintained with an appropriate stirrer which also aids oxygen transfer in aerobic processes. Low-speed impellers or the use of sparged air up draft tubes are typically used to aid mixing. Since all bioprocesses are exothermic, cooling is required using either an external jacket or, for very large vessels, internal cooling coils.

bioscrubbing The removal of odorous or toxic waste gases from biological processes. Foul-smelling sulphurous odours such as mercaptans, toxic wastes such as cyanide, and microorganisms in aerosol form are removed within a spray column in which the waste gas stream is fed. Within the column, finely sprayed water droplets flow down *countercurrent to the flow up of the waste gas.

biosphere The sphere of life for all living organisms on Earth. It extends several kilometres above the surface of the Earth and down to the bottom of the deepest oceans. On land, it extends only a few metres below ground surface.

biosynthesis The natural formation of complex biochemical products from simple biochemical molecules. An example is the synthesis of proteins from amino acids. It is also known as **anabolism**. The opposite is called **catabolism**.

Biot, Jean-Baptiste (1774–1862) A French professor of mathematics, physics, and astronomy, he made the first balloon ascent ever undertaken for scientific purposes in 1804 together with Joseph Louis *Gay-Lussac (1778–1850). He was particularly interested in the polarization of light and his observations laid the foundations of the polarimetric analysis of sugar. He evolved an experiment using a metal sphere and two metal hemispheres to show that there is no electric charge on the inside of a hollow charged conductor but instead it is all on the outside.

biotechnology A technology that uses biological processes for the development and industrial production of materials for medical and industrial purposes. The technology uses living organisms such as plants, animals, yeast, and bacteria, and uses techniques such as genetic modification and *recombinant DNA technology to improve production or produce new products. Examples include the production of beer, cheese, and wine, the development and production of vaccines, hormones, monoclonal antibodies, and antibiotics, as well as the production of energy and the recycling of waste.

Biot number A dimensionless number, Bi, used in non-steady-state heat transfer or mass transfer that relates the internal resistance to the flow of heat or mass transfer to the resistance of that flow at the surface of a solid body:

$$Bi = \frac{hl}{k}$$

where h is the film coefficient, l is the characteristic dimension as the half thickness of a slab or radius of a sphere or cylinder, and k is the thermal conductivity of the solid. It is named after the French physicist and mathematician Jean-Baptiste *Biot (1774–1862).

Birmingham Wire Gauge (BWG) A form of dimension used for the classification of the wall thickness of pipes and tubing using the units of decimal parts of an inch. It is also known as **Stubs' Wire Gauge**.

Black, Joseph (1728–99) A Scottish doctor and later professor of chemistry at the University of Edinburgh for 33 years. Unusually, he lectured in English rather than Latin, which was the normal practice at the time. His lectures involving many experiments were particularly popular and became a fashionable habit of Edinburgh society. James *Watt was one of his students, to whom he gave both money and ideas for his research. Black's work was largely on specific and latent heats. He distinguished between heat and temperature, found specific heats by the method of mixtures, and obtained the latent heat of water as it froze. He founded the first Chemical Society for his students.

black body Used in heat transfer to describe a body that absorbs all the thermal radiation falling upon it. A black body is noted for having an absorptivity of one and a reflectivity of zero. The emissive power of a black body is dependent on the temperature and proportional to the fourth power of the absolute temperature. **Black body radiation** is the

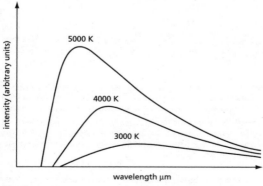

Fig. 8

*electromagnetic radiation that is emitted by a black body, which extends across the range of wavelengths. The intensity of energy rises to a maximum, the wavelength of which is dependent on the temperature (see Fig. 8). The wavelength decreases with increasing temperature. *See* STEFAN–BOLTZMANN CONSTANT.

black box A term used to describe a process or system that is considered only in terms of its inputs and outputs, and with no knowledge or understanding of internal operations. It is typically used to represent complex systems where the precise details or mechanism of the process or system is not known or fully understood. It uses empirical or semi-empirical mathematical models, experimental data, past performance data, and trends to describe or predict behaviour. It gets its name from the opaqueness of not seeing the internal operations. Examples include biological processes involving living organisms and the behaviour of catalysts in heterogeneous chemical reactions.

blanketing A process in which an inert gas such a nitrogen is used to fill the vapour space of a vessel containing a liquid to control its composition. It is used in a wide range of industries including the petroleum, pharmaceutical, and food industries. Carbon dioxide and argon gas are also used although carbon dioxide can tend to be reactive in some cases and argon is a relatively expensive gas.

blanking A method used to block pipelines to intentionally prevent the flow of materials or substances. It is used in maintenance, the closure of redundant pipelines, and process equipment, and usually consists of a plate bolted across the face of a *flange.

blast furnace A smelting furnace used for the continuous production of molten iron from iron ore, from which steel is made. It consists of a steel tower lined with refractory brick into the top of which iron ore, coke, and limestone are charged. Compressed hot air is blasted into the bottom. The chemical reaction of the oxygen in the air and the coke produces carbon monoxide. This then reacts with the iron ore to produce carbon dioxide and iron. The carbon dioxide then reacts with more carbon to produce more carbon monoxide. The limestone combines with the coke ash to form liquid *slag, which absorbs any sulphur in the iron, collects at the bottom of the tower, and floats on the surface of the molten iron. Both the molten iron and slag are periodically and separately tapped off. The hot gases are used as fuel for heating the incoming air. The cast or pig iron is used to produce steel in the *Bessemer process.

blast wave A pressure pulse moving outwards from the site of an explosion. It can be formed by a detonation, a rapid deflagration, or the sudden failure of a piece of process equipment containing a potential energy source that is released at a high rate. The **blast pressure** is the side-on overpressure and can have a destructive effect on buildings and structures.

bleach A disinfectant manufactured from caustic soda dissolved in water and reacted with the chlorine in either a batch or continuous process. The chlorine and caustic soda are made by the electrolysis of salt solutions. Industrial bleach includes calcium hypochlorite, chlorine, and hydrogen peroxide, and is used as a disinfectant and in the paper and textile industries. Domestic household bleach is an aqueous solution of sodium hypochlorite, $NaOCl$, made from caustic soda, chlorine, and water.

bleed The controlled removal of a small amount of material from a process as a *side stream that prevents an accumulation of undesired materials. It is also known as a **purge**.

blenders A type of mixer used for mixing powders. Powder blenders are broadly classified as vertical or horizontal mixers depending on whether the mixing impeller rotates on
a vertical or horizontal axis. Vertical mixers have a movable bowl in which the contents are
mixed by mechanical agitation. Paddle agitators are commonly used and the shape of the
impeller frequently conforms to the vessel walls. Planetary motion devices are commonly
used in vertical mixers in which the agitator revolves in a circle in addition to rotating on
its own axis. This ensures that the entire mixer volume receives a beating action and that
there are no dead spaces.

blending A process used in oil refining in which two or more hydrocarbon products are
mixed together to form a new product with desired and specified properties. The **blending
octane number** is a mix of hydrocarbons that provides an octane number greater than for
the pure fuel.

BLEVE *See* BOILING LIQUID EXPANDING VAPOUR EXPLOSION.

blinding The blockage of filters in which particles that become entrapped block other
particles to the point where the filter becomes blocked, the filter becomes inactive, and no
longer able to permit the passage of the flow medium, such as air.

block diagram A diagrammatic representation of a controlled process showing the relationships between the system variables. The controlled process is presented in blocks as
functional, non-interacting sections whose inputs and outputs are readily identifiable. The
blocks are connected in the same order as they appear in the physical controlled process.
The convention includes: lines, which represent signals as flows of information, material,
or energy; circular summing junctions, which represent an algebraic summation of the
input signals to that point (positive or negative); a branch point, which represents a division of a signal into more than one path without change; rectangles, which represent a
modification of the entering signal and describe the dynamic characteristics of the systems
that they represent, usually as conversions or transfer functions between the input and
output signals.

block flow diagram A schematic representation of an entire process or major part of a
process in which unit operations are symbolically represented as blocks in which process
material collectively enters for processing and products leave (see Fig. 9). Unlike a *process
flow diagram, a block diagram has limited information. No information is provided such
as flow compositions or details of process conditions within each unit operation. They are
useful at the design stage to present how a complex process fits together. It is also known as
a **schematic flow diagram**.

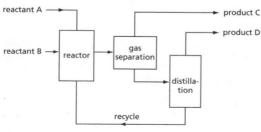

Fig. 9

block valve A type of valve used in a pipeline that provides a tight shut-off. It is typically used for *isolation purposes.

blowdown 1. The removal of oil and gas from a pipeline. **2.** The difference in pressure between that which opens a safety valve and that which closes it. **3.** A technique used to remove periodically accumulated solids in process vessels, boilers in particular, which would otherwise cause problems of deposition on heat transfer surfaces, foaming, and adverse effects on process performance. A **blowdown vessel** is used to receive a large amount of water and precipitated solids removed under boiler pressure.

blower A mechanical device designed to deliver a gas, usually air, with a low gauge pressure of less than 2 bar. The terms *'fan' and 'blower' are often used interchangeably although radial-flow fans are commonly referred to as blowers whereas axial-flow fans are referred to as fans.

blowoff valve A pressure release valve system used to prevent problems associated with fluid surge. It is used on turbocharged engines to prevent compressor surge and is usually vented to atmosphere. It is also known as a **dump valve**.

blowout A term used in offshore oil processing where there is a loss of control pressure of an oil reservoir resulting in an uncontrolled release of fluids to the surface. A **blowout preventor** is a hydraulically or mechanically actuated valve that is located at the wellhead and used to control pressure within the *well.

blue water gas *See* WATER GAS.

BO An abbreviation for **b**arrels of **o**il. It is a volumetric unit of crude oil and other petroleum products equivalent to 42 US gallons or 158.978 litres.

BOD 1. An abbreviation for *biochemical oxygen demand or *biological oxygen demand. **2.** An abbreviation for the *basis of design.

Bode plot A type of frequency response diagram used for analysing the frequency response of a system to a disturbance signal. It is the plot of the logarithm of the amplitude ratio with the logarithm of the phase angle measurements.

BOE *See* BARREL OF OIL EQUIVALENT.

boiler A heat exchange device used to raise the temperature or vaporize a liquid. The heat supplied by hot gases from combustion processes or by high-temperature heat transfer liquids such as oils. Many boilers are used to raise steam as a process utility. Steam itself is often used as the heat transfer medium in heat exchanger-type boilers such as *kettle reboilers.

boiler hotwell A container used for hot condensate water that is returned to a boiler and that has been settled out and filtered from impurities. The water is then fed to the boiler feed pump for reuse.

boiling The process of evaporation occurring throughout a liquid that occurs when the vapour pressure of a liquid is equal to the pressure above it. There are various types of boiling including natural convection from a heating surface, nucleate boiling, partial film boiling, and film boiling.

Boiling Liquid Expanding Vapour Explosion (BLEVE) The *explosion that occurs when a pressure vessel with a superheated flammable liquid or a liquefied flammable gas is heated and ruptures. It is the result of the rate of pressure build-up being greater than the rate of pressure relief due to venting. The pressure build-up, together with the reduced structural strength of the vessel caused by the external heating, increases the risk of explosion. At the point of rupture, the pressurized liquid discharges and immediately flashes to a vapour. The rapid rate of vapour expansion causes the explosion in the form of a blast wave. There may also be missile damage. Ignition of the expanding flammable vapour in air can result in a rapid combustion and create a *fireball.

boiling point The transition temperature at which a liquid state to the gaseous state occurs. The temperature corresponds to the point where the saturated vapour pressure of the liquid is equal to the system pressure. In a solution, the presence of a solute raises the boiling point of the solvent. In dilute solutions, the boiling point elevation can be used to determine the relative molecular mass of the solute.

boiling point-composition diagram A diagram that is used to present the relationship between the *boiling point and composition for two components at a given pressure (see Fig. 10). The x-axis ranges from 0 to 100 per cent for one component (i.e 100 per cent to 0 per cent for the other component) with the y-axis representing temperature. The diagram features two curves: the lower curve gives the boiling point temperature for the binary mixture, and the upper curve represents the composition of the vapour for that temperature. For an ideal mixture, the two curves coincide. However, deviations from *Raoult's law may show a maximum or a minimum where the composition of the liquid is the same as the vapour, and illustrates the existence of an *azeotrope.

boiling point elevation The increase in the temperature at which the transition from a liquid to the gaseous state occurs due to the presence of a dissolved substance. It is related to the mole fraction and the molecular weight of the dissolved molecules and ions.

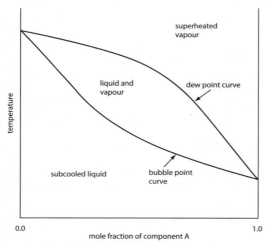

Fig. 10 Boiling point-composition diagram

boil-up ratio The ratio of vapour returned to the bottom of a distillation column to the bottom product removed.

Boltzmann constant (Symbol k) The ratio of the *universal gas constant, R, to *Avogadro's constant, N_A: $k = R / N_A = 1.380\ 658 \times 10^{-23}$ JK^{-1}. It may therefore be seen as the gas constant per molecule. It is named after Austrian physicist Ludwig Eduard Boltzmann (1844–1906).

Boltzmann equation An equation used in non-equilibrium statistical mechanics to calculate transport coefficients such as thermal conductivity. The equation describes a distribution function, f, which gives a description of the state and how it is changing with position at any time t. It has the mathematical form:

$$\frac{\partial f}{\partial t} + a \cdot \left(\frac{\partial f}{\partial v} \right) + v \cdot \left(\frac{\partial f}{\partial r} \right) = \left(\frac{\partial f}{\partial t} \right)_{coll}$$

where v is a velocity vector, a is the acceleration of particles between collisions and r a position vector. It was proposed in 1872 by Austrian physicist Ludwig Eduard Boltzmann (1844–1906).

bomb calorimeter An instrument used to determine the *heat of combustion of a substance. It typically consists of a steel chamber with a non-oxidizable material on the inside, with a tight-fitting screw lid. A known mass of sample of material is burnt in a platinum cup. Insulated platinum leads allow an electric current to be passed through a thin wire to ignite the material. The air in the chamber is replaced with pure oxygen, which is allowed to reach a pressure of around 20 atmospheres. The calorimeter is surrounded by water and once the combustion has begun, the rise in temperature recorded from which the heat of combustion is determined.

bond The linkage between atoms to form molecules. Most bonds involve the exchange or sharing of electrons. *See* CHEMICAL BOND.

Bond number A dimensionless number, Bo, used to characterize the shape of drops of liquid and relates the interactions of gravitational and surface tension forces, σ, in a capillary of diameter d:

$$Bo = \frac{(\rho - \rho')gd^2}{\sigma}$$

where ρ is the density of the droplet and ρ' is the density of the surrounding fluid. It is also known as the Eötvös number (Eo) and named after Hungarian physicist Loránd Eötvös (1848–1919).

Bond's Work Index A semi-empirical equation that relates the power consumption required for crushing and grinding solid particles such as rocks and ores into smaller particles. It is named after American mining engineer Fred Chester Bond (1899–1977).

bone seeker A radioactive isotope such as strontium-90 that is preferably deposited in the bones of humans and animals. Chemically, the isotope behaves in the same way as calcium and can actively displace calcium in the bone structure. The substance may enter the body through ingestion, inhalation, and through wounds.

BOPD An abbreviation for **b**arrels of **o**il **p**er **d**ay as a volumetric measure of crude oil produced by an oil well or field. The volume of a barrel is equivalent to 42 US gallons or 159 litres.

borosilicate glass A type of glass that contains a high level of boron. Boron is effective at absorbing neutrons emanating from radioactive isotopes. Borosilicate glass is also resistant to the process of *leaching and is therefore used in various nuclear fuel reprocessing applications such as in the process of *vitrification used to solidify high active liquid wastes.

bottleneck A term used to describe a process or system whose performance or throughput is limited by the action of an item of process plant such as the performance of a pump or volume of a vessel. The term originates from the phenomenon of a liquid pouring freely from a bottle being limited by the restricted area at the bottleneck. A management tool is **debottlenecking** and is used to identify and eliminate such limitations thereby increasing plant production. *See* CRITICAL PATH ANALYSIS.

bottoms The high-boiling liquid removed from the bottom of a *distillation process that is rich in the less volatile component. It is also known as **residue**.

botulinum cook A thermal food process that is used to destroy spores of the pathogenic bacterium *Clostridium botulinum*. It involves heating the food to 121°C and holding the temperature for three minutes which corresponds to twelve successful decimal reductions or 12-D. A decimal reduction is the time required to reduce a population of viable spores ten-fold. A 12-D process therefore corresponds to reducing 1 million million spores to a single surviving spore, which statistically is negligible, thereby rendering the food safe to consume. It is also known as a 12-D process.

boundary condition/value Used in solving differential equations that are used to describe natural phenomena such as the movement of mass or flow of heat or fluids; they are used to define the boundary of the region under consideration. The boundary condition or value is required to be specified for a solution to be reached. The boundary condition may be a physical entity, such as the velocity of a fluid at a surface, or the concentration of components at a fluid interface.

boundary layer The region between a surface or wall and a point in a flowing fluid over it where the velocity is at a maximum. Within this region, the movement of the fluid flow is governed by frictional resistance. By convention the edge of this region is assumed to lie at a point in the flow that has a velocity equal to 99 per cent of the local mainstream velocity. Within the boundary layer, which is laminar in flow, the transfer of heat and mass across it occurs only by molecular diffusion.

boundary slip A *boundary condition used in fluid mechanics. A fluid that flows over a surface of a solid can be assumed to be in the form of layers that are brought to rest at the surface, known as the *no-slip boundary condition. While the assumption is useful in macroscopic treatments and for finding analytical solutions such as the *Hagen–Poiseuille equation, the assumption breaks down when considering fluids at the molecular level. In such cases, a boundary slip condition occurs where the viscosity at the surface is different from the bulk viscosity of the fluid.

bound moisture The liquid that is held by a solid material within capillaries and cavities by physical or chemical adsorption and with a vapour pressure that is less than that of the pure liquid at the same temperature. The *unbound moisture is the moisture that is in excess of the equilibrium moisture content corresponding to saturation humidity.

Bourdon gauge An instrument used to measure gas or vapour pressure. It consists of a tube curved in a circular arc that has an elliptical cross section. One end is fixed while the other is linked to a mechanical pointer on a calibrated scale. The tendency for the tube to straighten out is measured by a scale calibrated for pressure. It was invented by Eugène Bourdon (1804–88).

BOV *See* BLOWOFF VALVE.

bowl centrifuge A type of continuous centrifuge used to separate liquids carrying suspended particles or droplets of a dispersed phase. It has a set of perforated conical discs that are used to separate the feed liquid into layers. The liquid is fed at the centre of the rotating disc and the suspended particles or heavier dispersion of liquid phase is removed radially.

Boyle, the Hon. Robert (1627–91) An Irish scientist noted for his experimental work on the pressure of gases. The fourteenth child and seventh son of the Earl of Cork, he was sent to Eton at the age of eight and overseas at eleven with a tutor. He settled in England and devoted his life to experimental science. He became an active member of the Invisible Society, which developed into the Royal Society.

Boyle's best-known works include the *Spring of the Air* (1660); he discovered that sound does not pass through a vacuum and established that air has weight. He observed the effect of altitude on pressure and the effect of pressure on the boiling point of liquids. He invented a type of thermometer and carried out many experiments on refraction, colour, electricity, relative densities, and the expansion of water when it freezes. He also defined the term 'element', distinguished between mixtures and compounds, and showed that a compound could have different properties from those of its constituents. He was the first to prepare, collect, and burn hydrogen and one of the first to isolate phosphorus in 1680.

Boyle-Charles's law The combination of *Boyle's law with Charles's law for an *ideal gas as:

$$\frac{p_1 V_1}{T_1} = \frac{p_2 V_2}{T_2} = nR$$

where R is the gas constant (8.314 J mol^{-1} K^{-1}) and n is the number of moles.

Boyle's law A law that states the relation between gaseous pressure and volume. It states that the volume of a fixed mass of gas is inversely proportional to its pressure at constant temperature. It was proposed in 1662 by Robert *Boyle (1627–91).

brake horsepowe *See* HORSEPOWER.

brass A yellowish alloy that consists of copper and zinc, and occasionally some other metals. The usual ratio is 2:1 copper to zinc. Brass is used for small bore pipes and some pipe fittings such as valves and couplings.

brazing A process used to join metal in which a filler metal is heated to a melting temperature and is then distributed between close-fitting parts to be joined together by capillary action. The filler metal is an alloy and is protected by a flux which is a chemical agent used to prevent oxides from forming while the work piece is heated. It also serves to clean any contamination left on the brazing surfaces.

break-even point A financial measure of the point that a process or business is economically viable. It is measured as the fraction of the production capacity of a process or business where the income is equal to the sum of the variable and fixed costs.

b

breakpoint chlorination A process used in water treatment in which chlorine is added to water to a point that the required level of chlorination has been satisfied. This corresponds to the amount of chlorine consumed before a freely available chlorine residual is produced. Public water supplies normally chlorinate beyond the breakpoint.

breakthrough 1. The point where an absorbate first appears in the fluid from an absorber. **2.** The maximum concentration of unwanted ions such as calcium ions that are left in a treated liquid in an ion exchange unit. **3.** The point where water or gas injected into an oil or gas reservoir used to maintain reservoir pressure breaks through to another production well.

Brent crude *See* WEST TEXAS INTERMEDIATE.

brewing A *fermentation process used to the production of beer, lager, and whisky. A major component used in beer and Scotch whisky process is barley. The barley is first converted to barley malt by soaking in water until germination. It is then dried in a kiln, followed by milling of the grain, and the starch extracted from the husks. Malting contributes to the flavour and provides the necessary enzymes to produce sugars from the starch. The barley malt is then mixed with water and yeast. When making beer, hops are added to provide bitterness. Mashing is a process involving heating the mixture to around 65°C for about an hour and used to convert starches to sugar; it takes place in a vessel called a mash tun. The mash liquid is known as wort. It is boiled to halt enzyme activity. Fermentation takes place in a washback which is a type of batch fermentation vessel and used to convert the sugar into alcohol. The beer is then filtered, pasteurized to kill any residual yeast, and bottled or canned. For whisky production, the wort is distilled in a *still and the distillate collected, stored, and matured in oak barrels. *See* WHISKY.

Bridgman, Percy Williams (1882–1961) An American physicist who studied the properties of matter under extremely high pressure. A graduate of Harvard, he remained there as professor from 1919 until retirement. He was awarded the Nobel Prize for Physics in 1946. He is also noted for his writings on the philosophy of science and studies of electrical conduction in metals and properties of crystals. He was president of the American Physical Society in 1942.

Brinell hardness number A measure of the hardness of metals and alloys. It involves pressing a small, hardened steel ball into the surface of the metal being tested by a known loading force. The number is the ratio of the mass of the load in kilograms to the area of the depression created by the ball in square millimetres. Typical numbers range from 60 (soft) to 800 (hard). It is named after Swedish metallurgist J. A. Brinell (1849–1925).

Brin process A process that was once used for producing oxygen through the heating of barium oxide in air to form barium peroxide. This was then heated to temperatures in excess of 800°C to produce oxygen:

$$2BaO_2 \rightarrow 2BaO + O_2$$

The reaction was discovered by the French chemists Joseph-Louis *Gay-Lussac (1778–1850) and Louis-Jacques Thénard (1777–1857). The process was particularly inefficient, and a major improvement was developed by Arthur and Leon Brin who found a way of removing the carbon dioxide with sodium hydroxide. The oxygen produced was used to be for limelight before more cost-effective methods of producing oxygen were discovered.

Britannia metal An alloy of tin with between 5 per cent and 10 per cent antimony and small amounts of copper, bismuth, lead, and zinc. It is used in bearings and tableware resembling pewter.

British thermal unit (Btu) The amount of energy required to increase the temperature of one pound of water at around 39.2 degrees *Fahrenheit by one degree Fahrenheit (454 g of water by 0.56°C at around 4°C). It is equal to around 1,056 joules but varies depending on the temperature of the water. As a unit of power, it is expressed as Btu per hour.

British thermal unit per hour An Imperial measure of the rate of release, gain, or flow of energy. In SI, one Btu per hour is equal to 0.293 W.

Brix A scale, expressed in degrees, that measures the percentage by weight of sucrose in water at a given temperature. It represents the concentration of a solution of pure sucrose in water expressed as parts by weight of sucrose per 100 parts by weight of solution.

bronze An alloy of mainly copper with tin and other elements such as phosphorous, manganese, aluminium, or silicon. Unlike *brass, which is an alloy of copper and zinc, and is malleable, bronze is hard and brittle. Being resistant to seawater, it is used in boats and ships for such things as propellers. Bronze is also used for valves, bearings, and other process machinery parts.

broth The aqueous growth medium used to support the growth of a microbiological culture in a *bioreactor. Also known as a **culture medium** or *growth medium, it contains all the necessary nutrients to sustain growth. The broth in a bioreactor is first inoculated with a sample of living microorganism, which then multiply consuming the nutrients. The rate of growth of the microbial cells can be determined by measuring the *turbidity of the broth or taking samples and determining the *cell dry weight.

Brownian motion The small, irregular, and continuous movement of very small particles suspended within a fluid. Particles with a diameter of less than one micrometre (1 μm) have a random movement caused by collisions with other particles. It is named after British botanist Robert Brown (1773–1858) who first noticed the phenomenon while studying pollen particles.

Brunner, Sir John Tomlinson (1842–1919) A British chemical industrialist and parliamentarian who worked in partnership with the German-born chemist Ludwig *Mond (1838–1909) to form the Brunner Mond & Co. Ltd chemical company. The company made alkali using the *Solvay process. He was a 1st Baronet and twice served as a Liberal Member of Parliament for the constituency of Northwich, Cheshire.

BSI An abbreviation for the British Standards Institution. It is responsible for the UK National Standards prefixed by BS as well as publication of standards in the UK prefixed by BS EN.

(⊕) SEE WEB LINKS
• Official website of BSI, with information about BSI standards.

BTL *See* BIOMASS TO LIQUIDS.

Btu *See* BRITISH THERMAL UNIT.

BTX A mixture of aromatics and includes benzene, toluene, and *xylenes. They are readily separated by distillation.

bubble cap column A *distillation column that is equipped with bubble cap trays, also known as **bubble trays**. These are flat perforated plates with short risers covered with inverted caps over the top. The caps have slots and allow vapour to rise through a liquid on the tray but prevent the liquid from draining back. This type of distillation column was common before the 1960s prior to the introduction of much cheaper *sieve plate column and valve tray designs.

bubble column A tall cylindrical vessel used for liquid-phase reactions using the sparging of a gas at the bottom to form bubbles within the liquid. The bubbles create the necessary turbulence for mixing. The surface area of the bubbles is important for mass transfer. An airlift column is a type of bubble column that has a concentric draft tube in which the bubbles rise outside the draft tube causing a liquid up-flow with the liquid flowing down the inside of the tube.

bubble flow A two-phase flow in which gas or vapour is present as bubbles in the liquid.

bubble point The temperature at which bubbles of vapour first appear on heating a liquid. For single component mixtures the bubble point and the *dew point are the same. For a mixture, the vapour will have a different composition from the liquid. Together with the dew point at different compositions these provide useful data when designing distillation systems. At the bubble point:

$$\sum_{i=1}^{n} y_i = \sum_{i=1}^{n} K_i x_i = 1.0$$

where K_i is the distribution or K-factor and is the ratio of the mole fraction in the vapour phase, y, to the mole fraction in the liquid phase, x, at equilibrium.

bubbling zone The location in a *fluidized bed in which a fluidizing gas is of sufficient pressure and flow to form rising pockets of gas or bubbles. The bubbling zone is where the bubbles grow by coalescence and rise to the bed surface where they break.

bubbly flow A two-phase flow regime in which the gas phase flows as bubbles dispersed in a flowing liquid. This type of flow tends to occur at low gas superficial velocities and high liquid velocities. The bubbles travel with a complex motion, each with a different velocity. The bubbles may coalesce and are generally of non-uniform size. In horizontal pipes, gravity tends to make bubbles accumulate in the upper part of the pipe except at very high liquid velocities when the intensity of the turbulence is enough to disperse the bubbles about the cross section.

Buckingham π theorem A theorem that describes how physically meaningful equations that describe observable phenomena involving n variables can be presented as an equation of $n-m$ dimensionless groups, in which m is the number of fundamental dimensions such as mass, M, length, L, and time, T. The theorem allows the dimensionless groups to be determined from the variables. The π theorem was first proved by French mathematician Joseph Bertrand (1822–1900) and is named after American physicist Edgar Buckingham (1867–1940). *See* RAYLEIGH'S DIMENSIONAL ANALYSIS METHOD.

bulk density The gross or apparent density of a material that includes void space. *See* APPARENT DENSITY.

bulk materials Process materials either used or produced in large quantities in a loose bulk form. Examples include ores, coal, grain, wood chip, sugar beet, and sugar cane. They are delivered to chemical plants by rail wagon, ship, and road tanker, are stored in *hoppers, *silos, or in stockpiles, and transported to the process by conveyor belts and bucket elevators.

bulk temperature The average temperature of a mass of substance or material.

bumping The violent boiling of a liquid caused by superheating such as bubbles of vapour that form at a pressure that is above the pressure of the liquid. Bumping can be prevented from occurring in the laboratory by placing small pieces of porous clay pot in the liquid in order to allow bubbles to form at the pressure of the liquid (i.e. atmospheric pressure for an open container).

bund A containment wall surrounding a storage tank or process vessel designed to contain its contents in the event of accidental leakage, spillage, or catastrophic failure. They are used to contain liquids that may be harmful, polluting, or explosive hazards.

Bunsen burner A type of burner used in laboratories that has a vertical metal tube through which gas is fed. There is a sleeve around a hole near the base used to regulate the amount of air to be mixed with the gas. Adjustment can provide a low-temperature smoky and luminous yellow flame through to an intense high-temperature flame with a blue inner cone. The burner is named after German chemist Robert Wilhelm Bunsen (1811–99).

buoyancy The upthrust on a body immersed in a fluid. The force is equal to the weight of the fluid displaced. *See* ARCHIMEDES' PRINCIPLE.

burn The capability of a material to undergo or cause to undergo combustion.

burn degree The severity of burns to the human body. Burns of the first degree show hyperemia (redness); second-degree burns show vesication (blistering); third-degree burns result in necrosis of the skin and underlying tissues in the form of charring.

burning rate The rate at which a defined mass of a solid or liquid burns and is measured in the direction normal to the surface. The **burning velocity** is the rate at which a combustion wave propagates into unburned gas. For pre-mixed flames the velocity depends only on the temperature, pressure, and composition of the cold gas. *See* FLAME SPEED.

bursting disc (rupture disc) A device used to prevent the unsafe over-pressurization or under-pressurization of a process vessel. In the event of excess internal pressure, the disc will rupture. It consists of a thin disc of corrosion-resistant metal whose thickness is determined by the pressure it is designed to contain. These are widely used in the petrochemical, nuclear, and pharmaceutical industries and are often used in combination with other *safety relief valves.

bushel An Imperial unit of capacity used for liquids and solid substances such as grain. In the UK, this is equal to 8 Imperial gallons. In the US, it is a unit of dry measure and equal to 64 US pints.

butex process *See* PUREX PROCESS.

butterfly valve A type of valve used to control the flow of a fluid in a pipeline; it has a circular body and a rotary motion disc closure member which is pivotally supported by its stem. There are various styles including eccentric and high-performance valves. They offer a low pressure drop and are suitable for dirty and viscous fluids, however, they are prone to inducing *cavitation and *choked flow.

BWG *See* BIRMINGHAM WIRE GAUGE.

by-product A substance that forms at the same time as the main or desired product during a chemical reaction. It is usually of secondary economic significance to the main product although many industrial by-products have significant economic value and commercial use in their own right. The product from the cracking of oil is petroleum. Many other by-products are also formed that have economic value such as paraffin, lubricating oils, and other distillates. In some cases, the revenues from by-products can exceed the main product, such as, the extraction of gold from porphyry (i.e. a coarse-grained type of igneous rock) deposits. *Compare* CO-PRODUCT.

caking The formation of powdered, ground, granular, or crushed solid material into a larger solid mass. This may be desirable as in forming blocks, tablets, and pellets, or undesirable as in the blocking of filters, pipes, drains, and vessels. The moisture content of granular materials such as sugar may affect the propensity to form a cake.

calandria A vertical tube through which heat is exchanged for the purposes of evaporation of a liquid inside. They are used in *evaporators, *thermo-syphon reboilers, and *nuclear reactors. Steam is often used as the heating medium.

calcining (calcination) A high-temperature process used for ores and other granular materials in the presence of air to bring about a thermal decomposition, phase change, or removal of a volatile component. A **calciner** is a high-temperature rotating oven or kiln. Rotating vacuum furnaces are also for calcining.

calculus A branch of mathematics that deals with the differentiation and integration of functions. That is, methods of calculation for solving problems in which an unknown is not able to be expressed as a number or a finite set of numbers, but instead is expressed as a function or system of functions. By treating continuous changes as if they consisted of very small step changes, *differential calculus can be used to find the rate at which something happens such as a particle accelerating as a change in velocity with time. The rate of change, dv/dt in this case, is called the derivative of v with respect to t. If v is a known function of time, t, then the acceleration at any instant can be found through the process of *differentiation. Integral calculus is the reverse mathematical process and is used to find the end result of a known continuous change. It is used to determine the infinitesimal change in a variable over a short period. The overall change is found by a process of summation called *integration. For example, it can be used to find the velocity of a particle, v, that has constant acceleration, a, over a period of time:

$$v = \int_{t_1}^{t_2} a \ dt$$

Integration is also used for finding the area under a curve and volumes, as well as solving problems involving the summation of infinitesimals. The mathematical techniques were developed independently by Sir Isaac *Newton (1642–1727) and Gottfried Leibniz (1646–1716).

calendering A method of producing plastic sheet and film. Plastic that is heated and softened is forced between two heated rollers with a fixed spacing apart. The thick continuous sheet is formed and further rollers reduce the thickness and can emboss the sheet if required. The process is used to produce sheet plastic products such as flooring and tape.

calibration A method of adjusting an instrument so that its reading can be correlated to the actual values that are being measured. Calibration is achieved using methods such as recognized standards or by experimentation. For example, the calibration of flow meters can be achieved by the *dilution method in which a concentrated extraneous material is added to the flow stream in the vicinity of the flow meter and its diluted effect analysed further downstream from which the actual flow can be determined. For a material balance, the actual flow can be determined and the flow meter calibration completed. Pressure gauges are calibrated against standards and checked periodically for any drift.

A **calibration chart** is a graphical representation of values read from an instrument. It is usually presented on the x-axis and the corrected value or quantity on the y-axis. An instrument should be calibrated with a sufficient number of data points to ensure a proper relationship between the measured and indicated value.

calorie The amount of heat required to raise the temperature of 1 gram of water by 1°C. The calorie has been replaced by the joule as an SI unit, for which the conversion of one calorie (1 cal) is equal to 4.186 joules. The **Calorie** (upper case C) is a largely obsolete unit of energy and confined to defining the energy in foods. It is equal to one kilocalorie (1 kcal) or 4,186 joules.

calorific value The heating value or heat of combustion of a combustible material in oxygen. Determined using a *bomb calorimeter, the calorific value is used as a measure of the energy content of a material such as a hydrocarbon fuel or foodstuff. The calorific values of fuels are usually expressed in joules per unit mass such as MJ kg^{-1}. Hydrocarbon gases such as methane are expressed per unit volume such as MJ m^{-3} with reference to *s.t.p. to allow for temperature and pressure variations.

calorimeter A device used to measure the quantity of heat produced by combustion. It consists of a chamber in which a specimen or sample is placed and is surrounded by an outer chamber containing water. Following combustion of the sample, the temperature change in the outer chamber is measured and the heat produced or transferred is calculated. *See* BOMB CALORIMETER.

candela (Symbol Cd) The SI unit of luminous intensity. It is the luminous intensity in a given direction of a source that emits monochromatic radiation of frequency 540×10^{12} hertz that has a radiant intensity in that direction of 1/683 watts per *steradian.

CANDU A type of Canadian heavy-water-moderated pressure tube natural uranium reactor. The name is derived from **Can**adian, **d**euterium (i.e. heavy water) and **u**ranium fuel.

cannibalized The extensive use of components and equipment parts from a redundant process or facility to modify or return to service another plant or item of process equipment.

canning A process of food preservation for long-term storage in which food is sealed in a container and then subjected to an elevated temperature and for a sufficient period of time so as to kill all spoilage and pathogenic microorganisms that may be harmful to human health. Various temperature-time combination exposures have been developed for destroying microorganisms. A 12-D process refers to the necessary time and temperature required to destroy the pathogenic microorganism *Clostridium botulinum* by a million million (10^{12}) times. It uses a steam retort operated at a temperature of at least 121°C for at least three minutes. *See* BOTULINUM COOK.

capacity 1. The working volume of a process vessel. For a liquid, it is less than the total volume as allowance is made for head space or overflow. **2.** The total production rate of a chemical plant, petroleum refinery, or petroleum production facility. **3.** The amount of electric charge stored in a battery.

capacity coefficient A dimensionless group used in the scale-up of centrifugal pumps:

$$C_Q = \frac{Q}{ND^3}$$

where Q is the flow, and N and D are the rotational speed and diameter of the impeller, respectively.

capacity ratio A parameter used to quantify the performance of a heat exchanger expressed as the ratio of the lower to higher heat capacity rate.

capex An abbreviation of **cap**ital **ex**penditure, it is the finance used to acquire equipment, machinery, and buildings. It tends to refer to significant levels of expenditure. It is also known as **capital spending** or **capital expense**. *Compare* OPEX.

capillary A tube that has a very small internal diameter. In a vertical glass capillary, the elevation of a liquid is due to *surface tension forces. In a porous body, moisture held within capillaries is removed by evaporation at the surface and therefore limits the rate of drying. The rate of evaporation is therefore controlled by the movement of moisture to the surface from within the capillaries, which is dependent on surface tension, pore size, and the density of the liquid.

capillary bound water A thin layer of water that is absorbed on to the walls of a *capillary. This is distinct from *free moisture, which is bound by a free meniscus.

capillary flow The flow of a liquid through the interstices of a porous solid by liquid-solid molecular attraction.

capillary flow number A dimensionless number, Ca, in which the viscous force is related to the surface tension:

$$Ca = \frac{\mu v}{\sigma}$$

where μ is the viscosity, v is the fluid velocity, and σ is the surface tension. It is used in the atomization of fluids and the two-phase flow in beds of solids. It is proportional to the ratio of the *Weber number to the *Reynolds number.

capital cost The total investment that is required for a process or item of equipment to be commercially operable. The capital costs, C, for equipment are often specified as a function of volume or capacity, Q, and can be scaled up using the power law relation:

$$C = \left(\frac{Q_1}{Q_2}\right)^n$$

where n is determined empirically and based on data for equipment of comparable design.

carbon assimilation A process that incorporates carbon from atmospheric carbon dioxide into organic material such as plants through photosynthesis.

carbonation A general name for processes that use carbon dioxide as the reactant and involve dissolving the gas in an aqueous solution. An example is the production of calcium carbonate by passing kiln gases through an aqueous suspension of calcium hydroxide.

carbon capture and storage (CCS) A post-combustion technology that is currently being developed to mitigate the effects of climate change linked to renewable energy systems. It involves extracting carbon dioxide from power plants, factories, and other industrial facilities before being expelled into the atmosphere. The captured carbon dioxide is then injected into a secure underground storage site.

carbon credit An international tradable certificate that shows that an industrial company or corporation has prevented the emission of one tonne of carbon dioxide or another form of greenhouse gas that is equivalent to one tonne of carbon dioxide from entering the atmosphere. Carbon credits are used as a way of mitigating the increase of greenhouse gases by capping the amount of gases that can be emitted. The intention is to encourage industries to seek ways of reducing the amount of emissions into the atmosphere. As commodities with a monetary value, they can be traded between partners and used as a way of lowering a company's *carbon footprint.

(⊕) SEE WEB LINKS
• Official website of Carbon Futures, with an explanation of how carbon trading works.

carbon cycle The series of processes in which carbon is exchanged between organisms and the atmosphere as carbon dioxide, biomass, and carbonates through the processes of photosynthesis, decomposition, and respiration.

carbon fibres The fibres of carbon used to provide strength to a wide variety of lightweight manufactured products such as sports equipment. There are several manufacturing processes: Rayon fibres with a high-modulus are charred around 300°C, carbonized between 1,000°C and 2,000°C, and heat-stretched at 3,000°C. Fibres made from polyacrylonitrile result in a product with lesser mechanical properties but improved yield and follow a similar process. They can also be made from molten coal and heavy petroleum hydrocarbons in which fibres are spun from the liquid in oxygen with heat treatment at 3,000°C.

carbon footprint A representation of the amount of emissions of carbon dioxide from an industrial process or organization. The calculation is based on the amount of energy that has been consumed. The carbon footprint is expressed as either tonnes of carbon per year or tonnes of carbon dioxide per year.

(⊕) SEE WEB LINKS
• Official website of the Carbon Trust.

carbonization A general name for processes used to produce gases, coke, and smokeless fuels from organic materials such as coal by heat treatment.

carbon offset A method used by industrial organizations to calculate their *carbon footprint and for offsetting their emissions against initiatives that are aimed at reducing the amount of carbon dioxide in the atmosphere. This includes the sequestering of carbon dioxide through the planting and growth of new trees, supporting renewable energy projects such as wind farms, and through buying *carbon credits in the emissions trading market.

A **carbon-neutral process** in one that is completely offset by the amount of carbon dioxide released.

carbon sequestration The process of removing carbon from the atmosphere by capture and long-term storage. In nature, trees, forests, and oceans capture carbon. Captured carbon from millennia is in the sequestered form of oil, gas, and coal located underground in oil and gas reservoirs and coal seams. Modern technologies are now being developed to capture carbon dioxide to reduce the amount of carbon in the atmosphere in order to reduce the *greenhouse effect and control global warming.

carbon tax A tax or surcharge levied on the sale of fossil fuels (i.e. oil, gas, and coal). The level of taxation is based on the carbon content of each fuel. It is designed to discourage the use of fossil fuels and reduce emissions of carbon dioxide into the atmosphere.

carboy A rigid cylindrical container used for storing and transporting process liquids. They vary in capacity and range from around 1 to 60 litres. They are generally made from polypropylene and are suitable for a wide range of liquids. They are often used for collecting waste solvents.

carburization A process used for the surface hardening of steel in which steel is heated in a gaseous, liquid, or solid carburizing medium that allows the absorption of carbon onto its surface. It typically uses temperatures of around 1,700°C and is followed by quenching.

carcinogen A substance or agent that is capable of producing cancer in humans. Chemical substances such as benzene and phenol are known carcinogens. Exposure to *ionizing radiation and *ultraviolet radiation is also known to cause cancer. Asbestos and tobacco particles in the lungs are also known to cause cancer.

Carman–Kozeny equation A semi-empirical relationship used to determine the pressure drop through a *packed bed of solids and permeability of porous media as:

$$\frac{\Delta p}{L} = \frac{180 \mu v (1-e)^2}{d_p^2 e^3}$$

where μ is the viscosity, v is the superficial velocity, e is the voidage, and d_p is the diameter of the particles. It is valid only for the laminar flow fluid through the bed up to Reynolds number of about one. It is named after Austrian physicist Josef Kozeny (1889–1967) who proposed the equation in 1927 and Philip Carman who subsequently modified it in 1938 and again in 1956.

Carnot, Nicolas Léonard Sadi (1796–1832) A French physicist who began his career as a military engineer before turning to scientific research. In 1824 he published a book *Reflections on the Motive Power of Fire*, which provided for the first time a general theoretical approach to understanding the conditions under which the efficiency of heat engines could be maximized. The thermodynamic *Carnot cycle eventually led to the concept of *entropy. He died aged 36 from cholera.

Carnot cycle A thermodynamic cycle used in a heat engine and comprises four distinct steps: isothermal expansion, adiabatic expansion, isothermal compression, and then adiabatic compression (see Fig. 11). According to the **Carnot principle**, the efficiency of a heat

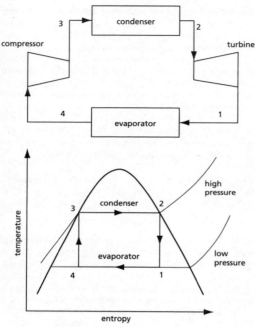

Fig. 11 Carnot steam cycle

engine depends on the temperatures at which it operates. The efficiency is the ratio of the work done, W, to the heat input, q_1

$$\eta = \frac{W}{q_1}$$

where, according to the *first law of thermodynamics, the work done is the difference between the heat in and heat out. The efficiency is therefore:

$$\eta = \frac{T_1 - T_2}{T_1}$$

For maximum efficiency, T_2 should be as small as possible and T_1 should be as high as possible. It was developed by Nicolas *Carnot in 1824.

carrier gas 1. An inert gas such as helium or unreactive gas such as nitrogen used in the chromatographic separation of gases. The carrier gas is the mobile phase and transports the gaseous substances to be separated through a long tube or column containing a packing material known as the stationary phase. The substances have different retention times and are detected as they are eventually eluted from the column. **2.** A gas used in a thermal spraying coating process in which melted or heated materials such as powders are sprayed onto a surface.

cascade control Two or more controllers working together to control a process. The output of the main (or master controller) is used as the *set point for the other (or slave controller). An example is the control of the output temperature of a *CSTR being used as the set point for the cooling jacket temperature. *See* FEEDBACK CONTROL.

cascade process A process that has a series of similar steps or devices having the same function. Each step or device may not provide sufficient chemical, biochemical, or physical change so further devices are therefore required to bring about a measurable or desirable change. For example, in the enrichment of the isotope uranium-235, the separation of uranium-235 from the heavier uranium-238 in the form of uranium hexafluoride (UF_6) gas in a single centrifugal separator is very low. Natural uranium contains 0.7 per cent of the uranium-235 isotope with the remainder being uranium-238. To achieve the sufficient level of enrichment consequently takes place in a cascade process involving many centrifugal separators operated in series in which the enriched gas forms the feed to the next separation.

Another example of a cascade process is where the conversion of a chemical reaction in a *continuous stirred-tank reactor (CSTR) is very low such that the continuous overflow from one can be fed into another reactor for further reaction and conversion. Where there are many CSTRs in series, the operation approximates to that of a *plug flow reactor.

case hardening 1. The formation of a layer in a solid being dried that resists the passage of moisture movement from the interior to the surface. In a food dehydration process, or paint or varnish drying, the surface hardens while much of the interior remains soft. **2.** The process of hardening the surface of steel that is used for tools and specialist purposes. It involves heating the steel in a hydrocarbon or by dipping the red-hot steel into molten sodium cyanide. A layer of nitrides can also be formed by the diffusion of nitrogen.

casing The body surrounding an impeller of a centrifugal pump or compressor.

cast iron A brittle and impure form of iron produced in a blast furnace. It is also known as pig iron and used in the manufacture of steel in a *Bessemer converter.

Castner–Kellner process A process once used to produce chlorine and sodium hydroxide by the electrolysis of sodium chloride solution. The electrolysis took place in a cell with a mercury cathode and graphite anode. The cell consisted of three compartments with a common bed of mercury and solution of sodium chloride above. The cathode was in the central compartment and anodes in the other two. It was invented independently by American industrial chemist Hamilton Young Castner (1858–99), who was working in the UK at the time, and Austrian chemical engineer Karl Kellner (1851–1905). The process was abandoned due to concerns over mercury pollution and replaced by various diaphragm-based electrolytic processes.

CAT An abbreviation for **c**arbon **a**batement **t**echnology. CAT is a group of technologies used to improve the efficiency of power plants with low-carbon alternatives such as biomass. *See* CARBON CAPTURE AND STORAGE.

catalysis The effect of a substance that, without itself undergoing change, aids a chemical change in other substances. The main forms of catalysis are: **homogeneous catalysis**, in which the catalyst is in the same phase as the reactants and products; and **heterogeneous catalysis**, in which the catalyst is in a different phase to the reacting system. A third, mainly biological form is **enzyme catalysis**, in which enzymes catalyze reactions. These are large complex organic molecules and do not form true solutions.

catalyst A substance that alters the rate of a chemical reaction without itself being consumed. While some catalysts are used in small quantities, such as the use of finely divided platinum in the decomposition of hydrogen peroxide, some catalysts are used in relatively large amounts such as in the Friedel–Crafts reaction:

$$C_6H_6 + C_2H_5 \rightarrow C_6H_5C_2H_5 + HCl$$

in which anhydrous aluminium chloride catalyst is required to be at least 30 per cent of the mass of benzene feed to be effective.

catalyst poison A substance that is absorbed more strongly than the reactants at the surface of a solid catalyst such that it prevents the catalyst from functioning. For example, in the *Haber process, sulphur compounds that may be present in the hydrogen gas feed can strongly bind on the iron catalyst preventing it from functioning.

catalytic coefficient A constant that relates the rate of reaction in a catalyzed reaction to the concentration of the catalyst. For a catalyzed reaction in which reactants are converted to products, the reaction rate is a function of the reactants. The proportionality or rate constant is dependent on the catalyst and the catalyst concentration. Where several catalysts are present, the overall rate constant is the sum of the products of the catalytic coefficients and catalyst concentrations.

catalytic converter A device used in the exhaust of internal combustion engines of motor vehicles to reduce the amount of harmful gases released into the environment. It consists of a chamber containing catalytic substances through which the exhaust gases flow. The catalyst converts unburnt hydrocarbons, carbon monoxide, and nitrogen oxide gases. Carbon monoxide is formed from the incomplete combustion of hydrocarbon fuel gases, and the nitrogen oxides or *NOx are formed from the high temperature reaction of nitrogen with oxygen in the air. The catalysts used to oxidize the unburnt hydrocarbons and the carbon monoxide, and convert the nitrogen oxides back to nitrogen, include platinum, palladium, and rhodium. Their performance is, however, adversely affected by some chemicals such as tetraethyl lead once widely used in some forms of petrol.

catalytic cracking A process that breaks down complex petroleum hydrocarbon molecules by the action of temperature and pressure in the presence of a catalyst. Zeolites and other minerals are used as the catalyst. Heavy oils can be converted into lighter products such as *LPG. The process takes place in a **catalytic cracker**, which is often abbreviated to the name *cat cracker.

catalytic distillation *See* REACTIVE DISTILLATION.

catalytic hydrocracking A refining process in which hydrogen and catalysts are used at a relatively low temperature and pressure to produce naphtha, diesel fuel, jet fuel, and high-grade fuel oil. *Hydrocracking is used for feedstocks that are particularly difficult to process by catalytic cracking or reforming because they may have a high sulphur content, or contain a high level of polycyclic aromatics or olefins.

catalytic reactor A vessel or item of process plant designed to contain a chemical reaction that requires the use of a catalyst to increase the rate of a reaction. The reactor may be operated batch-wise or continuously, depending on the particular requirements of the reaction, and charged with catalyst. The catalyst may be solid in the form of beads and either packed within tubes or free flowing. It can be either fluidized by the reaction mixtures or stirred to promote good heat and mass transfer.

catalytic reforming A process used to produce aromatic hydrocarbons by reforming straight-chain hydrocarbons in the C_6 to C_8 range from *naphtha or gasoline fractions into compounds containing benzene rings. Hydrogen is produced as a *by-product. The process is carried out at around 500°C and a pressure of 20 atmospheres in the presence of a catalyst. Various catalysts are used such as mixtures of platinum and aluminium oxide. Where platinum is used the process is known as *platforming.

catalytic rich gas process *See* CRG PROCESS.

catastrophic failure The sudden or complete failure of process equipment resulting in the rapid release and loss of process materials such as liquids and gases.

cat cracker An abbreviated form of *catalytic cracker and is a petrochemical refinery unit used to reduce large hydrocarbon molecules into smaller molecules. The conversion takes place at very high temperatures in the order of around 500°C in the presence of a catalyst.

cathode A negative electrode. In the process of *electrolysis in which electricity is passed through an electrolyte, positively charged ions (i.e. cations) are attracted to the cathode. *Compare* ANODE.

cathodic protection A method of preventing the rusting and corrosion of iron by using a reactive metal that makes the cathode of an electrochemical cell. Magnesium alloy bars are used for pipelines and iron structures, while zinc or an aluminium/zinc/indium alloy is used for the jacket legs of offshore platforms where the ions from the bars are formed in preference to the steel structure. It is therefore also known as **sacrificial protection** and the bars are called **sacrificial anodes**. Galvanized iron consists of a layer of zinc, where, like the magnesium ions, zinc ions are formed protecting the iron.

Cauchy number A dimensionless number, Ca, it is a measure of the ratio of the inertial to elastic forces in the compressible flow of fluids. It is the product of the density, ρ, and square of the velocity, v, to the bulk modulus of the fluid, E:

$$Ca = \frac{\rho v^2}{E}$$

causticization A process used for converting a solution of soda (sodium carbonate) with lime (calcium hydroxide) into sodium hydroxide and calcium carbonate as:

$$Na_2CO_3 + Ca(OH)_2 \rightarrow 2NaOH + CaCO_3$$

The process operates at around 90°C with an excess of lime and the precipitated calcium carbonate separated. Developed in the nineteenth century, the process was used after the *Leblanc process and before the *Castner–Kellner process.

Cavendish, the Hon. Henry (1731–1810) An English scientist and millionaire grandson of the Duke of Devonshire, he devoted his life to scientific investigation. A skilled experimenter, the drawing-room of one of his three London houses was a laboratory, with a forge in the next room and an observatory upstairs. He invented many new methods in both chemistry and physics. He determined many specific heats, the freezing point of mercury, and the relative density of gases. He also discovered the dielectric constant and 'weighed the Earth'. He was the first to investigate hydrogen, to synthesize water (which until then was thought to be an element), and to analyse air. In 1785 he made nitric acid by sparking nitrogen with oxygen. He also refined many methods such as drying gases for the correction of gas volumes for temperature and pressure. The French scientist Jean-Baptiste *Biot (1774–1862) said that he was: '*le plus riche de tous les savants, et probablement aussi le plus savant de tous les riches*'. The Cavendish Laboratory in Cambridge is his national memorial and contains much of his apparatus.

cavitation The destructive collapse of vapour in a liquid in localized regions of high pressure. Vapour will form in any liquid when the pressure is less than the vapour pressure at the liquid temperature. The possibility of this happening is much greater when the liquid is in motion, particularly in pumps on the suction side where velocities may be high and the pressure correspondingly reduced. Having been formed, the vapour bubbles travel with the liquid and eventually collapse with explosive force giving pressure waves of high intensity. In pumps this collapse, or cavitation, occurs on the impeller blades causing noise, vibration, and erosion of the blades, which then have a typically pitted appearance similar to corrosion. Another sign of cavitation is a rapid decrease in delivered head and efficiency. As a remedy, a throttle valve is placed in the delivery line and when any symptoms of cavitation are noticed the valve is partially closed. This restricts the throughput and hence lowers the velocity. Cavitation is more likely to occur with high-speed pumps and hot liquids. One way of ensuring that cavitation does not occur is to ensure that the available *net positive suction head (NPSH) exceeds the required NPSH.

CCR An abbreviation for **c**entral **c**ontrol **r**oom where process plant information is transmitted, logged, and controlled. The CCR is located in a safe part of the plant and is often protected against fire and explosion.

CCR platforming A **c**ontinuous **c**atalytic **r**egeneration process that uses platinum as the catalyst to convert straight-chain aliphatic hydrocarbons into aromatic hydrocarbons and hydrogen. It is a widely used form of the *platforming process.

CCS *See* CARBON CAPTURE AND STORAGE.

CEFIC An abbreviation for **C**onseil **E**uropéen des **F**édérations de l'**I**ndustrie **C**himique, it is an organization based in Brussels representing around 30,000 chemical companies, which account for about one-fifth of the world's chemical production. Founded in 1959, it comprises corporate, national federation, and business members and has responsibility for logistics, policy, international legislation, research and innovation, and means of public communication on chemical substances and their production. CEFIC is a member of the *International Council of Chemical Associations (ICCA).

(((⊕))) SEE WEB LINKS
• Official website of Conseil Européen des Fédérations de l'Industrie Chimique.

cell 1. A simple living organism such as a bacterium or yeast. Bacteria reproduce by binary division whereas yeast cells reproduce through budding. They are used as biocatalysts in biochemical reactions such as fermentation, converting sugars into alcohol and carbon dioxide and used in the food and brewing industries. **2.** A sealed chamber used in the nuclear reprocessing industry in which a process takes place. There is no access by process operators during normal operation. Control of the process is therefore carried out remotely. **3.** A system in which two electrodes are in contact with an electrolyte. In a *voltaic cell, electrochemical energy is produced by a chemical reaction that takes place between two electrodes made from different metals in an electrolyte consisting of salts or acidic substances. A voltaic cell is also known as a galvanic cell. *See* BATTERY.

cell dry weight A measure of the concentration of microorganisms in a bioreactor. The cell dry weight is obtained by completely drying a small but defined volume of the culture medium and weighing the dried material remaining. The **cell wet weight** can be measured more quickly but not as precisely, and is around four or five times greater than the cell dry weight. Measurement of the *optical density or absorbance of the culture medium is a fast and relatively easy process used to determine the cell density as a direct correlation which typically uses wavelengths between 548nm to 600nm. Most determinations are carried out using *off-line analysis, although absorbance measurements can be carried out *in situ*.

cell homogenizer A device used for the disruption of cell walls and membranes of microorganisms in order to release intercellular material. It involves forcing a volume of the living cells under a very high pressure through a small orifice onto an impact ring. Disruption of the cell wall and membrane occurs due to the high shear forces involved. *See* LYSIS.

cell output rate (COR) The production rate of microorganisms in a continuous stirred-tank bioreactor operating with a fixed volume. A product of the cell concentration and dilution rate, the cell output rate ranges from zero where the bioreactor is operated on a batch basis with no feed of substrate, to a maximum value. Where the rate of fresh substrate is increased further, washout of the microorganisms occurs since the rate of flow out exceeds the rate of growth.

Celsius, Anders (1701–44) A Swedish professor of astronomy who, in 1742, introduced the idea of using 100 divisions between the fixed points of melting ice and boiling water on a mercury thermometer. He called the upper fixed point 0 and the lower fixed point 100. In 1750 Martin Strömer (1707–70), with whom Celsius had worked, inverted the scale.

Celsius scale A temperature scale devised in 1742 by Swedish astronomer Anders *Celsius (1701–44). Celsius used the reference points of the freezing point of water (0°C) and boiling point of water (100°C) divided equally by 100 degrees. The Celsius scale officially replaced the earlier *centigrade scale in 1948. The conversion from degrees Fahrenheit is:

$$^{\circ}C = \frac{5}{9}\left(^{\circ}F - 32\right)$$

Degrees Celsius has the same magnitude as the *kelvin. Note that 0°C is equal to 273.15K.

cement A widely used material in the construction industry which, when mixed with sand and aggregates, is used in the construction of buildings, bridges, and roads. The raw materials in its manufacture are limestone, silica-rich clay, gypsum (used to control the hardening process), and other materials such as sand or slag from a blast furnace used to adjust the mix. They are finely ground and burnt in a kiln at 1,500°C to form clinker. The clinker is formed due to reactions and the evaporation of water, and results in the formation of carbon dioxide. Cement is chemically unstable. When water is added, it cures gradually hardening.

cementation A metallurgical process in which a metallic surface is impregnated with another surface. It involves heating the metal such as steel and dipping it into molten sodium cyanide. It is used to harden the steel. *See* CASE HARDENING.

CEng Post-nominal letters used after someone's name to indicate that the person is a chartered engineer registered with the Engineering Council in the UK. A chartered chemical engineer would additionally include the grade of member or fellow of the Institution of Chemical Engineers with the post-nominal letters MIChemE or FIChemE.

centigrade scale A former scale for representing temperature now officially called the *Celsius scale.

centipoise A unit for viscosity in the c.g.s. system and equal to one-hundredth of a *poise. 1 cP is equal to 0.001 Pa s. *See* VISCOSITY.

centistokes A unit of *kinematic viscosity in the c.g.s. system where one cSt is equal to $10^{-6} m^2 s^{-1}$.

centreline A reference point along the axis of a tube, pipe, or cylindrical vessel of uniform circular cross section, or along the axis of a rotating shaft.

centrifugal decanter *See* DECANTER.

centrifugal fan A machine used to develop large amounts of air or a gas at low gauge pressure and consists of a rotating shaft with blades or fan. The air is fed axially into the fan and is moved to the periphery by centrifugal force. Variants include forward-curved, backward-curved, and straight-bladed fans. Mechanical efficiencies may be as high as 80 per cent.

centrifugal force A force that acts radially outwards from the centre of a circular path. The opposite is the **centripetal force**, which is the force that acts towards the centre of the circular path.

centrifugal pump A mechanical device used to transport fluids by way of an enclosed impeller rotating at high speed. This commonly used device involves the fluid being fed in at the centre or eye of the impeller and thrown out in a roughly radial direction by centrifugal action. The large increase in kinetic energy that results is converted into pressure energy at the pump outlet by using either an expanding volute chamber or a diffuser. The latter is more efficient but more expensive. There are considerable variations in impeller design, but almost all have blades, which are curved, usually backwards to the direction of rotation. This arrangement provides the most stable flow characteristic. The head developed depends not only on the size and rotational speed of the pump, but also on the volumetric flow rate.

Centrifugal pumps are suitable for handling fluids with a wide range of properties including fluids with suspended solids. They are also capable of operating when the delivery line is blocked. They have low capital and maintenance costs, and easy fabrication in a wide range of corrosion-resistant materials. However, they have an inability to develop high heads unless multiple stages are used. They are also not self-priming. The pump must therefore be full at the point of start-up, which is achieved either manually, or requires ancillary equipment. They have a high efficiency over only a limited range of conditions and are not particularly suitable for highly viscous fluids. They are also prone to *cavitation in which the pressure of the fluid falls below the vapour pressure resulting in the formation of bubbles of vapour which collapse with a violent effect within the pump at a region of high pressure. This results in noise, vibration, and eventually damage to the impeller. It is avoided by ensuring that the available *net positive suction head exceeds the required net positive suction head.

centrifuge A mechanical device consisting of a rotating bowl, basket, or cones or blades, used to separate solid particles from a liquid or a suspension of immiscible droplets in a liquid by virtue of having a difference in density. The particles are usually small and have a similar density to the liquid, such that separation by gravity is too slow or not even possible. By applying a centrifugal force, separation can be achieved more easily. Centrifuges are used in the oil industry to separate droplets of oil from water streams. High-speed centrifuges are used in biotechnological processes such as separating spent yeast cells from fermented beer. *See* CYCLONE SEPARATOR.

CERN An abbreviation for the Conseil Européen pour la Recherche Nucléaire, which is the European Laboratory for Particle Physics located on the border of France and Switzerland near Geneva. As an intergovernmental research centre founded in 1954, it has been used for research into high-energy particle physics, and is now the home to the Large Hadron Collider. This is a 27-km particle accelerator.

(⊕) SEE WEB LINKS
• Official information website of CERN, accessible to the public.

cetane number A number used as a measure of diesel fuels that represents the percentage of hexadecane (i.e. cetane) in a mixture with 1-methylnaphthalene that has the same ignition characteristics as a diesel fuel being tested in a standard diesel engine. It is therefore an indication of the ease of self-ignition. The **cetane index** is a number calculated from the average boiling point and the density of a petroleum fraction in the diesel fuel boiling range, which estimates the cetane number of the fraction.

CFC *See* CHLOROFLUOROCARBON.

CFD *See* COMPUTATIONAL FLUID DYNAMICS.

CFM An Imperial volumetric rate of flow of gases and air expressed as cubic feet per minute. It is typically used in the design and specification of ventilation systems.

CFPD An Imperial volumetric measure of gas flow expressed as cubic feet per day. The flows are usually large and often given as **MMSCFPD** or **m**illions of **s**tandard **c**ubic **f**eet **p**er **d**ay.

c.g.s. system A system of units that uses centimetre, gram, and second as the base units. A great deal of early scientific work was done using the c.g.s. system. Derived units were given names and some are still in use today, such as the *poise for viscosity, *dyne for force, and *erg for energy or work. The c.g.s. system has largely been replaced by *SI units.

chain reaction A self-sustaining chemical reaction or nuclear process. In a *nuclear reactor, the neutron-induced fission of uranium-235 results in the release of neutrons, which in turn cause the fission of further uranium-235 atoms. In chemical chain reactions, free radicals act as intermediates in the overall chemical reaction.

chamber process *See* LEAD CHAMBER PROCESS.

change of phase A change from one physical state to another (i.e. solid, liquid, and gas). The transition is accompanied by a change in energy. While heat is absorbed or evolved, there is no temperature change. Examples include the melting of a solid to a liquid, the freezing of a liquid to a solid, the boiling or evaporation of a liquid to a vapour, the condensation of a vapour to a liquid, and the sublimation of a solid to a vapour.

channel A trough used for the passage or transport of liquids, usually with a free surface. That is, there is a gas or vapour such as air above the surface of the liquid. The cross section of the channel is usually square, trapezoidal, or semi-circular.

characteristic curve A diagram presenting the delivered head or pressure, total power consumption, and efficiency against volumetric flow rate for a centrifugal pump. The shape of the curve tends to show a fall or decrease in the delivered head or pressure with increasing volumetric flow rate while the power consumption increases with flow rate. The efficiency of the pump, which is expressed as the ratio of the work delivered by the fluid to the work input to the pump, rises with increasing delivered flow to a maximum known as the *best efficiency point, and then decreases thereafter.

charge The quantity or load of material fed into a process, the quantity contained within a storage vessel, or the amount of fuel fed into a burner.

charge hand A senior process operator whose responsibility is below that of a *foreman.

Charles, Jacques Alexandre César (1746–1823) A French chemist and physicist best known for his discovery of *Charles's law. He made the first hydrogen balloon ascent in 1783, which was sponsored by the Académie des Sciences. He prepared the hydrogen, filled the balloon, and with an assistant, Nicolas-Louis Robert (1760–1820), rose to a height of over 500 m. He also experimented with atmospheric electricity.

Charles's law A law that states the relation between the volume of a gas and its temperature. It states that the volume of a given mass of gas is directly proportional to its absolute temperature at constant pressure. It was proposed by Jacques *Charles (1746–1823) although he never published it. *Gay-Lussac (1778–1850), who had made the discovery in 1802, acknowledged that 'Citizen Charles' had remarked on it fifteen years earlier.

chartered chemical engineer A person who is both academically and professionally qualified in chemical engineering. A chartered chemical engineer has a proven ability to work at a high level without supervision to solve complex engineering problems, develop new or existing technologies through innovation, creativity, and change. He or she may be involved in pioneering or promoting advanced designs and design methods, work on new and more efficient production techniques, marketing and construction concepts, engineering services, and management methods. A chartered chemical engineer is also engaged in technical and commercial leadership. The person is entitled to use the post-nominal letters *CEng after his or her name, and will also be a member of the *Institution of Chemical Engineers.

(( )) SEE WEB LINKS
• Official website of the Engineering Council.

chart recorder An electromechanical device used to record data from several inputs using a paper and coloured pens. Strip chart recorders have a long continuous strip of paper that is fed at a constant speed, whereas circular chart recorders have a rotating disc of paper, which requires periodic replacement. Chart recorders were once a common feature of control rooms and used to record the history of process plant operations, but have now been largely superseded by electronic computer capture.

check valve A non-return valve used to control the direction of flow of a fluid such as the entry to and exit from a reciprocating pump. It is designed to close automatically in the event of reversed flow. The design consists of a flap or ball, which seals the pipe when the pressure caused by flow is in the reverse direction, but lifts or opens when in the required direction.

CHEMECA A major annual international conference of the Australian and New Zealand community of chemical engineers and industrial chemists held under the auspices of *IChemE in Australia, the *Royal Australian Chemical Institute, and the Society of Chemical Engineers New Zealand, the Institution of Chemical Engineers in New Zealand. The 2012 conference took place in Wellington, New Zealand.

(⊕) SEE WEB LINKS
• Official website for CHEMECA conference 2012.

chem eng An abbreviation or loose term for chemical engineering. A **chem-enger** is a slang term for a student of chemical engineering.

chemical amount *See* AMOUNT OF SUBSTANCE.

Chemical & Engineering News **(C&EN)** A weekly magazine published by the American Chemical Society that provides professional and technical information to chemical engineers. Founded in 1923, it includes topical information on news, research employment information, business and industry news, government and policy, education, and special reports. It is printed and also available online.

(⊕) SEE WEB LINKS
• Online access to *Chemical & Engineering News*.

chemical bond A mechanism by which atoms are held together to form molecules. There are various types of chemical bond and these include the attraction of opposite charges and the formation of stable configurations through electron-sharing. The main forms of bonds are covalent, ionic, metallic, and hydrogen bonds. Valence governs the number of bonds that an atom can form.

chemical engineering A branch of engineering that deals with the design, construction, and operation of processes and plant that involve physical, chemical, and biological change for the conversion of raw materials into useful products on an industrial scale. The principal operations include mixing, reaction, and separation.

Chemical Engineering **1.** A series of six textbooks co-written and authored by John *Coulson (1910–90) and Jack *Richardson (1920–2011) first published in 1954. **2.** A monthly magazine published in the US covering all aspects of chemical engineering and aimed at professional chemical engineers, academics, and associated professionals in the manufacturing, chemical, and allied industries.

(⊕) SEE WEB LINKS
• Online access to *Chemical Engineering* magazine.

chemical equation The use of symbols to present a chemical reaction. The equation provides a mathematically balanced number of atoms present in the reaction such that the left-hand side presents the reactants and equals the right-hand side, which presents the products. An example is the combustion of methane in oxygen (reactants) to form carbon dioxide and water (products):

$$CH_4 + 2O_2 \rightarrow CO_2 + 2H_2O$$

chemical equilibrium A thermodynamic equilibrium between the reactants and products in a reversible chemical reaction. The balance in a reversible chemical reaction in which two chemical reactions proceed at the same rate but in opposite directions. That is, the rate of the forwards reaction is balanced by the rate of the backwards reaction. For example, in the reaction $wA + xB \leftrightarrow yC + zD$ the reactants A and B form the products C and D, which can also return to form A and B. For the balanced reaction at equilibrium, the *equilibrium constant is the ratio of the molar concentrations of the products to the reactants given by:

$$K_c = \frac{[C]^y [D]^z}{[A]^w [B]^x}$$

The square brackets indicate equilibrium concentrations.

chemical formula The use of symbols to present a chemical compound. The symbols represent atoms. Subscripts are used to represent the number of atoms. An example is sulphuric acid, H_2SO_4, which comprises two atoms of hydrogen, one atom of sulphur, and four of oxygen.

chemical hazard Any chemical substance or chemical process that presents a hazard to humans, property, or the environment, and which may realize its potential through fire, explosion, or toxic or corrosive effects.

chemical industry A major economic manufacturing sector responsible for the development and mass production of chemicals. Early nineteenth-century processes included the mass production of alkalis and, in particular, soda ash, caustic soda, and bleach. By the end of that century, various metals were being extracted from ores in large-scale processes, along with sulphuric and nitric acid, explosives, and fertilizers. The *Haber process in the early twentieth century led to the production of fertilizers and radical changes in agricultural practices through the fixation of nitrogen. After the First World War, the rapid rise of the automotive industry resulted in the demand for petroleum and petroleum products including thermoplastics. The Second World War spurred the development of chemical products such as synthetic rubber, and the decades that followed led to new plastics such as polyester, polyvinylchloride, and polypropylene. The first decades of the twenty-first century are marked by the development of pharmaceutical production and with new and innovative food products.

chemically combined water Moisture that is chemically combined either in the form of hydroxyl ions (OH^-) or in molecular compounds such as hydrates. Chemically combined water is not usually removed from biological materials during drying processes.

chemical oxygen demand (COD) A standard method used for the indirect measurement of the amount of pollution that is not able to be oxidized biologically in a sample of water. The test procedure is based on the chemical decomposition of organic and inorganic contaminants dissolved or suspended in water. The test indicates the amount of water-dissolved oxygen consumed by the contaminants; the higher the chemical oxygen demand, the higher the amount of pollution in the sample. The quantity of oxidant consumed is expressed in terms of its oxygen equivalence and is expressed in mg L^{-1} of oxygen.

chemical plant An industrial factory that processes chemicals on a large or industrial scale. Raw materials undergo chemical, physical, or biological transformation into useful products within process equipment through the processes or *unit operations of mixing, reaction, and separation.

chemical potential (Symbol μ) The change in *Gibbs free energy, G, with respect to the amount, n, of component in a mixture $\partial G/\partial n$ with temperature, pressure, and the amounts of the other components being constant. The components are in equilibrium when their chemical potentials are equal.

chemical process A general term used for the industrial manufacture of useful chemical products from raw materials through chemical, physical, or biochemical transformation. The transformation may involve a single step or many steps within *unit operations such as mixing, reaction, and separation, in either batch or continuous forms of operation. A chemical process can be presented schematically as either a *block flow diagram or as a *process flow diagram. Block flow diagrams use blocks to represent the unit operations with arrows to represent the flows in and out, whereas process flow diagrams provide more detail of the unit operations and include process piping, the main items of equipment, control valves, and operational information such as temperature, pressure, and flows.

chemical reaction The chemical conversion of elements and compounds known as *reactants into other substances or *products that involves the breaking and formation of *chemical bonds. The chemical reaction can be *exothermic with the liberation of heat, or *endothermic in which heat is absorbed during the reaction. Other effects of a chemical reaction can include colour change, the formation of solids such as precipitation, the liberation of gas, light, and sound. All chemical reactions are theoretically reversible although some are deemed to be irreversible since the reverse reaction rate is negligibly small, such as the combustion of fuel in oxygen.

chemical reaction equilibrium The balance in a reversible chemical reaction in which two chemical reactions proceed at the same rate but in opposite directions. For example, in the esterification of fatty acids with an alcohol, an ester and water begin to be formed as products as the concentration of reactants become depleted, eventually reaching a balance point where the products break down and revert back into the reactants. *See* CHEMICAL EQUILIBRIUM.

chemical reactor *See* REACTOR.

chemical substance A material with a definite chemical composition. Elements, compounds, and alloys are chemical substances.

chemical vapour deposition (CVD) A coating process that uses a reactant gas or vapour that decomposes onto the heated surface of a material depositing a solid element or compound. The process takes place in a chamber and the deposition forms coatings that are corrosion resistant and durable. CVD is used to provide coatings in the production of integrated circuits, anti-oxidation coatings of refractory materials, and hard cutting-tool coatings of titanium carbide, and other similar materials.

chemical warfare The use of chemicals as weapons. These include poisons, nerve gases, defoliants, and herbicides; mustard gas was used during the First World War. Their deployment was prohibited by the Geneva Convention in 1925; however, their production is not prohibited. The defoliant Agent Orange was used by the US during the Vietnam War.

chemisorption The force of attraction of gases and vapours by chemical forces to a solid surface of a substance rather than by physical or van der Waals adsorption. It is dependent on the nature of the surface and is restricted to definite sites on the surface.

chemistry The study of chemical elements, the compounds that are formed from elements, and their reactions. There are many specialist fields of chemistry including organic, inorganic, physical, and biochemistry.

chemostat cultivation *See* CONTINUOUS CULTIVATION.

Chernobyl A major nuclear accident that occurred on 26 April 1986 at the Ukrainian nuclear power plant. Following a power output surge, an emergency shutdown was attempted. With an even greater power output surge, the reactor vessel ruptured. A series of explosions then followed, destroying the reactor. With the graphite moderator then exposed, a fire began, with the release of radioactive material into the environment. Many people were killed as the result of exposure to ionizing radiation, and many more were to suffer cancer-related deaths. The disaster had far-reaching implications for nuclear power and safety in the nuclear industry.

Chézy formula A semi-empirical formula that relates the rate of discharge of liquid in an open *channel to its dimensions, slope, and surface roughness as $Q = CA\sqrt{mi}$ where C is the Chézy coefficient, A is the cross-sectional area of the channel, m is the *mean hydraulic diameter, and i is the slope of the channel. The formula was devised by French engineer Antoine Chézy (1718–98) who was responsible for designing a canal system to supply water to Paris. The Chézy coefficient was developed further in 1890 by Irish engineer Robert Manning (1816–97). *See* MANNING FORMULA.

chilling A thermal process in which thermal energy is removed from a substance to reduce the temperature to below ambient but without a phase change. It is widely used in the food industry to reduce the temperature of perishable foods such as meat, fish, and dairy products to extend shelf life by slowing down the microbiological processes of enzymes and microorganisms present.

Chilton, Cecil Hamilton (1918–72) An American chemical engineer who co-edited the *Chemical Engineers' Handbook* with Robert H. Perry (1924–78). He was senior advisor at the Battelle Memorial Institute, Columbus, Ohio. Having completed the 5th edition, he died following heart surgery before its publication in 1973.

Chilton, Thomas Hamilton (1899–1973) An American chemical engineer noted for his pioneering work in chemical engineering practice. A professor of chemical engineering, he studied heat transfer, fluid flow, distillation, and other aspects of chemical engineering, and developed the *Chilton–Colburn analogy. During the Second World War he worked on the Manhattan Project. He was president of the American Institute of Chemical Engineers in 1951.

Chilton–Colburn analogy A widely used analogy from heat, momentum, and mass transfer analogies. Also known as *j-factors, they are used to determine an unknown transfer coefficient when one of the other coefficients is known. It applies to fully developed turbulent flow in pipes, and relates mass and heat transfer coefficients, and friction factors. It was proposed by and named after American chemical engineers Thomas H. *Chilton (1899–1973) and Allan P. Colburn (1904–55).

CHISA An international chemical engineering congress held in the Czech Republic, which began in the Czech city of Brno in 1962. It is aimed at advancing chemical engineering research, development, and practice. The word CHISA originates from the Czech abbreviation for 'Chemical Engineering, Chemical Equipment Design and Automation' and is now a form of trademark for large meetings that have emphasized European collaborations.

chlorination 1. A chemical reaction in which chlorine is added to a compound. **2.** A process for purifying water for drinking or for disinfecting water such as in swimming pools.

chlorofluorocarbon (CFC) A substance, usually an alkane, in which all the hydrogen atoms have been replaced with chlorine and fluorine atoms. Developed for use as refrigerants, aerosol propellants, solvents, and in the manufacture of foam packaging materials, they are chemically inert and unreactive. However, because of these qualities, they are known to diffuse into the upper atmosphere where photochemical reactions result in the reaction with the protective ozone layer of the Earth. Their manufacture and use has therefore been discouraged beginning with the Montreal Protocol in 1987.

choked flow A condition in which a fluid becomes limited in its flow or 'choked' and is not able to be increased further. For a fluid flowing through an orifice or small hole in a pipe, the increase in velocity corresponds to a decrease in pressure, known as the venturi effect. However, a point is reached in which the rate of flow will not result in any further decrease in pressure, thereby limiting flow. The choking of gases occurs when the velocity leaving the orifice approaches sonic velocity i.e. at a Mach number of one. This results in shock waves that restrict flow causing the choking effect. The deliberate choking of gases is sometimes useful for limiting the rate of flow to processes. For liquids, the decrease in pressure below the vapour pressure results in partial flashing and *cavitation, with the formation of vapour effectively limiting flow. It is also known as the **critical flow** and is important in process safety, particularly in terms of the rate of release of material from a vessel or vent when depressurizing.

CHP *See* COMBINED HEAT AND POWER.

Christmas tree An assembly of valves and fittings used to control the pressure and flow in oil and gas wells. It is located on top of the wellhead and provides the controlling mechanism for the isolation of wells. It has many other functions including permitting the injection of chemicals into wells and pressure relief. Playing an essential role in an *emergency shutdown (ESD), a basic form of assembly has several manual gate valves with typically four or five valves being arranged in a crucifix arrangement. The name is derived from its resemblance in both shape and decoration to a Christmas tree.

churn flow A two-phase flow regime in a pipe or tube characterized by the oscillatory transition from a continuous liquid phase to a continuous and predominant gas phase. It occurs when gas bubbles coalesce and liquid becomes entrained in the bubbles. At high gas velocities, **Taylor bubbles** that have been formed in *plug flow break down into an unstable pattern in which there is a churning or oscillatory motion of liquid. Churn flow may be regarded as a breaking-up of plug flow with an occasional bridging across the pipe by the liquid phase. At high gas flow rates, it may be considered as a degenerative form of *annular flow with the direction of the film flow changing and large waves being formed on the interface for which the term **semi-annular flow** is occasionally used.

CIMAH An abbreviation for Control of Industrial Major Accident Hazards, which were UK regulations issued in 1984. Since 1999 they have been superseded by the *COMAH regulations.

CIP An abbreviation for: **1.** Cold isostatic pressing. **2.** Constant injection pressure. **3.** *Clean-in-place. **4.** Carbon in pulp.

circulating pump A pump used to circulate a process liquid from, and back to, a process. Circulating pumps can be used as a form of mixing or in the prevention of suspended particles settling.

cladding The tight-fitting surrounding material used to contain *nuclear fuel. Its purpose is to protect the fuel against chemically active agents and to prevent the release of fission products into cooling water, particular when the fuel is stored in water prior to reprocessing.

Clapeyron–Clausius equation *See* CLAUSIUS–CLAPEYRON EQUATION.

clarification A general name for processes used to remove suspended matter from a solution. It includes the processes of *filtration, centrifugation, and *sedimentation. *Compare* THICKENING.

clarifier 1. A large tank used to remove suspended matter from a solution under the influence of gravity. It has a continuous input and output flow. **2.** A device used for the removal of suspended particles from a liquid in order to reduce or remove the turbidity. It may be either a filter or a centrifuge.

clarifying agent A soluble component added to a liquid used to remove turbidity. Examples include gelatine and pectinases used to clarify wine and beer.

Claude, Georges (1870–1960) A French chemist and physicist noted for his study of gases at different pressures. He discovered that acetylene is very soluble in acetone and gave rise to a method for storing the gas. His research on rare gases obtained by the *liquefaction of air was developed into the invention of neon signs. Two processes are known by his name: the *Claude process for liquefying gases by a series of processes including cooling by expansion while performing work against a piston; and the *Claude process for the manufacture of ammonia.

Claude process 1. A method used for the liquefaction of gases in which a highly compressed gas is cooled by expansion in an expansion engine. This is followed by further cooling in a heat exchanger, and finally cooled by the *Joule–Thomson effect as it passes through an expansion valve to a lower pressure. The liquid obtained in the last expansion is withdrawn in which the remaining gas is used as the cooling medium in the heat exchanger before it is recompressed and returned to the process. The use of the expansion engine to recover some of the energy used in compressing the gas is a refinement of the earlier *Linde process and also makes the liquefaction process more rapid. **2.** A process for the manufacture of ammonia involving the electrolysis of water to produce hydrogen which is then burnt in air, thus converting oxygen into water and leaving nitrogen. The nitrogen and more hydrogen are then mixed, compressed to 750 bar, and passed over a catalyst at 500°C. It is named after Georges *Claude (1870–1960).

Clausius, Rudolph Julius Emmanuel (1822–88) A German physicist noted for formulating the second law of thermodynamics in 1850 independent of Lord *Kelvin. He introduced the concept of *entropy, and also contributed to electrochemistry and electrodynamics. He held teaching posts in Berlin and Zurich before taking a post at Würzburg in 1869.

Clausius–Clapeyron equation A relationship between the change in saturated vapour pressure and the latent heat of vaporization when there is a change of state:

$$\frac{dp}{dT} = \frac{\lambda}{(V - v)T}$$

where λ is the latent heat of vaporization, p is the saturated vapour pressure, T the absolute temperature, V the specific volume of the vapour at T, and v is the specific volume of the liquid. Assuming that the *ideal gas law applies, then the equation can be written as:

$$\frac{d\ln p}{dT} = \frac{L}{RT^2}$$

The equation can also be used for the transition of a solid to a liquid. The equation is named after German physicist Rudolph *Clausius (1822–88) and French engineer Benoît Paul Émile Clapeyron (1799–1864).

Claus process A two-stage process used for the removal of sulphur from natural gas or crude oil. In the first, hydrogen sulphide is partially oxidized using air to form sulphur dioxide. In the second, the sulphur dioxide is reacted with the hydrogen sulphide in the presence of a catalyst at 300ºC to form elemental sulphur and water vapour:

$$2H_2S + 3O_2 \rightarrow 2SO_2 + 2H_2O$$
$$SO_2 + 2H_2S \rightarrow 3S + 2H_2O$$

It is named after its inventor, chemist Carl Friedrich Claus, who developed the process in 1883 while working on ways of recovering sulphur from the waste calcium sulphide produced in the *Leblanc process. He originally used iron ore and later bauxite as the catalyst.

Clean Air Act An Act of the UK Parliament introduced in 1956 to reduce the level of air pollution, in particular, by controlling the use of coal as a heating medium in homes. The Act was a major landmark in environmental protection in the UK and was responsible for reducing the level of smoke pollution as well as harmful sulphur dioxide emitted from domestic fireplaces into the environment. The Act was also responsible for the relocation of power stations away from urban areas, and also included an increase in the height of chimneys for more effective dispersion of emissions. The Act was amended in 1993.

In the US, the Clean Air Act was introduced in 1963 as a federal law to control air pollution. The Environmental Protection Agency is responsible for enforcing the regulations and for protecting the public from airborne contaminants that are hazardous to health arising from domestic and industrial sources. There have been subsequent amendments to the Act, particularly to address airborne pollution such as *acid rain and effects of ozone depletion.

SEE WEB LINKS
• Official website of HM Government managed by the National Archives to publish all enacted legislation in the UK.

clean-in-place (CIP) A fully or partly automated technique used to clean and sanitize closed process equipment after use and before reuse. Used throughout the food and biochemical industries, it avoids the time-consuming process of dismantling equipment and manual cleaning, or where cleaning by other means is too difficult due to restricted access. The equipment to be cleaned is equipped with nozzles with supply and return pipes to and from a CIP kitchen. This involves the preparation of the necessary chemicals and wash water, and heat exchangers. The cleaning solution is pumped through the equipment often as a spray through the nozzles. A CIP programme typically involves a pre-rinse with water, circulation with a cleaning solution, an intermediate rinse, disinfection, and a final rinse with water. Cleaning agents used in the food and drink industry include alkalis such as sodium and potassium hydroxide, sodium carbonate, acids such as nitric acid, phosphoric acid, citric acid, and gluconic acid. Formulated cleaning agents containing chelating agents include EDTA, NTA, phosphates, polyphosphates, phosphonates, as well as surface active agents.

climbing film evaporator *See* RISING FILM EVAPORATOR.

closed loop control A system or process being controlled in which the controlled variable is measured and the result of this measurement is used to manipulate one of the process variables, such as steam flow to a heat exchanger, for example.

closed system A system in which there is no transfer of material across the system boundary. Differential material and energy balances are used to describe closed systems. A batch process takes place in a closed system.

close packing A packing arrangement of spherical particles, such as catalysts, in which there is a minimum amount of space around them. For each spherical particle, there are twelve other particles in contact. The *voidage is the space between the spheres.

cloud point The temperature of a liquid at which dissolved solids precipitate giving a cloudy appearance. It is used in the petrochemical industry to measure the point at which wax forms in diesel fuels. It is used to indicate the point that such liquids will deposit wax onto surfaces causing the blockage of pipes.

coagulation The joining of colloidal particles to form a larger mass of particles. Coagulation occurs when an agent is added to a colloidal solution in which the ions change the ionic strength of the solution, and therefore destabilize the colloid. For example, alum is used to remove proteins in beer and wine which otherwise give cloudiness. Alum and iron (III) sulphate are used as coagulation agents in sewage treatment.

coal equivalent A measure of the energy within a fuel that is equivalent to the energy contained within coal. Although coal itself is variable in composition and calorific value, a standard of one coal equivalent corresponds to 7,000 kilocalories (≈ 29.3 MJ). For examples, 1.0 kg of fuel oil is equal to 1.52 kg coal equivalent; 1.0 m^3 of natural gas is equal to 1.35 coal equivalent; 1 kg of uranium-235 is equal to 27×10^6 kg coal equivalent.

coalescence The forming together of droplets of liquid that are dispersed within another liquid, such as droplets of oil within water. Coalescence can be improved by adding agents that reduce the surface tension of the droplets. A **coalescer** is a type of separation vessel used to separate emulsions. Baffles and filters are used to cause droplets to coalesce. Electrostatic coalescers use electric fields to cause the coalescence. Coalescers are used in the separation of oil and water, particularly in the offshore and onshore petroleum industries.

coal gas A *fuel gas once produced from the heating of coal in the absence of oxygen. Once used extensively in the nineteenth and early part of the twentieth centuries, coal gas typically contained hydrogen, methane, and a significant amount of toxic carbon monoxide. The *by-products of the production were coke and coal tar, which is a residue containing benzene, naphthalene, and other organic compounds. The availability of *natural gas in the 1970s led to the decline in its use.

cocurrent flow An arrangement in which the flows of two separate process streams are fed into a process in the same direction for the purpose of carrying out heat and mass transfer. It is typically used for heat-sensitive materials and in packed gas absorption columns where the chemical reaction in the liquid phase tends to be rapid. It is generally inefficient as a flow arrangement, since once equilibrium between the process streams has been reached, no further transfer takes place. It is also known as **parallel flow**. *Compare* COUNTERCURRENT FLOW.

coefficient 1. A number or symbol used in an algebraic expression that multiplies an unknown quantity. For example, in the expression $y = ax^2 + bx + c$, a is the coefficient of x^2 and b is the coefficient of x. **2.** A measure of a specified property under specified conditions. For example, the *coefficient of discharge for a discharging vessel through an orifice is the ratio of the actual to theoretical rate of flow and takes into account friction losses, etc.

coefficient of contraction A dimensionless number expressed as the ratio of the minimum flow area or *vena contracta* to the flow area for a fluid discharging through an orifice. It is often difficult to measure directly but can be determined indirectly from the *coefficient of discharge and coefficient of velocity which are more readily measured.

coefficient of discharge A dimensionless number expressed as the ratio of the actual to the theoretical flow rate of a fluid discharging through an opening or restriction. The coefficient is used as an indication of the recovery of energy following through the opening or restriction. Where there is full recovery and no permanent energy loss, the coefficient is equal to 1. For example, for a well-designed *venturi meter the coefficient may be in the order of 0.95 to 0.98 signifying very good energy recovery, whereas an *orifice plate meter may be as low as 0.6 at high flow rates. Orifice plate meters, however, are popular since they are considerably cheaper than venturi meters to fabricate and install.

coefficient of expansion The increase in the physical dimensions of a material due to the effects of temperature. It is expressed as a fraction of the original dimension per degree temperature rise. The coefficient may relate to either linear or cubic expansion.

coefficient of performance (COP) A coefficient used in *air conditioning, *refrigeration, and *heat pump cycles as a measure of the performance of the thermodynamic cycles. In refrigeration, the COP is the ratio of the duty of the condenser to the work input to the compressor, whereas in a heat pump the COP is the total heat output as a ratio of the heat equivalent of work required to produce the heating effect. The difference in COP between a heating and cooling system is due to the heat reservoir of interest being different. For a system where cooling is of interest, the COP is the ratio of the heat removed from the cold reservoir to the input work, whereas for a heating system, the COP is the ratio to input work of the heat removed from the cold reservoir together with the heat added to the hot reservoir by the input work. The COP of a refrigeration system usually varies between 3.0 and 9.0 depending on the refrigerant, head and suction pressure of the compressor, and the superheated conditions.

coefficient of volumetric expansion *See* THERMAL EXPANSION.

coherent units A system of units that are obtained by the multiplication or division of *base units without numerical factors. The SI system is a coherent system. For example, the newton is equal to one kilogram metre per second squared ($kg\ m\ s^{-2}$), while velocity is metres per second ($m\ s^{-1}$).

coke A porous material produced from the carbonization of coal in which all the volatile materials have been driven off. It is used in a *blast furnace.

Colburn j factor (Symbol j_H) A semi-empirical equation used for heat transfer in turbulent flow with *Reynolds numbers ranging from 5,000 to 200,000 inside long tubes and defined as:

$$\left(\frac{h}{c_p G} \right) \left(\frac{c_p \mu}{k} \right)^2 \left(\frac{\mu_w}{\mu} \right)^{0.14}$$

and is equal to $0.023\,Re^{-0.2}$. It applies over a range of *Prandtl numbers from 0.6 to 120, but should not be used for Reynolds numbers below 6,000 or for molten metals that have unusually low Prandtl numbers.

cold shot process A technique used to control the temperature in an exothermic reaction in which cold fresh feed is added to the reaction mixture in a *tubular flow reactor or a cascade of *continuous stirred-tank reactors (CSTR). It is used to overcome limitations in conversion due to chemical equilibria. It also avoids the need for a heat exchanger. It is typically used for high-pressure processes such as ammonia synthesis. This reduces the total volume of the reaction vessel since interstage heat exchangers are not required.

cold work A method of carrying out a task in a hazardous area using a tool or item of equipment that does not provide a source of ignition.

Collier, John Gordon (1935–95) A British chemical engineer who was director-general of the Central Electricity Generating Board (CEGB). He began his career at Harwell United Kingdom Atomic Energy Authority (UKAEA), before leaving in 1983 to join the CEGB, but he returned in 1987 to become its chairman. After its breakup he became the first chairman of Nuclear Electric. He was a Fellow of the Royal Society and president of the Institution of Chemical Engineers in 1993. The John Collier Medal is awarded biennially and jointly by the Royal Academy of Engineering, the Royal Society, and the *Institution of Chemical Engineers.

colloid A suspension of particles whose size lies within the range of 1 nm to 1 μm and dispersed within a liquid medium. Hydrophilic colloids consist of thermodynamically stable water-soluble macromolecules such as gelatine and starch. Hydrophobic colloids consist of insoluble particles in a finely divided state suspended in water and are thermodynamically unstable. Colloids often have a kinetic stability due to a surface-charge repulsion effect between particles.

column A tall cylindrical process vessel whose height is considerably greater than its diameter and used for unit operations such as distillation, absorption, and various forms of gas–liquid and liquid–liquid extraction processes. It usually allows liquids to descend under gravity and contact rising gases or vapours, or liquids of lesser density in which there is intimate contact between the two to allow equilibrium to be reached. Some types of columns are empty in which the liquids are sprayed in the form of droplets, while others have internal features such as baffles, plates, or packing materials used to promote the intimate contact between the ascending and descending materials. *See* TOWER.

COMAH An abbreviation for Control of Major Accidents and Hazards. These UK regulations under the Health and Safety at Work Act require the operator of a process plant or site that contains more than a defined amount of hazardous chemicals to provide a safety case report for the process and its operations, to demonstrate that it can be operated in a safe and environmentally acceptable manner. The regulations have control of manufacturing sites with major pollution potential and superseded the earlier *CIMAH regulations in 1999.

combined feed ratio The total quantity of a reactant fed to a chemical *reactor expressed as the ratio of the fresh feed to other feeds including any recycled feed.

combined heat and power (CHP) The use of a heat engine or power station to simultaneously generate both electricity and useful heat. The use of low-grade thermal energy such as for municipal district heating is an effective way of raising the overall efficiency of an engine or power station.

combustible A material that is capable of burning under normal conditions. For a material to combust there must be a sufficient supply of oxygen and an ignition source. A **combustible liquid** is a liquid that is capable of combustion, such as a hydrocarbon or alcohol.

combustion The rapid thermal oxidation of a fuel with the production of heat and light. Complete combustion is also known as **stoichiometric combustion**. The products of combustion of a hydrocarbon are water vapour and carbon dioxide. The presence of carbon monoxide indicates incomplete combustion.

comminution The break-up and particle-size reduction of solid materials into smaller particles and fragments by the process of crushing, grinding, pulverization, attrition, impact, or by chemical methods. It is used to break up ores prior to *flotation.

commissioning A final and thorough check of an installed process plant or item of equipment to ensure that it is fully operable as intended. All aspects of the process or equipment are tested individually and collectively. Prior to commissioning, a site acceptance test is carried out. At the end of the working life of a process, the process and its equipment are decommissioned and taken out of service. *See* DECOMMISSIONING.

common logarithms *Logarithms that use base 10.

common rail A type of fuel injection system used in modern automotive diesel engines. The injection nozzles to each cylinder are supplied by a common fuel line. A single pump is used to supply the fuel at a very high pressure of around 1,000 bar. The fuel injection valves open and close automatically and in sequence.

competitive inhibition *See* ENZYME INHIBITION.

complex medium *See* GROWTH MEDIUM.

complex number A number that has a real and an imaginary part of the form $x + jy$ where x and y are real, and $j = \sqrt{-1}$. It can also be written in polar form as $r\cos\theta + jr\sin\theta$ where r is the modulus and θ is the argument. Complex numbers are represented on an *Argand diagram, and are useful in the study of the stability of controlled chemical processes.

component A constituent in a mixture that is defined as a phase or as a chemical species. *See* GIBBS' PHASE RULE.

composition The parts of which something is made up. The chemical composition of a substance is made up of its elements. The composition of an *alloy is made up of *elements whereas the composition of an ore is made up of elements and compounds. The composition of a flow stream to or from a process, or a chemical reaction itself is made up of the various components involved. *See* PROCESS VARIABLE.

compound A substance of uniform composition throughout its bulk and containing two or more elements in a state of chemical combination. Compounds are formed when elements react and are chemically joined. Unlike a *mixture, a compound can only be separated into its components by chemical reaction.

compressed gas A gas or a mixture of gases at a pressure greater than atmospheric pressure. Compressed gases occupy a smaller volume than their uncompressed state and can therefore be conveniently stored and transported. Compressed air is used to power certain types of machines and is also used as *instrument air.

compressibility The fractional reduction in the volume of a substance with applied pressure. The **compressibility factor** is a measure of the compressibility of a gas, z, and used as a multiplier to adapt the *ideal gas law for non-ideal gases:

$$z = \frac{pV}{RT}$$

where p is the pressure, V is the volume, R is the universal gas constant, and T is the absolute temperature.

compressible fluid A fluid in which the density changes with applied pressure. The compressibility of liquids is negligible in comparison with gases and vapours. The **isothermal compressibility** of a gas is the change in volume per unit volume or density for a unit change in applied pressure given as:

$$c = \frac{-1}{V}\left(\frac{\partial V}{\partial p}\right)_T = \frac{-1}{\rho}\left(\frac{\partial \rho}{\partial p}\right)_T$$

Isothermal compressibility coefficients are frequently used in oil and gas engineering, transient fluid flow calculations, and in the determination of the physical properties of substances.

compressor A type of machine used to increase the pressure of a gas or vapour by reducing its volume. It is used to compress a gas or vapour in a vessel, to raise its pressure, and for its transport under pressure through pipelines. Compressors are broadly classified as those that are dynamic and those that are positive displacement in design and operation.

Centrifugal compressors are dynamic-type compressors and consist of various types of rotating impeller used to increase the velocity, which is converted to pressure through a diffuser. They are widely used in the chemical, oil, and gas industries. Axial-flow compressors consist of blades or airfoils mounted on the rotating shaft that compresses the flow along the axis of the shaft rather than radially. They are used for generating compressed gases with a high velocity such as in natural gas compression.

Reciprocating compressors are positive displacement-type compressors and consist of a piston attached to a crankshaft to compress gas that is drawn into a cylinder. Many reciprocating compressors are multistaged and can generate high pressures in excess of 2,000 bar. Rotary screw compressors are also positive displacement-type compressors and consist of two meshing rotors. The compressor on a domestic refrigerator is a hermetically sealed compressor used to compress the evaporated refrigerant vapour for condensation, throttling, and recycling back to the evaporator. They commonly consist of a scroll-type compressor although spindle-type compressors are also used.

computational fluid dynamics (CFD) A set of numerical methods and algorithms used to solve and analyse problems involving complex fluid flow behaviour. The problems require many computations that are required to be performed simultaneously using high-speed computers. The problems involve the simulation of the interaction of liquids and gases with surfaces defined by *boundary conditions. The *Navier–Stokes equations are used to define most CFD problems involving a single phase as a liquid or a gas, but not both. For problems involving fluid mechanics, the solvers are based on the finite volume method. For 2-D or 3-D problems, the geometry of interest is first defined as an area or volume and divided into discrete control volumes or cells known as a mesh. The flow into these cells obeys the general laws of conservation for mass, momentum, and energy as algebraic equations. The boundary conditions are specified and in the case of problems involving transient behaviour, the initial conditions are also defined. All the equations are then solved iteratively. The solution is then presented for visual analysis and interpretation. There have been many developments in CFD methodologies and many are available as commercial software packages that can be applied to the study of complex fluid flow systems.

computer-aided design (CAD) The use of computers for the efficient engineering design of process equipment and its layout. This permits prototype designs to be checked, altered, analysed, and tested prior to proceeding to fabrication and construction.

computer control The use of computers to control a process in which analogue signals of measured *process variables such as pressure, temperature, level, etc. are converted to a digital signal that is then manipulated according to a model of the process. The digital output from the computer is then converted back to an analogue signal to make the controlled adjustments to the process. The advantage of computer control is that it is able to process large volumes of data at high speed obtained in real time from around the process compared with traditional analogue controllers.

concentrate 1. The action of intensifying the purity or strength of a material, such as by the action of adding or removing a substance. For example, a solution of salt water can be concentrated by the evaporation of the water. **2.** A product of concentration.

concentration A quantitative measure of the relative amount of a component in a mixture. Concentration is often expressed as a mass and mole fraction. Other forms include volume fraction, molarity, molality, parts per million or billion on either a mass or volume basis, mass per unit volume, mass or weight per unit mass or weight, and activity.

concentrator A device used to increase the concentration of one component dissolved or suspended in another, usually by the removal of the latter. An evaporator is an example of a concentrator in which water as the solvent is removed from a solution of salt, leaving it more concentrated.

concentric Having the same centre.

conceptual process design A work activity performed by engineers at an early stage to evaluate in broad terms the technical feasibility of new and existing processes, as well as process redesigns based on existing feed materials. The work activity examines the thermodynamic feasibility of process routes and the *process variables required, and assesses the broad issues of chemical and process production, which includes information on process costs and material selection. The use of *heuristics and process simulation using computers are useful tools to provide rapid information before committing resources to a more detailed design using tools such as *computer-aided design (CAD).

It is important to define the necessary codes and standards to which a design will adhere. This permits effective communication with other engineering disciplines and equipment vendors, and also allows for appreciation of the full extent of the design. Standards commonly used include British Standards Institution, the *American National Standards Institute, the *American Petroleum Institute, the Deutscher Nornenauschuss, and the International Standards Organization.

conclusion A proposition made at the end of an argument upon what the argument set out to prove. It is based on evidence and facts, and not conjecture.

concurrent *See* COCURRENT.

condensate 1. A liquid obtained from the cooling of a vapour below its saturation temperature, or from a vapour–gas mixture cooled below the *dew point. **2.** A term used to describe liquid drops of light hydrocarbons in *natural gas.

condensation The change in state from a vapour to a liquid or a solid accompanied by the release of energy known as the heat of condensation, which has the same magnitude as the heat of vaporization at the same temperature. Condensation is the opposite of *evaporation. The direct change in state from a vapour to a solid phase is called **deposition.**

A vapour may condense on a cold surface by either *film condensation in which condensate forms a continuous layer of liquid that flows over the surface, or *dropwise condensation in which condensate forms at nucleation sites where droplets tend to coalesce and may form into rivulets that flow under the influence of gravity. Although dropwise condensation is associated with higher transfer coefficients, it is difficult to maintain in practical situations.

condensation reaction A type of chemical reaction in which two molecules combine to form a larger molecule with the elimination of a smaller molecule such as water, hydrogen chloride, or methanol. In a polymerization reaction, similar or different monomers form a long-chain polymer with the release of water molecules, such as in the formation of polyester. The formation of a peptide from two amino acids also involves the elimination of a water molecule.

condensation temperature *See* DEW POINT.

condenser A type of heat exchange used to cool a vapour at constant pressure to a temperature that is sufficiently low as to change the state from a vapour to a liquid, and to carry away the heat from the vapour–liquid mixture. The latent heat is removed using a coolant. The coolant evidently increases in temperature but it is the phase change action that is the important function.

Condensers fall into two classes: those that condense vapour using a coolant separated in a *shell and tube device, and those called **contact condensers** in which coolant and vapour are mixed and leave together in a single stream. They are used with distillation columns to produce reflux to control top temperatures and product quality. They are also used for steam turbines to produce condensate and to maximize the energy from the steam.

conditioner A tank used in mineral flotation into which chemicals are added and allowed sufficient time to absorb on to the particles before flotation.

conduction A mode of *heat transfer in which thermal energy is transmitted through a substance from a region of high temperature to a region of lower temperature. Within gases and liquids, the thermal energy is by collisions between atoms and molecules to those with lower kinetic energy. The rate of heat transfer for steady-state thermal conduction through a slab is given by *Fourier's law:

$$q = \frac{kA}{x}\Delta T$$

where q is the rate of heat transfer, k is the *thermal conductivity, x is the thickness of material, A is the area perpendicular to the direction of heat flow, and ΔT is the temperature difference along the path of heat transfer.

cone and plate rheometer An instrument used to measure the rheological properties of fluids (see Fig. 12). It consists of a fixed flat surface with a rotating cone above and a sample of the fluid sandwiched between them. The cone just touches the flat surface. The rotational speed and tapered gap defines the shear rate. The torque on the rotating cone that resists the motion defines the characteristic shear rate. The surface can be heated or cooled to determine the rheological properties as a function of temperature.

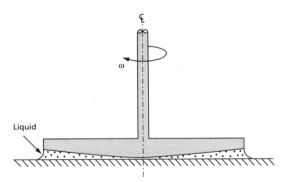

ω

Liquid

Fig. 12

confined space Any enclosed or partially enclosed space having restricted access and egress that may present itself as a form of trap and be life-threatening.

conflagration A large destructive fire.

conformation The spatial arrangement of atoms in a molecule responsible for its shape. There are many possible arrangements of atoms that can be interconverted by rotation about a single bond in a molecule. The conformation of proteins determines their function as in the case of enzymes as biocatalysts and proteins that have therapeutic properties such as insulin.

constant 1. A quantity that does not vary such as temperature, pressure, or level in a process. **2.** A number used in a mathematical relationship that is multiplied by a variable such as $y = ax$ where a is a constant. **3.** A fundamental constant used in formulae or calculations such as π or g. **4.** A fixed value, c, that is added to an indefinite integral such as:

$$\int x\,dx = \frac{x^2}{2} + c$$

constant-boiling mixture A mixture of components that boils giving a vapour with the same composition as the liquid. Separation of the components by evaporation or distillation is not possible. The mixture is also known as an *azeotrope.

constant molar overflow An assumed condition in which the number of moles of a mixture of volatile liquids evaporating at any point in a system or process, such as on the plate of a distillation column, is equal to the number of moles condensing at the same point. That is, on the plate, the liquid and the vapour flow rates remain constant.

constant rate drying The point in the drying of a solid material in which the rate of evaporation per unit area of the drying surface remains constant. It is dependent on the humidity of the drying air, the mass transfer coefficient, and the velocity of the drying air.

constant variable *See* CONTROLLED VARIABLE.

construction cost The financial expense incurred in the building of process plant and buildings by a contractor for the labour, construction materials, and equipment, as well as services and *utilities. It also includes profits and overheads. It does not include the cost of the site or fees for architects or construction engineers.

contact angle The angle between a solid surface and the surface of a liquid in contact with it. It can be used to determine the *surface tension of a liquid.

contact process A major industrial process used for the manufacture of sulphuric acid. Sulphuric acid is widely used with large quantities involved in the making of rayon, the refining of petroleum, and the manufacture of fertilizers, pharmaceuticals, dyes, and paints. The process involves the exothermic reaction of combining sulphur dioxide and oxygen to produce sodium trioxide, SO_3. The sulphur is produced by roasting ores such as iron pyrites, FeS_2. Excess oxygen is used and the process is controlled at as low a temperature as possible. Platinum catalysts were once used. However, due to their susceptibility to poisoning, especially by arsenious oxide, As_4O_6, finely divided vanadium and vanadium pentoxide catalysts are preferred since these are less susceptible to poisoning although less efficient. The sulphur trioxide is not absorbed in water due to an unmanageable mist of sulphuric acid droplets that is produced, but instead the cooled gases are passed up a tower down which 98 per cent sulphuric acid flows. The water in this acid forms the acid $H_2O + SO_3 = H_2SO_4$. The adsorption can be carried further to produce fuming sulphuric acid or oleum:

$$H_2SO_4 + SO_3 = H_2S_2O_7$$

The process was invented in 1831 in the UK by P. Phillips Jr. and is named after the German *kontaktverfahren* meaning 'catalytic process'. It was once in competition with the *lead chamber process.

containment The prevention of a hazardous material from being released into the environment or beyond a defined boundary. Codes of practice or legislation are often used to define the boundary within which such materials must be contained, such as vessels, chambers, cabinets, glove boxes, nuclear cells, and other similar facilities. Examples include the containment of radioactive process materials and biological substances such as pathogens.

contamination The unwanted or undesirable presence of a substance that may have an undesirable effect such as be infectious or harmful in some way to humans or the environment. It may be in the form of deposition, absorption, or adsorption of radioactive material, or involve biological and chemical agents such as in food, soil, skins. Bacteria in drinking water, radioactive material in soil, oil in water, or water in oil, are all examples of **contaminants**. *Compare* POLLUTION.

continuity equation An equation that describes the total transport of a conserved quantity moving from one place to another within a pipe, process, or system. It applies to the transport of mass, energy, momentum, and all other natural quantities. Expressed as a simple balance equation, the transport into a system is equal to the transport leaving. There is neither accumulation nor depletion within the system under steady conditions. The conserved quantity is usually defined with reference to a flow area.

continuous cultivation The operation of a *bioreactor that involves the continuous addition of fresh sterile media and its withdrawal at the constant rate. The media contains all the necessary nutrients required to promote biological growth of the microorganisms within the bioreactor. The volume of the bioreactor can be kept constant by an overflow weir arrangement. The *dilution rate is used to characterize the operation which is defined as the ratio of the flow of feed to the volume of the bioreactor. **Chemostat cultivation** involves the continuous feed of substrate in which at least one nutrient is limiting. **Turbidostat cultivation** involves maintaining a constant biomass concentration by varying the dilution rate.

continuous phase A phase such as liquid within which another phase is dispersed such as suspended gas bubbles, droplets, or solid particles. A colloid has two or more phases in which one is dispersed in another. *Compare* DISPERSED PHASE.

continuous process A process in which raw materials are continuously fed and in which processed materials continuously leave at the same rate. As an *open system, both material and energy are transferred across the system boundary. In contrast with a *batch process, the process equipment volumes are smaller and the operating costs lower. Continuous processes are used for high-quality production. Automated control is necessary to ensure that process conditions of flows, temperatures, and pressures are maintained at all times. A **semi-batch process** has certain elements within the continuous process that operate as a batch process, such as the removal of moisture by a molecular sieve, but the overall effect is a continuous production.

continuous stirred-tank reactor (CSTR) A type of idealized chemical reactor vessel used to contain a chemical reaction in which liquid reactants continuously flow into the reactor and products continuously flow out such that there is no accumulation within the reactor. By assuming perfect mixing of the reactants within the reactor by using stirring, the composition of the reaction does not vary with either position or time, and the output composition of the material is therefore assumed to be the same at the composition at all points within the reactor. It is also known as a *back-mix reactor.

continuum A region of material space through which properties such as temperature, density, and composition vary in a mathematically continuous manner.

contractor A company or person that undertakes a contract to provide a service, labour, or materials. A subcontractor is a company or person that is assigned by the contractor to undertake part of the contract. A contract is a legally binding document. Many chemical companies use contractors with specialist expertise to undertake chemical engineering design, construction, and commissioning of chemical plant as well as to undertake plant maintenance and revamp projects.

control charts Graphical tools used in statistical and quality control of a process to represent the state of control. They are used to indicate whether changes are required to be made to control parameters, as well as identify the parameters that require control. They are also used to predict the control of the process. A process that is stable but lies outside desired control limits requires understanding of the causes for control. They are also known as **process-behaviour charts** or **Shewhart charts** after American engineer Walter A. Shewhart (1891–1967) who devised them in the 1920s.

controlled variable A process variable that is not permitted to change unpredictably. It is also known as a **constant variable** as it is not expected to change its value. In process control, the controlled variable, for example, may be the flow of material leaving a heat exchanger that is required to be maintained at a constant temperature, and is achieved by adjusting the flow of the heat transfer medium through the heat exchanger.

controller A device used to regulate a *process variable in a controlled process. The purpose is to receive an input signal of the difference between a measured value and the desired value, and provide an output signal that is then used to control the process. The controller has a reference input or *set point. This is the desired value for the process measurement signal and is transmitted to the controller. The controller measures the difference or **error** between the set point and the measurement signal. The error is manipulated by the controller to provide the **controller output,** which corrects a valve position to reduce the error towards zero. In a *block diagram, the controller is represented by the summing junctions and the control modes block. The main forms of control are proportional, integral, and derivative control. The simplest type is on/off control. To reduce the frequency of switching on and off, which can lead to excessive wear and tear, a neutral zone is used. For example, if temperature is being measured and is rising within the neutral zone, the controller stays on, whereas if the temperature is falling within the neutral zone then the controller stays off. *See* PID CONTROL.

controller tuning The technique of selecting the optimum *controller settings used to control a process. The *Ziegler–Nichols tuning method is a way of selecting the controller settings.

control loop A part of a process control system. *Open loop control involves human operator intervention, whereas *closed loop control is an automated system in which the output signal to the process is compared to a defined *set point.

control mode A type of control action such as *proportional control, *integral control, or *derivative control. *See* PID CONTROL.

control rod A neutron-absorbing rod that is used to control the reactivity variations of a *nuclear reactor. Cadmium and boron are neutron-absorbing materials used for control rods.

control room An operations centre that receives all the process plant information. The information includes details of temperature, pressure, flows, levels in process vessels, and concentrations, etc. and is often presented on an array of computer display screens that represent the process. Process alarms indicate significant deviations from the expected values, and are monitored by process control room operators. The control rooms in hazardous chemical plants are often located a safe distance from the plant itself and are also used as a safe place of refuge. On offshore oil and gas platforms, the control room is shielded and protected from fire and explosion.

control valve A device used to control the rate of flow of a process material through a pipe. It is actuated either electronically or pneumatically in which either a stem or diaphragm changes the position of a plug in a seat either restricting or opening the passage of flow. As the final control element in a control system, it is therefore responsible for changing the value of the manipulated variable to the output signal from the controller. *Pneumatic control valves are either air-to-open or air-to-close. The application depends on safety consideration based on the impact of supplied air failure. An equal percentage valve characteristic is used to describe a type of control valve flow characteristic in which there are equal increments of valve plug movement for the change in flow rate. The change in flow rate with respect to movement is small when the valve plug is near its seat, and high when the valve plug is nearly wide open. A linear control valve characteristic is where the controlled flow is directly proportional to valve travel and is often used with distributive control systems or programmable logic controllers. A quick-opening valve characteristic provides a maximum change in flow rate at low movement of the stem and plug. It is used for on/off control.

control valve actuator A pneumatic or electrically powered device that supplies the force and motion necessary to open or close a *control valve.

convection A mode of heat transfer caused by the movement of currents within a fluid as the result of different localized densities. It is a combination of *conduction within the fluid and energy transport due to fluid motion, which is either by a natural flow of density currents or by a forced fluid flow, known as **natural convection** and **forced convection**, respectively. In natural convection, the movement of fluid is due to gravitational effects in which heated fluid has less density and rises, allowing cooler and denser fluid to descend. This results in a circulating flow. In forced convection, a pump or fan is used to circulate the fluid.

convective mass transfer The mass transfer between one substance or phase and another caused by simultaneous convection and molecular diffusion. It can involve the mass transfer between a fluid in motion and a surface, which may be either a solid or an immiscible liquid. The **convective mass transfer coefficient** relates the molar mass flux of a species to the concentration difference between the boundary surface concentration and the concentration of the diffusing species in a moving fluid. The coefficients k_G and k_L refer to the gas and liquid phases, respectively, and are related to the properties of the fluid, the dynamic characteristics, and the geometry of the system. They are often presented as a product with area through which mass transfer takes place as $k_G a$ and $k_L a$. In this form, they are useful when considering gas transfer as bubbles, particularly in biological systems for oxygen transfer since oxygen is usually a limiting factor in a *bioreactor.

convergence The approach to a limit of a sequence or series. It is usually the solution of a non-linear problem. Iterative solutions to complex problems solved by computers often seek convergence to a solution. The opposite is **divergence**.

conversion A measure of the completeness of a chemical reaction. It is often presented as the fraction of a particular reactant consumed by the chemical reaction. The **conversion per pass** is a measure of the limiting reactant that is converted in a chemical reactor and recycled for combination with fresh reactant feed. Not all reactions are complete within the reactor, and in many cases, unreacted reactants are separated from products and recycled for further opportunities for reaction.

conversion factor A number used as a multiplication factor to convert a quantity expressed in one set of units to those of another.

converter The reaction vessel used in the *Bessemer process in the production of steel.

conveyor A mechanical device used to transport efficiently large quantities of solid materials from one place to another. It is often used to handle bulky loose materials such as crushed ores, coal, and grain. Used in a wide number of industries, there are many types commonly used. The choice depends on the application such as bottles, pharmaceuticals, foods, packaging, etc. For example, belt conveyors consist of a belt moved by a series of rollers, with the materials resting on the belt.

coolant A fluid used in a process as a heat transfer medium for extracting heat from one place and transferring it to another. Water is widely used as an effective coolant in heat exchangers due to its abundance, low cost, and high heat capacity. In a nuclear reactor, the coolant is used to remove heat from the *core. High-pressure water is used in pressurized water reactors to prevent boiling. Liquid metal-cooled reactors use molten sodium. Mercury has also been used in the past. Gases have been used as coolants; for example, carbon dioxide was used in Magnox and AGR nuclear reactors.

cooler A heat exchange device used for reducing the temperature of a process stream or product but not necessarily with a change in phase.

cool flame A weak luminous hydrocarbon flame of fuel-rich air mixture. It usually has a temperature below 500°C.

cooling jacket An outer cover or surrounding of a process vessel or pipe to contain and transport a heat transfer medium or *coolant in order to reduce the temperature of the process material in the vessel or pipe, or to maintain it at a low temperature. *See* JACKET.

cooling tower A device or structure used to condense steam or reduce the temperature of water used as a cooling medium in a process for reuse. Cooling towers are either natural draught or forced draught in design and operation. Natural-draft cooling towers are large structures that contain packing material with a high specific surface area down which the water to be cooled trickles and cascades, contacting with cool air that is drawn up through the packing by convection. The cooled water collects at the bottom of the tower and is returned to the process for reuse and a make-up of water is added to account for loss by evaporation. Forced-draft cooling towers use fans to pass the cooled air through the packing. Although they have a higher operating cost, they are comparatively smaller and more compact than natural-draught cooling towers. In mechanical-draft cross-flow cooling towers, the air flows horizontally across the downward-flowing water. They therefore have a shorter path for the air to flow and allow a greater flow of air for the same power demand as counter-flow forced-draft cooling towers.

cooling water A supply of water that is used to remove heat from a process that operates at a temperature of less than 100°C. The water, which may be chilled or at ambient temperature, is used as a heat transfer medium or *coolant, and contacted with a hot process stream either by direct contact or directly such as through the walls of a *heat exchanger. After being heated, the water may be reused by removing the heat gained through a *cooling tower, with a make-up to allow for loss from evaporation. Depending on the quantities required and the type of application, the water may be freshwater taken from a river, such as used in a nuclear power station, or saltwater from the sea although there are associated corrosion issues. The deposition of dissolved salts can also lead to fouling issues.

COP *See* COEFFICIENT OF PERFORMANCE.

copper A valuable metal noted for its high electrical conductivity, malleability, suitability as an alloy with other metals, and resistance to corrosion. It is extracted from various ores by crushing and ball milling, followed by *flotation separation to raise the copper content, and remove unwanted minerals. Smelting with sulphur then produces copper and iron sulphide. This is melted in a reverbatory furnace in which air (oxygen) is added to produce sulphur dioxide, iron, and copper. The copper is then electrolytically refined to produce commercial copper. Copper has many uses in the process industry in addition to electrical wiring. It is also used for small bore pipes and for copper stills in the *whisky industry.

co-product A substance formed at the same time as the main or desired product during a chemical reaction that has equal or comparable economic significance. In the mining and extraction of ores, many of the elements recovered are of similar economic significance to one another. *Compare* BY-PRODUCT.

copyright The exclusive legal right to produce copies and to control a published literary work granted by law for a specified period.

COR *See* CELL OUTPUT RATE.

core The part of a *nuclear reactor where the fission chain reaction takes place. A **core meltdown** is the uncontrolled reaction within a nuclear reactor in which the core cooling fails, leading to the nuclear fuel heating up due to *radioactive decay of the fission products to the point that the fuel melts. The cooling system may fail due to a major leakage in the nuclear reactor cooling circuit with the simultaneous failure of the emergency cooling system. *See* CHERNOBYL.

Coriolis flow meter A non-invasive type of flow meter used to measure the mass flow of a fluid through a pipeline. It is based on the Coriolis effect and involves diversionary loops of pipe through which the fluid passes. As the fluid moves through the loop, the fluid momentum changes and rotates, exerting a force on the loop causing it to twist and vibrate. The extent of the twisting and vibration effect gives an indication of the rate of mass flow. It is named after French physicist Gaspard de Coriolis (1742–1843) who first used the term for describing the movement of fluids in rotating systems.

corollary *See* THEOREM.

correlation coefficient (Symbol R) A statistical number that represents the linear relationship between two variables or groups of variables (X and Y). The correlation has a value between -1 and +1. A positive value represents a positive linear relationship, whereas a negative value represents a negative relationship. A value of 0 represents no relationship between the variables. It is calculated from:

$$R = \frac{n\sum XY - \left(\sum X\right)\left(\sum Y\right)}{\sqrt{\left(n\sum X^2 - \left(\sum X\right)^2\right)\left(n\sum Y^2 - \left(\sum Y\right)^2\right)}}$$

where n is the number of values. The correlation coefficient requires that the relationship between the two variables is linear. Where this is known not to be the case, then the correlation coefficient is not useful.

corresponding states A condition in which fluids are compared at temperatures and pressures that are fractions of their corresponding critical properties. That is:

$$T_r = \frac{T}{T_c}; p_r = \frac{p}{p_c}$$

where the subscript r refers to the reduced state and subscript c refers to the critical state. In practice, the law of corresponding states is valid only near the critical point since all substances deviate from the law. Within groups of similar substances that have a similar form of intermolecular interaction, the deviations are often relatively small such that the properties of otherwise little-studied substances can be determined with confidence based on corresponding states.

corrosion The unwanted wastage of metallic materials due to reaction with the environment. The effect includes the loss of strength of material, a change in appearance, change in surface heat transfer and fluid flow properties, contamination, seizure, electrical contact failure, leakage, and general surface damage. Corrosion rates are determined to a large extent by the chemical nature of the process stream and its pressure and temperature; due account must be taken of the flow conditions, and how they interact with the ongoing chemical processes. The electrochemical corrosion process of *rusting involves the oxidation of iron to form a hydrated iron oxide that occurs in the presence of both water and oxygen, and is particularly damaging to process equipment and support structures. *Compare* EROSION. In some cases, the corrosive action of the environment can be reduced through the use of chemicals known as **corrosion inhibitors**. Cathodic corrosion inhibitors include oxygen scavengers such as sodium sulphite used in enclosed systems such as boilers, and ions such as Ca^{2+}, Mg^{2+}, HCO_3^-, and Zn^{2+}, which form insoluble precipitates at the cathodic (alkaline) surface. Anodic corrosion inhibitors are used to encourage oxidized passive films of surfaces using an oxidizing agent such as chromates (Cr^{VI}).

COSHH An abbreviation for **C**ontrol **o**f **S**ubstances **H**azardous to **H**ealth. These are regulations used in the UK for the safe use of chemicals. Originally established in 1988, there have been subsequent amendments. Where chemicals are used, such as in a laboratory, a COSHH assessment must first be completed to identify the controls required for their safe usage.

cost estimation A method of determining the capital cost of process plant and equipment. There are four recognized ways this is done: 1. *Rules of thumb provide approximations to the order of magnitude of cost. They are useful as a rough guide but are prone to major error. 2. The use of cost curves provide estimates based on similar process plants or equipment, and may involve *scale-up factors. 3. Multiplication factors can be applied to different types of equipment, such as heat exchangers and pumps etc., that might be involved in a process. 4. Definitive estimates require the use of detailed materials and equipment, the direct and indirect costs. Although time-consuming due to the level of detail involved, it is the most accurate method.

Couette flow A type of flow in which a fluid is sandwiched between two parallel plates, one of which is stationary and the other is moving at some constant velocity. For a Newtonian fluid, the velocity gradient is linear between the plates. It is named after French physicist Maurice Marie Alfred Couette (1858–1943).

Couette rheometer An instrument used to measure the rheological properties of fluids. It consists of a cup within which fits a cylindrical bob (see Fig. 13). A sample of the

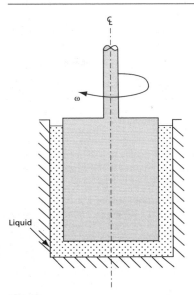

Fig. 13

fluid being tested is sandwiched between them. Either the cup is held and the bob rotated or, more rarely, the bob is held and the cup rotated. The rotational speed and distance between cup and bob define the shear rate. The torque to resist the motion defines the characteristic shear rate. The cup can be heated or cooled to determine the rheological properties as a function of temperature.

coulomb (Symbol C) The SI unit of electric charge and is equal to the charge transferred by a current of one ampere in one second. It is named after the French physicist Charles-Augustin de Coulomb (1736–1806).

Coulson, John Metcalfe (1910–90) A British chemical engineer and academic who co-authored a series of six textbooks, *Chemical Engineering*, with Jack *Richardson, that were first published in 1954. A twin to his brother Charles, who was a noted chemist, John Coulson gained his first degree from Cambridge and his PhD in chemical engineering from Imperial College, London. He was the first head of chemical engineering at the University of Newcastle where he remained until his retirement.

countercurrent flow The flow of two fluids in contact with each other or separated by a surface but flowing in opposite directions. In a distillation or adsorption column, the flow of liquid descends while the flow of vapour ascends, meeting in intimate contact. In a heat exchanger, the two fluids travelling in opposite directions are not in direct contact with each other, but exchange heat from one to the other through the tube walls. This arrangement is more efficient than *cocurrent flow, in which the flows are in same direction.

covalent bond A type of chemical bond that involves the electrostatic attraction between a pair of atoms through the sharing of electrons and the positive nuclei. This bond was discovered by American chemist Gilbert Newton Lewis FRS (1875–1946).

Cowles process A process used for the production of aluminium alloys from aluminium ores (bauxite). The ores are reacted with carbon in the form of charcoal in an electric furnace, together with the metal used to form the alloy, which is usually copper. Another metal is required in the process, since without it, the product would result in the formation of aluminium carbide. The process is named after its American inventors, the Cowles brothers: Alfred H. Cowles (1858–1929) and Eugene H. Cowles (1855–92).

CPA *See* CRITICAL PATH ANALYSIS.

cracker The chemical reactor in which the catalytic cracking of high molecular weight hydrocarbons to small molecules takes place. It is also known as a *cat cracker.

cracking A process in which high molecular weight hydrocarbons are broken down into lower molecular weight products by the effect of high temperature in the presence of an alumina-silica catalyst. The process is used to produce gases such as methane, ethane, propylene, and propane that are the subsequent raw materials used for the manufacture of a wide range of products including of plastics, detergents, textiles, and agricultural chemicals. *Thermal cracking uses high temperatures and pressures to break the molecular bonds to form smaller molecules. *Catalytic cracking uses a catalyst to assist in the breakdown of the molecules. In a *fluidized bed catalytic cracker, the catalyst is present as a bed of very fine particles which is agitated by the vaporized hydrocarbons as they pass up through the bed.

creep The continuous deformation of the structure of a solid material that is held under constant stress but below the *yield point. It is usually observed in metals that are held at elevated temperatures but may also occur over long periods at ambient temperatures. Horizontally supported pipelines at fixed points may eventually be seen to bow under their own weight.

CRG process (catalytic rich gas process) A catalytic process used to produce *fuel gas from *naphtha, which is a light petroleum distillate. The naphtha is reacted with steam over a nickel-based catalyst at a temperature of up to 650°C and pressure of 70 bar to produce a gas mixture that is rich in methane. Other gases in the product include carbon dioxide, carbon monoxide, and trace amounts of hydrocarbons. The process was superseded in the UK by the discovery of North Sea gas.

cricondenbar The maximum pressure at which two phases, such as a liquid and vapour, can coexist. A gas cannot be formed above this pressure irrespective of the temperature.

cricondentherm The maximum temperature above which liquid cannot be formed irrespective of the pressure.

critical damping *See* DAMPING.

critical dilution rate The highest possible dilution rate at which steady state is able to be attained within a constant volume *bioreactor such as a fermenter with continuous inflow and outflow. The dilution rate is the ratio of the flow of fresh feed to the volume of broth containing a viable population of microbial cells. At the critical dilution rate, the rate of microbial growth is insufficient to replenish the cells being washed out. Above the critical dilution rate, all the cells are eventually washed out.

critical dissolved oxygen concentration The lowest possible concentration of oxygen in a *bioreactor or biological system below which oxygen becomes the limiting substrate for growth. Under normal operating conditions, the level of oxygen supplied must be sufficient to ensure that the microorganisms are able to function metabolically. A dissolved oxygen electrode is used to monitor the level of dissolved oxygen in the medium to ensure that sufficient oxygen is being supplied, which is usually in the form of sparged bubbles of air or oxygen.

critical flow *See* CHOKED FLOW.

criticality accident An accident occurring in the nuclear industry resulting in the release of dangerous levels of gamma and neutron radiation, and energy from uncontrolled *nuclear fission reactions.

critical mass The smallest amount of a fissile material required to sustain a nuclear fission chain reaction. In a *nuclear reactor, the chain reaction is controlled in order to produce thermal energy, which raises steam used to produce electricity. The fissionable material of an *atomic bomb is divided into portions, each less than the combined critical mass. When they are brought together at the moment of detonation, their combined mass exceeds the critical mass.

critical moisture content The moisture of a solid material at the point when the *constant rate drying period changes to the *falling-rate drying period. It is also used to represent the point between the *bound moisture and *unbound moisture content.

critical path analysis (CPA) A management tool used to manage complex projects. It uses all the necessary information about the project as individual activities and their inter-relationship, and the time required to complete them. The mathematical analysis allows the project manager to identify the critical path, which determines those activities that must be completed and the time taken to reach a target date. It was originally devised in the US as a visual planning technique to manage large-scale military projects and has been adapted to other project management applications, in particular engineering and construction projects.

critical point The temperature and pressure of a substance at equilibrium when two phases become identical and form a single phase. The **critical state** is the condition in which the density of both the liquid and vapour phases of a substance are the same as occurs at the critical point.

critical pressure The minimum pressure required to liquefy a substance at its *critical temperature.

critical properties The properties of substances at their *critical point. Critical properties include critical temperature, critical pressure, critical volume, and compressibility factors. They are used to determine the properties of liquids and gases.

critical temperature The temperature above which a gas cannot be liquefied by pressure alone.

critical velocity The velocity above which the flow of a fluid no longer continues to be streamline but becomes turbulent.

critical volume The volume occupied by one mole of a substance at its *critical temperature and *critical pressure, known as the critical state. The SI units are $m^3 \, mol^{-1}$.

CRO An abbreviation for **c**ontrol **r**oom **o**perator, this is a person who forms part of a team based primarily in the control room of a process plant. The CRO is responsible for monitoring the screens and displays that present real-time data of the process, as well as taking appropriate action when process variables deviate beyond the expected limits. This is normally brought to the attention of the CRO by flashing displays and audible alarms.

cross-current separation A stage-wise liquid–liquid separation technique in which fresh solvent is added to each stage that progressively removes the solute (see Fig. 14). The number of stages and the amount of solvent needed is determined by a material balance at each stage. Experimental data is required to determine the effectiveness of each separation stage.

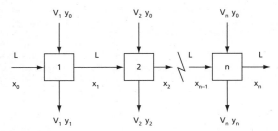

Fig. 14

crossflow filter A type of continuously operated filter configured such that the feed slurry flows across the face of the filter so as to prevent the build-up of solid material on the filter surface, but allows some of the liquid to pass through the filter.

crossflow plate A plate or tray used in a distillation column in which liquid flows enter from one side and flow across and out of the other (see Fig. 15). In a *distillation column the liquid is retained on the plate and prevented from descending through the perforations by the rate of upflow of vapour, bubble caps, or valve seals. A weir is used to control the depth on the plate over which the liquid descends to the plate below via the downcomer.

cross-sectional area The area of pipe or process vessel cut perpendicular to its axis. The cross-sectional area is the area through which material passes. For a circular cross section, the area related to diameter by $a = \dfrac{\pi d^2}{4}$. The volume of the pipe or vessel, V, is the cross-sectional area multiplied by the length $V = al$. The average velocity, v, of the movement of a fluid through a pipe is the volumetric flow, Q, divided by the cross-sectional area $v = \dfrac{Q}{a}$.

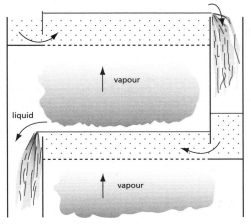

Fig. 15

crucible process An early process used to purify steel by heating steel in a graphite crucible and pouring the melt into a cast-iron mould. It is also known as the *Huntsman process. The method was used until it was superseded by the *Bessemer process.

CRUD A term used in the nuclear reprocessing industry to describe unidentified materials; it is an abbreviation of **c**orrosive **r**adioactive **u**ndetermined **d**eposit. It is usually used in the description of precipitation processes involving fission products.

crude distillation The separation or *fractionation of *crude oil into separate components or groups of components known as fractions in an distillation column called a

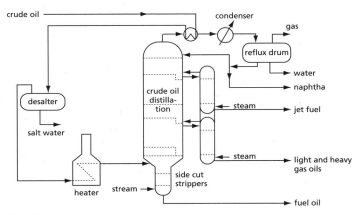

Fig. 16

fractionator operating at atmospheric pressure (see Fig. 16). It is the first step in the petrochemical refinery process in which the crude oil is first desalted and heated to around 370°C in a furnace. It is then separated into fractions according to their boiling points such as butanes, unstabilized naphtha, heavy naphtha, kerosene, and topped crude. This allows subsequent processing units to have *feedstocks that meet particular specifications. *Vacuum distillation is used to separate the heavier portion of the crude oil into fractions that would require higher temperatures to vaporize it at atmospheric pressure and cause *cracking to occur. The fractionator consists of a tall cylindrical column containing perforated trays upon which liquid sits and is in intimate contact with vapour rising up the column through the perforations. Unlike conventional distillation, the column does not feature a *reboiler. Instead, steam is introduced below the bottom tray to strip out any remaining gas oil and to reduce the partial pressure of the hydrocarbons. Reflux at the top of the column is provided by condensing the overhead vapour and returning a portion to the top. Side cut strippers are used to withdraw a liquid sidestream, which contains lower boiling point components, known as light ends. Steam is added and the vapour vented back into a higher position in the column.

crude oil A viscous black liquid composed of a mixture of hydrocarbons formed over millions of years through the gradual decay of buried aquatic animal and vegetable matter. It collects in vast underground pockets in sedimentary rock at depths ranging from a few metres to several kilometres. It is brought to the surface by drilling and pumping operations. It is then refined into useful products in *petroleum refineries.

crude unit A petrochemical refinery processing unit in which initial crude oil distillation takes place to make the first *cut. Lighter hydrocarbon products are further refined in a *catalytic cracker or *reforming unit. Heavier hydrocarbon products are further distilled by *vacuum distillation.

crusher A mechanical device used to reduce large quantities of coarse rock, ore, or other solid bulk material to smaller and more manageable sizes for further processing such as extraction. The main types of crusher include *jaw crushers, gyratory, and smooth-roll crushers that operate by compressing the material. Toothed-roll crushers tear the material apart as well as crushing it.

cryodesiccation *See* FREEZE DRYING.

cryogenic process A very low-temperature process used for separating substances, such as nitrogen and oxygen in air, that are gases at normal ambient conditions. Foods can also be preserved at very low temperatures using solid carbon dioxide or liquid nitrogen.

cryogenic pump A vacuum pump used to produce very low pressures of 10^{-6} Pa by the condensation of gases onto the surface kept at low temperatures using liquid hydrogen or liquid helium. It is possible to reduce the pressure further to 10^{-13} Pa with the combined used of a diffusion pump.

crystal A pure and homogenous solid form of matter in which the atoms, molecules, or ions are arranged in a regular and orderly three-dimensional array or lattice. A crystal structure is formed by the regular arrangement of atoms, molecules, or ions in which they have the same angle between their faces. However, depending on the conditions, different faces can grow at different rates, leading to slightly irregular-appearing shapes. *Compare* AMORPHOUS.

crystal bar process *See* VAN ARKEL–DE BOER PROCESS.

crystallization The process of forming crystalline substances from vapour, solutions, or melts. For example, snowflakes are formed from water vapour, ice is formed from water, and alum crystals from a saturated solution. A crystal may form from molecules, atoms, or ions by coming together randomly to form a cluster. An embryo is formed when sufficient particles have come together to form a solid phase. If *supersaturation is sufficient, the embryo may grow into a nucleus, which if it gains more particles, will grow into a crystal. The crystallization of petroleum fraction contaminants is used to remove wax and other semi-solid substances from heavier fractions. The removal of wax from lubricating oils involves mixing the oil with solvents and then cooling to a temperature of around -20ºC to cause the wax to crystallize. This is then separated.

crystallizer A vessel used to contain and bring about the process of *crystallization. This involves nucleation and crystal growth within a supersaturated solution.

crystalloid A substance dissolved in a solvent that can pass through a *semi-permeable membrane.

CSTR *See* CONTINUOUS STIRRED-TANK REACTOR.

culture medium *See* BROTH.

cumene process A process used for the production of phenol from benzene. The process, developed in the 1940s, involves reacting benzene and propene (propylene) vapour over a phosphoric acid catalyst at high temperature and pressure to produce cumene (isopropyl benzene):

$$C_6H_6 + CH_3CH:CH_2 \rightarrow C_6H_5CH(CH_3)_2$$

The cumene is then oxidized in air and reacted with dilute acid to yield phenol.

cupellation process A process used to separate lead and other base metals from noble metals such as gold and silver by blowing hot air over the surface of the molten metal held in a shallow refractory dish known as a cupel. The lead oxidizes to lead monoxide, floats to the surface, and is then removed.

cupola furnace A brick-lined furnace known as a cupola used for the conversion of pig iron into iron castings. It involves charging the furnace with coke and igniting it. Air is forced up beneath a charge of heated material to raise the temperature required to bring about the conversion. The molten iron picks up carbon as it descends under gravity and is collected at the bottom.

Curie, Marie (née Marja Sklodowska 1867–1934) A Polish-born scientist noted for her work on ionizing radiation. Taught by her father, she was not accepted by Warsaw University but at the age of 24 she instead went to study in Paris. There she married Pierre Curie (1859–1906) four years later. Together they discovered radium and polonium. They extracted less than a gram of radium from eight tonnes of pitchblende, which was noted for its high *radioactivity. In 1903, together with Henri *Becquerel, the Curies received the Nobel Prize for Physics. However, Pierre was killed in a street accident three years later. She took his place as professor of physics at the Sorbonne, and in 1911 she was awarded the Nobel Prize for Chemistry, becoming the only woman to receive two Nobel prizes, and for different sciences.

curie (Symbol Ci) A former non-SI unit of radioactivity named after Marie *Curie. It was originally defined as the volume of radon gas in equilibrium with 1 g of radium-226. It is now associated with the quantity of a radioactive isotope that decays at a rate of 3.7×10^{10} disintegrations per second. The *becquerel (symbol Bq) is the SI-derived unit of radio-activity where one Bq is equal to one disintegration per second. The millicurie (mCi) and microcurie (μCi) are also used.

curing A process used for hardening polymers by causing the cross-linking of polymer chains. The curing process can involve the use of heat, ultraviolet radiation, and chemical additives. The curing process of rubber is known as *vulcanization.

customary units A system of measurement used in the US. These are largely similar to the British Imperial units.

cut A division in separating distilled products of differing compositions or purities from a distillation process based on composition or temperature. In the continuous fraction-ation of *crude oil, the *naphtha cut, which contains a number of different hydrocarbon compounds, has a boiling point range of around 35°C to 200°C. In the batch distillation of *whisky, the differential compositions of alcohol are termed the *foreshots and **feints** whose composition is determined by *specific gravity.

CVD *See* CHEMICAL VAPOUR DEPOSITION.

cyanide process A process used to extract gold from crushed rock. It involves dissolv-ing the rock in aqueous sodium or potassium cyanide in the presence of air. The gold is converted to aurocyanide, which is then reduced back to gold with zinc in the following reactions:

$$4Au + 8KCN + 2H_2O + O_2 \rightarrow 4KAu(CN)_2 + 4KOH$$
$$2KAu(CN)_2 + Zn \rightarrow 2Au + K_2Zn(CN)_4$$

The process is also known as **cyanidation**.

cycle A series or sequence of periodic changes in which a system moves away and returns to its expected or normal condition or position. *See* THERMODYNAMIC CYCLES.

cyclic system *See* SYSTEM.

cyclone separator A device used to separate particles from air or a gas stream. Particles in sizes that typically range from 5 to 200 μm enter the separator with a high velocity, and are swirled around the circular chamber as a vortex under centrifugal action, and descend to the bottom of the separator for collection. The separated particle-free air or gas leaves the top of the separator. Cyclone separators are commonly used after a spray drying pro-cess to collect the dried product.

cylinder A geometric shape that has a uniform circular cross section and length. The length is greater or equal to the diameter. Examples of cylindrical vessels include drums and tanks, and may be either vertically or horizontally positioned along their axis. Columns are also cylindrical and mounted vertically, whereas gravity separators such as those used to separate oil, gas, and water on offshore platforms, are mounted horizontally.

Czochralski process A process used for growing single crystals. It involves melting a material in a crucible. A single crystal of the material is lowered onto the surface and drawn slowly upwards producing a cylindrical crystal known as a boule. The process is used to grow single crystals that have a high value, such as crystals of silicon and germanium used in the semi-conductor industry, and pure metals, such as platinum, silver, and gold, and synthetic gemstones. It is named after the Polish chemist Jan Czochalski (1885–1953) who invented the process in 1916.

DAC *See* DIGITAL-TO-ANALOGUE CONVERTER.

Dalton, John (1766–1844) A British chemist and physicist noted for his pioneering work on chemistry. From a poor Quaker family, he left his village school at the age of 11 and then taught at there from 12 to 14; from 15 he was a farm labourer. However, he began making scientific studies and even wrote a paper on his own colour-blindness. For 57 years Dalton kept a meteorological diary with over 200,000 observations. He began lecturing in Manchester and by 1800 was secretary of the Manchester Philosophical Society. He measured the rise in temperature of air when compressed and suggested in 1801 that all gases might be liquified by compressing and cooling them. After much work on vapour and gas pressures, he stated his law on partial pressures (*see* DALTON'S LAW) and applying Newton's idea of atoms in gases to his own work, formulated the law of Multiple Proportions in 1804. By 1808 he had a clear theory and published his famous Atomic Theory, which forms the basis of modern chemistry today.

Dalton's atomic theory A theory of chemical combination first postulated in 1803 by British chemist and physicist John Dalton (1766–1844). It includes the postulates that elements are made of individual particles (atoms); that atoms of the same element are identical and that different elements have different types of atoms; that atoms can be neither created nor destroyed; and that so-called 'compound elements' are formed when different elements join together to form molecules. He proposed symbols for the different elements that were later replaced by the present notation for chemical elements.

Dalton's law A law that states that the total pressure of an ideal mixture of gases or vapours is equal to the sum of the partial pressures of its components. The partial pressure is the pressure that each component would exert if it was present alone. The law was formulated by John *Dalton (1766–1844).

Damköhler number A dimensionless number, Da, used to relate the rate of a chemical reaction to certain phenomena that occur in the reaction. There are several types of numbers that are used whose definition depends on whether the chemical reaction is related to momentum transfer or to heat transfer. For example, for an nth *order of reaction, the Damköhler number is:

$$Da = kC_o^{n-1}t$$

where k is the rate constant, C_o is the initial concentration of reactant, and t is the reaction time. It is used to indicate the extent of conversion within continuous flow reactors. In continuous chemical processes, the Damköhler number is defined as the ratio of the chemical rate of reaction to the convective chemical reaction time. A second Damköhler number is defined as the ratio of the chemical reaction rate to the rate of mass transfer:

$$Da_{II} = \frac{kC_o^{n-1}t}{k_g a}$$

where k_g is the mass transfer coefficient and a is the interfacial area. There are two further Damköhler numbers related to heat transfer. They are all named after the German chemist Gerhard Damköhler (1908–44).

damping The gradual reduction or suppression of the oscillatory behaviour of a *process variable such as temperature, pressure, and liquid level in a controlled system or process. Damped harmonic oscillators satisfy the second-order differential equation of the form:

$$\frac{d^2x}{dt^2} + 2\varsigma\omega_o\frac{dx}{dt} + \omega_o^2 = 0$$

where ω_o is the undamped angular frequency and ς is the damping ratio. Critical damping ($\varsigma = 1$) is where response to a change is abrupt, in which there is no overshoot of the set point. Under-damping ($0 < \varsigma < 1$) occurs when there is overshoot, while over-damping ($\varsigma > 1$) occurs when the response is slower than critical damping. An undamped system ($\varsigma = 0$) corresponds to its natural resonant frequency (ω_o).

Danckwerts, Peter Victor (1916–84) A British chemist awarded the George Cross for bomb disposal work in the Second World War. Noted for his application of simple science to complex industrial applications, he was Shell professor in chemical engineering at Cambridge University and executive editor of *Chemical Engineering Science*.

Danckwerts' surface renewal theory A conceptual mass transfer model used to describe the transfer of a substance from a liquid to a gas. It assumes that an element of the surface interface comprises a mosaic of elements of various ages. Each element has a random chance of being replaced by another element from the bulk of the liquid. A feature of the model is that a simple mathematical solution is used for complex cases involving chemical reactions. It was formulated by Peter V. Danckwerts (1916–84).

dangerous substance A chemical substance used or present in the workplace that can cause harm to people due to its physical or chemical properties if not correctly controlled. Such substances include liquids, vapours, gases, and dust in the form of solid particles or fibrous materials that can form an explosive mixture with air. Examples include paints, solvents, reagents, varnishes, flammable gases, such as LPG, dusts and particulates from foodstuffs, machining and sanding operations that can cause harm through fire or explosion. In the UK, the *DSEAR Regulations 2002 of the *HSE (Health and Safety Executive) are intended to protect workers and members of the public from dangerous substances.

Darcy, Henry Philibert Gaspard (1803–58) A French engineer who specialized in fluid hydraulics. In France, he was responsible for many significant hydraulic projects including the construction of a remarkable pressurized water network system in his birth town of Dijon. He developed a way of calculating head loss due to friction, which with further modification by the German mathematician and engineer Julius Weisbach (1806–71) became known as the Darcy–Weisbach equation. The unit of permeability, the darcy, is named after him.

Darcy–Weisbach equation An equation used in fluid mechanics to determine the pressure or head loss due to friction within a straight length of pipe for a flowing fluid:

$$\Delta p_f = f\frac{\rho v^2}{2}\frac{L}{d}$$

or in head form:

$$h_f = f \frac{L}{d} \frac{v^2}{2g}$$

where f is the **Darcy friction factor**, L and d are the pipe length and inside diameter, and v is the average velocity of the fluid. It is also known as the Darcy–Weisbach or **Moody friction factor** whose value depends on the nature of the flow and surface roughness of the pipe. Note that this friction factor is four times greater than the *Fanning friction factor. The value of the friction factor can be obtained from various empirical equations and published charts such as the *Moody diagram.

data Information that is obtained and accumulated from a process or system. It can be in the form of chemical information, measurements, observations, and numerical information used by computers, etc. A **database** is a collection of organized data stored in a computer. The data is available to users and used for various purposes, added to, deleted, or updated as required. The data can include chemical analyses, financial data, process flow information, accumulated history of process information, personnel records, etc. **Data processing** is the sequence of operations that is performed on data to extract information or to achieve some form of order. It usually refers to the use of computers to handle large amounts of data rapidly. **Data mining** is the structured organization of data used to identify relationships within large amounts of computerized data. It is used in various fields of science and engineering, and involves looking for existing or new patterns in data or sequences, as well as forecasting data from patterns within data.

datum A reference or benchmark point from which other measurements are taken. For example, it may be an arbitrary elevation from which other elevations are measured. For fixed offshore installations, the datum is taken as the seabed and not the sea level, which, due to the tide, is variable.

Davis, George Edward (1850–1906) Regarded as the founding father of chemical engineering, he began his career as an apprentice bookbinder. He decided to study chemistry at the Slough Mechanics Institute while working at a local gas works. He spent a year at the Royal School of Mines in London (later Imperial College) before moving to work in the chemical industry around Manchester. Identifying the main features in common to all chemical factories, his *A Handbook of Chemical Engineering* was published in 1901. Working as an alkali inspector, he was responsible for implementing the Alkali Act of 1863 recognizing the importance of air pollution and the need for environmental protection. He also published a series of twelve lectures given in 1888 at Manchester Technical School and helped to define chemical engineering as a distinct discipline.

Deacon, Henry (1822–76) A British chemist and industrialist who founded a major chemical factory in Widnes, Lancashire. He filed a patent for an improved process for the manufacture of sulphuric acid in 1853. He later filed many more patents which included alkali manufacture. In 1870, he invented an improved method for the manufacture of chlorine and hydrochloric acid that used copper chloride as a catalyst. The *Deacon process is named after him.

Deacon process A catalytic process that was once used to convert hydrochloride gas to chlorine gas used in the manufacture of bleaching powder. The bleach was used in the textile and paper industries, and the process was able to reduce the production of hydrochloric acid as a waste product. It involved reacting hydrochloric acid and oxygen in the form of air at a temperature of between 400°C and 450°C in the presence of a copper chloride catalyst:

$$4HCl + O_2 \rightarrow 2Cl_2 + 2H_2O$$

The process was invented by British chemist Henry *Deacon (1822–76) as a way of using the hydrochloric acid from the *Leblanc process. The process was later replaced by an electrolysis process for making chlorine from brine.

deactivation A complete or partial reduction in the reactivity of a substance such as an enzyme or poisoning of a catalyst.

dead band A term used to control a process representing the range through which an input signal can be varied without initiating a response. The dead band is often described as a percentage of the *span.

dead-leg A vertical section of pipe filled with stagnant process liquid that normally has no flow.

dead time The interval of time in a controlled process in which a *process variable begins to change after an input change or stimulus. It is also called a *delay.

de-aeration The removal of air, oxygen, or gas from a substance. This can be achieved by disengagement, by using agents to strip out the gas, by raising the temperature to achieve evaporation or boiling, or by using a reduced pressure.

dealkylation process A catalytic process used for the removal of alkyl groups from molecules. The process is carried out in an atmosphere of hydrogen and is therefore also called **hydrodealkylation**. An example is the formation of benzene from toluene:

$$C_6H_5CH_3 + H_2 \rightarrow C_6H_6 + CH_4$$

Dean number A dimensionless number, Dn, used for flow in curved channels. It is a modified form of the *Reynolds number used to characterize the flow and heat transfer of fluids particularly through helical coils as:

$$Dn = \text{Re}\left(\frac{d}{d_C}\right)^{0.5}$$

where d is the tube's inside diameter and d_C is the diameter of the coil. It is, in effect, the ratio of the centrifugal force to the inertial force of the fluid.

death phase The interval of time that follows the stationary phase in the batch culturing of microorganisms in a bioreactor and where the number of viable cells begins to fall. This is due to complete consumption of available substrate such that the microorganisms then die.

Deborah number A dimensionless number, De, used to classify the rheological behaviour of a fluid, which is able to elastically store energy. It is defined as the ratio of the fluid characteristic time to the process or observed time, where the characteristic time is the stress relaxation time, τ:

$$De = \frac{\tau}{t}$$

Values of De that approach zero indicate liquid behaviour, while values that approach infinity have solid-like behaviour. It was coined by Israeli scientist Markus Reiner (1886–1976) from the Bible (Judges 5:5) as 'The mountains melted from before the Lord ...'.

debottlenecking An analysis used to identify and find engineering solutions to the limiting part or parts of a process that restrict the way it operates. *See* BOTTLENECK.

de-butanizer A continuously operated distillation column used to remove butane as distillate from a mixed feed of hydrocarbons with the heavier components leaving the bottom of the column. Lighter hydrocarbons may also leave the top of the column with the butane. To ensure that the light hydrocarbons can be boiled and condensed in the liquid phase, the distillation is operated at a gauge pressure of around 10 bar.

decanting The process of separating a liquid from a settled solid suspension or from a heavier immiscible liquid in a vessel known as a **decanter**. This uses either gravity to bring about the separation or centrifugal force such as in a **centrifugal decanter**. The separation of the upper liquid layer can be achieved by careful pouring into a separate vessel or by siphoning of the upper layer. It is also known as **decantation**.

decay The spontaneous transformation of one radioactive nuclide into another radionuclide, or into another energy state of the same nuclide with the emission of one or more particles or protons. The decay of N_o number of nuclides to give N number of nuclides after a period of time t is given by $N = N_o \exp(-\gamma t)$ where γ is known as the **decay constant**.

decay ratio An oscillatory response to a controlled system defined as the ratio of successive peaks and troughs above and below the final steady-state value. *Quarter damping is regarded as the optimal decay ratio in which the magnitude of successive peaks or troughs is one-quarter of those of the preceding peaks or troughs.

DECHEMA An abbreviation for Gesellschaft für Chemische Technik und Biotechnologie, which is the German Society for Chemical Engineering and Biotechnology. Founded in 1926, it is based in Frankfurt and is a non-profit-making organization with over 5,000 members. It aims to support developments in chemical technology, biotechnology, and environmental protection.

(⊕) SEE WEB LINKS
• Official website of DECHEMA.

deci- (Symbol d) A prefix used in the metric system to denote one-tenth. For example, 0.1 metre is the same as 1.0 decimetre (dm).

decimal reduction time (Symbol D) The time required to reduce the population of viable microbial spores ten-fold in the thermal sterilization process of foods that may be contaminated with harmful spores. Twelve successive reductions is known as a *botulinum cook or a **12-D process**. This is deemed to be sufficient to reduce statistically any potential spores of the highly pathogenic microorganism *Clostridium botulinum* to insignificant levels. This bacterium produces a highly toxic toxin which is destroyed in the cooking process.

decision tree analysis An engineering procedure used in complex multistage decision-making problems that involves planning and organizing decisions. It takes into account how choices are made at earlier stages, the outcomes of possible external events that determine the types of decisions, and events at later stages of that sequence. The decision-making tree represents, in an organized way, the decisions and events that introduce uncertainty, as well as possible outcomes of all those decisions and events. The nomenclature involves squares that represent decisions. Lines from squares represent options that can be selected. Circles show various circumstances that have uncertain outcomes. Lines from circles denote possible outcomes of that uncontrollable circumstance.

decommissioning The procedure of closing down a chemical process or nuclear facility to a point that permits the release of the property for demolition and site clearance. In nuclear power and reprocessing plants, decommissioning begins by closing down the facility, followed by reducing residual radioactivity to a low level. In nuclear power plants, the process begins with the removal of the nuclear fuel, coolant, and radioactive process waste. The three recognized stages of decommissioning involve safe enclosure, partial removal with safe enclosure, and the complete removal of materials.

decomposition **1.** A chemical reaction in which a compound breaks down into more simple compounds or into elements. **2.** The chemical breakdown of organic matter into its constituents by the action of bacteria and other organisms. **3.** The factorization of an integer into other integers that are second- or third-degree (squares or cubes).

decontamination A process that involves the complete removal of hazardous or harmful substances from the surfaces of process pipes, vessels, and equipment. Various cleaning methods are used including acids, rinsing, and the use of steam.

decrepitation A cracking sound produced when certain crystals are heated. This is caused by the loss of water of crystallization during changes in the structure of the crystals.

deep shaft process A biotechnological process used to treat and purify domestic sewage and industrial biodegradable effluents. It consists of a tall vertical loop-type bioreactor with a height of between 30 m and 150 m that is usually installed below ground level. Compressed air is introduced into the downflow leg to drive the circulation of the liquid medium containing microorganisms. As the liquid rises up the upflow section, the hydrostatic pressure decreases, and the bubbles add to the mixing and aeration. This type of bioreactor is used where there is a high oxygen demand such as in the processing of activated sludge. The sludge can therefore have a higher density than conventional *activated sludge systems. It is also known as a **deep shaft airlift fermenter**.

de-ethanizer A continuously operated distillation column used for the removal of ethane as distillate from a mixed feed of light hydrocarbons. Any methane also leaves the top of the column along with the ethane while heavier components leave the bottom. To ensure that the light hydrocarbons can be boiled and condensed in the liquid phase, the distillation is operated at a gauge pressure of around 14 bar.

defined medium A growth medium used for the culture of microorganisms in which the composition and concentration of all the chemicals is known. It is also known as synthetic medium.

deflagration A type of subsonic explosion as a result of a combustion-type chemical reaction in which the *shock wave arrives before the combustion reaction is complete. The flame front advances outwards, whereas the reaction front advances into the unreacted substance. *Compare* DETONATION.

deformation The change in shape of a body due to shrinkage, stretching, or torsion but does not involve breaking up or destruction of the body.

degassing The release of dissolved, absorbed, or adsorbed gases from a liquid or solid. Degassing is often achieved in reduced pressure or vacuum systems.

degradation 1. The reduction of complex molecules into smaller, simpler molecules through natural processes (biogradation) or synthetic processes. **2.** An organic chemical reaction in which a compound is converted into a simpler compound. For example, the action of certain enzymes can bring about the degradation of proteins to amino acids.

degree 1. A division on a *temperature scale. **2.** An angle equal to 1/360th of a circle. **3.** The highest power of a polynomial. For example, the quadratic equation $ax^2 + bx + c = 0$ has a degree of 2. **4.** The highest power to which a derivative of a differential equation is raised. For example:

$$\left(\frac{d^2 y}{dx^2}\right)^4 + \left(\frac{dy}{dx}\right)^3 = 0$$

is a second order differential equation of the fourth degree.

degrees absolute *See* ABSOLUTE.

degrees of freedom The number of independent variables that are needed to describe fully the equilibrium state of a system or process. In defining the thermodynamic equilibrium for a system involving components and phases, the thermodynamic degrees of freedom are the minimum number of variables (temperature, pressure, and composition) that must be stated to define the system completely. In the use of statistics, the degrees of freedom refers to the number of values that can be varied in a calculation determined from the difference between the number of independent values (sample size) and the steps (which is usually one). It is used frequently in linear regression and *analysis of variance calculations, and in some other calculations involving the sum of squares.

dehumidification The process of decreasing the water vapour or humidity of air or a gas through the processes of condensation at a cold surface, diffusion of an absorbing agent, chemical reaction, or heating.

dehydration The process of removing water from a substance. *See* DRYING.

dehydrogenation A chemical reaction involving the removal of hydrogen from a compound. In organic molecules such as straight-chain hydrocarbons or fatty acids, single carbon–carbon bonds are converted into double bonds by the removal of hydrogen atoms.

deionization A process used to purify water by the removal of mineral ions such as the cations calcium, copper, iron, and sodium, and chloride and sulphate anions. The process uses ion exchange resins that exchange the hydrogen and hydroxide ion for dissolved minerals and then form water. Deionized water is used for many purposes such as the cooling water in nuclear reactors. It is also used as an alternative to distilled water in the laboratory. With the absence of ions, deionized water does not conduct electricity.

De Laval, Carl Gustaf Patrik (1845–1913) A Swedish mechanical engineer who worked on the design of steam turbines and machinery used in the dairy industry. After gaining his doctorate in 1867, he invented a number of machines used in the dairy industry, including a cream separator that involved a spinning container powered by a steam engine. The centrifugal action permitted the lighter cream to separate to the inside with the heavier milk moving to the outside. His most notable achievement was the development of the impulse steam turbine in 1882 that could operate at 42,000 rpm. It was used in many industries through special reduction gearing mechanisms allowing high-speed turbines to drive slow-speed shafts such as propellers for ship propulsion. He was a member of the Royal Swedish Academy of Sciences and was elected to the Swedish parliament. Together with Oscar Lamm (1848–1930), he founded the Alfa Laval company in 1883.

delay A commonly used term instead of *dead time in process control.

del factor The ratio of the number of viable contaminating microorganisms to the initial number in a sterilization process. It is used as a measure of the effectiveness of a sterilization process in which the population of viable microorganisms is reduced due to a combination of both temperature and time.

deliquescence The absorption of moisture from the atmosphere by hygroscopic solids. The extent of the effect produces a concentrated liquid solution of that solid. Examples include sodium chlorate, sodium hydroxide, and calcium chloride.

delivery pressure The pressure of a fluid at the outlet of a pump.

delta (Symbols Δ and δ) As a capital, it represents the difference between two values such as the difference in pressure between two points, Δp. As a lower case, δ commonly represents a partial derivative used in partial differential calculus.

de-methanizer A continuously operated distillation column used in petrochemical refineries to remove methane as distillate from a mixed feed of light hydrocarbons. Any hydrogen will also leave from the top of the column while ethane and other heavier components leave from the bottom. To ensure that the hydrocarbons can be separated in the liquid phase, the distillation process is typically operated at a gauge pressure of around 24 to 28 bar and temperature of $-60°C$.

demister A device used to remove a fine dispersion of liquid droplets from a gas or vapour. In the processing of natural gas, demisters are often called *scrubbers.

denature 1. To produce an irreversible structural change in a protein or nucleic acid that results in the total or partial biological inactivation. Denaturation can be caused by certain chemicals, high temperature, very low and high pH, ionizing radiation, and UV light. **2.** To add a poisonous or unpleasant smelling substance to ethanol to make it unsuitable for human consumption.

dendrite A crystal that has grown into a tree-like structure.

denitrification A natural process involving a chemical reaction in which nitrates in soil are reduced to molecular nitrogen.

denominator The bottom part of a mathematical fraction. Used in ordinary division operations, it is the number or expression by which the numerator is divided. For example, in the fraction ½, 2 is the denominator and 1 is the numerator. The denominator is the divisor.

dense gas A gas that has a density greater than that of air at the same temperature. A **dense gas cloud** is a release of a gas-and-air mixture that is heavier than the surrounding air. The density of the cloud may be heavier due to its temperature being below ambient.

densitometer An instrument used to measure the density of a substance by measuring the volume for a given mass.

density The mass of a substance per unit volume. The SI units are kg m^{-3}.

dependent variable *See* VARIABLE.

dephlegmator A condenser used in a distillation process in which only part of the vapour is condensed and used as reflux. The rest of the vapour is used as the top product. A dephlegmator is also known as a *partial condenser.

depleted uranium Uranium mostly consisting of the non-fissionable isotope uranium-238, obtained as a *by-product of enriching natural uranium in a nuclear reactor, or obtained from nuclear reprocessing plants. Being radioactive with a *half-life of many thousands of years, it therefore needs to be stored for indefinite periods of time. It has a very high density of 19,000 kg m^{-3} and has found use in various military applications such as coatings for armour-piercing missiles.

deposition *See* CONDENSATION.

depression of freezing point The reduction of the point of freezing of a pure liquid when another substance has been dissolved in it.

DePriester charts *Nomographs that present the complex relationships between pressure, temperature, and *K-factor for various light and heavy hydrocarbons. They are used to determine the *bubble point and *dew points of hydrocarbon mixtures and were first published in 1953 as an improvement on earlier charts known as Kellogg charts.

de-propanizer A continuously operated distillation column used to remove propane as distillate from a mixed feed of hydrocarbons with the heavier components leaving the bottom. Lighter hydrocarbons may also leave the top of the column with the propane. To ensure that the light hydrocarbons can be separated by distillation, the distillation column is typically operated at a gauge pressure of around 10 bar.

depth The distance downwards from a reference location or plane. For example, the depth within a liquid is the distance below the surface, while the depth of a mineral ore mine can refer to the horizontal distance to the back of the mine.

derivative 1. A compound that is derived from another compound and retains its general structure. **2.** A function derived from another by the application of differentiation and partial differentiation. A second-order derivative is a derivative of a derivative.

derivative action control A mode of control used to control a process in which the controller output is proportional to the rate of change of the process variable or process error. *See* PID.

derived units An acceptable unit that is defined from base *SI units. In the SI system, there are seven base units: kilogram (kg) for mass, metre (m) for length, second (s) for time, kelvin (K) for temperature, mole (mol) for the amount of substance, ampere (A) for electrical current, and candela (cd) for luminous intensity. The newton (N) is a derived unit of force where 1 N is equal to 1 kg m s^{-2}. The pascal (Pa) is the derived unit for pressure where 1 Pa is equal to $1 \text{ kg m}^{-1} \text{ s}^{-2}$. *See* BASE UNIT.

derrick The tall metal tower on an oil or gas drilling platform above a well. It is used for lifting and lowering tubes and tools down into the well.

desalination The removal of salt from seawater used to provide potable water for drinking or irrigation purposes. Desalination is only used where there is a cheap source of energy and where there is a distinct shortage of fresh water. Desalination methods used include reduced pressure evaporation, freezing, reverse osmosis, electrodialysis, and ion exchange.

desalting The process of removing mineral salts from water-containing oil.

desiccation The process of drying and removing the moisture within a material. It involves the use of a drying agent known as a **desiccant**. Desiccants that function by adsorption of moisture include silica gel and activated alumina, while chemical desiccants that function by the reaction with water to form hydrates include calcium chloride and solid sodium hydroxide. A **desiccator** is a container used for drying substances or for keeping them dry and free of moisture. Laboratory desiccators are made of glass and contain a drying agent such as silica gel.

design codes (design standards) Published standards required for equipment and working practices within the chemical and process industries that represent good practice and define the level or standard of design. Developed and evolved over many years and based on tried and tested practices, there are a number of national standards organizations and institutions that provide published standards for design, materials, fabrication, and testing of processes and equipment. These include the British Standards Institute (BSI), the Institute of Petroleum (IP), the American Petroleum Institute (API), the American National Standards Institute (ANSI), the American Society of Mechanical Engineers (ASME), the American Society for Testing and Materials (ASTM), and the American Iron and Steel Institute (AISI). In Europe, there has been a steady move towards harmonization of national standards with the formation of the Euronorm (EN) engineering standards.

design project A distinguishing feature of degree programmes in chemical engineering that are accredited by the *Institution of Chemical Engineers. The design project features near the end of the degree programme and involves the complete design of a chemical process from raw materials through to the production of a product that is defined in terms of an output flow rate and quality. Typically working in teams, students are required to pool their accumulated knowledge in chemical engineering to meet the specifications, which involves the production of a *flowsheet, material balances, and energy balances, with unit operations that typically include mixing, chemical reaction, and separation. It also includes process economics, safety, environmental issues, start-up, *HAZOP, and the preparation of a *P&ID.

desorption The removal of absorbed atoms, molecules, or ions from the surface of a solid material. It is the reverse of absorption and can be achieved by the use of heat.

desublimation *See* SUBLIMATION.

desulphurization A process used to remove sulphur from crude oil and refined petroleum products, and also the removal of hydrogen sulphide from natural gas. It involves a catalytic reaction in the presence of hydrogen. The use of bacteria to remove sulphur from crude oil has also been developed. It is important to remove sulphur as it poisons catalysts and results in harmful sulphur dioxide in the combustion of fuels. Also known as **hydrodesulphurization**, the process involves a high temperature and pressure reaction in the presence of a nickel or alumina catalyst impregnated with molybdenum.

detection limit The ability of an instrument or an analytical technique to detect and measure the lowest quantity of a substance, or some other variable such as temperature or pressure. In analytical work, confidence limits are usually stated for the level of detection.

detonation A supersonic combustion process in which the reaction front advances into unreacted material with a flame front or shock front and reaction products travelling in the same direction. The resulting *blast wave is initially characterized by a very high peak pressure acting over a very short period of time. As the wave travels outwards from the source, the pressure decays. *Compare* DEFLAGRATION.

deuterium (Symbol D) Heavy isotope of hydrogen with a mass about double that of ordinary hydrogen and is sometimes known as heavy hydrogen. A **deuterated compound** is a compound in which some or all of the hydrogen-1 atoms have been replaced by deuterium atoms. Heavy water is water in which the hydrogen atoms are replaced by deuterium oxide (D_2O). It is found naturally in water in very small amounts, and can be separated and concentrated by *fractional distillation or by *electrolytic separation; it is used as a moderator in the nuclear industry due to its ability to reduce the energy of fast-moving neutrons.

deviation The departure from a desired or expected process value. Monitoring departures from expected operating conditions forms the basis of *process control.

devitrification The loss of amphorous structure of glass as a result of crystallization. It is the opposite of *vitrification. It occurs very slowly in nature, which is why glassy rocks are rarely found, having turned to crystalline structures over time.

Dewar, Sir James (1842–1923) A Scottish chemist who was the first to liquify hydrogen in 1898 and later succeeded in solidifying it. He studied the magnetic properties of liquid oxygen and ozone, and the phosphorescence of substances at low temperatures. He invented the *Dewar flask for keeping liquids at very low temperatures.

Dewar flask A container used for storing hot or cold liquids that are able to maintain their temperature independently from their surroundings. The heat loss is kept to a minimum by using a container held within a container separated by a vacuum. The thin walls of the inner container are made of glass (or steel in the case of bigger containers). The inner surface of the glass vessel is silvered to reduce loss through radiation. It was devised around 1872 by Sir James *Dewar, and is also known by its trade name Thermos flask.

dewaxing A process used to remove *wax from processed lube oil to allow it to flow at ambient temperature. Either refrigeration is used to crystallize the wax with the use of solvents to dilute the oil to allow filtration of the wax from the oil; or selective *hydrocracking is used to crack the wax to form light hydrocarbons.

dew point The temperature at which drops of condensate first appear on cooling a condensing vapour. When the *relative humidity of the vapour is 100 per cent, the wet and dry bulb temperatures are the same. When the relative humidity is less than 100 per cent, the wet bulb is less than the dry bulb temperature.

diafiltration A type of *ultrafiltration membrane separation process involving a *semi-permeable membrane to remove salts and microsolutes from a solution. These are small molecules that are separated from larger molecules, which are retained as the *retentate. Unlike *dialysis, which uses osmotic pressure to drive the solutes across the membrane, diafiltration uses an external force such as pressure. The process is typically used to reduce the level of salts in solutions containing proteins, peptides, nucleic acids, and other biomolecules from biochemical processes.

dialysis The separation of molecules in a liquid by differences in their ability to pass through a *semi-permeable membrane using osmotic pressure. Large molecules such as glucose or amino acids in an aqueous solution can be separated by diffusion across the membrane in which they move from an area of high concentration to one of low concentration. Smaller solutes and water are able to pass through the pores of the membrane and retain the larger molecules. The cells of living organisms are semi-permeable. The kidneys in the body are used for the excretion of nitrogenous waste. An artificial kidney also uses dialysis for the same purpose. *Compare* DIAFILTRATION.

diameter (Symbol d) The distance across a circular plane figure at its widest point. The internal diameter of a pipe of uniform cross-sectional area is twice the internal radius of the pipe.

diamond-anvil cell A device used for producing extremely high pressures. Using a screw arrangement, high pressures of up to a megabar (1 Mbar or 10^{11} Pa) can be generated within a small chamber or cell that contains two high-quality diamonds through which samples of materials under investigation can be observed optically. It is used to study the effects of materials under very high isostatic pressures.

diaphragm pump A type of reciprocating pump used to transfer liquids in which a flexible diaphragm is flexed to and fro. The liquid is drawn into a chamber and expelled as the diaphragm flexes. Non-return or *check valves ensure that the liquid flows in the desired direction. The diaphragm provides a barrier between the liquid and the moving mechanical parts of the pump. It is a comparatively low-maintenance sealed pump and useful for transporting liquids where leakage may be a concern. They also have the ability to stall if run against too high a pressure. A **double diaphragm pump** is a variation that has two diaphragms that are flexed to and fro separating a wetted chamber and used where it is essential to fully safeguard against leakage. They are therefore used in the nuclear reprocessing industry, such as in the transfer of radioactive solutions.

diaphragm valve A type of valve in which a flexible membrane is used to restrict the rate of flow. The membrane is usually made from a flexible natural or synthetic rubber. Diaphragm valves are typically used for fluids that contain suspended solids.

diauxic growth The growth of a microorganism that first assimilates a limiting substrate and, on eventual depletion, it metabolically adjusts itself to assimilate another limiting substrate. For example, in the batch culture of yeast, sugar is first assimilated with the production of carbon dioxide and ethanol. Once the sugar has been depleted, the yeast then has a further period of growth in which the ethanol is used as the limiting substrate. *See* GROWTH CURVE.

Diesel, Rudolph (1858–1913) A German inventor who invented the diesel engine that was first publicly exhibited in 1898. He built a factory and spent the rest of his life perfecting and constructing his engine. While on a night steamer crossing the English Channel to England, he fell overboard and was drowned.

Diesel index An empirical measure of the ignition quality of a diesel fuel defined in terms of the *API gravity and aniline point of the fuel. The higher the number, the better the ignition quality.

differential calculus A branch of mathematics that is concerned with the behaviour of functions at a point and involves the evaluation of the derivative at that point, written as dy/dx or f(x). The derivative gives the slope or gradient of the tangent to the function at x. The overall process of obtaining the derivative is called **differentiation**, and the reverse is called **integration**.

differential distillation A batch distillation process used to separate components in a liquid mixture that does not involve reflux. As the charge is boiled, the first distillate is richest in the more volatile component. As the distillation proceeds, the vaporized product becomes leaner. The distillate can therefore be collected into fractions or cuts of differing purities. *See* WHISKY.

differential equation A mathematical equation expressed as a *derivative of one variable with respect to another. The order of the differential equation is the order of its highest derivative. There are many types of differential equation and each has its own method of solution. The simplest form of differential equation has separable variables enabling each side of the equation to be integrated separately. For example, the rate of change in level, dH/dt, in a vessel of uniform cross-sectional area, A_t, containing a liquid allowed to drain freely through an orifice of cross-sectional area a_o located at the bottom of the tank can be given by the first order differential equation with separable variables:

$$A_t \frac{dH}{dt} + C_d a_o \sqrt{2gH} = 0$$

The *degree of a differential equation is the highest power to which a derivative of a differential equation is raised.

differential pressure The difference in pressure between two points. The difference may be due to frictional pressure resistance of a flowing fluid, blockage, and obstacles, and for some types of flow meters, it can be used to determine flow rate. The differential pressure can be measured using devices such as *manometers. A **differential pressure cell** is widely used to measure the flow rate in pipes in which the differential pressure measured across an orifice is converted to a pneumatic, electrical, or mechanical signal for transmission to a local controller. *See* DP CELL.

differential scanning calorimetry (DSC) A thermal technique used for the analysis of substances in which heat is electrically added or removed to change the temperature, thereby allowing enthalpy changes due to thermal decomposition to be accurately studied.

differentiation A mathematical process used to find the rate at which one variable changes with respect to another. The result of differentiation of a function of y = f(x) is dy/dx or f′(x) and called the *derivative. On a graph of f(x), dy/dx is the slope of the tangent to the curve y = f(x) at that point. In general, the rule for differentiation is that where $y = ax^n$ than

$$\frac{dy}{dx} = anx^{n-1}$$

Tables of standard differentials are used for other forms of equation.

diffuser A widening section of a duct or channel used to carry air or a gas, which is shaped in such a way that its kinetic energy is converted to pressure energy.

diffusion The process of movement in which molecules of one substance move and penetrate other substances. In a mixture of gases, the rate of diffusion of one gas into another is inversely proportional to the square root of its densities (*see* GRAHAM'S LAW). In liquids, the diffusion of a solute through a solvent to produce a solution of uniform concentration is slower. Diffusion in solids occurs much more slowly.

diffusion coefficient, diffusion constant (Symbol D) A proportionality constant between the molar flux as the result of the molecular *diffusion of one substance in a mixture and the driving force. The driving force is a concentration gradient across the mixture. The greater the value, the greater the diffusion of the substance into the other. The use of subscripts such as D_{AB} represents the diffusivity of substance A in substance B. The diffusivity of gases is dependent on temperature, pressure, and the nature of the gas components. The diffusivity of liquids, however, varies appreciably with concentration. For gases, the coefficients are either obtained experimentally or are based on formulae that are based on *kinetic theory. For liquids, empirical correlations are generally used. The SI units are $m^2\,s^{-1}$. Typical values for the diffusivity of gases is in the order of $10^{-5}\,m^2\,s^{-1}$ and for liquids in the order of $10^{-9}\,m^2\,s^{-1}$. *See* FICK'S FIRST LAW OF DIFFUSION; FICK'S SECOND LAW.

diffusion flame The combustion of an unmixed fuel in oxygen that results in a laminar flame in which the propagation of the flame is governed by the diffusion of the fuel and oxygen (air). A candle flame is an example of a diffusion flame.

diffusivity Another name for *diffusion coefficient.

digester A vessel used to produce cellulose pulp from wood chips. The wood chips are reacted with chemicals and heated under pressure to remove lignin. The digester is the first step in the process of making paper.

digital signal Used in the computer control of processes, it is a discrete value at which an action is carried out. As a binary signal using the notation of 0 or 1, it is often used to represent on or off.

digital-to-analogue converter (DAC) The electronic hardware used in the control of processes that converts a *digital signal to an analogue signal such as electrical voltage or current.

dilatant A non-Newtonian fluid in which the apparent viscosity increases with shear rate. That is, it thickens when it is being sheared. It can be described by:

$$\tau = k\gamma^n$$

where τ is the shear force, γ is the shear rate, and n has a value greater than unity. Examples include titanium dioxide suspensions, cornflour and sugar suspensions, and cement aggregates.

dilation (dilatation) An increase in volume.

diluent A substance that is added to dilute a solution or mixture; to reduce the strength of a solution by the addition of water or other solvent. Diluents are used to alter the viscosity of a solution in order to meet specifications, such as the addition of naphtha or condensate to heavy oils to meet pipeline specifications for transportation. In solvent extraction, diluents are inert solvents such as kerosene in the *purex process to control the extraction capacity of the organic layer.

dilution The volume of solvent in which a given amount of solute is dissolved.

dilution method A method used to calibrate flow meters. It involves the addition of a concentrated extraneous material whose presence can be quantitatively determined by an analytical technique. The extraneous material may or may not be one that is already present as an impurity and is diluted into the flow of fluid whose flow rate is being measured. Examples are salt in water, ammonia in steam, and chlorine in oxygen. The procedure is to add steadily a known amount of extraneous material to the flowing process fluid over a known period of time. If the addition is made upstream of the meter, the latter can act as a mixing device during the period before samples of the fluid are taken downstream of the meter and then analysed. For the addition of extraneous material added upstream of the meter this can be expressed mathematically as:

$$Q_i C_i + Q_a C_a = C_o (Q_i + Q_a)$$

where Q_i is the flow rate of the process fluid, Q_a is the flow rate of the extraneous material, C_i is the concentration of the extraneous material in the process fluid, C_a is the concentration of the added extraneous material, and C_o is the diluted downstream concentration. This equation can be rearranged to calculate the flow rate of the process fluid as:

$$Q_i = Q_a \left(\frac{C_a - C_o}{C_o - C_i} \right)$$

If the addition of extraneous material is after the meter, then the meter measures Q_i and not $Q_i + Q_a$.

dilution rate (Symbol D) A term used in the continuous operation of a fixed volume vessel supplied with a constant flow stream of material and an equal amount leaving, and defined as the ratio of volumetric flow rate, Q, to volume of the vessel, V:

$$D = \frac{Q}{V}$$

In the continuous cultivation of microorganisms in a bioreactor containing viable cells, fresh nutrients and limiting substrate promoting growth are supplied while inhibiting metabolic products are withdrawn at the same rate. A material balance for cells, X, and limiting substrate, S, for an ideally mixed bioreactor is:

$$\frac{dX}{dt} = \mu_s X - DX$$
$$\frac{dS}{dt} = D(S_o - S) - \frac{\mu_s X}{Y_{X/S}}$$

where S_o is the concentration of supplied substrate, μ_s is the growth rate, and $Y_{X/S}$ is the yield coefficient. The SI unit is s^{-1}.

dimensional analysis A method used to check an expression or a solution to a problem used to describe an observable phenomenon. There must be dimensional consistency between the variables and the phenomenon. The relationship of the variables are considered in terms of their dimensions, such as mass, length, time, heat, and temperature, and are regrouped into dimensionless groups. These groups, by virtue that they have no scale of size, mass, or time, can then be used to study the effects of any terms in any situation independent of its application. Dimensionless groups are therefore important in both scale-up and scale-down of processes in chemical engineering.

dimensionless number (or group) A group of variables, which collectively, have no dimensions. They are used to describe observable phenomena such as friction in pipes, the behaviour of fluidized and packed beds, and heat transfer from surfaces. Examples of dimensionless numbers in fluid flow and heat transfer include the *Reynolds number and the *Fourier number, respectively. They are particularly useful for scale-up and scale-down in chemical engineering and can be obtained through *dimensional analysis.

dimensions 1. The size of an area or a body, such as a rectangle or a cylinder that is measured in terms of breadth, height, and diameter. **2.** The product or quotient of the basic physical quantities, raised to the appropriate powers, in a derived physical quantity. The basic physical quantities are mass, length, and time from which derived quantities can be formulated. For example, force is defined as the product of mass [M] and acceleration $[LT^{-2}]$ to give $[MLT^{-2}]$. Pressure is defined as force per unit area $[L^2]$ to give the derived units of $[ML^{-1}T^{-2}]$. **3.** The number of coordinates needed to define a line, shape, or solid. A rectangular area is two-dimensional and a solid is three-dimensional.

direct-acting controller A type of control device used to control a process in which the output signal from the controller increases with increasing measured (input) value. It has a negative gain. *Compare* REVERSE-ACTING CONTROLLER.

directional solidification A process used to cool a liquid metal or alloy to form a solid casting in which the solidification is allowed to occur progressively along the length of the casting. The process is used to overcome shrinkage problems in alloys, and is also used to remove impurities from the metals in which the impurities collect at the liquid–solid interface. The end piece containing the impurities can then be cut off and removed for recycling.

disc-and-doughnut An arrangement of plates used in distillation and adsorption columns in which alternate discs and rings (doughnuts) are used to allow liquid to cascade from a disc onto a ring and down onto another disc. This arrangement allows a good contact of descending liquid with ascending vapour, and also has the benefit of a low pressure drop up the column.

disc centrifuge *See* BOWL CENTRIFUGE.

discharge head The *delivery pressure from a pump expressed as a head rather than as a pressure for which the SI unit is metres of the fluid being pumped.

discontinuous phase The phase in a dispersion or emulsion, which is dispersed as particles or droplets in the continuous phase.

dispersion The mixture of one substance dispersed within another as bubbles, drops, or particles. A **dispersed phase** is the dispersion of one phase within another such as suspended gas bubbles, droplets, or solid particles dispersed in a liquid. Fog is the dispersion of water droplets in air, a colloid is a dispersion of suspended particles, and margarine consists of water dispersed in a hydrogenated fat phase. *Compare* CONTINUOUS PHASE. **Dispersed flow** is a two-phase flow of liquid in a vapour or gas in which the liquid is almost entirely entrained as fine droplets.

dissociation A reversible chemical reaction in which ionic compounds break down into smaller particles. For example, in the reversible dissociations of:

$$2HI \Leftrightarrow H_2 + I_2$$

the dissociation constant is constant and is given by:

$$K_d = \frac{[H_2][I_2]}{[HI]^2}$$

The brackets denote the equilibrium concentrations. For gases, this can be expressed in partial pressures.

dissolve The ability to make or become a solution.

distance–velocity lag The *dead time of a signal in a controlled system between the measurement of a variable in a process and making the appropriate controlled adjustment, and seeing some effect due to that adjustment arising solely from the finite speed of the propagation of the signal. For example, it may arise from the measurement of temperature of a flowing fluid in a pipe located downstream some distance from a heat exchanger being controlled. Most in-line chemical analysers such as gas chromatographs take time to return a value once a process sample has been taken and can lead to a dead time.

distillate The liquid or vapour from the top of a *distillation process that is rich in the more volatile component. **Distillate fuel** is a light group of oils extracted by crude oil distillation. They include diesel oil, light heating oil, and heavy gasoils that are used as a feedstock for other petrochemical products. Diesel oil is distilled at between 180°C and 380°C and used in diesel engines.

distillation The separation of liquids by virtue of their difference in boiling points. The process consists of boiling the liquid mixture and condensing and collecting the vapour. It is used to purify liquids and to separate liquid mixtures into its constituent components. It is the most widely used method for the separation of miscible liquids and takes place in a *distillation column.

distillation column A tall vertical cylindrical vessel used for the process of *distillation. Hot vapour rises up the column, which is brought into intimate contact with cooled liquid descending on stages or trays for a sufficient period of time so as to reach equilibrium between the vapour and the liquid. The vapour rises up from the tray below through perforations in the tray, and the liquid on the tray flows over a weir to the tray below. In this way, the more volatile component increases in concentration progressively up the column. In continuous distillation, fresh feed is admitted at the tray corresponding to the same composition. Below the feed point, the section of column is known as the stripping section, while above is known as the rectifying section. A *reboiler heat exchanger is used to boil the bottom product and produce vapour for the column. A *condenser is used to condense some or all of the vapour from the top of the column. A small portion of liquid is returned to the column as *reflux. The height of the column is an indication of the ease or difficulty of separation. For example, an ethylene splitter in a petrochemical refinery used to separate ethylene from ethane, which have close boiling points, requires many trays and the column is very tall. The width of the column is an indication of the internal vapour and liquid rates. In batch distillation, a charge of feed for separation is fed to the still and allowed to be boiled. The vapour is rich initially in the more volatile component or components. Scotch *whisky distillation uses copper stills and operates as a single equilibrium stage. Multistage rectification is used in small production units where a pure product is required.

distillation train A sequence of distillation columns used to separate components from a multicomponent feed. Each column is required to perform a particular separation of either a pure component or a cut between two components. For example, in the separation of four components ABCD in a mixture in which A is the most volatile and D the least, then the five possible separation sequences requiring three columns are:

Separation	Column 1	Column 2	Column 3
1	A:BCD	B:CD	C:D
2	A:BCD	BC:D	B:C
3	AB:CD	A:B	C:D
4	ABC:D	A:BC	B:C
5	ABC:D	AB:C	A:B

Where it is required to separate a larger number of components, the number of possible separation sequences becomes much larger according to the relationship

$$N = \frac{(2n-2)!}{n!(n-1)!}$$

where N is the number of sequences and n is the number of components:

Components (n)	4	5	6	7	8	9	10
Sequences (N)	5	14	42	132	429	1430	4862

Due to the difficulty in seeking the best train sequence, enumerative or heuristic approaches are used to rapidly identify the best train.

distilled water A form of water that has been purified by the process of distillation such that it is free from dissolved salts and other compounds. *See* DEIONIZATION.

distillery A facility or establishment that involves the process of *distillation. Scotch *whisky is produced in a distillery in which malted barley is fermented to alcohol (ethanol) and then distilled to a higher concentration. An industrial distillery uses other grains as the raw material, such as maize or wheat, to produce *neutral spirit that is also used for making gin and vodka.

distributed control system (DCS) A general name for control systems that are used to control processes characterized by multiple controllers distributed throughout the process and connected by networks for the purpose of communication and monitoring data. They are connected to sensors and valve actuators and can use proportional, integral, and derivative control, as well as perform neural network and fuzzy-logic control, and be connected to a *human–machine interface.

distribution coefficient, distribution ratio A coefficient used in solvent extraction to quantify the extent of extraction between phases. It is evaluated as the concentration of the solute in the organic phase divided by the concentration in the aqueous phase. The distribution ratio can be a function of temperature and concentration of chemical species amongst other factors. It is also referred to as the *partition coefficient.

disturbance A change in the expected operating conditions of a process that requires control. The disturbance can be intentional or unintentional, and can be in the form of a step change in a process variable such as temperature, or a ramped change, etc.

Dittus–Boelter correlation A dimensionless equation used in heat transfer for forced convection. For a fluid with turbulent flow in a pipe of circular cross section:

$$Nu = 0.023 \, \mathrm{Re}^{0.8} \, \mathrm{Pr}^n$$

where the Nusselt number is given by:

$$\mathrm{Nu} = \frac{hd}{k},$$

*Reynolds number by:

$$\mathrm{Re} = \frac{\rho v d}{\mu},$$

and Prandtl number by:

$$\mathrm{Pr} = \frac{\mu c_p}{k}.$$

The exponent n has a value of 0.4 for the heating of a fluid and 0.33 for the cooling of a fluid. This shows that doubling the forced flow of the fluid does not double the heat transfer coefficient, but raises it by a factor of 1.75 times. *Compare* NATURAL CONVECTION.

divider Used in *flowsheets to represent a process input stream, which is split into two or more output streams that have the same composition with no phase change or reaction taking place. It is also known as a *splitter.

dominant dead time process Used to describe a process being controlled where the *dead time is larger than the *lag time. A **dominant lag process** is used to describe a process being controlled where the lag time is greater than the *dead time. The majority of controlled processes are dominant lag types and includes most temperature, level, flow, and pressure control loops.

dose The exposure to chemicals or ionizing radiation with hazardous effect on the human body. Solid and liquid toxic effects may be presented as LD_0, LD_{10}, or LD_{50}, describing the dose that will produce 0, 10 per cent or 50 per cent chance of fatality, respectively. The values are usually milligrams per kg body weight. Another form of dose is the OEL (Occupation Exposure Limits). The TLV is the *Threshold Limit Value for eight hours per day, while the STEL is the Short Term Exposure time for fifteen minutes. For ionizing radiation, the dose is a measure of the extent to which the human body has been exposed. The **absorbed dose** is the amount of energy per unit mass. The SI unit is the *gray or the rad where 1 rad = 0.01 gray. The **dose rate** is the exposure to ionizing radiation expressed as the ratio of dose with time.

dosing pump A pumping system that is used to supply fluids to a process from time to time. They often form part of a control system such as the need to maintain the pH in a bioreactor with the controlled addition of acids and alkalis. Other dosing examples include chlorine dioxide being added to drinking water and power station water, and chlorine gas for swimming pool disinfection. *Peristaltic pumps are often used for small systems such as laboratory bioreactors, and *diaphragm pumps for larger applications.

double-acting A term used to refer to both sides of a reciprocating pump or engine being used for compression or expansion, respectively.

double block and bleed The use of two block or isolation valves on a length of pipe for the purposes of isolation with a bleed valve between them. It is used to prevent a process fluid from reaching an item of equipment such as a pump requiring maintenance. The pump can therefore be taken out of service without having to drain the entire length of pipe.

double bond *See* CHEMICAL BOND.

Dow, Herbert Henry (1866–1930) A Canadian industrial scientist who founded the Dow Chemical Company in 1897. He is noted for the invention of the *Dow process for extracting bromine using electrolysis to oxidize bromide to bromine. He had a major impact in the breaking of a cartel of European companies in chemical manufacture at the time.

downcomer A duct in an item of process equipment, such as a distillation column or evaporator, used to channel liquid from one location, such as a plate, down to the one below.

Downs process A process used in the extraction of sodium by the electrolysis of a *eutectic mixture of sodium chloride and calcium chloride at 580°C. The **Downs cell** has a central graphite anode and is surrounded by a cylindrical steel anode. Released chlorine is led away through a hood over the anode while molten sodium is formed and collected at the cathode. A small amount of calcium chloride is added, which lowers the melting point and the sodium chloride is kept molten electrically. Additional sodium chloride is added as it becomes exhausted. The process was invented by J. C. Downs in 1922.

downstream A generic term to mean the stream of material for processing that has already passed through a process. *See* UPSTREAM.

downstream processing A general term for production facilities and unit operations that use process materials derived from another process. They are often in an unchanged state, such as crude oil and natural gas, or products from a fermentation process, and require separation, concentration, reaction, and purification. Downstream processes include petrochemicals, oil refining, biological, pharmaceutical, and fertilizer production facilities. **Downshore processing** are onshore-based processing facilities.

downtime The time that a process or item of equipment is not operating and unavailable for use. It may be out of action due to a fault, failure, or routine *maintenance, such as cleaning and repairs.

Dow process A process used to extract magnesium from seawater. Lime (calcium hydroxide) is used to react with magnesium chloride in the seawater to produce magnesium hydroxide and calcium chloride. The hydroxide is precipitated and filtered. It is then treated with hydrochloric acid, dried, and melted at 710°C. Electrolysis is used to produce the magnesium along with chlorine gas.

DP cell An abbreviation for **d**ifferential **p**ressure cell. A widely used device used to measure the pressure difference between two points, such as a process fluid flowing through a pipeline or across an item of equipment such as a heat exchanger. They are used to measure flow rate, level, or depth of fluids in vessels, and the status of equipment for the presence of fouling, amongst many other uses. By detecting and measuring the pressure difference between two points, such as across an orifice plate, a transducer then converts the signal into an electrical signal. This signal is then used to adjust the rate of flow by sending an appropriate control signal to open or close a valve, for example. *See* DIFFERENTIAL PRESSURE.

drag coefficient (Symbol C_D) The resistance to movement of a particle immersed in a fluid. For a small particle with a particle *Reynolds number (Re_p) less than unity, the drag coefficient predicted by *Stokes's law is:

$$C_D = \frac{24}{Re_p}$$

drain valve A valve located on process equipment such as vessels and tank, and pipelines to discharge liquids as and when required.

drift The change in the output–input signal relationship of a controlled process over a period of time.

drilling mud A mixture of water and additives used in drilling oil wells to cool the drill-bit, remove rock cuttings and transport them up to the surface. It also prevents the well wall from collapsing and maintains sufficient pressure at the bottom of the well to avoid a *blowout.

driving force The difference in physical properties that causes the movement of mass or heat from one place to another. In heat transfer, the driving force is the temperature difference or gradient between two points in which heat flows from a region of high temperature to a region of lower temperature. In mass transfer, the driving force is the difference in concentration of a substance or the partial pressure.

dropwise condensation A type of condensation that occurs at a cold surface that is not wetted by the condensate. *Compare* FILM CONDENSATION.

drum filter A type of filtration device that consists of a horizontal drum covered by a filter cloth. The drum rotates in a bath that is continuously fed with the process solution to be filtered. A vacuum is applied to the inside of the drum and the solid suspension collects on the surface of the rotating drum as a cake as the filtrate is drawn through. The filter cake is removed from the surface by a scraper or knife. Alternatively, a string arrangement is used in which a series of endless belts run around the drum and a separate cylinder. The strings continuously lift off the cake. Drum filters are used in the preparation of certain catalysts.

dry basis 1. A method of representing the moisture content of a substance in which the amount of water is taken as a ratio of the amount of substance. The moisture content of a very wet substance on a dry basis can be above 100 per cent. **2.** In the analysis of gases leaving a process, such as in refining and petrochemical operations, the gases can be defined on a dry basis in which steam is not included in the analysis. *Compare* WET BASIS.

dry bulb temperature A temperature measurement taken from a dry bulb thermometer. Together with the *wet bulb temperature, which uses a soaked wick surrounding the bulb of a thermometer, the humidity of air can be determined.

dryer A device used to reduce the moisture from a solid material. The most common type involves the passing of heated air with a low humidity over the moisture-bearing solid, thereby causing evaporation and removal of the moisture. The drying process is carried out until a desired moisture content is achieved. Other types of batch dryer include the use of desiccant and hygroscopic materials, which absorb the moisture from the air causing a drying effect of the material to be dried. Refrigerated dryers use a reduced temperature form of moisture removal, while membrane dryers use a *semi-permeable membrane to remove moisture.

dry ice Solid carbon dioxide. Instead of melting, it sublimes at $-78°C$ (195 K) at standard pressure. It is used as a refrigerant or cooling agent. It is also used in the preservation of foods such as ice cream and has the advantage that it leaves no residue after sublimation. It is also used in fog machines to produce dramatic fog effects, and in plumbing to freeze water to form a plug in a pipe to allow repairs to be made. It is also used to remove warts by freezing the infected skin, and is also used to attract biting mosquitoes and midges. Dry ice is produced by either carbon dioxide capture or manufacture from another process such as fermentation, and pressurized or refrigerated until it liquefies. When the pressure is reduced causing some of the carbon dioxide to vaporize, there is a lowering of the temperature of the liquid resulting in solidification. The solid carbon dioxide is then compressed into blocks or pellets.

drying The removal of moisture or a liquid from a solid in a process that involves simultaneous heat and mass transfer. The heat is transferred for evaporation by a combination of conduction, convection, and radiation. The moisture may evaporate directly from the surface of the non-porous solid or be transferred from within the body of a porous solid by *diffusion or capillary flow to the surface.

drying rate The speed at which the moisture content of a material is reduced to a lower moisture content. The drying rate is dependent on the humidity and temperature of the drying vapour as well as the properties of the material such as its porosity. There are many theoretical and empirical methods used to estimate the drying rate. The *constant rate drying period is where the rate of evaporation is constant due to continual surface replacement of moisture from within the materials, while the *falling-rate period is determined by the gas phase mass transfer or rate of heat transfer within the material.

dryness The condition of a substance being dry or having a low quantity of water or liquid. The **dryness fraction** is the amount of moisture in a vapour expressed as the ratio of the mass of moisture in the vapour, to the total mass of the vapour. The dryness fraction of *wet steam ranges from 0 (totally dry) to 1, in which it contains the maximum possible amount of moisture.

dry process *See* WET PROCESS.

dry scrubbing A process used to remove acid gases from flue gases that use sorbent materials such as hydrated lime or soda ash to remove sulphur dioxide and hydrochloric acid gas. They are also used to remove undesirable odours and corrosive gases from wastewater treatment plants, and use activated alumina as the sorbent material to remove hydrogen sulphide as well as mercaptans amongst others. In **spray dryer absorbers**, the flue gases are contacted with a finely atomized alkaline slurry where the acid gases are absorbed and react to form solid salts and are removed.

DSC *See* DISTRIBUTED CONTROL SYSTEM; DIFFERENTIAL SCANNING CALORIMETRY.

DSEAR An abbreviation for **D**angerous **S**ubstances and **E**xplosive **A**tmospheres Regulations 2002 of the UK's Health and Safety Executive (HSE). The regulations are intended to protect people from dangerous substances by putting particular duties on employers and self-employed people to protect workers in the workplace as well as members of the public. The regulations require that employers identify the substances that present a risk and put control measures in place to remove the risks or to control them.

(((⊕))) SEE WEB LINKS
• Official website of the Health and Safety Executive, DSEAR regulations.

Dubai crude *See* WEST TEXAS INTERMEDIATE.

duct A pipe, tube, or enclosed channel for transporting gases or vapours. They are square, rectangular, triangular, or circular in cross section, and used in air conditioning for transporting heated air and ventilation purposes. The name is derived from the Latin *ductus* meaning both 'leading' and 'aqueduct', formed from *ducere* meaning 'to lead'.

Dulong–Petit law A law proposed by the French physicists Pierre Louis Dulong (1785–1838) and Alexis Thérèse Petit (1791–1820) for the molar heat capacity of a crystal. The law states that the molar heat capacity of a solid element is approximately three times the *universal gas constant, which is about 25 J mol^{-1} K^{-1}. They proposed the law from experimentation before the kinetic theory of gases had been established. The law is only approximate and applies to elements with simple crystal structure.

dump valve *See* BLOWOFF VALVE.

duplex process An early process used in the production of steel in which one furnace begins as another is finished. Iron is converted to steel in a Bessemer furnace in which the molten product is then transferred to an arc furnace to oxidize the impurities. **Duplex** is a group of stainless steels that have near-equal amounts of austenite and ferrite. They are noted for their high strength and resistance to corrosion resistance, and are used for process equipment.

duplex pump A type of reciprocating pump that operates with two pistons or plungers. The pistons operate out of phase in which one pump begins discharging as the other has just finished its stroke and returns to fill. The result is a near-continuous flow. The rate of flow is the product of the frequency of the strokes and the displacement per stroke.

DuPont, Pierre Samuel (1870–1954) An American chemist and industrialist who worked for the E. I. du Pont de Nemours and Company, and becoming its president from 1915 to 1920. He is noted for transforming the American explosives producer into the well-known diversified chemical producer.

Duralumin A trade name class of strong lightweight aluminium alloys containing copper, magnesium, manganese, and occasionally silicon. Their strength and lightness of weight are useful in aircraft fabrication.

dust Dry powders with a particle size of less than 500 micrometres.

dust cloud explosion The rapid combustion of a dispersion of combustible airborne dust particles within a *confined space. Dust can arise from many processes and include materials such as coal, flour, metals, custard powder, and sawdust, and form an explosive suspension in air (oxygen). Within the confined space, the dispersion of particles collectively have a very high surface area to volume ratio. Typical ignition sources include open flames, electrical static discharge, or arcing in process equipment, hot surfaces, and friction.

duty The power requirement for a machine or item of process plant such as a heat exchanger. The duty of a heat exchanger is dependent on the amount of liquid to be heated, cooled, vaporized, or condensed. The duty of a centrifugal pump is dependent on the flow delivered and pressure generated. The SI units for duty are watts or more usually kW or MW.

duty point The maximum possible delivery that can be achieved by a particular centrifugal pump to meet the pressure drop demand of a system (see Fig. 17). It is represented as the crossover point for the pump characteristic curve with the system characteristic on a pressure-flow rate curve. The system characteristic is parabolic since the pressure drop through the pipes is proportional to the square of the rate of flow.

dynamic analysis A method used to control an inherently unstable and dynamic system or process. It involves mathematically describing the system or process in terms of rate equations for materials, energy, and momentum, and determining how the associated variables change with time in order to restore controlled conditions to a disturbance.

dynamic equilibrium *See* EQUILIBRIUM.

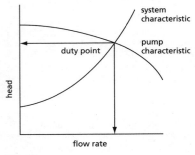

Fig. 17

dynamic error The difference between the actual process parameter and the instrument reading in a controlled process following a *disturbance.

dynamic response The behaviour of the output of a controlled process or item of equipment as a function of the input with respect to time.

dynamics The study of mechanics concerned with the motion of bodies due to the action of forces. *Compare* STATICS.

dynamic similarity A method used to study systems that are geometrically similar such as two stirred tanks, or two piping geometries, in which viscous effects and interfacial phenomena occur, and where the *Reynolds, Weber, and *Froude numbers are the same for both systems. It is useful in the design and interpretation of experimental observations and for the *scale-up of equipment.

dynamic viscosity *See* VISCOSITY.

dynamite A high explosive of nitroglycerine mixed with an inert absorbent, compacted into a cylindrical shape and wrapped in paper. It was invented by the Swedish chemist Alfred *Nobel (1833–96) and was originally formulated from nitroglycerine-absorbed kieselguhr. Once used for military purposes, modern forms of dynamite are used for blasting, mining, and quarrying, and contain sodium or ammonium nitrate with the nitroglycerine soaked into absorbent materials. It is not to be confused with *TNT.

dyne The c.g.s. unit of force. It is the force required to give a mass of one gram an acceleration of $1 \, cm \, s^{-2}$. 1 dyne = 10^{-5} newton.

e The irrational number that has the value of 2.718 281 828 . . . It is calculated from $(1+1/n)^n$ where n tends to infinity. It is used as the base of natural or Naperian logarithms and exponential functions involving e^x.

EA *See* ENVIRONMENT AGENCY.

Eadie–Hofstee plot A graphical method used to obtain a straight line from experimental data from enzyme kinetics (see Fig. 18). It involves forming a plot of V/S versus V in which S is the substrate concentration at which the velocity v is observed. The gradient of the line is equal to $-K_m$ and the intercept on the y-axis is equal to the maximum velocity V. It is named after Canadian biochemist George Sharp Eadie (1895–1976) and B. H. J Hofstee who developed the plot in 1942 and 1959, respectively. *See* MICHAELIS–MENTEN KINETICS.

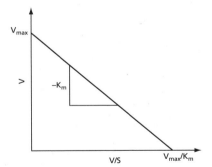

Fig. 18

ECHA An abbreviation for the European Chemicals Agency, which is a regulatory authority within the European Union on the safe use of chemicals. Its aims are to assist chemical manufacturers, importers, and users to comply with legislation and promote the safe use of chemicals. It also provides information on chemicals that are of concern to human health and the environment through authorization, restriction, and labelling of chemicals.

(SEE WEB LINKS
• Official website of the European Chemical Agency.

economic pipe diameter The diameter of a pipe that gives the minimum overall cost for any specific rate of flow of a fluid (see Fig. 19). The total costs of a pipe comprise the capital and operating costs. The capital cost increases with pipe diameter while the operating costs decrease with increasing diameter since the pumping costs to overcome frictional pressure resistance decreases. Where the cost can be related to diameter, the economic pipe diameter can be found mathematically as the turning point by differentiating the total cost with respect to diameter.

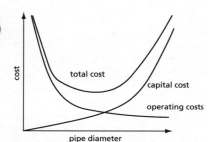

Fig. 19

economic potential The economic viability of a process expressed as the difference between the revenue gained from sales of a manufactured product or products and the cost of the raw materials used. It does not take into consideration capital or operating costs.

economizer A type of *heat exchanger used to raise the efficiency of a steam boiler. It involves first heating the feed water to a steam boiler by the hot flue gases from the boiler. By using some of the heat content of the flue gases, the steam boiler operates more economically since it uses less additional fuel to separately preheat the water feed.

eddy A bulk movement in a stream of fluid which doubles back on itself as a whirlpool or *vortex. *See* TURBULENCE.

EFCE *See* EUROPEAN FEDERATION OF CHEMICAL ENGINEERING.

effectiveness A parameter used to quantify the performance of a heat exchanger, expressed as the ratio of the actual heat transfer in a heat exchanger to the maximum heat that could be transferred by an infinitely long countercurrent single-pass heat exchanger.

effectiveness factor A term used in heterogeneous catalysis defined as the ratio of the actual reaction rate to the reaction rate without diffusion resistance.

efficiency A measure of the effectiveness of a machine or process expressed as a ratio of the energy or power output or delivered to the energy or power supplied. For example, the efficiency of a compressor is the ratio of the energy consumed in isoentropic compression to the actual power consumption.

effluent The liquid or solid waste stream from a process. It can contain contaminated process waste materials such as solvents, acids, alkalis, suspended solids, water from washing operations, and sewage. There is strict legislation governing the release of effluent into the environment. Effluent treatment plants are therefore designed to treat the effluent from a process to comply with the legislation prior to safe release.

efflux The discharge of a fluid from a pipeline or process vessel. The flow in is called the *influx.

effusion The flow of gases under pressure through small holes such as a gas escaping from a cracked weld on a process vessel. The rate of flow is inversely proportional to the density. *Compare* DIFFUSION.

Einstein, Albert (1879–1955) A German-born mathematical physicist who took Swiss nationality in 1901. He originated the theory of relativity and also explained important results from the quantum theory of energy. He explained the variation of specific heat with temperature and also developed a law of the photoelectric effect. He later produced a complete theory to account for Brownian motion. Einstein resigned his German professorship with the rise of the Nazi regime, and worked first at Oxford University and then in Princeton University. He was awarded the Nobel Prize for Physics in 1921.

Einstein theory of specific heat A theory presented by Albert *Einstein that proposes that the specific heat of solids is a consequence of the vibrations of the atoms in a lattice structure. The theory correctly shows that when the temperature tends towards absolute zero, the specific heat of solids tends to zero.

ejector A type of compressor in which gas or vapour is entrained by injecting a high-velocity jet of gas or vapour. A *steam jet ejector uses live steam as the entraining high-velocity jet in which part of the kinetic energy is converted to pressure energy.

elastic collision An idealized assumption used in the *kinetic theory of gases in which the kinetic energy of atoms before collision is the same as the kinetic energy after collision. Perfect elastic collisions occur unless the kinetic energy is converted into another form of energy. For molecules other than monatomic molecules, the kinetic energy is converted into vibrational energy and rotational energy.

elasticity A fundamental property of certain materials to retain energy during deformation caused by an applied force, with the release of the energy to allow the material to return to its original shape. Certain *non-Newtonian fluids exhibit elasticity. For example, elasticity is seen as the energy that is retained in a fluid once a shear stress is removed causing a reversal of flow.

elastomer A substance or material that has the ability to return to its original shape after an external force has been applied and removed. Natural and synthetic rubber or related rubberoid materials are elastomers.

elbow A short section of curved pipework, usually at 90°.

ELD *See* ENGINEERING LINE DIAGRAM.

electrical resistance *See* RESISTANCE.

electric-arc furnace A type of furnace used to melt metals for the production of alloys such as steel in which the heat generated is in the form of an electric arc. The arc is either formed between an electrode and the metal, known as a direct-arc furnace, or is formed between two electrodes and the heat radiated into the metal, known as an indirect-arc furnace.

electrochemical equivalent (Symbol z) The mass of an element released from a solution of its ions in electrolysis by one coulomb of charge. *See* FARADAY'S LAWS OF ELECTROLYSIS.

electrochemistry A branch of chemistry concerned with chemical reactions that involve electricity. The electrical energy may generate electricity such as in a *fuel cell or *battery, or may initiate a chemical process such as in *electrolysis.

electrode A conductor of electricity through which an electric current enters or leaves an *electrolyte. The *anode is the positive electrode and the *cathode is the negative electrode.

electrodeposition A process involving the deposition of one metal onto another by *electrolysis such as *electroplating. An example is the electrodeposition of copper for on-chip wiring of integrated circuits.

electrode potential The potential difference between an electrode and the solution in a *half-cell. This is not able to be measured directly but instead standard electrode potentials E^* are defined by measuring the potential relative to a standard hydrogen half-cell using molar solution at 25°C. By convention, the cell is written with the oxidized cell first. Published tables are used that list standard electrode potentials. For example, the standard electrode potential of the half-cell $Cu|CuSO_4$ is 0.34 volt, and the standard electrode potential of the half-cell $Ag|AgNO_3$ is 0.80 volt. The e.m.f. of a cell $Cu|CuSO_4| AgNO_3|Ag$ is therefore 0.46 volt.

electrolysis The removal of ions from a solution by the passage of electric current in which the solutions are separated in separate chambers by a *semi-permeable membrane. It requires a source of direct current, which is conveyed to the electrolyte by electrodes. The electrode connected to the positive pole is called the *anode; the electrode connected to the negative pole is called the *cathode. The current enters the electrolyte by the anode and leaves by the cathode.

electrolyte An ionic compound which, when molten or in solution, can conduct an electric current with decomposition. Conversely, a non-electrolyte is a compound that will not conduct electric current. Typical electrolytes include acids, bases, and salts, and their solutions, which are also termed electrolytes.

electrolytic cell A device or cell in which *electrolysis occurs. It is sometimes abbreviated and simply called a cell.

electrolytic corrosion A corrosion of metals that occurs as a result of electrochemical activity.

electrolytic refining An electrolytic process used to obtain pure metals. For example, copper can be obtained by using impure copper as the *anode and pure copper as the *cathode. Using a solution of copper sulphate, electricity is passed through the *electrolytic cell resulting in copper being deposited on the cathode. The impurities either remain in solution or are collected as a *sludge.

electrolytic separation A process used to separate isotopes using electrolysis. It was once used to separate deuterium and hydrogen from water since hydrogen is more readily formed at the cathode than deuterium causing an enrichment of the water with deuterium oxide or *heavy water.

(⊕) SEE WEB LINKS
• Official website of the Norsk Industriarbeider Museum, Vemork (Norwegian Industrial Workers Museum).

electromagnetic radiation A form of energy in the form of electromagnetic waves, which are oscillating electric and magnetic fields at right angles to one another from the point of propagation. It includes radio waves, infrared radiation, visible light, ultraviolet radiation, x-rays, and gamma rays, which travel at the speed of light $(2.9979 \times 10^8$ m s$^{-1})$ in a vacuum but slower through materials.

electrometallurgy Processes involving metals and electricity such as the separation of metals from their ores and the plating of metals. *See* ELECTROPLATING.

electron A negatively charged subatomic particle with a mass of about 1/1840 of the mass of a hydrogen atom; together with protons and neutrons in the atomic nuclei it makes up an atom. The electron was identified as being a particle by British physicist Sir Joseph John Thomson FRS (1856–1940) for which he was awarded the Nobel Prize for Physics in 1906, by measuring the angles through which the cathode rays were deflected by known magnetic and electric fields. He succeeded in determining a value for the ratio of charge to mass, which was found to be constant irrespective of the nature of the cathode. The charge was determined around 1910 by American physicist and Nobel laureate for Physics Robert Millikan (1896–1953) to be 1.6×10^{-19} coulombs. The sharing of electrons is the main cause of *chemical bonds.

electro-osmosis The movement of a liquid through a porous material in which an electrical potential has been placed. It is used in dewatering and drying operations, and for materials that are normally difficult to dry such as jelly-like substances. It is used in *fuel cells, where protons move from the anode to the cathode through a proton exchange membrane together with water molecules.

electroplating A process that involves coating one metal with another using *electrodeposition. Within an *electrolytic cell, the item to be coated is made the *cathode, and the other metal used for the plating is made the *anode. The technique is used to provide coatings that provide resistance to corrosion or for decorative purposes.

electrostatic precipitation The process of removing small electrostatically charged particles from a gas stream, such as *ash from combustion processes, using highly charged plates or tubes, and of the opposite polarity. An **electrostatic precipitator** is the apparatus used to bring about the separation in which the charged plates or tubes collect the particles and are cleaned periodically, either by strongly tapping the plates, or by surface cleaning with water. They are typically used to clean the flue gases from coal-fired power stations and can operate effectively at high temperatures and pressures. They are expensive to operate and can tend to be inefficient. They were first commercialized by the American physicist and inventor Frederick Gardner Cottrell (1877–1948) in 1907.

element 1. A substance that is not able to be decomposed into simpler substances by any known chemical process. All the atoms in an element have the same number of protons or electrons but the number of neutrons may vary. All the elements are listed in the *periodic table of which 93 are naturally occurring. **2**. A small part of a system used as the basis for an analysis of an entire system. For example, an element may be a portion of the fluid in a pipe or vessel upon which a force or energy balance is performed to determine the fluid behaviour within the entire pipe or vessel. **3.** A component of a control device or system.

elementary particle A fundamental particle from which all matter is composed and includes electrons, neutrons, and protons.

elementary reaction A simple form of chemical reaction that takes place in a single step with a single transition state and does not involve intermediates.

elimination reaction A chemical reaction that involves the decomposition of a molecule into two smaller molecules with one being much smaller than the other. An example is the chemical reaction used for synthesizing alkenes and alkynes by removing hydrogen from saturated hydrocarbons.

Ellingham diagram A graphical representation of the dependence of temperature on the stability of compounds, particularly for the reduction of metal oxides and sulphides. The diagram is used to predict the equilibrium temperature between a metal, oxygen, and the metal oxide, and therefore the reduction of an ore to the metal. It is, in effect, a graphical form of the second law of thermodynamics as a plot of the *Gibbs free energy (ΔG) for the oxidation reaction with absolute temperature. The diagram is named after British chemist Harold Johann Thomas Ellingham (1897–1975).

elution A process used to wash components of a mixture through a chromatography column in which absorbed material (absorbate) is removed by washing with a liquid (**eluent**). The **eluate** is the solution consisting of the absorbate dissolved in the eluent.

elutriation The process of separating suspended particles in a liquid by the upward flow of liquid such that the smaller particles with insufficient buoyancy are washed out or **elutriated**. It is used for separating particles into different size fractions.

emergency shutdown (ESD) The rapid and safe shutdown of a process plant or item of equipment due to a serious deviation in plant operation. Critical valves shut to isolate sections of the process. Other valves may be opened to depressurize vessels or rapidly discharge contents of reactors to quench tanks. Emergency shutdowns may occur due to changes in process conditions causing unstable or unsafe operating conditions, a failure in the control system, operator intervention causing unsafe conditions, plant and pipe failure, or some other external event such as an electrical storm.

emissions The discharge of materials into the environment emanating from an industrial or domestic source. It includes solids such as particulates, liquids, gases and vapours, and noise. The exhausts from the internal combustion engine in cars and other vehicles gives emissions of particles, water vapour, and carbon dioxide, and other gases.

emissive power (Symbol E) The rate at which electromagnetic radiation energy is emitted per unit-time, per unit-area of the surface spanning all wavelengths and in all directions. The emissive power is proportional to the absolute temperature to the fourth power, $E_b = \sigma T^4$, where subscript 'b' refers to *black body radiation and σ is the *Stefan–Boltzmann constant with a value of $5.67 \times 10^{-8}\,\mathrm{Wm^{-2}\,K^{-4}}$.

emissivity (Symbol ε) The ratio of the rate of emission of radiant energy from an opaque body due to the temperature, to the rate of emission by the same area from a black body at the same temperature. The emissivity of a *black body therefore has a value of 1.0 while a perfect reflector of radiant energy has a value of 0. The emissivity is dependent on the material, its surface characteristics, and its temperature.

empirical Derived from or relating to an experiment or observation rather than from established *theory. It can refer to an equation, formula, curve, analysis, or a number.

empirical formula The simplest chemical formula that expresses the composition of a compound by mass. It may not necessarily be the same as the molecular formula. For example, the empirical formula of ethane is CH_3 and, as its relative molecular mass is 30, its molecular formula is C_2H_6.

emulsification The dispersion of two immiscible phases within one another. The emulsification of foods is one of the most complicated unit operations since the nature of the final product varies greatly depending on the method of preparation. The method of addition of components and the rate of addition can significantly affect the emulsion quality. Oil-in-water emulsions can be produced in impeller-agitated vessels operating at high rotational speeds, colloidal mills, or high-pressure valve homogenizers. Continuous processing may be achieved by *in-line mixers, which consist of a high-speed rotor inside a casing, into which the components are pumped and subjected to high shear.

emulsion A colloidal suspension of one liquid dispersed within another. The dispersed phase has droplet sizes usually less than 1 mm. Butter is an example of a water-in-oil emulsion, while mayonnaise is an example of an oil-in-water emulsion. Surfactants or emulsifiers are surface-active agents and used to stabilize emulsions. Detergents possess both hydrophilic and hydrophobic parts to their molecules to behave as emulsifiers. In the offshore oil industry, emulsions form at the interface of water and oil in crude oil gravity separators. Sufficient hold-up time is used to separate the emulsion, or alternatively, surface-active agents are used to encourage separation.

end-of-pipe Used to describe technologies that aim to reduce the emissions of scrubbers, smoke stacks, and catalytic converters on vehicle exhausts that are considered to be pollutants.

endothermic reaction A chemical reaction that absorbs heat from its surroundings in order for the reaction to proceed. Such reactions have a positive enthalpy change and therefore do not occur spontaneously. *Compare* EXOTHERMIC REACTION.

energy The capacity or ability of a system to do work. It may be identified by type as being kinetic, potential, internal, and flow, or by source such as electric, chemical, mechanical, nuclear, biological, solar, etc. Energy can be neither created nor destroyed, but converted from one form to another. It can be stored as *potential energy, nuclear, and chemical energy, whereas *kinetic energy is the energy in the motion of a body defined as the work that is done in bringing the body to rest. The *internal energy is the sum of the potential energy and kinetic energy of the atoms and molecules of the body. Like work, energy has the SI units of joules.

energy balance An accountancy of the energy inputs and output to a process or part of a process, which is separated from the surroundings by an imaginary boundary. All energy forms are included in which the energy input across the boundary must equal the energy output plus any accumulation within the defined boundary. Where the conditions are steady and unvarying with time, the energy input is equal to the energy output. The most important energy forms in most processes are kinetic energy, potential energy, enthalpy, heat, and work. Electric energy is included in electrochemical processes.

Engel process A process invented in the nineteenth century for the production of potassium carbonate from potassium chloride. The process has two steps and first involves the production of soluble salt $MgKH(CO_3)_2.4H_2O$, known as Engel's salt, using carbon dioxide which is passed through a suspension of magnesium carbonate in aqueous chloride. This is followed by its decomposition by hot water and magnesia to form the product and insoluble hydrated magnesium carbonate:

$$3MgCO_3 + 2KCl + CO_2 + 5H_2O \rightarrow 2MgKH(CO_3)_2.4H_2O + MgCl_2$$
$$2MgKH(CO_3)_2.4H_2O + MgO \rightarrow 3MgCO_3.3H_2O + K_2CO_3$$

There were several variations of this process but it was abandoned in the 1930s for economic reasons.

engine A machine used for transforming one form of *energy to another, such as chemical or electrical energy to *kinetic energy or mechanical work. The development of Thomas Newcomen's (1664–1729) steam engine, together with James *Watt's (1736–1819) improvement with a separate condensing chamber, brought about the Industrial Revolution in which steam energy was converted into useful work to operate machinery including the development of the steam locomotive engine.

engineering The study of the design, construction, and operation of mechanical, electrical, and chemical systems, processes, and devices. Derived from the Latin *ingenium* meaning 'invention', the engineering disciplines that we know today originated from military engineering that involved machines of war. Civil engineering was therefore the first engineering discipline to be identified for the design of structures such as bridges, buildings, and roads for civilian and non-military purposes. The establishment of mechanical engineering was also to follow in the early nineteenth century. The discipline of *chemical engineering was developed during the Industrial Revolution through the development of industrial-scale chemical plants outside of the realms of applied chemistry. By 1908, the *American Institute of Chemical Engineers was founded some fourteen years before the *Institution of Chemical Engineers in the UK in 1922. A person who practises engineering is known as an **engineer**. In the UK, a professional *chemical engineer may be qualified as a *chartered chemical engineer or incorporated engineer by the *Institution of Chemical Engineers. An engineer applies the laws of physics and mathematics to solve problems, leading to improvement in the wellbeing of society, while using available resources responsibly, safely, and ethically, and without harm or damage to the environment.

engineering line diagram (ELD) A diagrammatic representation of a process. Also known as an **engineering flow diagram**, it features all process equipment and piping that is required for start-up and shutdown, emergency, and normal operation of the plant. It also includes identification numbers, identifiers for the materials of construction, diameter and insulation requirements, direction of flows, identification of the main process and start-up lines, all instrumentation, control, and interlock facilities, key dimensions and duties of all equipment, operating, and design pressure and temperatures for vessels, equipment elevations, set pressures for relief valves, and drainage requirements.

enhanced oil recovery (EOR) One of a number of methods used to increase the amount of crude oil that can be recovered from an oil reservoir. These include the use of water, gas, chemical, and steam injection into the reservoir. Gas injection is the most commonly used method and involves the use of natural gas, carbon dioxide, or nitrogen to displace the oil. The method is dependent on the temperature and pressure of the reservoir and composition of the oil within it. Carbon dioxide is miscible in light oil and can alter the viscosity and surface tension of the oil. The use of chemical injection involves dilute alkaline solutions to alter the surface tension and enhance oil recovery. Surfactants are also used. Steam injection also reduces the viscosity of the oil and can also partially vaporize it, aiding recovery.

enrichment The process of raising the amount of a substance in another. In nuclear reprocessing, enrichment involves raising the amount of one radioactive isotope in a mixture of isotopes. The enrichment of uranium-235 from a mixture of uranium-238 involves converting the uranium isotopes into uranium hexafluoride and separating them in a *gas centrifuge process. This involves using an array of centrifuges in which the heavy atoms are separated from the lighter atoms by centrifugal forces. The separation is dependent on the difference in mass between the isotopes being separated.

enthalpy (Symbol H) The thermal energy of a substance or system with respect to an arbitrary reference point. The enthalpy of a substance is the sum of the *internal energy and the flow of energy, which is the product of the pressure and specific volume:

$$H = U + pV$$

The reference point for gases is 273 K and for chemical reactions is 298 K. For foods and refrigerants the reference temperature is 243 K (-40°C).

enthalpy balance A form of energy accountancy for a process in which the stream energies to and from the process are expressed as enthalpies. At steady state, the total enthalpy into a process is equal to the total enthalpy out. Where there is an inequality, there is either a loss or an accumulation of material with an associated loss or increase in enthalpy. An enthalpy balance is used to determine the amount of heat that will be generated in the process or that needs to be removed to ensure that the process operates safely and to specification.

enthalpy-concentration diagram A chart that presents the thermodynamic data for non-ideal mixtures of substances. It is used to calculate the heat of mixing when two pure components are combined to form a non-ideal mixture such as sulphuric acid and water.

entrainer A substance that is added to a homogenous *azeotrope to convert it to a heterogenous azeotrope that can then be readily separated by *distillation. A high concentration is usually needed and is only justified when the improvement in the *relative volatility offsets the extra investment, replacement, and recycle costs involved in the purchase, heating, and pumping of the entrainer. *Azeotropic distillation is useful for separations where the overhead component is present in small amounts. Entrainers should ideally be cheap, readily obtainable, non-corrosive, non-toxic, unreactive, thermally stable, and easily recoverable. As well as being able to form an azeotrope, it is required to have a low latent heat of vaporization. *See* EXTRACTIVE DISTILLATION.

entrainment The capture of bubbles, drops, or particles from one phase to another from which it is being separated.

entrance and exit losses The irreversible energy loss caused when a fluid enters or leaves an opening, such as into or out of a pipe into a vessel. Where there is a sudden enlargement, such as when a pipe enters a larger pipe or vessel, eddies form and there is a permanent energy loss expressible as a head loss as:

$$H_{exit} = \frac{v^2}{2g}\left(1 - \frac{a}{A}\right)^2$$

where v is the velocity in the smaller pipe, a is the cross-sectional area of the smaller pipe, and A is the cross-sectional area of the larger pipe. For a considerable enlargement the head loss tends to:

$$H_{exit} = \frac{v^2}{2g}$$

With a rapid contraction, it has been found experimentally that the permanent head loss can be given by:

$$H_{entrance} = k\frac{v^2}{2g}$$

where for a very large contraction $k = 0.5$.

entropy (Symbol S) The extent to which energy in a *closed system is unavailable to do useful work. An increase in entropy occurs when the free energy decreases or when the disorder of molecules increases. For a *reversible process, entropy remains constant such as in a friction-free *adiabatic expansion and compression. The change in entropy is defined as:

$$dS = \frac{dQ}{T}$$

where Q is the heat transferred to or from a system, and T is the absolute temperature. However, all real processes are irreversible, which means that in a closed system there is a small increase in entropy.

Entropy can be regarded as a measure of disorder. That is, the higher the value, the higher the level of disorder. A closed system tends towards higher entropy and therefore a higher disorder. For example, two layers of coloured balls in a box shaken will generate a level of randomness and disorder, and will not return to their original state without the intervention of more work in their separation. *See* FIRST LAW OF THERMODYNAMICS.

entry length The distance from the entry plane of a pipe or tube in which a fluid flows that corresponds to 99 per cent of the *centreline velocity of the fully developed velocity profile. That is, the point where fully turbulent or laminar flow has fully formed. The type of flow regime can be determined from the *Reynolds number.

ENVID An abbreviation for **env**ironmental **id**entification, it is a systematic and wide-ranging structured hazard analysis tool used to identify the environmental hazards at an early stage in process design and development. It is conducted like a *HAZID except that it focuses on identifying environmental issues.

Environment Agency (EA) A non-departmental public body that is responsible for the protection of the environment in England and Wales. Its primary role is to protect and improve the environment by being an environmental regulator to protect and regulate activities that can cause harmful pollution. *See* SEPA.

(⊕) SEE WEB LINKS
• Official website of the UK Environment Agency.

Environmental Protection Agency (EPA) A governmental agency in the US that is responsible for setting standards and monitoring policies for environmental pollutants.

(⊕) SEE WEB LINKS
• Official website of the US Environmental Protection Agency.

enzyme A protein molecule that catalyzes biochemical reactions. It has a complex molecular structure and its molecular shape or conformation plays an important role in its function as a catalyst. Enzymes as biocatalysts promote biochemical reactions and involve a substrate molecule, which is converted into products. As with all catalysts, enzymes lower the activation energy of the biochemical reaction and increase the rate of reaction. Every type of enzyme is specific to a particular biochemical reaction. For example, maltose and sucrose are disaccharides but are hydrolyzed by different enzymes, with the former by an enzyme called maltase and the latter by invertase. The molecular structure of the enzyme is complex and, being a protein, is formed from long folded chains of amino acids. The substrate undergoing reaction binds to an active site on the enzyme in a *lock and key mechanism to form a short-lived enzyme–substrate complex which dissociates to form the product. Enzyme reactions operate most effectively at a certain optimum temperature and within a narrow pH range. The names of most enzymes end with –*ase*, such as amylase, which is used to break down starches into simple sugars. Enzymes are denatured with excessive temperature, very low or high pH, and certain chemicals.

enzyme inhibition A reduction in the rate of an enzyme-catalyzed biochemical reaction by a substance or inhibitor. **Competitive inhibition** occurs when the conformation of the inhibitor mimics or resembles the substrate and thus prevents the active site from being available to form an enzyme–substrate complex. In **non-competitive inhibition**, the inhibitor binds to the enzyme–substrate complex or part of the enzyme such that the enzyme is unable to catalyze the reaction.

enzyme kinetics The study of the behaviour of enzyme-catalyzed reactions. The rate at which an enzyme-catalyzed reaction proceeds is determined from the disappearance of the substrate or the formation of the product. Simple kinetic models such as *Michaelis-Menten kinetics can often be used to determine the rate of the reaction.

EOR *See* ENHANCED OIL RECOVERY.

EPA *See* ENVIRONMENTAL PROTECTION AGENCY.

equation of state A relationship that links the pressure, volume, and temperature of an amount of a substance. It is used to determine thermodynamic properties such as liquid and vapour densities, vapour pressures, fugacities and deviations from ideality, and enthalpies. Various equations of state have been developed to predict the properties of real substances. Commonly used equations of state include the *ideal gas law, *virial equation, *van der Waals' equation, *Peng-Robinson, *Soave-Redlich Kwong, and *Lee-Kesler equations. Cubic equations of state are relatively easy to use and are fitted to experimental data. The van der Waals equation is comparatively poor at predicting state properties. The Lee–Kesler model, which is based on the theory of corresponding states and uses reduced temperature and pressure, covers a wide range of temperatures and pressures.

equilibrium A condition or state in which a balance exists within a system, which may be physical or chemical. A system is in equilibrium if it shows no tendency to change its properties with time. **Static equilibrium** occurs if there is no transfer of energy across the system boundary, whereas **dynamic equilibrium** is when transfer occurs but the net effect of the energy is zero. *Thermodynamic equilibrium occurs when there is no heat or work exchange between a body and its surroundings. *Chemical equilibrium occurs when a chemical reaction takes place in the forwards direction, when reactants form products at exactly the same rate as the reverse reaction of products revert to their original reactant form.

equilibrium constant A reversible process, chemical or physical, in a closed system will eventually reach a state of equilibrium. The equilibrium is dynamic and may be considered as a state at which the rate of the process in one direction exactly balances the rate in the opposite direction. For a chemical reaction, the equilibrium concentrations of the reactants and products will remain constant providing the conditions remain unchanged. For the homogenous system:

$$wA + xB \leftrightarrow yC + zD$$

the ratio of the molar concentrations of products to reactants remains constant at a fixed temperature:

$$K_C = \frac{[C]^y [D]^z}{[A]^w [B]^x}$$

where K_C is the equilibrium constant and the square brackets indicate equilibrium concentrations. The relationship is known as the **equilibrium law**. For example, for the *Haber process for the synthesis of ammonia, nitrogen is reacted with hydrogen as:

$$N_2(g) + 3H_2(g) \leftrightarrow 2NH_3(g)$$

The equilibrium constant is expressed as partial pressures as:

$$K_C = \frac{[NH_3]^2}{[N_2][H_2]^3} = \frac{p_{NH_3}^2}{p_{N_2} \cdot p_{H_2}^3}$$

equilibrium distillation *See* FLASH VAPORIZATION.

equilibrium law *See* EQUILIBRIUM CONSTANT.

equilibrium moisture The balance in which the rate of loss of moisture from a system or material is equal to the rate of gain from its surroundings. The **equilibrium moisture content** of a substance is the smallest possible moisture content that can be achieved in the presence of a gas of given pressure, temperature, and humidity. It is also dependent on the nature of the solid and is usually expressed on a dry basis. The *free moisture content is the difference between the *total moisture content and equilibrium moisture content.

equilibrium ratio (Symbol K) The ratio of the mole fraction in the vapour phase of a component in a mixture, y, to the mole fraction in the liquid phase, x, at equilibrium:

$$K_A = \frac{y_A}{x_A}$$

It is a function of both temperature and pressure. The *relative volatility, α, is less dependent on temperature and pressure than the equilibrium constant where for an ideal mixture of two components, A and B:

$$\alpha_{AB} = \frac{K_A}{K_B}$$

equimolar counter-diffusion The process of diffusion of one component (A) into another (B) in which their respective molar fluxes are equal but in opposite directions. The diffusivity of A in B is therefore the same as B in A in which the total pressure within the process remains the same throughout. The process of steady-state equimolar counter-diffusion in an ideal gas mixture can be described by *Fick's law for which the concentrations of A and B are expressed in terms of their partial pressures.

equivalent hydraulic diameter An alternative value used in calculations in place of the diameter of a pipe in turbulent flow calculations when the pipe or duct cross-section is not circular. It is equal to four times the flow area divided by the wetted perimeter. It does not apply to laminar flow.

equivalent length A method used to determine the pressure drop across pipe fittings such as valves, bends, elbows, and T-pieces. The equivalent length of a fitting is that length of pipe that would give the same pressure drop as the fitting. Since each size of pipe or fitting requires a different equivalent length for any particular type of fitting, it is usual to express equivalent length as so many pipe diameters, and this number is independent of pipe. For example, if a valve in a pipe of diameter, d, is said to have an equivalent length, n, pipe diameters, then the pressure drop due to the valve is the same as that offered by a length, nd, of the pipe. Values are determined experimentally.

Erbar–Maddox correlation An empirical method used for the design of distillation columns that relates the number of ideal stages for a given separation and reflux ratio to the minimum number at total reflux and the minimum reflux ratio. The minimum reflux ratio corresponds to an infinite number of stages to bring about separation. It is named after American chemical engineers John H. Erbar and Robert N. Maddox.

erg A unit of work in the *c.g.s. system in which one erg is the force when one *dyne of force is exerted upon a body through a distance of 1 cm. This is a very small unit and one joule is preferred in which one erg is equal to 10^{-7} joules.

Ergun equation An equation used to determine the pressure drop per unit length of a fixed bed of particles such as catalyst at incipient gas velocity, v:

$$\frac{-\Delta p}{L} = \frac{150(1-e)^2 \mu v}{\phi e^3 d^2} + \frac{1.75(1-e)\rho v^2}{\phi e^3 d}$$

where $-\Delta p/L$ is the pressure drop over the depth of bed, e is the bed voidage, d is the mean particle diameter, ρ is the fluid density, μ is the fluid viscosity, and φ is the *sphericity. The incipient point of fluidization corresponds with the highest pressure drop at the minimum fluidization velocity. It is named after Turkish-born American chemical engineer Sabri Ergun (1918–2006) who developed the equation in 1952.

erosion The physical removal of material from a surface by mechanisms that exclude chemical attack. The usual phenomenon that causes erosion is impingement by either liquid droplets or entrained solid particles. If there are no corrosive substances present, then in many cases, the most common mechanism for material damage due to erosion is impingement by solid particles. *Compare* CORROSION.

erosion-corrosion The process of metal surface damage and removal due to the combined action of chemical species and mechanical attack. The mechanical damage is by the removal of an otherwise protective film or surface coating by repeated bombardment of either liquid droplets or solid particles. This exposes the reactive metal substrate to chemical attack, and the rate of material damage is frequently many times higher than the base corrosion rate.

error 1. The difference between a measured or indicated value and the expected or true value. The error can be used to control a system or process while **statistical error** can be used as a measure of uncertainty. For example, reading the level of liquid in a vertical glass manometric leg used to measure the pressure in a process may only be possible to the nearest 1 mm. A level of 30 cm should therefore be written as 30 ±0.1 cm, which means a level of between 29.9 and 30.1 cm. **Random error** is a form of error that can neither be predicted nor accounted. **Systematic error** is due to a fault that requires correcting. For example, a thermometer that reads 10°C less than the actual temperature means that all recorded temperatures are 10°C less than the correct value. **Human error**, blunders, and mistakes can lead to disastrous consequences. **2**. Used in process control of process, the error is the difference between a desired or set point value and the actual and controlled process variable. The control of a process aims to minimize the error. Various forms of process control strategies are used to achieve this. *See* PROCESS CONTROL

ESD *See* EMERGENCY SHUTDOWN.

estimation A solution to a problem that is deemed to be sufficiently close to the right answer and that has involved some rational thought or calculation. It is not a precise value and the level of error may not be known.

ethylene (ethene) A colourless and flammable hydrocarbon with the formula C_2H_4 (m.p. -169°C; b.p. -103.7°C). It is widely used in the chemical industry to produce organic chemicals such as ethanol and ethanal, as well as ethylene dichloride, ethylbenzene, and ethylene oxide. It is polymerized to form polyethylene. Ethylene is made by the cracking of hydrocarbons derived from petroleum and is the simplest form of alkene.

Eulerian fluid dynamics A branch of fluid dynamics that considers the velocity of a fluid at fixed points. It is named after Swiss mathematician and physicist Leonhard Euler (1707–83).

Euler number A dimensionless number, Eu, that represents the relationship between pressure drop due to friction and inertial forces in a moving fluid in a system such as in a pipeline.

$$Eu = \frac{\Delta p}{\rho v^2}$$

where Δp is the pressure drop due to friction, ρ is the density, and v is the velocity. It is named after Swiss mathematician and physicist Leonhard Euler (1707–83).

European Congress of Chemical Engineering (ECCE) A biennial conference on chemical engineering, now held concurrently with the European Congress of Applied Biotechnology (ECAB). The ninth ECCE was held in The Hague, the Netherlands, in April 2013. The ECCE conference series is held under the auspices of the *European Federation of Chemical Engineering (EFCE), an association run jointly by the *Institution of Chemical Engineers, *DECHEMA, and the French Society of Process Engineers.

European Federation of Chemical Engineering (EFCE) A professional organization formed in 1953 representing over 100,000 chemical engineers and scientists from over 30 countries within Europe. It has nineteen working parties in specialist areas covering all areas of chemical engineering.

((⊕)) SEE WEB LINKS
• Official website of the European Federation of Chemical Engineering.

eutectic mixture A mixture of substances having the lowest freezing point of all mixtures of the substances. The **eutectic point** is the lowest temperature at which a solution can exist and that three condensed phases can co-exist (see Fig. 20). Low melting-point

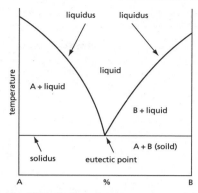

Fig. 20 Eutectic point

alloys are usually eutectic mixtures. For example, the freezing point of pure lead and pure tin are 327°C and 232°C, respectively, whereas a mixture with 62 per cent lead (soft solder) has a **eutectic point** of 183°C.

eutrophication A biological process in which there is excessive algal growth in water due to an excess of nutrients and in particular phosphates and nitrates. The eventual decomposition of the algae depletes the water of available oxygen, resulting in a sterile body of water. While it is a natural process, it is assisted by the release of chemicals into water as a form of *pollution.

evaporation The change of state from a liquid to a vapour at a temperature below the boiling point of the liquid. Evaporation occurs at the surface of the liquid in which the kinetic energy is sufficient to enable it to leave the surface as vapour. The average kinetic energy of the molecules in the liquid is therefore reduced and there is consequently a reduction in temperature resulting in **evaporative cooling**. *See* BOILING.

evaporator A heat exchanger device used to concentrate a solution by boiling. The solution is heated until vapour is released and removed, leaving behind a more concentrated solution. A drying process may then follow evaporation and low pressures may be used for heat-sensitive materials such as milk and other thermally labile foods. Several evaporators or 'effects' can be connected in series as a *cascade process to obtain a desired evaporation with the benefit of good energy utilization. They may be operated as feed-forward or feed-backward. Cold feeds use a backward process because the boiling temperature in the last effect is lower and consequently the required temperature for evaporation is less than that required in the first evaporator. *See* MULTIPLE EFFECT EVAPORATION.

event The occurrence of something happening. It is a term used in *risk analysis and *risk assessment, and used in an evaluation of the probability or likelihood of the event occurring and its consequences. A risk analysis seeks to estimate the risks associated with the event and the risk assessment seeks to make decisions to reduce or mitigate the risks.

excess air A supply of air to a combustion process that exceeds the stoichiometric requirement of oxygen. In practice, a supply of 1.2 times greater than that which is stoichiometrically required is often used to ensure a good level of combustion.

excess flow valve A type of flow valve that permits the flow of a fluid in either direction but prevents excessive flow, in which case, the valve closes automatically.

excess reactants The reactants consumed in a chemical reaction that exceed the stoichiometric proportion. Excess oxygen or *excess air are usually added to a combustion process to ensure good combustion efficiency. The fuel itself is the *limiting reactant. Excess reactants are normally expressed as per cent excess, such as 100 per cent excess, which means that twice as much of one of the reactants is added as is theoretically necessary for complete reaction.

exchange capacity The number of ions that can be retained or exchanged per unit volume by a particular *ion exchange resin within an ion exchange unit. It is generally measured in milliequivalents per gram of resin. The exchange capacity depends on the number of exchange sites and the type of resin.

exergy The maximum amount of useful thermodynamic work that can be extracted from a system in a given surrounding. This can be mechanical work such as running a pumping or lighting a bulb with electrical energy. For example, there is more exergy in an ice cube in a room at ambient temperature than a cup of water in the same room. It applies to kinetic, potential, chemical, nuclear, magnetic, and electric energy. **Exergoeconomics** is the study that combines exergy analysis with process economic evaluation to identify the cost of operating a system in terms of materials in, products out, and the cost of exergy destruction. It is used for the analysis of many processes in terms of the amount of work that can be lost or needs to be applied.

exothermic reaction A chemical reaction that liberates heat. No energy input is required for the reaction to proceed. It has a negative enthalpy change and therefore under the appropriate conditions the reaction will occur spontaneously. Chemical reactors used to contain exothermic reactions therefore require cooling facilities to remove the excess heat that is generated and to maintain a constant temperature. *Compare* ENDOTHERMIC REACTION.

expansion An increase in the volume of a substance for a given mass. This may be due to a decrease in pressure or an increase in temperature of a gas, or due to a swelling effect of moisture in a porous material. When water cools, the density decreases until it reaches 4°C where it has a minimum volume and maximum density. Cooled further and turned to ice, its volume increases and density decreases to below that of liquid water, which is why ice floats and can lead to burst pipes in winter.

expansion bend A loop of pipe placed in a straight run of pipeline to allow for the possible thermal expansion of the pipeline, thereby avoiding the build-up of excess temperature-induced stress. Examples include horseshoe bends and bellows. It is also known as an **expansion joint** or **expansion loop.**

experimental design A statistical procedure used to evaluate the influence of process variables on the outcome from a process. The purpose is to determine the key information such as a process optimization quickly and cheaply. The simplest approach is to adjust one variable at a time, such as changing the temperature, and evaluating the effect before adjusting another variable. Various protocols have been developed in which combinations of all the important variables can be considered simultaneously. The adjustments follow a defined matrix. A simple approach and early method known as response surface methodology (RSM) was developed by British statistician George Edward Pelham Box, FRS, with K. B. Wilson in 1951. It was designed to explore the relationships between several process variables and one or more response variables. It uses two levels (coded as $x=-1$ and $x=1$)

for each process variable such as high and low in a full-factorial design. It involves all combinations of all levels of the variables. For example, in a two-level full factorial experimental design with two variables, there are four treatments. In this case, the number of treatments is equal to 2^n. The effects of the two variables and their interactions can be described by the response surface in the form:

$$y = \beta_o + \beta_1 x_1 + \beta_2 x_2 + \beta_{11} x_1^2 + \beta_{22} x_2^2 + \beta_{12} x_1 x_2$$

Other experimental designs have been developed and involve more than two levels, and also involve the use of replicates used to evaluate the statistical error. Fractional factorial experimental designs use part of the full factorial experimental design and cut down the number of trials needed.

explicit A mathematical function that contains no dependent variables. *Compare* IMPLICIT.

explosion A rapid, violent, and uncontrolled release of energy as a result of a chemical or nuclear reaction, which causes a *blast wave or *shock wave. This is a sharp pressure pulse. An explosion is also accompanied by noise, heat, and light. An explosion as the result of a chemical reaction can be due to a deflagration in which the reaction front moves at less than sonic velocity, whereas a detonation is a chemical reaction that is extremely rapid and the reaction front moves into the unreacted material greater than sonic velocity, resulting in considerable blast damage.

explosion limit The highest or lowest concentration of a flammable gas or vapour in air or oxygen that will propagate a flame when ignited.

explosion-proof equipment The equipment in an enclosure that is capable of withstanding an internal explosion of a specified gas or vapour, and of preventing possible ignition of a surrounding flammable atmosphere.

explosion suppression A method, device, or system to effectively extinguish an explosion.

explosive A substance capable of a sudden high-velocity reaction with the generation of high pressure. High-energy explosives generate detonations. An **explosive atmosphere** is a mixture of air and one or more *dangerous substances in the form of gases, vapours, mists, or dusts in which, after ignition has occurred, combustion spreads to the entire unburned mixture. More generally, an **explosive mixture** is a combustible-oxidant mixture that is potentially explosive or capable of propagating flame.

exponent A number or symbol placed as a superscript to a number, expression, or quantity that represents the power to which it is to be raised. That is, the number of times that number, expression, or quantity is to be multiplied by itself. Exponents follow simple laws:

Multiplication: $x^a x^b = x^{a+b}$
Division: $x^a / x^b = x^{a-b}$
Power of a power: $\left(x^a\right)^b = x^{ab}$

Where the exponent is zero, the number or variable is equal to 1, i.e $x^0 = 1$.

exponential decay The decrease in a quantity at a rate that is proportional to its value:

$$\frac{dN}{dt} = -kN$$

For example, the decrease in the viability of contaminating microorganisms during the thermal sterilization of foods in the canning process follow this form of decay where the number of surviving microorganisms after time, t, is:

$$N = N_O e^{-kt}$$

where N_o is the initial number and k is the decay constant. The reciprocal of the decay constant is known as the *time constant. A special case is the radioactive decay of a substance known as the *half-life. This is the time taken for a substance to fall to half of its original value ($N = 0.5N_O$):

$$t_{1/2} = \frac{\ln 2}{k}$$

The half-life of a particular isotope is constant and independent of the amount of starting material.

exponential growth The growth of a microorganism in a culture that is unrestricted in terms of nutrients or substrate such that there is a repetitive doubling. The increase in population is:

$$\frac{dN}{dt} = kN$$

Integrating from an initial population N_o at time t=0, gives $N = N_o e^{kt}$. In practice, the nutrient or substrate is limiting and the growth curve is not indefinitely exponential but sigmoidal or S-shaped.

exposure The amount of a toxic or harmful substance to which a person is exposed. The exposure may be in the form of ingestion, absorption, or inhalation, and may involve both concentration and time of exposure. It also applies to ionizing and thermal radiation. *See* DOSE.

expression A mathematical variable, function, or combination of constants, variables, or functions.

exsiccant A material capable of drying or being dried. The material is dried in an **exsiccator**. It is applied mainly to paints and varnishes.

extended surface A surface used for heat transfer in which projections are attached, such as fins or pins. These attachments are used to increase the available area of heat transfer on the side that has the lowest surface heat-transfer coefficient.

extensive variable *See* INTENSIVE VARIABLE.

extent of reaction The ratio of the number of moles of a substance consumed or produced in a chemical reaction to its stoichiometric coefficient.

extinguishing agent A substance used to put out a fire by cooling the burning material, inhibiting the chemical reaction, and/or blocking the supply of oxygen. Examples include water, carbon dioxide, foam, and sand.

extract The solvent-rich product stream leaving a *liquid–liquid solvent extraction process. The extract carries an extracted material.

extraction A separation process in which a component is selectively removed by chemical or physical means such as the pressing of seeds to remove oils. Liquid–liquid extraction or *solvent extraction is a separation process in which a component is selectively transferred from one liquid phase to another, such as the extraction of uranium and plutonium from a fission product containing nitric acid solution into a kerosene diluent containing the solvent tri-butyl phosphate.

extractive distillation The separation process used for azeotropic mixtures in which a solvent is added, leaving the bottom of the column carrying one component with it, allowing the other component to leave at the top of the column (see Fig. 21). Extractive distillation is more widely used than *azeotropic distillation since there is a wider choice of agent and there is no need for close matching of volatilities. Azeotropic entrainers must usually boil within 5–20°C of the other components. There is also a lower heat requirement since the solvent is not vaporized. There is also a wider choice of operating conditions. The solvent/feed ratio, for example, is not critical, unlike an entrainer/feed ratio, and there is easier recovery of the agent. Solvents are cheap, readily obtainable, non-corrosive, non-toxic, unreactive, thermally stable, and easily recoverable, as well as being readily miscible, and have a low volatility. The solvent is added a few trays down from the top of the column with the trays above the solvent feed point functioning as a solvent recovery section to remove traces of solvent from the overhead vapour A. The solvent leaves the bottom of the main column together with component B, from which it is usually readily separated in a solvent-recovery column in view of its presumed low volatility. The solvent is recycled to the main column with a make-up stream being provided to allow for any solvent losses. Examples include the separation of paraffins (A) from toluene (B) using phenol, and isobutane (A) from butane (B) using furfural.

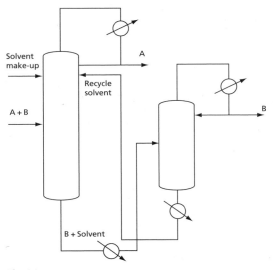

Fig. 21

extractor A device used to selectively remove components from one phase to another, such as a solid or a liquid to another liquid, by leaching or solvent extraction.

extraneous material A substance that may appear intentionally in another substance as an impurity. A method of calibrating flow meters involves the use of a controlled amount of extraneous material that is intentionally added to the process stream either upstream or downstream of the flow meter. A sample is then taken downstream and analysed from which the rate of flow of the process stream is determined, and relevant calibration adjustments made to the flow meter. Examples include the addition of dyes to water, brine to seawater, the addition of radionuclides and oxygen in chlorine.

extrapolation A method of obtaining data that falls outside a range of known data that assumes the estimated data can be obtained by extending the known data.

extrusion A process in which a soft material is forced under pressure through a restricted orifice or opening in a die to produce a continuous flow of material in a desired shape or form. The material is conveyed under pressure using an Archimedes screw. Either a single or twin screw auger is used that is generally powered by an electric motor. Heated *billets of aluminium can be extruded to form sheets, rods, and tubing. Plastics and polymers can also be extruded. The first extruders were developed in 1879 for wire coating. The first hot-melt-extrusion extruders for thermoplastics were developed in Germany in 1935.

Eyring equation An equation used to describe the kinetics of chemical reactions and relates the reaction rate to temperature in the form:

$$k = \frac{k_B T}{h} e^{\frac{-\Delta G^{\ddagger}}{kT}}$$

where k_B is the Boltzmann constant, T is the absolute temperature, h is the Planck constant, $\Delta G^{\ddagger}$ is Gibbs free energy. The equation is based on the transition state theory and is of a similar form to the *Arrhenius equation. The equation is based on the statistical probability that a chemical reaction will take place once the system has reacted the activated state. There are similar equations used to describe transport phenomena such as viscosity. The Eyring equation is named after Mexican-born American chemist Henry Eyring (1901–81) who derived it.

factor 1. A number that can be multiplied by another number. A common factor is a number by which another number is divided exactly. It is also known as a divisor. For example, 12 can be divided by 1, 2, 3, 4, 6, and 12. **2.** An activity that contributes to a particular situation or effect. For example, greenhouse gas emissions are one of the factors that contribute to global warming.

factorial (Symbol !) The product of all the positive whole integers being considered. Note that 0! has a value of 1. For example, 5! is $0 \times 1 \times 2 \times 3 \times 4 \times 5 = 20$.

factorial experimental design A statistical procedure used to study the effects of process variables on a process and their collective effect on an independent variable. It is used when the effect of the combination of the variables is complex and involves the study of the various combinations of the variables. For example, temperature, pressure, and the concentration of reactants can have an effect on the yield of a desired product. By selecting high, medium, and low levels for each variable, the number of separate experiments that would be required to study fully the effects on the product would be $3^3 = 27$. These would be randomly sequenced and possibly repeated in duplicate or triplicate. The test results on yield would then be interpreted by a regression analysis and analysis of variance to determine the influence of each of the variables. Where there are many variables that give too many combinations to be practically considered, a fractional factorial experimental design is used. This allows for a fraction of the full factorial experimental design to be used that can lead to the same outcome, although the level of statistical uncertainty is correspondingly increased. *See* EXPERIMENTAL DESIGN.

factorization The mathematical process of changing an algebraic or numerical expression from a sum of terms into a mathematical product. For example, the equation $2x^2 - 3x - 2 = 0$ can be factorized to $(2x+1)(x-2)$, thereby making it easier to solve for x. Since the product of the two factors is zero, then the solutions are $x = 2$ and $x = -\frac{1}{2}$.

Factory Acts A series of early Acts of Parliament in the UK designed to limit the number of hours worked by women, children, and young people, and to improve working conditions. They were initially aimed at those working in the textile industries and later applied to all industries, in order to protect workers and to compel industry owners to operate lawfully and with accountability. There were a number of Acts passed over the course of the nineteenth century. The Factory Act of 1844, for example, was responsible for reducing the hours of work for children aged between eight and thirteen and stipulated that they must begin after 6 am and end before 9 pm. Women and young persons aged between fourteen and eighteen were to work no more than twelve hours a day. Fines could also be imposed on factory owners. Today, workplace health and safety is governed by the Health and Safety at Work Act 1974. *See* HSE.

facultative anaerobe A living organism, such as certain bacteria and yeast, that can adapt its metabolism to enable it to survive and grow in either the presence or absence of oxygen. Yeast can grow aerobically producing more cells and carbon dioxide, which is useful in the production of bread, or anaerobically producing carbon dioxide and ethanol as well as more living cells, which is useful in the production of beer, lager, and wine.

Fahrenheit scale A temperature scale invented in 1714 by German physicist and instrument-maker Gabriel Daniel Fahrenheit (1686–1736). Fahrenheit was the first person to make practical thermometers using mercury instead of spirits of wine (alcohol). He calibrated his thermometer between two *fixed points for which he used the eutectic point of salt water as 0°, which was the lowest temperature he could obtain in the laboratory, and the temperature of a healthy human which he originally called 12° and later 96°. He found that on his scale, pure ice melted at 32°F and steam at normal atmospheric pressure was 212°F. The conversion to the *Celsius scale is given by:

$$^\circ C = \frac{5}{9}(^\circ F - 32)$$

fail-safe A type of control used in situations such that a malfunction or unsafe condition results in a device, item of equipment, or process safely shutting down. An example is a gas burner where the fuel supply is automatically shut off in the event that the air (oxygen) supply should inadvertently cease. In contrast, a **fail-to-danger** is a fault or failure of a device or item of equipment that renders the plant in a dangerous condition such that a control system is unable to respond to provide the necessary protection.

Falconbridge process A selective leaching process used to extract copper and nickel from a sulphide ore containing nickel, copper, and other precious metals that has been roasted to remove the sulphur known as *matte. The nickel is removed using hydrochloric acid and recovered as nickel chloride crystals. The residue is then leached with sulphuric acid to dissolve the copper.

falling film evaporator A type of tubular heat exchanger in which the feed solution for evaporation falls under the influence of gravity as a film down the walls of the tubular heat exchange surface. The solution separates into liquid and vapour near the bottom of the tube. It is used in concentrating temperature labile solutions such as fruit juices.

falling-rate drying The period of drying in which the rate of change of moisture decreases with time. The falling-rate period follows the *constant-rate drying period. For non-porous solids, the rate of drying changes non-linearly with time and is limited by molecular diffusion of bound water.

fallout The release of radioactive material into the environment either from an accidental release from a nuclear installation such as a power plant, or from a nuclear explosion. The *Windscale nuclear accident in 1957, *Chernobyl in 1986, and *Fukushima Daiichi in 2011 led to wide-scale contamination of the environment. The most harmful isotopes to human health include iodine-131 and strontium-90. Both isotopes are taken up by animals such as grazing cows and transferred to humans through the consumption of milk. Iodine-131 accumulates in the thyroid gland while strontium-90 accumulates in the bones.

fan A mechanical device used to move air, gas, or vapour at low gauge pressure. It consists of blades on a rotating shaft. Fans are classified as being axial or radial, and refer to the direction of flow with respect to the rotation of the fan. The efficiency of a fan is the ratio of the power output to the power input resulting in the flow delivered. *Compare* BLOWER.

Fanning friction factor A dimensionless friction factor used to determine the frictional pressure drop resistance of a moving fluid in a pipe or over a surface. The friction factor is the proportionality constant in which the wall shear stress, τ_w, of a fluid at a surface is proportional to the kinetic energy of the fluid per unit volume

$$\tau_w = f\frac{\rho v^2}{2}$$

where ρ is the density and v is the fluid velocity. It is named after American engineer John Thomas Fanning (1837–1911)

FAR *See* FATAL ACCIDENT RATE.

farad (Symbol F) The SI unit of capacitance and is the capacitance of a capacitor where if it were charged with one coulomb there would be a potential difference of one volt between its plates. The farad is too large for most applications, so it is usually expressed in microfarads (μF).

Faraday's laws of electrolysis Laws that were defined by Michael Faraday (1791–1867) to describe the process of electrolysis. The first law states that the mass of a given element liberated during electrolysis is directly proportional to the magnitude of the steady current consumed during the electrolysis and to the time for which the current passes. The second law states that when the same quantity of electricity is passed through different electrolytes, the masses of the different substances liberated are directly proportional to the masses of the substances that require one mole of electrons (1 faraday) for neutralization. The transfer of 1 mole of electrons corresponds to the passing of approximately 96,500 coulombs of electricity, which is known as the Faraday constant. A coulomb is equivalent to the passage of 1 ampere for 1 second.

fast reactor A type of high-temperature nuclear reactor that has no moderator and uses a liquid metal coolant such as sodium. They are used as either fast breeder reactors in which fissile material such as uranium enriched with plutonium-239 is used as the fuel and is able to produce more plutonium-239 at a rate faster than it uses; or as converter reactors in which *fertile material such as uranium-238 is converted into plutonium-239 by the absorption of a neutron to form uranium-238 and which then decays to plutonium-239. Fertile material is a nuclide that can absorb a neutron to form fissile material.

fatal accident rate (FAR) The number of fatal accidents in a particular industry occurring within a period of 10^8 hours of activity. This is the period of time that corresponds with approximately the working life of a group of a thousand workers. For the chemical industry the FAR is between 4 and 5. For the agricultural industry it is 10, while for the mining and construction industries the FAR is 12 and 64, respectively.

fatigue The failure of materials such as metals under cyclic applications of an applied stress.

fault tree analysis A method used to calculate the probability of an event occurring based on the probabilities or frequencies of its component parts. The fault tree is a diagram used to provide a model of the interactions between the various components in a system to identify how failure may occur and to provide an estimate of the likelihood or frequency of failure. It is used to plan maintenance and to avoid plant and equipment failure.

FCCU *See* FLUIDIZED CATALYTIC CRACKER UNIT.

F-distribution A statistical probability distribution used in the analysis of variance of two samples for statistical significance. It is calculated as the distribution of the ratio of two chi-square distributions and used, for the two samples, to compare and test the equality of the variances of the normally distributed variances.

fed batch bioreactor A type of bioreactor in which microorganisms are cultured in such a way as to overcome growth limitation due to the initial substrate concentration. The substrate is therefore added to the growing cell culture either on a step-wise or continuous basis. A constant flow of substrate can be used to maintain a steady growth rate or exponential growth. Alternatively, the addition can be adjusted to maintain a constant substrate concentration but requires good and rapid measurement of the substrate in the medium.

A substrate-limited fed batch bioreactor uses a continuously supplied substrate with a low concentration. The required amount of substrate supplied can be checked periodically by stopping the supply and monitoring the effect on dissolved oxygen concentration.

FEED *See* FRONT END ENGINEERING DESIGN.

feedback control A closed-loop method of controlling a process in which information about the controlled variable is fed back to the input and compared against a desired value. The difference between the two signals is called the *error or *deviation. The feedback can be accomplished by a human operator as in **manual control**, or by the use of instruments as in **automatic control**. In a **negative feedback control**, the applied counteracting disturbance is motivated by the difference between the desired value and the actual value of the controlled variable. For example, in the manual control of the heating of a liquid in a vessel with steam, an operator periodically measures the temperature of the liquid; if this temperature is below the desired value, the steam flow is increased by opening the valve slightly. In the automatic control of the vessel, a temperature-sensitive device is used to produce a signal, which may be electrical or pneumatic, and is proportional to the measured temperature. This signal is fed to a controller, which compares it with some desired value or set point. Where there is a difference in signals, the controller changes the opening of the steam control valve to adjust the temperature.

feedforward control A method of process control in which a disturbance is detected before it enters the system for which the controller calculates the required counter-acting disturbance. Process disturbances are measured and compensated for without waiting for a change in the controlled variable to indicate that a disturbance has occurred. Feedforward control is useful where the final controlled variable is not able to be measured. The necessary equations are solved by the controller relating all the process variables, such as steam flow, liquid output temperature, etc., which are usually designated as the process model. Perfect models and controllers are rare so a combination of feedback and feedforward control is more desirable.

feed point The location in a process where a stream of material is first introduced. In a distillation column this is located between the rectifying and stripping sections in which the feed is fed onto the **feed plate**.

feed rate The amount of material fed to a process per unit time. The rate may be expressed on a volumetric, mass, or molar basis depending on the process.

feedstock The main raw material used in a process that is converted into a product or products. In a petroleum refinery, the feedstock is crude oil.

feints The final distillate fraction that is collected from a Scotch *whisky *batch distillation once the desired mid-cut composition has been collected. It is lean in alcohol and returned to the next distillation batch.

Fenske equation A shortcut method used to find the minimum number of theoretical trays in a distillation column required for the separation of two *key components. It is calculated from the mole fraction, x, of the key components in the distillate, and in the bottoms where the relative volatility of the key components is reasonably constant. For a binary mixture of components a and b, the minimum number of trays is:

$$N_{min} + 1 = \frac{\ln \left(\frac{\left(\frac{x_a}{x_b} \right)_D}{\left(\frac{x_b}{x_a} \right)_B} \right)}{\ln \alpha_{ab}}$$

where N_{mn} is the total minimum number of theoretical trays, σ_{ab} is the relative volatility and the subscripts D and B represent the distillate and bottoms, respectively. The method was proposed in 1932 by American professor of chemical engineering Merrell Robert Fenske (1904–71).

fermentation A biotechnological process that involves the anaerobic respiration of certain microorganisms, in particular yeast. Forming the basis of baking, wine, and beer production, alcoholic fermentation comprises a series of biochemical reactions in which sugars are consumed as a substrate to produce ethanol and carbon dioxide in the reaction:

$$C_6H_{12}O_6 \rightarrow 2C_2H_5OH + 2CO_2$$

Industrial applications of fermentation include the production of alcohol and glycerol from yeast, lactic acid, acetone, and butyl alcohol from various bacteria, and citric acid, antibiotics, and vitamins from mould fermentations.

fermenter A vessel used to contain the biochemical process of *fermentation. It is typically made of steel although smaller vessels are made of glass. It can be operated as a batch process, a fed batch process, or continuously. The fermentation process is usually sufficient to create natural turbulence and mixing. However, the extent of mixing can be enhanced using a mechanical stirrer such as an impeller. To avoid a mechanical stirrer, an air-lift or loop fermenter uses an internal or external draft tube up which a small flow of air is admitted to circulate the fermenting medium. In the *whisky industry, the vessel is called a **washback** and is traditionally made of wood.

fertile material A radionuclide that can absorb a neutron to form fissile material. The fissile material used in fast reactors such as uranium-238 is converted into plutonium-239 by the absorption of a neutron to form uranium-238, which then decays to plutonium-239.

FIChemE The post-nominal letters used after someone's name to indicate that he or she is a chartered chemical engineer and a fellow of the Institution of Chemical Engineers. Fellow is the highest category of membership and represents a person with significant leadership and management responsibility within an organization.

Fick, Adolf Eugen (1829–1901) A German physiologist who began his studies in mathematics and physics before pursuing a doctorate and career in medicine. In 1855 he introduced *Fick's first law of diffusion, which governs the diffusion of a gas across a fluid membrane. In 1870, his law of diffusion led to a method for measuring cardiac output in what is known as the Fick principle.

Fick's first law of diffusion A law concerning the movement of mass by the process of diffusion. The law states that a species diffuses in the direction of decreasing concentration:

$$J = -D\frac{\partial C}{\partial x}$$

where J is the *diffusion flux defined as the amount of substance per unit area per unit time, D is the *diffusion coefficient or diffusivity, C is the concentration, and x is the position. The negative sign shows that the flux is driven in the direction of increasing position. The diffusion coefficient is proportional to the square of the velocity of the diffusing particles, which depends on the temperature, viscosity of the fluid, and the size of the particles, according to the *Stokes–Einstein equation. The SI units for the diffusion coefficients are $m^2\ s^{-1}$. The law was proposed in 1855 by Adolf *Fick (1829–1901) and is analogous to the relationships discovered around the same time by other eminent scientists, such as Darcy's law for fluid flow, Ohm's law for the transport of charge, and *Fourier's law for heat transfer. Diffusion processes that do not follow Fick's laws are known as **non-Fickian**.

Fick's second law A law that predicts how diffusion causes the concentration of a diffusing species to change with time and given by:

$$\frac{\partial \phi}{\partial t} = D\frac{\partial^2 \phi}{\partial x^2}$$

where ϕ is the concentration per volume, and x is the position. The law can be derived from *Fick's first law of diffusion and a mass balance of the diffusing species by assuming that the *diffusion coefficient, D, is constant.

field operator A process plant operator who is based on the chemical plant rather than in the central control room.

film A layer, usually thin, of a gas or liquid on a surface. In convective heat transfer, the heat transfer rate is increased by agitation, reducing the thickness of the film. Boiling and condensation heat transfer is related to the film characteristics. *See* CONVECTIVE MASS TRANSFER.

film badge A lapel badge worn by workers in the nuclear industry and in businesses that use ionizing radiation. It is used to measure and record the accumulative exposure to harmful exposure to beta particles, gamma rays, and X-rays. The badge consists of a photographic film, covered with various metals and thicknesses, which are usually lead and tin.

film boiling A type of liquid boiling in which a layer of vapour is formed on a very hot heat transfer surface causing an insulating effect. The heat transfer from the surface into the bulk of the liquid is due to radiation rather than convection and conduction.

film condensation A type of condensation that occurs on a cold surface that forms a continuous film of liquid condensate. Most *condenser type heat exchangers operate using film condensation. *Compare* DROPWISE CONDENSATION.

film model A mathematical model used to describe the mass transfer between two immiscible liquids in which mass transfer takes place as molecular diffusion through stationary layers between the interface and the bulk material. The model assumes that the interface is maintained at an equilibrium condition.

filter A device used to remove and separate particles from gases or liquids. The particles may be undesirable such as contaminants in an air stream as in the case of ventilation, or as extraneous material in food processing. The size of the pores or mesh size of a filter determines the size of the particles retained. Filters are classified as roughing, medium efficiency, high efficiency, ultra-high efficiency, and absolute. Strainers are a type of coarse filter. Filter cloths are used to retain particles from liquids and are typically made from cotton, glass, and synthetic woven fibres or mats. *See* FILTER PRESS.

filter cake A layer of particles formed on the surface of a *filter that have compacted to form a solid cake during the process of *filtration.

filter press A device used to carry out the process of *filtration for separating particles from a liquid suspension or *slurry. It consists of a series of filter cloths with a mesh size sufficient to withhold particles, and through which the suspension to be filtered can pass. A *plate and frame filter press consists of alternating plates and frames with a filter cloth between each. The suspension is fed under pressure and the filtrate passes out through channels embedded within the plates. They can be operated as a batch operation or continuously, and are suited for the filtration of large volumes of liquid carrying suspended solids. Once the *filter cake has built up on the filter cloth, the filter press is taken out of service, and the cake removed from the cloth.

filtrate The clear liquid separated from the process of *filtration, leaving behind the suspended particles.

filtration A process in which a suspension of solid particles in a liquid or a gas are separated. The process uses a *filter, which is a material used to entrap the particles. The rate of filtration can be increased by using a vacuum to draw the gas or liquid through, such as in the process of *vacuum filtration, or the use of pressure to force the liquid through the filter, such as *ultrafiltration.

final moisture content The *moisture content of a substance at the point of leaving a *dryer or drying process. The moisture content is expressed on either a *dry basis or *wet basis.

fine chemicals Chemicals produced industrially in relatively small quantities and of high purity, specification, or composition. For example, dyes, pigments, and pharmaceuticals are considered to be fine chemicals.

fines Small particulates ranging in diameter from 1 to over 400 micrometres. They may be carried by gases and accumulate in undesirable places. They can be removed by *filters, centrifugal separators, *cyclones, *electrostatic precipitators, and washing, etc.

finite difference A mathematical approximation method used for solving first- and second-order differential equations. It is based on determining the gradient from two adjacent points separated a finite distance apart. For example, the forward difference for a first order derivative is given by:

$$\frac{d}{dx}f(x_i) = \frac{f(x_{i+1}) - f(x_i)}{\Delta x}$$

The approximation therefore requires the difference in value of the function separated by Δx. It is used for complex functions where an analytical solution may not be possible or too cumbersome.

fire The rapid thermal oxidation (combustion) of a fuel source resulting in heat and light emission. There are various types of fire, classified by the type of fuel and associated hazards. In the US, the National Fire Protection Association (NFPA) classifies fires and hazards by types of fuel or combustible in order to facilitate the control and extinguishing of fires: Class A are ordinary combustibles such as wood, cloth, paper, rubber, and certain plastics; Class B are flammable or combustible liquids, flammable gases, greases, and similar materials; Class C is energized electrical equipment; Class D are combustible metals, such as magnesium, titanium, zirconium, sodium, and potassium.

(⊕) SEE WEB LINKS
• Official website of the National Fire Protection Association in the US.

fireball A phenomenon that occurs as the result of the deflagration of a cloud of combustible vapour but which does not result in a blast wave. The burning cloud may lift off the ground and form into the shape of a mushroom cloud. Combustion rates are extremely high and may exceed one tonne per second.

fire point The lowest temperature at which a flammable liquid gives off sufficient vapour to produce a sustained combustion after removal of the ignition source.

fireproof A condition in which a structure, equipment, wiring, controls, or piping is capable of functioning under the most severe conditions of fire likely to occur at its location. A **fire retardant** is a substance or treatment that reduces the combustibility of a material. It can be used in *personal protective equipment or coated onto process equipment and structures to protect them against severe heat for a limited period of time. **Fire prevention** involves the measures that are needed to prevent an outbreak of fire. **Fire protection** involves the engineering design features and systems that are required to reduce the damaging and harmful effects of a fire.

fire suppression system A method, device, or system used to detect a fire or ignition source, and to extinguish the fire in sufficient time so as to prevent structural damage and/or debilitation of personnel.

fire triangle A way of illustrating the three *factors necessary for the process of combustion which are fuel, oxygen, and heat. All three are needed for combustion to occur. A fire can therefore be prevented or extinguished by removing one of the factors. That is, a fire is not able to occur without sufficient amounts of all three.

first law of thermodynamics A law that is applied to the conservation of energy in which the change in internal energy, ΔU, of a system is equal to the difference in the heat added, Q, to the system, and the work done, W, by the system:

$$\Delta U = Q - W$$

When considering chemical reactions and processes, it is more usual to consider situations where work is done on the system rather than by it.

Fischer–Tropsch process A process used for producing hydrocarbons from carbon monoxide and hydrogen. Named after German chemist Franz Fischer (1852–1932) and

Czech Hans Tropsch (1839–1935), the process was invented to produce motor fuel in Germany during the Second World War. The process involves passing the reactants over a nickel or cobalt catalyst at 200°C. The general reaction is:

$$nCO + (2n+1)H_2 \rightarrow C_nH_{2n+2} + nH_2O$$

The resulting complex mixture of hydrocarbons is then separated into various fractions. The process also results in the formation of alcohols, aldehydes, and ketones.

fissile material A radionuclide of an element that undergoes nuclear fission either spontaneously or through being irradiated by neutrons. Examples include uranium-235 and plutonium-239 used in nuclear reactors. Fissile materials are also used in nuclear weapons.

fission 1. The division of a living cell into new cells as a mode of reproduction. **2.** In a nuclear reaction, fission is the splitting of a heavy atomic nucleus spontaneously or as the result of an impact of another particle with the release of energy, and possibly causing a chain reaction within the mass of the element. Nuclear fission is the process used in nuclear reactors and atomic bombs. **Fission products** are nuclides that are produced by nuclear fission or the radioactive decay of nuclides by fission. Examples include krypton-85, strontium-90, and caesium-137.

fittings Connections and couplings used in pipework and tubing. The type of fittings used depends largely on the wall thickness as well as in part on the properties of the pipes and tubes including welds, flanges, and screw fittings. Fittings include elbows, bends, reducers, and branches.

fixed bed A stationary and immovable layer of particulate material through which a fluid passes. The particulate material may typically be a catalyst, absorbent, or ion exchange resin. *Compare* FLUIDIZED BED.

fixed costs The costs of a process that include the capital cost repayments, scheduled maintenance costs, and overheads such as personnel facilities, administration, laboratory services, etc. The fixed costs of a process are independent of the rate of production. The sum of the fixed and *variable costs is the total cost of the process. *See* PROCESS ECOMONICS.

fixed point A data point that can be accurately located or reproduced to enable it to be used as a reference point. For example, fixed points are used for calibrating temperature scales such as the triple point of water. *See* FAHRENHEIT SCALE.

flaking A process for extracting chemical products as solid flakes from a liquid solution. An internally cooled rotating drum is partially immersed in the liquid; a thin film of product solidifies on the surface, and is removed by a scraper. It is widely used in the food industry.

flame The visible effect of the combustion of gases. A flame may be luminous or non-luminous, and is the result of the burning of soot particles and gaseous products. A **flame front** is the boundary between the burning and unburnt parts of a combustible mixture, such as a vapour and air.

flame proofing A surface treatment or impregnation of wood products, textiles, and other materials with **fire retardant** chemicals. These are flame-inhibiting chemical compounds used to reduce the flammability of a product or a structure over which it has been applied.

flame propagation The spread of a flame from one place to another in a combustible material, especially in a combustible vapour–air mixture.

flame resistant The property of a material that does not conduct a flame or continue to burn when an ignition source is removed.

flame speed The velocity of a propagating flame relative to the observer. The **flame speed rate** is the propagation velocity of a flame over a surface of combustible material.

flame temperature The heat intensity of a flame.

flame trap A device used to prevent a gas flame from entering a pipe or process vessel containing flammable material. It is typically made from a screen, mesh, or metal gauze, and located in the vent pipe to a storage vessel.

flammability limits The maximum and minimum concentration of a combustible gas or vapour in air or oxygen that is capable of propagating a flame at a specified temperature and pressure with the aid of an ignition source such as a spark or open flame. The concentrations are generally expressed as the per cent fuel by volume. Above the *upper flammable limit (UFL), the mixture is too rich to burn, while below the *lower flammable limit (LFL), the lack of oxygen does not allow the fuel to burn. Any concentration between these limits can result in ignition or explosion. For example, the LFL for hydrogen in air is 4 per cent $^V/_V$ and UFL is 75 per cent $^V/_V$. Concentrations below or above these limits therefore do not result in ignition or explosion.

The bounds of the flammability envelope expand with increased temperature and pressure. That is, the minimum oxygen level and the LFL decreases and the UFL increases. For example, diesel oil has a very low vapour pressure in which a fuel–air mixture at room temperature is too lean. That is, it is below the lower flammable limit. Once the oil is warmed to above its *flash point it will ignite.

flammable A substance or material that has the ability to support combustion and be capable of burning with a flame. It is easily ignited or highly combustible. The term is more widely used than *inflammable as this is often confused with *incombustible, which means an inability or lack of ability to combust. A **flammable liquid** is a liquid that has the capability of catching fire. In the US, the National Fire Protection Association defines a flammable liquid as a liquid that has a flash point below 100°F (37.8°C) and a vapour pressure not exceeding 40 psia (2.72 bar) at that temperature.

flange A pipe fitting at the end of a length of pipe that consists of flat disc or ring of metal used to attach another length of pipe which has an identical flange. The flanges are bolted together and a gasket is compressed between them. The flanges are attached to the pipe by welding or brazing. A blind flange or blank flange is a flange with no opening and used to close a pipe. Slip-on, weld neck, socket weld, and screwed flanges are types of flange to be found at the end of a length of pipe.

flaring The burning of unwanted gaseous hydrocarbons in a petroleum refinery or on an offshore installation. Once widely used, it is now largely used during the testing of a new well, and when small amounts of unwanted or surplus refinery gas have accumulated, or when there is insufficient pipeline or vessel capacity to transport or store it. A **flare stack** is an installation used to provide the safe disposal of gaseous hydrocarbons produced during the oil refining process. A **flare boom** is used on offshore oil platforms, angled away and extending a safe distance from the deck to permit the safe burning of the gases.

flashback The propagation of a *flame from an ignition source back to a supply of flammable gas or liquid.

flash drum A vessel used for the rapid separation of a mixture into a liquid and vapour by *flash evaporation caused by a sharp drop in pressure. The liquid leaves cooled from the bottom and the vapour from the top at the *saturation temperature of the liquid.

flash drying A rapid form of *drying of particulates using hot gas or air. The particulates either descend through a current of hot gas or air, or are transported pneumatically in which their *moisture content is reduced.

flash evaporation/vaporization The rapid separation of a saturated liquid of a single component into a vapour and a liquid when it passes through a throttling valve to a lower pressure. As a single-stage vapour–liquid separation process, the liquid mixture is partially vaporized by heating and reducing the pressure, allowing the vapour to form at the expense of the liquid adiabatically. The vapour is allowed to reach equilibrium with a residual liquid, and the resulting vapour and liquid phases are separated and removed. It may be operated as either a batch or continuous process. It is also known as **equilibrium distillation**. The process in which both the liquid and vapour are cooled to the saturation temperature of the liquid at the reduced pressure, which forms the basis of vapour compression cycles used for *refrigeration. The flash process is isenthalpic since it involves no work, and is also known as an *adiabatic flash.

flash fire A fire that spreads with considerable rapidity. It occurs when a fuel source such as a flammable vapour and air are mixed in sufficient amounts and ignited. The combustion is rapid and characterized by a rapidly moving flame front with a high temperature. The flame passes through the mixture at less than sonic velocity and with negligible overpressure. *See* FLAME.

flash freezing A rapid freezing technique used for foods in order to retain food quality. Pioneered by American inventor Clarence Birdseye (1886–1956) in 1922, it involves a freezing process that causes the water content to freeze in a very short time, preventing the formation of ice crystals that have a destructive effect on biological cell wall structure. It is used for many foods such as fish, meat, and fruits, and is also known as *quick freezing.

flash point The lowest temperature at which a volatile liquid will produce a sufficient amount of vapour to ignite in air at a given pressure. The rapid combustion occurs in the form of a momentary flash. Flash point data is important for the safe storage and transportation of volatile liquids. *Compare* AUTOIGNITION TEMPERATURE.

Flixborough disaster A major incident that occurred in the UK in 1974. It involved an explosion at the Flixborough Works of Nypro (UK) Limited near Hull, killing 28 people and injuring 36 others, together with extensive damage beyond the factory including 2,000 houses and 200 shops and factories. The incident took place on 1 June 1974 and was the result of a massive leak and subsequent deflagration of cyclohexane gas from a temporary bypass in a chain of six reactors. The reactors were used for the oxidation of cyclohexane to form cyclohexanone and cyclohexanol by the injection of air in the presence of catalyst. The cyclohexane and cyclohexanol were in subsequent stages of the process for conversion to caprolactum, which is a raw material for the production of Nylon 6. At the time, the report from the Court of Enquiry was the most comprehensive inquiry into a disaster in the chemical industry, and was crucial in the development of *Loss Prevention in the UK.

float glass process A process for making sheets of glass and involves molten glass being fed onto a shallow tank containing molten tin. The less dense glass floats on the tin and spreads to form a sheet. The process is carried out in a slightly reducing atmosphere to prevent oxidation of the molten tin. The molten glass starts at around 1,050°C and is cooled to around 650°C before being lifted off the tin.

floating head heat exchanger A type of shell and tube heat exchanger in which the tube sheet assembly is independent and free to move within the shell or within the shell cover. This allows for differential expansion of the tubes. Since the floating tube plate is smaller in diameter than the shell, removal for cleaning and general maintenance is relatively straightforward.

floating production systems (FPS) A floating oil platform that contains all the equipment associated with a fixed installation. They are used in conjunction with sub-sea wellheads to exploit moderate to deep-water oil fields. They are used for small or isolated oil fields or where the water is too deep for a conventional fixed platform. The term is used generally to mean any vessel associated with oil production or oil storage, and includes *FSO, *FSU, and *FPSO.

floating roof The roof of a tank used to store volatile liquids. The roof is allowed to float on the surface thereby reducing the vapour volume and vaporization, and therefore loss of the material by evaporation.

floating, storage, and off-loading unit (FSO or FSU) A floating vessel into which processed oil is pumped from a fixed platform that neither has any storage capacity of its own, nor is connected to a sub-sea pipeline. A **floating, production, storage, and off-loading unit** (FPSO) is a floating oil platform with a storage capacity that is not connected to a sub-sea pipeline. The oil from sub-sea wells is processed and stored aboard the vessel before being off-loaded into shuttle tankers.

flocculation A process that is used to separate particles in a liquid. It is applied to particles in a colloid that aggregate into larger clumps or flocs by weak electrostatic forces. The ability of particles to flocculate relies on a surface charge on the surface of the particles. Reagents or **flocculants** can be added to the liquid to bring the particles together. The clumps are then able to settle under gravity or be filtered more easily. The particles can be dispersed again back into the liquid with agitation. Flocculation is used in the *water treatment industry and for removing yeast from beer at the end of fermentation.

flooding An excessive build-up of liquid in adsorption columns or on the plates of a *distillation column. It is due to high vapour flow rates up the column. In distillation columns this is caused by high heating rates in the reboiler. *Compare* WEEPING. The **flooding point** is a condition in a packed column such as an absorption column which receives a counter-current flow of gas at the bottom and a liquid descending under gravity from the top where there is insufficient liquid hold-up in the packing for mass transfer to take place effectively. The liquid therefore descends to the bottom of the column without mass transfer. The rate of flow through the packing for effective mass transfer is controlled by the pressure drop across the packing material.

flotation A separation process in which suspended particles in a liquid are carried to the surface by air or gas bubbles attached to them. The air or gas is blown as bubbles through the liquid and the particles are removed from the surface. It is widely used in the mining industries for the separation of minerals, such as copper, lead, and zinc, and *gangue. It is also used for the separation of oil from oil sands. A **flotation agent** is a surface-active substance added to improve the separation process by promoting the attachment of the particles to the bubbles. Other reagents that may be added can improve the stability of the foam to facilitate the removal of the froth of solids. *See* ORE FLOTATION.

flow The movement of a fluid under the influence of an external force such as gravity or a pump.

flowline A pipeline that carries materials from one place to another. In the offshore industry, a flowline is a pipeline that carries oil on the seabed from a well to a riser. On a *process flow diagram, the flowline is indicated by a line entering and leaving a vessel or unit operation. An arrow indicates the direction of flow.

flow meter A device used to measure the flow of process fluids. Flow meters are broadly classified into those that are intrusive and those that are non-intrusive to the flow of the fluid. Flow meters include differential pressure meters, positive displacement meters, mechanical, acoustic, and electrically heated meters. The measurement of the flow of process fluids is essential not only for safe plant control but also for fiscal monitoring purposes. It is important to select correctly the flow meter for a particular application, which requires a knowledge and comprehension of the nature of the fluid to be measured and an understanding of the operating principles of flow meters.

flow rate The movement of material per unit time. The material may be a gas, liquid, or solid particulates in suspension or combinations of all of these, and expressed on a volumetric, mass, or molar basis. The volumetric flow rate of material moving through a pipe is the product of its average velocity and the cross-sectional area of the pipe.

flow regimes The behaviour of a combined gas and liquid flow through a channel, duct, or pipe can take on an almost infinite number of possible forms. There are many descriptions used to define the possible flow patterns and there is often confusion through the subjective way in which flow patterns are characterized. In general, depending on the conditions of flow of the two phases, one phase may be considered to be the continuous phase while the other is the discontinuous phase. An example is the flow of a mist or fine dispersion of liquid droplets in a gas phase. The smaller the liquid droplet, the higher are the surface tension effects. Distortion of the discontinuous phase causes the shape to become non-spherical. There is a tendency for the liquid phase to wet the wall of the pipe and for the gas phase to congregate at the centre. An exception to this is in evaporators such as in refrigeration where nucleate boiling occurs on the pipe surface resulting in a vapour film or bubbles forming at the surface with a central core of liquid. The flow of fluids through pipes and over surfaces can be broadly described as being:

1 Steady flow in which flow parameters at any given point do not vary with time.
2 Unsteady flow in which flow parameters at any given point vary with time.
3 *Laminar flow in which flow is generally considered to be smooth and streamline.
4 *Turbulent flow in which flow is broken up into eddies and turbulence.
5 Transition flow, which is a condition lying between the laminar and turbulent flow regimes.

Flow regime maps are charts representing the various flow regimes and *flow patterns that are possible for two-phase gas–liquid flow in both horizontal and vertical pipes and tubes. There are many types of flow regime map that have been developed. The simplest form of map involves a plot of superficial velocities or flow rates for the two phases with the most widely used generalized flow regime map for horizontal flow having been developed by

Baker in 1954. More complex maps plot the volume fluxes, mass fluxes, momentum fluxes or similar quantities for both the liquid and gas or vapour flows. The maps are populated with experimental data in which lines are drawn to represent the boundaries between the various distinguishable regimes of flow. These include stratified flow, intermittent flow, annular flow and bubble flow. Maps may also include identification of other regimes including slug, plug, wavy and annular flow. The boundaries between the various flow patterns are due to the regime becoming unstable as it approaches the boundary with the transition to another flow pattern. As with the transition between laminar and turbulent flow in a pipe, the transitions in a flow regime are unpredictable. The boundaries are therefore not distinctive lines but loosely defined transition zones. A limitation of the maps is that they tend to be specific to a particular fluid and pipe.

flowsheet A schematic diagram or representation of a process illustrating the layout of process units and their functions linked together by interconnecting process streams. The development of a flowsheet involves the process of synthesis, analysis, and optimization. The heat and material balances are solved using thermodynamic properties and models. An economic analysis is also completed as well as a safety and environmental impact assessment. The choice of equipment and their interconnectivity are optimized along with the choice of operating parameters such as temperature, pressure, and flows. Steady-state flowsheet computer software packages are frequently used to develop flowsheets.

flue gas A mixture of gases produced as a result of *combustion that emerge from a stack or chimney. The gases contain *smoke, particulates, carbon dioxide, water vapour, unburnt oxygen, nitrogen, etc. An *Orsat analysis is a reliable way to determine the composition of the flue gas and the efficiency of combustion although it has largely been replaced by other techniques.

fluid A substance that offers no resistance to change of shape by an applied force. Liquids, vapours, and gases are fluids while solids are not. However, solid particles may be made to behave as fluids when they are dispersed in liquids, vapour, or gases. *Pneumatic conveying is a technique used to move solid particles such as wheat and catalyst particles and involves mixing the particles in a strong current of air. *See* FLUIDIZED BED.

fluid bed *See* FLUIDIZED BED.

fluidics The use of both the flow and pressure of fluids in pipes to perform sensing and control functions in devices that have no moving parts. Although much slower than electronic devices, they are not influenced by temperature, magnetic fields, and ionizing radiation.

fluidization The process of suspending solid particles in an upward flow of a fluid and is widely used to promote good heat transfer and mass transfer in *fluidized beds. In **particulate fluidization** involving liquids, each particle behaves individually and collides with others remaining a certain distance apart. As the velocity of fluidization is increased, the bed expands. It is used in backwashing of filter beds and ion exchange resin beds. In **aggregated fluidization** involving gases, similar conditions exist up to the point of incipient fluidization. At higher velocities, flow passes through the bed in the form of bubbles and the bed expands due to the volume of the bubbles.

fluidized bed A vessel or chamber in which solid particles are suspended in an upwards flow of a gas (see Fig. 22). The buoyant solid particles therefore behave as though they were in a liquid state. Used extensively in the chemical industry, fluidized beds provide excellent

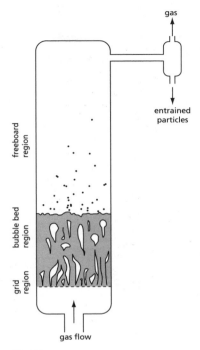

Fig. 22

mixing, heat transfer, and mass transfer characteristics. They are used in catalytic reactions where powdered or pelleted catalysts have a high specific surface area. They are also used in furnaces in which coal is combusted in a hot bed of ash or sand through which air is passed. Fluidization permits lower temperatures to be used thereby avoiding the production of polluting oxides of nitrogen. The behaviour of a fluidized bed depends on the particle size and the fluidizing gas velocity. When fine particles are fluidized at low gas velocities, the bed expands but without the formation of bubbles. At higher velocities, at the so-called bubbling regime, there are three distinct zones in the bed:

• The grid zone is located at the bottom of the bed and corresponds to gas penetrating the bed and is dependent on the types of grid used.
• The bubbling zone is where bubbles grow by coalescence and rise to the bed surface where they break.
• The freeboard zone is where some particles are carried above the bed surface and are elutriated from the system while others are returned back to the bed.

fluidized catalytic cracker unit (FCCU) A unit operation used in petrochemical refineries to convert the heavy components in crude oil into useful products such as fuels for motorized vehicles. They consist of a catalytic reactor front end section in which cracking takes place, reducing long-chain hydrocarbon molecules into smaller short-chain molecules. In the light or back end section, separation takes place where the cracked molecules are fractionated through a series of columns to produce fuel gas, fuel oil, LPG, diesel, and gas oil.

fluid mechanics A branch of engineering science that concerns the behaviour of fluids when subjected to changes in pressure, the effects of frictional resistance, the flow through pipes, ducts, and restrictions, and the production of power. An understanding of the behaviour of fluids is critical in the cost-effective design and efficient operation of chemical plants, and includes the development and testing of theories devised to explain various phenomena. **Fluid statics** deals with fluids in an equilibrium state of no shear stress, whereas **fluid dynamics** considers fluids in relative motion. Many of the fundamental principles of fluid mechanics were established in the seventeenth and eighteenth centuries by scientists such as *Bernoulli, *Newton, and Euler. Many of today's fluid mechanics problems are complex, non-linear, three-dimensional, and transient. High-speed and powerful computers are increasingly used to solve complex problems particularly in *computational fluid dynamics (CFD).

fluoridization A process used in the water industry that involves the addition of fluoride to public water supplies for the purposes of reducing the risk of dental decay. It uses a highly soluble fluoride compound whose action in the mouth is to reduce the demineralization of tooth enamel.

(((⊕))) SEE WEB LINKS
• Official website of the British Fluoridation Society.

fluorination A chemical reaction that involves the introduction of fluorine atoms to a molecule such as in the production of uranium hexafluoride:

$$UO_2 + 3F_2 \rightarrow UF_6 + O_2$$

The densities of the isotopes of uranium-235 and uranium-238 hexafluoride are appreciably different such that they can then be separated by centrifugation for the purposes of *uranium enrichment.

flux 1. The rate of flow of a substance expressed as the mass or energy per unit cross-sectional area. The SI units are $kg\,m^{-2}\,s^{-1}$ or $W\,m^{-2}$. **2.** A substance used in soldering to inhibit oxidation. *See* BRAZING. **3.** A substance used in the smelting of metals for the removal of impurities or reaction products as *slag. *See* SMELTING.

FMEA (failure modes and effects analysis) A technique or procedure used to analyse the likelihood of failure within a process or system. It involves a process of review in assessing and quantifying risks; identifying those that are of the greatest concern to the overall process or system, and preventing problems before they arise, saving both time and money. The 'failure modes' are problems that need to be identified and the 'effects analysis' are the consequences of the problems, which may include loss of plant function, equipment damage, plant shutdown, injury, release of materials to the environment, etc.

foam An aggregation of gas or vapour bubbles that remains as a stable suspension. The liquid is the continuous phase and exists as bubble films. Foaming is often a problem with fermentation processes where it forms on the surface of the broth. The existence of foam can be controlled by the addition of chemicals such as glycol, which interrupts the stability of the foam. A **foam breaker** is a mechanical device used to break up the formation of foam. Foam breakers are sometimes used in fermentation vessels in which foaming may occur and otherwise overflow. They usually consist of high-speed spinning discs and are positioned in a region above a potentially foaming liquid.

foot A unit of length used in *f.p.s. engineering units. It has a length equal to 12 inches and one third of a yard. In science and engineering, the foot as a measure of length, and other foot-related units such as the foot-pound-force, has been replaced by SI units, where one foot is equal to 0.3048 m.

force (Symbol F) A physical agent that brings about a change in momentum. For a body of mass, m, travelling with a velocity, v, the momentum is mv. The agent that brings about a change in momentum is therefore:

$$F = \frac{d(mv)}{dt} = m\frac{dv}{dt}$$

The force is therefore directly proportional to the product of mass and acceleration of a body. That is $F = ma$. The SI unit is the newton (N).

forced convection The process by which heat is transported within a fluid by the movement of the fluid with the assistance of a fan, pump, or blower. *Compare* NATURAL CONVECTION.

forced vortex A rotating fluid due to the action of a rotating impeller or stirrer that creates a vortex. *Compare* FREE VORTEX.

foreman A senior or experienced process operator in a factory or chemical plant who has responsibility for a team of operators.

foreshots The first distillate fraction taken from a Scotch whisky *batch distillation. It has a high alcohol content and contains many volatile components not suitable for Scotch *whisky, some of which may be harmful. It is therefore recovered and redistilled. After the heart is then recovered, which is used for Scotch whisky, the *feints are finally recovered, which are lean in alcohol and also returned to the next distillation batch.

formula 1. A statement in algebraic form representing a mathematical relationship of variables. An *empirical formula is based on experimental results and not necessarily on the laws of physics. **2.** The use of symbols to present a chemical compound in which the symbols represent elements. Subscripts are used to represent the number of atoms. For example, the molecular formula for ethanol is C_2H_5OH, which is a molecular compound that has two carbon atoms, six hydrogen atoms, and one oxygen atom. *See* CHEMICAL FORMULA.

forward-curved fan A centrifugal or radial-flow fan that has blades on a rotating shaft such that they are arranged with the leading edge on the outer periphery.

forward mixing The propensity of part of a flowing fluid to accelerate through a system such as a pipe, vessel, or reactor, and to reach the exit point having effectively short-circuited the rest of the fluid. Forward mixing leads to reduced reactor efficiencies where some of the reactants that are expected to undergo a chemical reaction pass through unreacted. It is therefore a form of departure from ideality where the flow of fluids leave before the expected time. Conversely, some reactants will reside for longer periods, known as *back-mixing.

fossil fuel A naturally occurring organic hydrocarbon fuel formed millions of years ago from the decomposition of fossilized remains of plants, trees, and aquatic life to form coal, oil, and natural gas found in rock strata. With a high carbon and hydrogen content, they are energy-rich having captured the energy of the sun in prehistoric times. Fossil fuels are extracted from the ground and used for *power generation in power stations, as fuel for motorized vehicles, and for domestic heating. In power stations, the fuel is combusted to produce heat, which is used to produce steam from water that supplies steam turbines to drive a generator. Coal is the most widely used fossil fuel. Both coal and wood were burned in the earliest power stations. In 1920, crushed coal was first used thereby improving combustion efficiency. However, limited economical coal supplies and pollution concerns have led to a recent decline in its use. Oil and gas are also used as fuel in power stations. Natural gas, which is largely methane, produces less airborne pollution than either coal or gas. Natural gas provides a significant proportion of the UK's electricity and is combusted directly in a *gas turbine. As a finite and non-renewable energy source, fossil fuels produce carbon dioxide through combustion and are attributed to causing *pollution and the *greenhouse effect.

fouling A deposition of solid material on a surface causing resistance to heat and mass transfer. The material may be process materials, precipitates, and particulates that build up on the surface of heat exchangers, packing supports, and distillation trays, etc. Pipe scale, lime, carbon, gums, and other chemical deposits can restrict the passage of flow, reducing heat transfer, and increasing pressure drops. Extreme fouling can cause blockages or plugging. Periodically, the process equipment must be taken out of service and cleaned. A **fouling factor** is a number used in heat transfer calculations where fouling is likely to be an issue on a surface causing resistance to heat transfer. Scale, dirt, and other deposits may accumulate on the heat transfer surface resulting in additional resistance to heat flow. Fouling factors are usually specified to provide a margin of safety.

Fourier, Jean Baptiste Joseph (1768–1830) A French physicist and mathematician noted for his work in heat transfer and his work leading to the Fourier series. He made a notable contribution to the French Revolution and as a result was rewarded with a position at the École Normale Supériéure in 1795 and then a chair at the École Polytechnique. He accompanied Napoleon Bonaparte to Egypt in 1798 and was appointed to senior positions there before becoming cut off by the British fleet. Later returning to France, Fourier carried out his work on heat propagation. The Fourier transform and *Fourier's law are named in honour of him. He is also credited with the discovery of the *greenhouse effect.

Fourier number A dimensionless number, Fo, used in unsteady-state heat and mass transfer calculations that characterize the connection between the rate of change of temperature, the physical properties, and the dimensions of the body defined as:

$$Fo = \frac{\alpha t}{l^2}$$

where α is the thermal diffusivity, t is time, and l is the length through which conduction takes place. The Fourier number is used together with the *Biot number to solve unsteady-state heat transfer problems. In the case where the Biot number is less than 0.1, the time can be calculated from:

$$t = \frac{mc_p}{hA} \ln \frac{T_o - T_\infty}{T - T_\infty}$$

where T_o is the initial temperature and T is the temperature at the centre of the body. The analogous form of the Fourier number used for mass transfer is given by:

$$Fo = \frac{Dt}{l^2}$$

where D is the diffusivity. The Fourier number is named after the French physicist and mathematician Jean Baptiste Joseph *Fourier (1768–1830).

Fourier's law A relationship that states that the steady-state rate of heat transfer by conduction is proportional to the cross-sectional area perpendicular to the direction of flow and to the *temperature gradient of the path of conduction:

$$Q = -kA \frac{\partial T}{\partial L}$$

where k is the *thermal conductivity, A is the area, T is the temperature, and L is the distance through which heat flows. The negative sign signifies that heat flow occurs from hot to cold.

FPS *See* FLOATING PRODUCTION SYSTEMS.

f.p.s. engineering units A non-decimal British system of units based on the foot, pound, and second, formerly used in engineering. It also uses the thermodynamic Rankine scale (°R) for temperature. The gravitational unit of force is called the pound force (lb_f) and is defined as the standard gravitational field that exerts a force of one pound on a mass of one avoirdupois pound. The standard gravitational acceleration (g) is 32.174 ft s^{-2}. The system has now been replaced by *SI units.

FPSO *See* FLOATING, STORAGE, AND OFF-LOADING UNIT.

fracking An informal name for **hydraulic fracturing,** which is a method used to release and recover oil and natural gas from rock in natural underground reservoirs by fracturing the rock using a pressurized fluid. By injecting a fluid such as water, oil, or gas at high pressures into channels within the reservoir rock, new channels are created by fracturing the rock thereby increasing the release and recovery of the fossil fuels. A proppant such as sand is also used, which is intended to fill the cracks and prevent collapse of cracks. There is some concern that fracking results in noticeable seismic activity and that drinking water supplies from underground aquifers may become contaminated with oil.

fractional crystallization A process used to produce pure crystals that involves dissolving crystals in a small amount of a hot solvent and cooling the solution to produce crystals. The process is repeated until a required purity of crystal has been obtained.

fractionation The process of separating a mixture of components into distinct components or groups (fractions) of components having different properties such as size, boiling point, etc. It is applied to solids, liquids, isotopes, and particles in suspension. **Fractional distillation** is used to separate mixtures of liquids with different boiling points such as the separation of hydrocarbons in crude oil in a fractionating column or **pipestill**. The crude oil is heated to around 350°C until it vaporizes and enters the fractionating column. The vapour rises through perforated trays in the column condensing back down into distinct liquids. Light fractions such as kerosene and naphtha collect near the top while heavier fractions such as lubricants and waxes collect near the bottom.

Frasch process A production method used to recover sulphur from underground deposits. Named after German-born US chemist Hermann Frasch (1851–1914), it uses three concentric pipes in which superheated steam is passed down the outer pipe to melt the sulphur. Compressed air is fed down the inner pipe to force the molten sulphur up the middle pipe while steam in the outer pipe keeps the sulphur molten.

freeboard zone The location in a *fluidized bed of solid particles in which the fluidizing gas is sufficient to form bubbles within the particles. The freeboard zone is where some particles are carried above the bed surface and are elutriated from the system while others are returned back to the bed.

free convection The movement of a fluid caused by localized differences in density due to differences in temperature. The heated fluid due to a hot surface is less dense than the cooler and more dense fluid, and therefore rises, creating natural circulating currents. *Compare* FORCED CONVECTION.

free energy A measure of a system's ability to do work. The *Gibbs free energy provides an indication of the conditions under which a chemical reaction will occur and is given by $G = H - TS$ where G is the energy liberated or absorbed in the reaction process at constant temperature, T, and pressure, H is the enthalpy, and S is the entropy. Changes in Gibbs free energy, ΔG, are useful for indicating the conditions under which a reaction will occur. Negative values of ΔG indicate that the reaction will proceed spontaneously to equilibrium, while positive values indicate that the reaction will only occur if sufficient energy is supplied to force it away from equilibrium. The **Helmholtz free energy** is given by $F = U - TS$ where U is the internal energy. For reversible isothermal processes, the change in the Helmholtz free energy represents the useful work available. It is named after German physicist and physiologist Hermann von Helmholtz (1821–94).

free moisture content The moisture in a substance that is not held by hygroscopic forces and is the difference between the *total moisture content and the *equilibrium moisture content. It therefore indicates the excess mass of water that is removed prior to reaching equilibrium. It can be removed by mechanical methods such as by applying centrifugal force or pressing. It is sometimes defined as the moisture above the equilibrium moisture content and can also include *unbound moisture.

free surface The interface of a stationary liquid with a vapour or gas above.

free vortex A freely rotating liquid as it drains from an orifice at the bottom of a tank. It occurs with non-viscous liquids in which the tangential velocity is proportional to the inverse of the distance from the axis. The streamlines move freely in horizontal concentric circles with no variation of the total energy across them. *Compare* FORCED VORTEX.

freeze crystallization A process used to produce crystals in which heat is removed from a solution to form crystals of the solvent rather than of the solute. The product can be either melted crystals, as in water desalination, or the concentrated solution, as in the concentration of coffee extracts.

freeze drying A process used to remove a solvent by sublimation from the frozen state. The process is used to preserve or extend the shelf-life of heat-sensitive materials by removing the water (i.e the solvent) by rapid freezing and subsequent drying in a vacuum. Frozen water sublimes directly to the gas phase reducing the moisture content. The porous structure of the material is maintained by carefully applying heat. It is also known as *lyophilization or **cryodesiccation**. A **freeze dryer** is a device used to remove the moisture of a body in which a vacuum is maintained. In freezing foods, the vacuum is maintained such that the water is frozen in the product and the moisture is removed by sublimation of the ice to water vapour. Freezer dryers are used for drying heat labile foods and pharmaceutical products that would otherwise be damaged if higher temperatures were to be used.

freezing point The temperature at which a liquid turns to the solid state. The freezing point can be depressed by adding a solute.

frequency A statistical number of occurrences within a defined interval. Used in a *risk analysis, it is the number of occurrences of an *event happening per unit time.

frequency response An analysis used in process control to determine the output signal as the result of a disturbance in a system. It involves both the magnitude and phase angle of the output signal as a function of frequency in comparison with the input signal. The magnitude is typically measured in decibels and the phase angle is measured in radians. The frequency response of a system can be analysed by plotting the magnitude and phase on a chart. The three commonly used frequency response diagrams are the *Bode plot, the *Nyquist plot, and the *Nichols plot.

Freundlich adsorption isotherm An empirical adsorption isotherm proposed in 1909 by German–American physical chemist Herbert Max Finlay Freundlich (1880–1941) that relates the quantity of gas molecules adsorbed onto a surface at a constant temperature of the form:

$$\theta = kp^{\frac{1}{n}}$$

where θ is the measure of the sites occupied per unit area of surface measured as the ratio of the mass of adsorbate to the mass of adsorbent, p is the equilibrium pressure of the adsorbate, and k and n are empirical constants for a particular temperature. When $1/n$ approaches zero, the adsorption becomes independent of pressure. The isotherm is often applied to the adsorption from liquid solutions and is useful when considering the performance of heterogeneous catalysis.

frictional head The head required by a system involving a flowing fluid to overcome the resistance in pipes and associated fittings. The frictional head, h, is related to frictional pressure as:

$$h = \frac{p}{\rho g}$$

where ρ is the density of the fluid and g is the acceleration due to gravity.

friction factor A dimensionless number used in fluid flow calculations that accounts for the permanent energy loss as a fluid moves over a surface. It is related to shear stress at a surface and kinetic energy of the fluid per unit volume as:

$$\tau = f\frac{\rho v^2}{2}$$

Proposed by John Thomas Fanning (1837–1911), the Fanning friction factor for laminar flow in pipes and tubes is related to *Reynolds number as:

$$f = \frac{16}{\text{Re}}$$

For turbulent flow, numerous correlations exist for both smooth and rough-walled pipes. A number of charts have been prepared such as those by Moody, and by Stanton and Pannell, in which friction factor is correlated against Reynolds number for differing pipe surface roughness. It is important to note that this Fanning friction factor has a value of one-quarter of the Darcy friction factor.

Friedel–Crafts reactions Catalyzed chemical reactions that involve the alkylation and acylation reaction of aromatics. An example is the production of toluene from benzene using chloromethane in the presence of aluminium chloride catalyst.

$$C_6H_6 + CH_3Cl \rightarrow C_6H_5CH_3 + HCl$$

The reactions are named after French chemist Charles Friedel (1832–99) and American chemist James Craft (1839–1917).

front end engineering design (FEED) A conceptual study used for the development and analysis of process engineering projects. It defines the processing objectives and examines the various technical options.

froth A dispersion of a high fraction of gas or vapour in a liquid that readily separates at the surface of the liquid as bubbles that then disintegrate without forming a *foam.

froth flotation A process used to separate ore and *gangue using small bubbles of air or another gas generated inside a tank containing particles of the ore and gangue. Water is treated with a water-active agent (detergent), which allows either but not both materials to adhere to the bubbles. The bubbles rise and are skimmed from the surface.

Froude number A widely used dimensionless number, Fr, that represents the influence of gravity in the power relationship for fluid systems such as pumping, mixing in unbaffled tanks and reactors, and in determining the extent of fluidization of particles in a fluidized bed. For a fluidized bed it is expressed as:

$$Fr = \frac{v^2 d_p}{g}$$

where v is the minimum fluid velocity calculated over the entire cross-section of the bed, d_p is the particle diameter (or diameter corresponding to the surface mean diameter), and g is the gravitational acceleration. Since the viscosity of gases is low, the velocity is high and aggregative fluidization occurs for Fr greater than a value of 1. Particulate fluidization occurs for Fr less than 1. For stirred tanks it is expressed as:

$$Fr = \frac{N^2 D}{g}$$

where D is the diameter of the impeller and N is the rotational speed in rps. It is named after British engineer William Froude (1810–79) who developed laws for the resistance of the hulls of ships in water and in predicting their stability.

FSO (or FSU) *See* FLOATING, STORAGE, AND OFF-LOADING UNIT.

fuel 1. A substance that has a calorific value and is used for producing thermal energy by combustion. Examples include coal, oil, natural gas, wood, and *biofuels. Their controlled combustion is used to release the energy as heat, and conversion to mechanical motion in automotive vehicles, and conversion to electricity in power stations. Primary fuels are those obtained directly from nature such as methane, whereas secondary fuels are derived from primary fuels, such as coke. *See* COMBUSTION. **2.** A nuclear fuel is a substance capable of nuclear fission for the production of nuclear energy such as certain isotopes of uranium and plutonium. *See* NUCLEAR FUEL.

fuel cell An electrochemical device in which the energy of a chemical reaction is converted directly into a low-voltage, direct current electrical energy. The simplest fuel cell involves the oxidation of hydrogen to form water. Hydrogen gas is used as the fuel and fed to the porous anode, and oxygen is fed to the porous cathode. The two electrodes are separated by a hot alkaline electrolyte such as potassium hydroxide. The electrodes are porous to allow the two gases to react with the electrolyte. At the anode the hydrogen reacts with the hydroxide ions in the electrolyte to form water and the release of two electrons per hydrogen molecule. At the cathode, the oxygen reacts with the water to take up the electrons and form hydroxide ions. The flow of electrons from the anode to the cathode is via an external circuit as an electrical current:

Anode: $H_2 + OH^- \rightarrow 2H_2O + 2e^-$

Cathode: $1/2 O_2 + H_2O + 2e^- \rightarrow 2OH^-$

The overall fuel cell reaction of $H_2 + \frac{1}{2} O_2 \rightarrow H_2O$ is, in effect, the spontaneous cold combustion of hydrogen. However, unlike the combustion process, hydrogen and oxygen must be kept separate. Little heat is liberated in the fuel cell and instead the free energy is released directly as electrical energy. Fuel cells are therefore more efficient than combustion processes. However, fuel cells are more bulky than heat engines and require a continuous supply of gaseous fuels.

fuel element *See* FUEL ROD.

fuel gas A mixture of methane and hydrogen produced in petrochemical refineries. It is used as a fuel in gas turbines and furnaces such as for cracking, etc.

fuel rod A *nuclear fuel used in a nuclear reactor that has been compressed into pellets and sealed into metal tubes. There are various designs of rod depending on the type of nuclear reactor. The pellet is formed from an enriched form of the easily fissionable material such as the isotope of uranium-235, which has been converted to uranium dioxide powder. The powder is calcined, compressed into pellets, and loaded into the tubes. The rods used in nuclear reactors are made into an assembly called a **fuel element**. At the end of the life of the fuel rod in the nuclear reactor, once fissile impurities have built up, the rod is removed, and the unused fuel is recovered and reprocessed for reuse. This involves cutting the rod and recovering the fissile material inside. *See* NUCLEAR REPROCESSING.

fuel tank inerting A method used to prevent fire and explosion in tanks and vessels that contain flammable liquids by the use of inert or non-combustible gases such as nitrogen. The addition of the inert gas is used to exclude air and prevent otherwise flammable vapour and air mixtures from building up. *See* INERTING.

fugacity A thermodynamic function that represents a pseudo-pressure in place of the partial pressure of a real gas or gas mixture. The **fugacity coefficient** for gases is the ratio of the fugacity to the partial pressure. It is therefore equal to 1.0 for an ideal gas. For liquids, the fugacity is the mole fraction of the constituent multiplied by the fugacity of the pure substance at the temperature and pressure of the solution.

fugitive emissions The uncontrolled release of air pollutants from a process other than those via stacks and vents. Fugitive emission are often due to leaks, evaporation, and accidental releases.

Fukushima Daiichi A nuclear disaster that occurred at the Fukushima nuclear power plant in Japan following a major earthquake and tsunami on 11 March 2011. It led to a series of equipment failures resulting in a nuclear meltdown, explosion, and the release of radioactive material into the environment. The tsunami was responsible for breaking the connection to one of the boiling water reactors to the power grid, causing the reactors to overheat. The earthquake and flooding prevented a controlled shutdown. Seawater was used in an attempt to cool the reactors before some electrical power was eventually restored to cool the reactors. It was the biggest nuclear accident since *Chernobyl in 1986.

fully developed flow The point above a surface in which fluid flows in such a way that it has attained its final form and remains unchanged thereafter. The fluid in contact with a surface, such as in a pipe wall, is retarded to the point that it is theoretically brought to rest. Above the surface it increases in velocity to the point that it has reached the mean velocity of the bulk flow. *See* BOUNDARY LAYER.

fume A cloud or dispersion of very small airborne particles or droplets that generally have a diameter less than one micrometre (1 μm). The particles are caused by a chemical reaction, by the condensation of vapour, emitted through the process of calcination, through volatilization of liquids, by distillation, and sublimation. They are sufficiently small that they are visible as a cloud.

function A mathematical formula or algorithm that expresses an output value for a given input. For the form of notation $y = f(x)$, then y is a function of x in which x is the input or independent variable and y is the output or dependent variable. If x is known, then y can be determined.

fundamental constants Parameters that do not change such as the *universal gas constant, *Avogadro's constant, the Planck constant, and the speed of light. They are also known as *universal constants.

fundamental dimensions The primary dimensions used in a system of units. These are mass, length, and time denoted M, L, and T. It is convenient to include temperature (θ) and electrical charge (Q), although they are not strictly fundamental dimensions. Fundamental dimensions are used in *dimensional analysis.

fundamental units An independent set of units of measurement that forms the basis of a basis of units. In the *SI unit system (Système International d'Unités), there are seven fundamental units: kilogram, metre, second, kelvin, mole, ampere, and candela. Derived units are also acceptable in the SI system and include the units for pressure as newtons per square metre (Nm^{-2}) with base units $kgm^{-1}s^{-1}$. The pascal (Pa) is also acceptable. In British Imperial units, the units are foot, pound, second. Many process instruments are still to be found with these units. *See* F.P.S. ENGINEERING UNITS.

furnace A device used for the controlled combustion of fuel with air for the heating of solids, liquids, or gases. Furnaces operate at high temperatures and are used for melting metals, cracking hydrocarbons, and promoting strongly endothermic reactions.

fusel oil A liquid by-product arising from the fermentation of yeast and separated by distillation. It comprises a mixture of high molecular weight alcohols and also contains esters and fatty acids. It is sold and used by the paints, perfumes, and plastics industries.

fusion 1. The process of changing from the solid state to liquid state, as in melting, by the application and absorption of heat. The latent heat of fusion is the quantity of heat required to change a mass of solid to a liquid. The latent heat of fusion of changing ice to water is 333 kJ kg^{-1}. **2.** The combining of atomic nuclei of lighter elements to form nuclei of a heavier element.

fuzzy logic A form of logic used in the computer control of a process that allows for a degree of impression. In contrast to using truth values of 'true' or 'false', fuzzy logic uses terms such as 'fairly true' or 'more or less true'. These are then represented numerically within the range of 0 to 1 depending on the level of truth.

F-value The time in minutes required to bring about an acceptable level of sterilization of foods in a thermal treatment process such as canning. The F-value often has a subscript, which denotes the temperature at which it applied. For example, F_{121} is the time in minutes at 121°C.

Galilei, Galileo (1564–1642) An Italian scientist noted for his experiments and studies on gases, dynamics, and temperature measurement. He studied medicine at the University of Pisa aged 17. At the cathedral in Pisa, he timed the swinging lamp with his pulse and found that the oscillations took equal times. This discovery led to the development of the pendulum clock. He became professor of mathematics at Padua and studied dynamics. It is said that he demonstrated from the Leaning Tower of Pisa that bodies of different weights fell with the same acceleration and also that the path of a projectile is a parabola. He also showed that air has weight by weighing a vessel under ordinary conditions and then filling it by means of a pump with compressed air. He carried out experiments on inclined planes to test his theory of falling bodies. He also made the first thermoscope or temperature detector based on the expansion of air with rise in temperature. He was one of the first scientists to make a practical telescope and discovered the four largest moons of Jupiter in 1610, which are known as the Galilean moons. The Galilean telescope has a convex lens for objective and a concave lens for eye-lens and gives an upright image. His hand-made instruments were in demand all over Europe. He was later persecuted by the Church for supporting Copernicus' theory that the sun is the centre of the solar system.

Galileo number A dimensionless number, Ga, used in the study of viscous liquids. It is the relationship between the force due to molecular friction and the force of gravity in a flowing viscous fluid:

$$Ga = \frac{l^3 g \rho^2}{\mu^2}$$

where l is the characteristic dimension of length, g is the gravitational acceleration, and ρ and μ are the density and viscosity of the viscous liquid. Ga is related to the *Grashof number:

$$Gr = \beta \Delta T \; Ga$$

where β is the coefficient of bulk expansion $(1/T)$ and ΔT is the temperature difference. It is used in calculations involving thermal expansion such as in the design of heat exchangers. The Galileo number is also referred to as the **Galilei number** and is named after the Italian scientist Galileo *Galilei (1564–1642).

gallon A British Imperial unit of volume defined as the volume occupied by exactly ten pounds of distilled water of density 0.998 859 grams per millilitre in air of density 0.001 217 grams per millilitre. One gallon is therefore equal to 4.546 09 litres. In the US Customary system, one gallon is equal to 0.832 68 Imperial gallons or 3.785 44 litres. It is the volume occupied by 8.3359 pounds of distilled water and owes its origin to the Winchester or wine gallon.

galvanic cell *See* VOLTAIC CELL.

galvanization A metallurgical process in which zinc is deposited onto steel to provide galvanic protection, abrasion resistance, and resistance to corrosion. The method involves either a hot metal bath in which the steel object is immersed in the molten zinc providing a layer of zinc up to 150 μm thick, or the zinc is electrochemically deposited to form a layer up to 30 μm thick.

gamma radiation An energetic, short-wave *electromagnetic radiation. The waves are emitted as photons by the nuclei of radioactive atoms and have radiation levels that vary in energy between 10^{-15} and 10^{-10} joules corresponding to a wavelength of 10^{-10} to 10^{-14} m, and are highly penetrating. They are stopped by materials with a high density such as lead, concrete, and steel.

gangue The undesired minerals that are associated with ore. Gangue is mostly non-metallic and is separated as tailings. It has little commercial value.

Gantt chart A chart used in project management to provide an easy-to-interpret pictorial representation of the progress of a project. It also shows all the critical activities, which determine the overall timescale of the project. Gantt charts can also be used to monitor the progress of a project by signing off activities as and when they are completed. Major engineering projects may have a hierarchy of Gantt charts that provide a general overview and more detailed analysis of individual parts of the project.

gas A physical state of matter in which the molecules of a substance are free to move and which has neither a definite shape nor volume. An *ideal gas behaves as if the molecules occupy a negligible volume and that the collisions between the molecules are perfectly elastic. A *real gas allows for these deviations from ideal behaviour since the molecules themselves occupy a volume and there is a small amount of attractive force between them.

gas centrifuge process A separation process used to separate radioactive isotopes of uranium in which heavy atoms are separated from the lighter atoms by centrifugal forces. The separation is dependent on the difference in mass between the isotopes being separated. The uranium is prepared as uranium hexafluoride gas (UF_6) in which the lighter uranium-235 is separated from the heavier uranium-238. *See* ENRICHMENT.

gas constant *See* UNIVERSAL GAS CONSTANT.

gas formation volume factor The volume of gas within an underground reservoir at the reservoir conditions of temperature and pressure divided by the volume occupied by the gas at standard conditions of 298 K and 101,325 Pa. The factor is used to convert surface measured volumes to reservoir conditions.

gasification A process used for the production of *fuel gas from liquid or solid hydro-carbons. Examples include the production of *water gas, which is a mixture of carbon dioxide and hydrogen by the reaction of coke or coal with steam. *Producer gas, which is a mixture of carbon monoxide and nitrogen, can be produced from passing air over heated coke or coal. *Synthesis gas, which is a mixture of carbon monoxide and hydrogen, can be produced from liquid petroleum fuels.

gas laws The laws related to the temperature, pressure, and volume of an ideal gas. Boyle's law states that the volume of a fixed mass of gas is inversely proportional to its pressure at constant temperature and was proposed by Robert *Boyle (1627–91) in 1662. Charles's law states that the volume of a given mass of gas is directly proportional to its absolute temperature at constant pressure and was proposed by Jacques *Charles (1746–1823). The combination of Boyle's law with Charles's law for an ideal gas leads to the universal gas equation $pV = nRT$. The gas laws were established experimentally. However, the laws are only applicable to real gases at low pressures and high temperatures. Equations of state have been developed to predict the actual behaviour of real gases.

gas lift A method for transporting liquids in which compressed gas or air is introduced to the bottom of a long vertical and open leg containing a liquid to be transported. By reducing the overall density in the leg relative to the density of the liquid around the leg, a flow is induced up the leg. At the top of the leg, the liquid and gas are disengaged. It is used to raise oil from wells and also used in *air-lift reactors.

gasohol A liquid fuel comprising a blend of alcohol and gasoline. Ethanol-based gasohol is a blend of typically 10 per cent ethanol and 90 per cent gasoline in which the ethanol is produced from the fermentation of agricultural crops such as maize or sugar cane. Methanol and wood alcohol are also used in other blends. The fuel is used in certain types of combustion engines and noted for its high octane and anti-knock properties.

gas oil An oil refinery distillation stream in which the molecular weights and boiling points are higher than heavy naphtha (205°C). It was once added to city gas supplies to make it burn with a luminous flame, and hence its name. It is used for industrial heating, by off-road diesel vehicles and machinery, as a marine fuel, and as a fuel for rail locomotives.

gas scrubbing *See* SCRUBBING.

gas stripping *See* STRIPPING.

gas sweetening A process used to remove hydrogen sulphide and mercaptans from natural gas. Commonly used in petroleum refineries, the gas treatment uses amine solutions such as monoethanolamine. The process uses an absorber unit and a regenerator. The amine solution flows down the scrubber and absorbs the hydrogen sulphide as well as carbon dioxide from the upflowing gases. The regenerator is used to strip the amine solution of the gases for reuse. It is known as gas sweetening as the foul smell is removed from the gas.

gas turbine An engine that uses internal combustion to convert the chemical energy of a fuel into mechanical energy and electrical energy. It uses air, which is compressed by a rotary compressor driven by the turbine, and fed into a combustion chamber where it is mixed with the fuel, such as kerosene. The air and fuel are burnt under constant pressure conditions. The combustion gases are expanded through the turbine causing the blades on the shaft to rotate. This is then converted to electrical energy. Gas turbines are used in the process industries and on offshore gas platforms for electrical generation.

gas void fraction The fraction of the gas in a multiphase flow system. Knowledge of the gas void fraction is needed in identifying the type of flow regime and flow behaviour of multiphase flow mixture. It is also needed in the calculation of pressure drop and prediction of fluid residence time for pumping requirements and understand the changes in thermal properties along long pipes such as sub-sea pipelines.

Experimentally, the gas void fraction can be measured by quick-closing valves, capacitance, and optical probes, devices based on X-ray and gamma-ray attenuation, and local electrical conductivity. However, different techniques tend to give different values of the gas void fraction for the same flow. There is, therefore, an inherent uncertainty in the published data on gas void fraction.

gate valve A widely used device that regulates the flow in a pipe. It consists of a vertical moving section across the flow area. They are useful for on–off type flow control operations and provide a low pressure drop when fully open.

Gauckler–Manning formula *See* MANNING FORMULA.

gauge An instrument used to give a reading or a value.

gauge pressure The pressure of a system or fluid measured above the local atmospheric pressure. Atmospheric pressure is a variable quantity and standard atmospheric pressure is taken to be 101 325 N m^{-2}.

Gaussian distribution *See* NORMAL DISTRIBUTION.

Gay-Lussac, Joseph Louis (1778–1850) A French chemist and physicist noted for his two laws on gases and for his work on alcohol-water mixtures. He was professor of physics at the Sorbonne and later took a chair of chemistry at the Jardin des Plantes. Together with Jean-Baptiste *Biot (1774–1862), he made the first-ever hot-air balloon ascent for scientific purposes in 1804, reaching an altitude of over 6 km. He is also credited with recognizing iodine as a new element and suggested the name iodo, with the codiscovery of boron, and the terms 'pipette' and 'burette'.

Gay-Lussac's law A law stating that if the mass and pressure of a gas are held constant, then gas volume increases linearly as the temperature rises. When gases react, they do so in volumes that bear a simple ratio to one another and to the volume of the product, if it is a gas, temperature and pressure remaining constant. This is sometimes written as $V = kT$ where k is a constant dependent on the type, mass, and pressure of the gas, and T is the absolute temperature. For an ideal gas $k = nR/P$. It is named after *Gay-Lussac (1778–1850) who proposed the law in 1808, which led to *Avogadro's law.

Geiger, Hans Wilhelm (1882–1945) A German physicist with an interest in radioactivity and cosmic rays, and who worked with Ernest *Rutherford (1871–1937). He worked in Manchester and then in various centres in Germany. He is noted for the Geiger counter, an instrument he invented for detecting charged particles. Forms of the instrument are sufficiently sensitive to detect beta-particles (electrons) and gamma-radiation.

gel A pliable semi-solid mixture or viscous emulsion that solidifies or sets to a jelly upon cooling. Colloidal suspensions may convert from a gel to a sol by the addition of a solvent, increase in temperature, change in pressure, pH, or some other physical or chemical influence. The conversion may be reversible such as through *drying.

generator A machine used to produce electrical power. Electromagnetic generators are widely used to produce electrical power and are driven by water turbines, steam turbines, internal combustion engines, windmills, or other forms of mechanical energy. In power stations, generators produce alternating current (a.c. current) and are also known as **alternators**.

genetic engineering The general term for the artificial manipulation of the nucleic acids DNA and RNA to produce new, modified species of living cells. It involves the formation of new combinations of heritable material by the isolation of nucleic acid molecules by whatever means outside the living cell, into a virus or plasmid, and transferred into a host organism in which they do not naturally occur, but in which they are capable of continued propagation. It has been applied to microorganisms such as bacteria and yeast, and plants in which genes for antibiotic resistance can be transferred to the living organism, for example. It is also known as *recombinant DNA technology.

geometrically safe A term used to describe the geometry of an item of process plant used in the nuclear industry to contain fissile material. The geometry is designed in such a way that it is impossible for a nuclear chain reaction to accidently occur since the concentration of fissile material will always fall below that required for a self-perpetuating chain reaction. Examples include toroidal-shaped tanks and tubular-shaped tanks, known as *harp tanks.

geometric mean A type of mean value that is the average value of a set of n numbers, and calculated by multiplying them together, and extracting the nth root of the product. For example, the geometric mean of 10, 20, and 40 is $\sqrt[3]{10\times20\times40}=20$ and is less than the arithmetic mean, which is $(10+20+40)/3=23.33$.

geometric similarity Systems that have similar physical dimensions such that their ratio is constant. This is useful in the scale-up of process equipment.

geometric view factors The ratio of thermal radiation leaving a grey surface that is absorbed by another surface, to the total radiation leaving the surface were it to be a *black body. Charts are used for complex surface orientations such as the thermal radiation between adjacent rectangular surfaces in perpendicular or parallel planes.

geothermal energy A natural and renewable source of thermal energy extracted from regions of volcanic activity such as fumaroles, hot springs, and geysers. The heat is extracted by drilling a borehole down into the high-temperature porous rock, which ranges from just below the surface of the Earth in some places, to several kilometres down. Hot water and steam is either brought to the surface and used directly, or water is sent down where it is heated by conduction and returned to the surface. Geothermal energy is used for domestic or district heating, industrial processes, or for power generation in turbogenerators. Countries that use significant amounts of geothermal energy include the US, the Philippines, Iceland, and New Zealand. The district heating system in Reykjavik, Iceland provides heat for 95 per cent of the buildings.

Gibbs, Josiah Willard (1839–1903) An American mathematician and physicist noted for his work on thermodynamics. A professor at Yale University, he developed the theory of chemical thermodynamics and devised the concepts of *Gibbs free energy and *Gibbs' phase rule. He was also a founder of *statistical mechanics and was responsible for the introduction of *vector notation.

Gibbs free energy (Symbol G) A measure of the maximum useful work in a system that can be obtained, defined by:

$$G = H - TS$$

where T is the absolute temperature, H is the enthalpy, and S is the entropy. A quantitative measure as to how near or far a potential reaction is from occurring can be determined from the change in Gibbs free energy ΔG. Negative values of ΔG indicate that change can occur spontaneously whereas positive values indicate that energy is required to be supplied for the reaction to occur. It is also known as **Gibbs energy** or **Gibbs function** after American physicist Josiah Willard *Gibbs (1839–1903).The Helmholtz free energy is defined as $F = U - TS$ where U is the internal energy. For a reversible isothermal process, the change Helmholtz free energy represents the useful work available. It is named after German physicist and physiologist Hermann von Helmholtz (1821–94).

Gibbs' phase rule A method used to calculate the variance of a heterogeneous equilibrium of a non-reactive system or process, and to establish the number of independent variables. It is calculated from:

$$F = C - \Pi + 2$$

where F is the number of thermodynamic *degrees of freedom, C is the number of components, and Π is the number of phases. Detail of phase equilibria is important in mass transfer and problems are analysed in terms of the phase rule. If insufficient degrees of freedom are fixed, then the problem will be under-specified. If too many are chosen, the problem will be over-specified. For problems involving two components with two phases such as in binary distillation, then $F = 2$. If the pressure is fixed then only one variable remains that can be changed independently, such as temperature. The concentration of the liquid and vapour phases are then fixed. The phase rule does not apply to systems involving chemical reactions. It was deduced by American physicist Josiah Willard *Gibbs (1839–1903). A modified Gibbs' phase rule is used for reacting species:

$$F = n - \Pi + 2 - r$$

where n is the number of components and r is the number of independent chemical reactions. For example, in a gas reaction involving three components in which only two react (e.g. fuel with air), then there are three independent variables such as T, P, and y_1, which fixes y_2 and y_3.

giga- (Symbol G) A prefix denoting 10^9. For example, the output from a power station may be 1 GW (i.e. 10^9 watts).

Gilliland equation An empirical equation used to estimate the number of stages required in a distillation column. It uses the minimum number of stages as well as minimum and actual reflux ratios in the calculation. It is named after American chemical engineer Edwin Gilliland (1909–73) who was professor of chemical engineering at MIT.

Gilliland–Sherwood correlation A dimensionless equation used to determine the mass transfer in gas absorption and relates the Sherwood number, Reynolds number, and Schmidt number:

$$Sh = 0.023 \, Re^{0.83} \, Sc^{0.44}$$

It was developed in 1934 based on experimental data from wetted wall columns and is valid for Reynolds numbers between 2,000 and 35,000 and Schmidt numbers between 0.6 and 2.5.

global warming potential (GWP) The overall assessment of the effect on the climate by the release of one kilogram of a gas to carbon dioxide, which is used as the reference. A GWP of a so-called greenhouse gas is dependent on its lifetime within the troposphere that surrounds the Earth and the radiative efficiency, which is the amount of infrared radiation that the gas can trap and is dependent on the frequency at which infrared radiation is absorbed. The GWP of carbon dioxide is 1, whereas the GWP of methane is 25. The GWPs of CFC-11 and CFC-13 are 45 and 640, respectively. Their manufacture and use is no longer permitted. *See* GREENHOUSE EFFECT.

globe valve A device used to regulate the flow of a fluid in a pipe and consists of a flat disc that sits on a fixed ring seat. The disc is movable and allows flow through the valve.

glove box A gas-tight box used to handle either harmful substances or substances that may react with air. The box has a transparent window for viewing and full-length gloves used for the operator to handle the substances inside without direct contact. A *negative pressure is applied to the glove box to ensure that there are no inadvertent leaks or emissions. They are typically used in the nuclear and biomedical industries for handling radioactive substances and biological agents. The glove box can be filled with another gas such as an inert gas to exclude oxygen from the air and used to handle substances that may otherwise react with it. The glove box is therefore maintained at a pressure greater than atmospheric.

glowing combustion The thermal oxidation of solid material with light emission but without a visible flame.

GOLDOX A process that extracts gold from ore by the injection of oxygen into a cyanide solution; the name is an abbreviation of **gold ox**idation.

GOR The ratio of produced gas to produced oil from a well or reservoir. It is an abbreviation of **g**as to **o**il **r**atio.

GOSP An abbreviation for a **g**as **o**il **s**eparation **p**lant.

gradient The slope of a line on a *graph. In Cartesian coordinates, the general equation of a straight line is $y = mx + c$ where m is the gradient or slope and c is the intercept on the y-axis. The gradient at some point on a curve can be found from the derivative dy/dx at that point. The gradient is therefore the tangent to the curve.

Graetz number A dimensionless number, Gz. Used in fluid dynamics, it is used to characterize heat transfer by convection compared with the heat transfer by conduction in ducts as:

$$Gz = \frac{c_p \rho Q}{kL}$$

where c_p is the heat capacity, ρ is the density, Q is the flow rate, k is the thermal conductivity, and L is the length. It is named after German physicist Leo Graetz (1856–1941).

Graham, Thomas (1805–69) A Scottish chemist who was professor of chemistry at Glasgow University in 1830 and later moved to University College London in 1837. He is noted for his law for the diffusion of gases and as founder of the science of colloids based on his work between 1851 and 1861.

Graham's law of diffusion A law that states that the rates of diffusion of gases are inversely proportional to the square root of their densities under the same conditions. This was first proposed by Scottish chemist Thomas *Graham (1805–69) in 1829.

gram (Symbol g) A *fundamental unit of mass equal to one thousandth of a kilogram. It was formerly used within the *c.g.s. system and included units such as the **gram-molecule,** which has now been replaced by the *mole.

granulation A process used in the pharmaceutical industry to prepare powders for the production of tablets. The granulation process involves combining one or more powders to create bonds between the particles of the powder. These are formed by compression or by using a binding agent to produce a tablet with good hardness and a consistent quality. The machines used for granulation are called **granulators** that apply the necessary shear to combine the powders and the binding solution. The fluid-bed granulator uses air to elevate the powders while the binding solution is sprayed onto the particles.

graph A diagram that presents the relationship between numbers or quantities. A graph is normally drawn with coordinate axes at right angles. They are used to display and illustrate the geometric relationship between data. The dependent variable is plotted on the x-axis while the independent variable is plotted on the y-axis.

Grashof number A dimensionless number, Gr. It is used in the study of natural convection and represents the ratio of buoyancy forces to viscous forces:

$$Gr = \frac{\rho^2 g \beta l^3 \Delta T}{\mu^2}$$

where ρ is the density, g is the local acceleration due to gravity, β is the volume coefficient of expansion, l is the characteristic length, ΔT is the temperature difference, and μ is the viscosity. The product of the Grashof number and the *Prandtl number characterizes the convective heat transfer. It can be applied to laminar and turbulent regimes and for vertical and horizontal planes and cylinders, such as pipes. An analogous form of the Grashof number used in the study of mass transfer by natural convection is:

$$Gr = \frac{g \Delta \rho L^3}{\rho v^2}$$

where g is the acceleration due to gravity, $\Delta \rho$ is the density difference between two points, L is the characteristic length, and v is the kinematic viscosity of the fluid. It is named after German engineer Franz Grashof (1826–93).

gravitational acceleration (Symbol g) The attractive force towards the centre of the Earth that causes an acceleration of a falling body. That is, there is a change in velocity with time. Within a vacuum, the gravitational acceleration is equal to 9.806 65 m s^{-2}. The variation of gravity over the Earth's surface is negligible for most engineering work. *See* GRAVITY.

gravity The force that pulls a body towards the centre of the Earth. According to Newton's second law of motion, the weight of an object is the product of its mass and *gravitational acceleration $F = mg$. It varies with latitude and elevation above sea level. For precise calculations, the *gravitational acceleration is taken to equal to 9.806 65 m s^{-2}.

Suspensions of small particles can be separated from solutions by the force of gravity as in the process of *sedimentation. The process, however, tends to be slow. Centrifugation is a process used to apply a far higher 'gravitational' force to the particles, increasing the rate of separation.

gravity separation A process in which immiscible phases or particles in a liquid separate due to the influence of gravity. In the separation, the more dense phase or particles settle out. It is used in numerous chemical and mining processes.

gray (Symbol Gy) An SI derived unit for the absorbed radiation dose of ionizing radiation defined as the absorption of one joule of ionizing radiation by one kilogram of matter. It is named after British physicist Louise Harold Gray (1905–65) and replaces the c.g.s. unit of rad where 1 rad = 0.01 Gy.

green chemistry A term used to describe the steady move towards the development of environmentally acceptable chemical processes and products. It was first coined by the US Environmental Protection Agency in the 1990s and focuses on influencing education, research, and industrial practice.

greenhouse effect A phenomenon in which the Earth's atmosphere and surface is steadily heating up. It is caused by the ability of certain gases and particles in the Earth's atmosphere to trap infrared radiation from the sun reflected back from the Earth's surface more effectively than nitrogen and oxygen in air. The principal **greenhouse gases** are water vapour, carbon dioxide, methane, nitrous oxides, chlorofluorocarbons (CFCs), and ozone. Fossil fuel combustion is the main cause of carbon dioxide release, while methane is a by-product of agriculture and landfills. CFCs are even more potent but their use has been reduced in recent times.

The levels of carbon dioxide and other greenhouses gases have been steadily increasing over the past century. It is thought that the level of carbon dioxide is due to industrialization and the burning of fossil fuels has led to the greenhouse effect, contributing to global warming. *See* GLOBAL WARMING POTENTIAL.

grey body A body that does not absorb all the incident thermal radiation and emits radiation that is less than that of a *black body. It has a constant emissivity over all wavelengths and temperatures, and has a value less than 1.0.

grid zone The location in a *fluidized bed in which the fluidizing gas flow is sufficient to form bubbles. The grid zone is located at the bottom of the bed, corresponds to the gas penetrating the bed, and is dependent on the types of grid used.

grinding The process of breaking up particles into more particles or fragments. *Ball mills are typically used, which consist of a slowly rotating vessel on its horizontal axis containing material to be ground and steel or ceramic balls. Due to the cascading and tumbling action of the balls, the material is ground to a smaller particle size distribution.

gross drying rate The loss of moisture from a substance per unit time in a drying process but ignores the change in drying area.

gross economic potential The difference in the economic value of manufactured products and the value of the feed materials. Where this is positive, there is added-value in the process. Where it is zero or negative, there is no economic gain or business value from the process. It does not take into account the cost of equipment or operating costs.

growth curve Used in biochemical engineering to describe and illustrate the evolution of biomass over time. These are plotted to illustrate the change of microbial cell concentration, either as dry weight or total population, over a period of time. Features of the batch cultivation of microorganisms show a lag phase in which the microorganisms adjust to the growth medium, an *exponential growth phase as they rapidly multiply, a **quiescence phase,** and a death phase. A second growth curve may occur, known as **diauxic** or bi-phasic growth in which the microorganisms consume another nutrient, which may be an excreted product from the first phase such as ethanol. *See* LIMITING SUBSTRATE.

growth medium A liquid or a gel used to support the growth of a plant, animal, or microbial cell culture. Also known as a **culture medium**, it contains all the necessary nutrients to sustain growth. An undefined medium, also known as a **complex medium**, contains a rich source of amino acids, nitrogen, vitamins, and trace elements needed for growth, such as yeast extract. It is non-selective in that all species of microorganisms if present would be capable of growth. A defined medium, also known as **synthetic medium**, has a composition in which all the chemicals used are known. A **minimal medium** contains all the nutrients required for growth except for certain amino acids. They are used to select for or against recombinant DNA microorganisms. *See* SELECTIVE MEDIUM.

guess and check A problem-solving method used to obtain a solution by using conjecture to obtain the answer and then checking that it fits the conditions of the problem. The method is useful when there is no knowledge or information within a problem to reach a solution by alternative means. It is a widely used method and very useful for solving differential equations. *Compare* TRIAL AND ERROR.

guestimate A reasoned estimate that is made to solve a problem with insufficient data or information. It is a corruption of the words 'guess' and 'estimate', and is usually used for solving complex problems as a first trial in an iterative approach used in reaching a solution.

guideword A list of words used in a *HAZOP (hazard and operability) study to identify undesired deviations from the normal operating conditions that may present a hazard. It includes the words *none, more of, less of, part of, more than*, and *other*. In each case the word is applied to the effect of flow, temperature, and pressure.

Guldberg rule A rule stating that the ratio of the normal boiling point to the critical temperature of a substance is equal to a value of 2/3. It was proposed in 1890 by Norwegian chemist Cato Maximilian Guldberg (1836–1902). For simple molecules containing one or two atoms and discounting hydrogen atoms present, the ratio is 0.567, while for other inorganic molecules the ratio is 0.635.

Gurney–Lurie charts Charts used for determining thermal diffusion in materials. First published in 1923 by H. P. Gurney and J. Lurie, they plot the *Fourier number on the x-axis and a dimensionless temperature on the y-axis. The charts can be used to determine the temperature within materials that have a defined geometry such as slabs, cylinders, cubes, and spheres.

Haber, Fritz (1868–1934) A German chemist who, working with Carl Bosch (1874–1940), developed the *Haber process in 1908 in which ammonia was synthesized in the catalytic high-temperature, high-pressure reaction of hydrogen and nitrogen. He received the Nobel Prize in Chemistry in 1918 for this work. Being born a Jew, he left Germany in 1933 and worked at the Cavendish Laboratory in Cambridge.

Haber process A major industrial process for the synthesis of ammonia, which is principally used in the manufacture of fertilizers. The process involves passing a mixture of nitrogen and hydrogen at a volumetric ratio of 1:3 under high pressure in the presence of a heated catalyst: $N_2 + 3H_2 \rightarrow 2NH_3$. With a contraction of volume and being exothermic, *Le Chatelier's principle indicates that the yield of ammonia will be increased by pressure and decreased by a rise of temperature. However, a relatively high temperature is used otherwise the reaction is too slow. Most industrial plants vary in operation but, typically, finely divided iron catalysts are used with a temperature of 550°C and pressure of 250 atmospheres. The hydrogen is obtained from the *steam reforming of natural gas. The process was developed by Fritz *Haber (1868–1934) and Carl Bosch (1874–1940) in 1908, and is also known as the **Haber–Bosch process.** *See also* SYNTHESIS GAS.

Hagen, Gotthilf Heinrich Ludwig (1797–1884) A German physicist noted for his contribution to engineering particularly in the field of hydraulics. Qualified in civil engineering and mathematics, he worked as a civil engineer managing various engineering projects before turning to teaching in Berlin. He is credited with independently developing the *Hagen–Poiseuille equation.

Hagen–Poiseuille equation An equation used to determine the rate of flow of a Newtonian fluid with laminar flow expressed as:

$$Q = \frac{\pi}{8\mu} \frac{\Delta p}{L} R^4$$

where μ is the viscosity of the fluid, $\Delta p/L$ is the pressure drop per unit length of pipe, and R is the pipe inner radius. It is named after G. *Hagen and J. *Poiseuille who independently derived this equation in 1839 and 1840, respectively.

half-cell An electrode immersed in a solution of ions that forms part of a cell. *See* ELECTRODE POTENTIAL.

half-life The time taken for a substance to fall to half of its original value and is independent of the amount of starting material. A reaction that has a constant half-life is a first-order reaction. *Radioactive decay is an example of a first-order reaction. The half-life of a radionuclide is the time for half the nuclei to disintegrate. The time varies between radionuclides ranging from fractions of a second to millions of years. For example, the half-life of beryllium-16 is $2.0x^{-16}$ seconds and tellurium-128 is $7.2x10^{24}$ years. In biological systems, the half-life is the time taken to naturally excrete half of an absorbed substance from a body or an organ.

Hall–Héroult smelting process A continuous electrolytic process used to produce aluminium from alumina. The alumina is dissolved in a bath of sodium aluminium chloride called cryolite that contains alumina fluoride and calcium fluoride. The solution is heated to 950°C in a steel tank with a carbon liner. Carbon anodes are lowered into the solution with the liner being the cathode. Electrolytic action separates the alumina into liquid aluminium, which collects at the cathode, and oxygen at the anode, which combines with the carbon to form carbon dioxide gas. It was invented in 1886 simultaneously and independently by American chemist Charles M. Hall (1863–1914) and French scientist Paul L. T. Héroult (1863–1914).

halogenation A general name for a chemical reaction that involves the introduction of a halogen atom into a compound. The name of the specific reaction is named after the halogen such as bromination, chlorination, and fluorination. The halogenation of aromatics such as benzene involves electrophilic substitution. The halogenation of alkanes involves free radicals and high temperatures.

hammer mill A mechanical device used to shred solid materials such as grains and sugar cane in order to aid extraction of starches and sucrose. It consists of a rotating shaft upon which hinged hammers are fixed. The shaft is contained within a shell through which solid material is fed and emerges from the other end in a shredded form.

hardness The resistance of a material or substance to crushing, indenting, or abrading. Hardness is measured on the *Moh scale of hardness of 1 to 10, representing soft (talc) to very hard (diamond). Hardness is also a measure of the applied force with time.

harp tank A geometrically safe tank used to store radioactive liquids that are capable of fission if the emitting neutron concentration reaches a critical condition. The tanks consist of long slender tubes and join in a manifold at either end much like the strings of a harp. They are used in *nuclear fuel reprocessing.

hastelloy A widely used alloy of nickel, molybdenum, and chromium used for process equipment. It provides good resistance to wet chlorine, hypochlorite bleach, ferric chloride, and nitric acid.

HAZAN An abbreviation for **haz**ard **an**alysis, it is a technique used to assess the probability of a hazard occurring and for determining the subsequent consequences.

hazard A biological, chemical, or physical agent or situation that is reasonably likely, or has the potential, to cause illness or injury to humans, damage to property, damage to the environment, or some combination of these if it is not controlled. The product of the consequence and frequency of that situation occurring, hazards are associated with fire and explosion, pollution, chemical reactions, toxicity, mechanical failure, corrosion, and nuclear radiation. Hazards may arise due to the relaxation of management control, fatigue, carelessness, boredom and complacency, a loss or changes in operational knowledge, ageing and poorly maintained equipment, design modifications, and the abuse of trips. Hazards can be controlled through elimination, control of containment, and the controlled reduction in the likelihood or frequency and effect. They may be reduced through design standards, control of work, inspection, maintenance, safety reviews, control of ignition sources, use of detection devices, fireproofing, and effective operational and emergency response procedures.

hazard analysis The identification of undesired events that result in a hazard. The analysis seeks to evaluate the likelihood of the undesired events and the harmful consequences of the hazards. A *HAZID and a *HAZOP are both forms of hazard analysis.

hazardous energy Any form of energy in a form that can cause harm to a human. It includes ionizing and thermal radiation, electrical, electromagnetic, mechanical, and chemical energy. A **hazardous substance** is a chemical material which, due to its chemical properties, constitutes a hazard. It may realize its potential through fire, explosion, toxic, or corrosive effects.

HAZID An abbreviation for **haz**ard **id**entification, it is a systematic and wide-ranging structured hazard analysis technique used to identify the potential hazards at an early stage in process design and development. It is applied once details of process materials and flows presented in a *flowsheet have been quantified. HAZID considers both internal and external events such as weather effects, and seeks to broaden the hazard understanding of those involved by encouraging lateral thinking. Whereas *HAZOP is cause-driven, HAZID is consequence-driven. HAZOP can accept a conclusion that an event will not occur, while HAZID assumes that an event will occur and consequently requires a rigorous analysis of the sequence of events for it to occur.

HAZOP An abbreviation for **haz**ard and **op**erability, it is a systematic and structured hazard evaluation technique used to identify the potential failures of equipment or plant systems that may otherwise become hazards and present potential operating problems. The aim is to eliminate or minimize the probability of an incident from occurring and the severity of consequence arising from that incident. It uses a multidisciplinary team-based approach to consider what can go wrong, the causes, consequences, frequency of occurrence, measures for prevention, and justification of the associated costs of prevention. The study uses *guidewords for the possible deviations from an intended design or operation of a process. These include *none, more of, less of, part of, more than*, and *other*, and in each case is applied to the effect of flow, temperature, and pressure. It also includes the effects of impurities, changes in phase and composition, extra phases being present, and activities that can happen, as well as normal operation such as maintenance, purging, access, start-up, shutdown, alternative operation mode, and failure of plant services. The HAZOP team works systematically on a particular point in the process and considers a particular parameter such as flow, pressure, temperature, level, heat, and deviation. The causes and consequences are recorded along with any required action to improve the safety or operability of the process. On completion, the next guideword at the same node is considered until all the nodes with their parameters and deviations have been considered. Finally, the costs of recommended changes to the design or plant are reviewed and justified.

head The pressure of a liquid expressed as the equivalent height that a column of the liquid would exert. The head is related to pressure, p, density, ρ, and gravitational acceleration, g, as:

$$h = \frac{p}{\rho g}$$

See HYDROSTATIC HEAD.

head coefficient A dimensionless group used in the scale-up of centrifugal pumps:

$$C_H = \frac{gH}{N^2 D^2}$$

where H is the head developed, g is the gravitational acceleration, and N and D are the rotational speed and diameter of the impeller.

head end The first process step used in *nuclear fuel reprocessing. It involves all the process stages of mechanical sectioning of fuel elements in preparation for chemical extraction. This includes feeding the fuel elements into a sectioning machine that cuts the fuel element bundles into roughly 5cm-long pieces. The rod pieces then fall into concentrated nitric acid to dissolve the uranium, plutonium, and fission products. The cladding material of the fuel rods is resistant to the acid. *Compare* TAIL END.

header 1. A large tank, reservoir, or hopper used to maintain a gravity feed to a process or to provide a static fluid pressure to a process or item of equipment. It is located above the process and feeds down into the process or equipment under the influence of gravity. **2.** A manifold pipe connected to smaller pipes.

headspace The space above a liquid in a vessel filled with gas or vapour. A space is usually left in sealed liquid storage tanks to allow for safe thermal expansion of the liquid.

health physics A branch of physics that is concerned with the study of the hazards of ionizing radiation and other aspects of atomic physics, and its harmful effects on human life. It has particular application in the protection of industrial workers in the nuclear power and *nuclear fuel reprocessing industries, as well as medical workers and researchers in related fields. It considers monitoring and establishing safe and permissible levels of radiation dosage, the disposal of radioactive contaminated waste, and measures for shielding against radiation exposure. *See* DOSE.

heat A form of thermal energy transferred from one body to another driven by a difference in temperature between the two bodies. The thermal energy of a body in thermodynamic equilibrium with its surroundings involves the internal movement of molecules and atoms. That is, it has internal energy made up of the kinetic and potential energies of the atoms and molecules. When a body changes its temperature, its internal energy changes in which heat is absorbed from the body and work is done simultaneously on the surroundings.

heat balance *See* ENERGY BALANCE.

heat capacity (Symbol C) The amount of heat Q transferred to a substance resulting in a change in temperature ΔT as $C = Q / \Delta T$. The SI units are $J\,K^{-1}$.

heat engine A machine used to convert heat into work. The heat is the result of combustion of a fuel. In an internal combustion engine, fuel is burnt inside the engine. A steam turbine is an example of an external combustion engine in which steam is raised outside the engine. Some of the steam's internal energy is then used to do work inside the engine. Heat engines operate as a cycle. The most efficient is the *Carnot cycle. However, this is an idealized cycle and not realized in practice.

heat exchanger A device used to transfer heat from one fluid to another without the two streams coming into contact with one another (see Fig. 23). Temperature is the *driving force for heat transfer to take place in which one of the fluids is either heated or cooled.

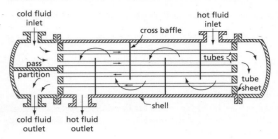

Fig. 23

There are many types used in the chemical and process industries. Examples include condensers, kettle reboilers, evaporators, shell and tube heat exchangers, and plate heat exchangers. The choice of heat exchanger depends on the heating or cooling requirement, the thermophysical properties of the fluids, and economics.

heat exchanger network (HEN) A technique based on *pinch analysis to identify the minimum requirements of energy, heat transfer area, and the number of units for a given process at the pinch point. The technique is applied in advance of a detailed design to determine the financial saving through process heat integration in maximizing heat recovery and reducing external utility loads. A network of heat exchangers is optimized through minimization of the total annual cost of energy costs and capital costs.

heat flux The transfer of heat energy from one place to another per unit time per unit cross-sectional area over which heat transfer takes place. Heat transfer calculations are based on the area of heating surface. In the case of tubes or pipes, heat fluxes may be based with reference to either the inside or the outside areas. While the choice is arbitrary, it must be stated as the magnitude of the heat flux is not the same for both. The SI units are $W\,m^{-2}$.

heating medium A fluid used to transport heat. It may be a liquid, a vapour, or a mixture of the two. Oil, water, and steam are often used as the heating media in many processes.

heat loss The unrecoverable dissipation of thermal energy that flows from a body to a place of lower temperature. The amount of heat flow, or loss, is dependent on the temperature difference of the body to the environment at the lower temperature, the exposed area, and the surface heat transfer coefficient. Heat is lost through the walls of pipes that transport hot fluids. The heat loss can be reduced by surrounding the pipe with an insulating material that has a low thermal conductivity. *See* INSULATION.

heat of absorption The heat released when a substance is absorbed into another substance at constant pressure.

heat of adsorption The heat released by a gas when it is adsorbed onto the surface of a solid material at constant temperature.

heat of atomization The heat required to atomize one mole of a substance into atoms.

heat of combustion The heat of reaction for the complete combustion of a mole or unit mass of a fuel in oxygen at the standard state of 25°C and 101.3 kPa. For the combustion of a hydrocarbon, the carbon is converted to carbon dioxide and the hydrogen is converted to water vapour. For fuels that contain sulphur and nitrogen, the combustion products are sulphur dioxide and nitrogen. It is also known as the **standard enthalpy of combustion**. *See* CALORIFIC VALUE.

heat of condensation The heat released when a substance is converted from a vapour to a liquid and both have the same temperature.

heat of crystallization The heat absorbed when a substance crystallizes in a solution.

heat of evaporation see HEAT OF VAPORIZATION.

heat of formation The energy released or absorbed when the elements at the standard state of 298 K and 101.3 kPa units produce one mole of a compound. The heat of formation may provide evidence of the structure of the compound. Tables of values are available for compounds in the standard state. It is also known as the **standard enthalpy of formation**.

heat of fusion The energy required to melt a substance under isothermal conditions.

heat of mixing The energy required to form a mixture from its components under isothermal conditions.

heat of reaction The energy absorbed or evolved when one mole of product is formed by a chemical reaction at the standard state of 298 K and 101.3 kPa. It can be calculated from the difference in the sum of the *heats of formation of the products and reactants in a reaction carried out in stoichiometric proportion.

heat of solution The energy liberated or absorbed when one mole of a substance is completely dissolved in a solvent.

heat of vaporization The energy required to evaporate a liquid under isothermal conditions; also known as **heat of evaporation**.

heat pump A device used to transfer heat from a cold source to a place or reservoir at a higher temperature by expending mechanical energy, such as in *air conditioning. Unlike a *refrigeration cycle, the heat pump is used to raise the temperature of the reservoir and not to cool it. It operates with an adiabatically compressed gas as the working fluid such that its temperature rises so that heat is given out to the reservoir. The fluid then condenses to a liquid. It is then expanded into an evaporator where it absorbs heat converting it to a vapour again for adiabatic compression.

heat transfer The movement of thermal energy through a body or across a space as a result of the difference in temperature. The three modes of heat transfer are *conduction, *convection, and *radiation. Heat is generated in chemical processes from a number of sources, including *exothermic reactions, mechanical agitation, and adiabatic processes. Heat is removed by cooling water through cooling coils, external cooling jackets, and heat exchangers. Heat may need to be added to *endothermic reactions, which require heat to proceed. For example, high-temperature heat is supplied directly to cracking processes by the combustion of fuel in the form of thermal radiation to heat tubes within which the endothermic reaction takes place, such as in the cracking of dichloroethane to form vinyl chloride monomer and hydrogen chloride.

heat transfer coefficient *See* OVERALL HEAT TRANSFER COEFFICIENT.

heat treatment The process of applying heat to a material to bring about a physical, chemical, or biological change. It is used in the manufacture of glass and metallurgical processes, and can involve heating to very high temperatures as well as chilling to very low temperatures. The heat treatment processes commonly used include *annealing, case hardening, tempering, and quenching. The use of heat and time can modify the grain structure of metals leading to improved strength, machinability, and electrical conductance.

heavy chemicals Bulk chemicals that are manufactured and used in industry in which the production rate is of the order of thousands of tonnes per day. They are often used as the starting materials and feedstocks for other products, such as sulphuric acid that is used in the petroleum, steel, and agrochemical industries. Chemicals produced in large amounts include fertilizers, cement, paints, and plastics. *Compare* FINE CHEMICALS.

heavy key *See* KEY COMPONENTS.

heavy phase The liquid with the higher density in a liquid–liquid extraction process and sits below the *light phase.

heavy water Water in which the hydrogen atoms are replaced by deuterium oxide (D_2O). *Deuterium is a heavier isotope of hydrogen. Heavy water has slightly different properties than H_2O with a melting point of 3.8°C and boiling point of 101.4°C. It is found naturally in water in very small amounts and can be separated and concentrated by *fractional distillation or by *elecrolytic separation. It is used as a moderator in the nuclear industry due to its ability to reduce the energy of fast-moving neutrons.

hecto- (Symbol h) A prefix used to signify one hundred times. For example, a hectolitre- (hl) is a unit of volume equal to 100 litres.

height equivalent to a theoretical plate (HETP) The height of packing required in a distillation column or an absorption column that is able to provide the same change in composition in the liquid or vapour phase as one theoretical plate. It can be found from the total height of the packed bed divided by the height of a *theoretical stage. This is a device that has perfect contact between the liquid and vapour phases such that both streams leave in equilibrium. The HETP effectively divides the column into equal portions whereas the *height of a transfer unit (HTU) is a continuous function.

height of a transfer unit (HTU) A dimensionless number used as a measure of the performance of a process separation column or unit such as a distillation column, adsorption column, or cooling water tower. It relates the height of the column or packing that gives the same change in liquid or gas composition as one transfer unit. The height of a transfer unit is therefore calculated from the ratio of the height equivalent to a theoretical plate (HETP) to the number of transfer units (NTU).

Heisler charts A set of graphical plots used to determine the time taken for thermal penetration by heat conduction into a solid body by heating or cooling at its surface. The plots are prepared for standard geometric shapes such as slabs, cylinders, and spheres with the *Fourier number on the x-axis and a dimensionless temperature on the y-axis. The lines represent the reciprocal of the *Biot number. They are named after M. P. Heisler, who compiled them in 1947.

Helmholtz free energy *See* FREE ENERGY.

hemisphere The surface bounded by a sphere and a plane through the centre of the sphere. The ends of cylindrical pressure vessels and columns have domed or hemispherical ends in order to spread the applied stresses.

HEN *See* HEAT EXCHANGER NETWORK.

Henry's law A law that states the effect of change of pressure on the solubility of a gas in a solvent. The law takes the form that the mass of a given gas which saturates a given solvent is directly proportional to the pressure of the gas at equilibrium, provided that the temperature is constant and that the gas does not react chemically with the solvent:

$$p_A = Hx_A$$

where p_A is the partial pressure of component A, H is Henry's law constant, and x_A is the mole fraction of the component in the solvent. It is used for many gases to describe the relationship between the concentration of a gas dissolved in a liquid and the equilibrium partial pressure. It does not apply to the aqueous behaviour of very soluble gases, and slight inaccuracies also arise because gases do not obey Boyle's law exactly. The SI units for H are pascals per mole fraction. It was formulated in 1803 by English chemist William Henry (1774–1836).

HEPA filter A very efficient type of filter that is able to entrap very small particles. An abbreviation of **h**igh **e**fficiency **p**articulate **a**ir filter, it is required to trap in excess of 99.97 per cent of particles with a diameter of 0.3 microns. HEPA filters are made from high-density mats of glass fibres and used in processes where very clean conditions are required such as the nuclear and pharmaceutical industries.

hermetically sealed A term used to describe something that is fully sealed and without leaks. The compressor on a domestic refrigerator is hermetically sealed. The compressor typically has a steel outer shell that seals the vapour of the refrigerant into the system such that there are no leaks. The packaging of certain foods is hermetically sealed to ensure that no potentially contaminating air can enter and spoil the food.

Herschel–Bulkley fluid A type of fluid that exhibits non-Newtonian behaviour and can be described by a three parameter rheological model as:

$$\tau = \tau_o + k\gamma^n$$

where τ is the shear stress, τ_o is the yield stress, k is a constant, γ is the shear rate, and n is a power law exponent. Drilling fluids are examples.

hertz (Symbol Hz) The SI unit of frequency where one hertz is equal to one cycle per second. It is named after German physicist Heinrich Rudolf Hertz (1857–94) who discovered radio waves.

Hess's law A law that states that the total heat resulting from a chemical reaction is dependent only on the initial and final states of the reactants and is independent of the reaction route. It is the essential basis of calculations in thermochemistry and has both experimental and theoretical justification. The law is an aspect of the general law of conservation or the first law of thermodynamics. For example, the chemical route of converting calcium oxide (CaO) to calcium chloride ($CaCl_2$) follows the two routes:

$$CaO(s) \quad \rightarrow \quad \underset{\Delta H_1 = -192.5 \text{ kJ}}{+2HCl(aq)} \quad \rightarrow \quad CaCl_2(aq) + H_2O(l)$$

$$\downarrow \qquad\qquad\qquad\qquad\qquad\qquad\qquad \uparrow$$

$$\underset{\Delta H_2 = -63 \text{ kJ}}{+H_2O(l)} \qquad\qquad\qquad\qquad \underset{\Delta H_4 = -117 \text{ kJ}}{+2HCl(aq)}$$

$$\downarrow \qquad\qquad\qquad\qquad\qquad\qquad\qquad \uparrow$$

$$Ca(OH)_2(s) \quad \rightarrow \quad \underset{\Delta H_3 = -12.5 \text{ kJ}}{+H2O(l)} \quad \rightarrow \quad Ca(OH)_2(aq)$$

$$\Delta H_1 = \Delta H_2 + \Delta H_3 + \Delta H_4$$

The law is named after Swiss-Russian chemist Germain Henri Hess (1802–50) who pioneered thermochemistry.

heterogeneous mixture A mixture of two or more different phases. For example, oil and water form a liquid heterogeneous mixture. Solid particles suspended in a liquid is an example of a solid and liquid heterogeneous mixture. A blend of silver and gold filings is an example of a heterogeneous mixture of elements. *Compare* HOMOGENEOUS.

HETP *See* HEIGHT EQUIVALENT TO A THEORETICAL PLATE.

heuristics A procedure based on a *rules of thumb used to provide a shortcut solution to a complex problem rather than an *algorithm. In the case of separating a multicomponent mixture, the rules of thumb are based on a general understanding of the factors that determine distillation cost. This includes the capital costs that comprise the column itself, as well as reboiler and condenser costs, and operating costs mainly associated with the heat flux from the bottom to top of the column. If all separations are equally difficult the whole cost of the train would be minimized if the sum of all the flows from the bottoms and tops of columns were minimized. This is the case if each column produces approximately equal quantities of material from top and bottom. The fewer times that any component in large excess is distilled, the less cost in treating it will be involved. Any arrangement that sets about removing components in large excess early in the train are likely to be favoured. Highly corrosive components should also be removed early as they are likely to cause damage to equipment. For separations of unequal difficulty, it is clearly not favoured to treat a large quantity of material with a difficult separation. It is preferable to delay difficult separations to later in the train when smaller quantities are being dealt with. It is also beneficial not to separate components at all, if separation is not needed. The following heuristics are therefore applied:

1 Remove highly corrosive components first.
2 Favour equimolar splits.
3 Split by widest difference in relative volatility first.
4 Take a single pure component from the top of the column.
5 Remove components in large excess first.
6 Do not separate components that are acceptable as mixed products.

The set of heuristics, however, may not suggest the same decision in each case.

hierarchical control The use of computer control to oversee the control of other computers that are themselves used to control a process.

Higbie's penetration theory A conceptual approach to mass transfer at the interface between a liquid and a gas. The gas–liquid interface is assumed not to be at steady state but is continually being refreshed by liquid from the bulk below. A theory assumes that an element of liquid reaches the interface for a fixed time depending on the hydrodynamic conditions and arrives with a particular concentration. The residence time is assumed to be short enough for the transferred component not to penetrate the element far enough to affect the bulk concentration. It was proposed by R. Higbie in 1935.

highly active waste The radioactive residues remaining after nuclear reprocessing of spent nuclear fuels and which contain a high level of fission products. These are not able to be further processed and are therefore stored in managed environments for very long periods of time. This includes the process of *vitrification in which the waste is converted to a glassy material and stored in containers in geologically stable shafts deep below the surface of the Earth. Over the period of storage, the fission products decay. The rate of decay depends on the *half-life of the fission products.

high-pressure processing (HPP) A non-thermal food-processing technique used to extend shelf life of certain products as well as to create new food textures. Depending on the type of food, it typically uses isostatic pressures in the order of 300 to 800 MPa held for between 1 and 30 minutes. It has the benefit that important food qualities such as flavour, nutrition, and colour are not affected since the covalent bonds of food components including saccharides, vitamins, lipids, and pigments are able to resist the effects of high pressures. Pioneered by Percy *Bridgman, commercial developments began in Japan around 1989. Today many foods worldwide are increasingly being processed by this technique, although the major limitation is the expensive pressure vessels required. It is also known as **ultra-high-pressure processing.**

Hill plot A graphical procedure used to fit experimental enzyme kinetic data to an S-shaped or sigmoidal curve that deviates from *Michaelis–Menten kinetics. It involves plotting $\log V/(V-v)$ as a function of the substrate concentration, S, where V is the maximum velocity, and v is the observed velocity. The straight line has a gradient that indicates the number of interacting sites on the enzyme or enzyme complex.

Hinchley, John William (1871–1931) A British chemical engineer and one of the founders of the *Institution of Chemical Engineers. He was the first professor of chemical engineering at Imperial College, London whose aims and ideas for the development of chemical engineering were embodied in his teaching.

histogram A statistical graph that uses the height of a column to represent the number of occurrences or frequency in a particular class of results that arise from an experiment or sample.

Hock process A process used to produce phenol and acetone from cumene (isopropyl benzene). The process first involves using air in the oxidation of cumene liquid to form cumene hydroperoxide:

$$C_6H_5CH(CH_3)_2 + O_2 \rightarrow C_6H_5C(CH_3)OOH$$

It is then followed by a reaction with sulphuric acid or phosphoric acid:

$$C_6H_5C(CH_3)OOH \rightarrow C_6H_5OH + (CH_3)_2CO$$

The process is named after the German Heinrich Hock who invented it in 1942.

Hofstee plot A graphical method used in enzyme kinetics to obtain a straight line from experimental data. It involves forming a plot of V/S versus V in which S is the substrate concentration at which the velocity v is observed. The gradient of the line is equal to $-K_m$ and the intercept on the y-axis is equal to the maximum velocity V. Also known as the Eadie–Hofstee plot, it is named after Canadian biochemist George Sharp Eadie (1895–1976) and B. H. J. Hofstee who developed the plot in 1942 and 1959, respectively. *See* MICHAELIS-MENTEN KINETICS.

holding tank A vessel used for temporary storage. The capacity and materials of construction depend on the process fluids being temporarily held. Examples of uses include process water, intermediate chemicals, waste fluids, and sewage.

hold-up The amount of material contained within a process vessel operating continuously. It is applied to liquids and solids such as materials contained within distillation columns and furnaces. **Liquid hold-up** is used in two-phase flow in which the liquid phase is retained in a system at any given time when the gas phase is moving faster than the liquid phase. **No-slip hold-up** is used for two-phase flow in which both liquid and gas phases are moving with the same velocity. The hold-up can be determined by measuring the gas fraction using capacitance probes, optical probes, electrical conductivity measurements, and devices based on x-ray or gamma ray attenuation.

holley-mott A type of *mixer-settler used particularly for solvent recovery in the *nuclear fuel reprocessing industries. With liquids flowing under gravity and with no moving parts, the design allows for the intimate contact of two liquids promoting mass transfer. They are used to separate spent fuels using the *purex process.

hollow fibre membrane A semi-permeable membrane in the form of tubes with inner and outer diameters ranging in the order of 0.5 mm to 1.0 mm. They are used for both liquid and gas separations, depending on the *selectivity of the pores and membrane materials.

hollow fibre bioreactor A type of bioreactor in which *hollow fibre membranes are used to transport fresh nutrients or oxygen to the medium, or to remove products or waste, or to immobilize biocatalysts such as enzymes and mammalian cells. The hollow fibres are porous semi-permeable membranes and packed into bundles within a cylindrical container. The pore size is sufficiently small to retain cells but sufficiently large to permit the passage of nutrients and waste materials. The hollow fibre bioreactor can be operated continuously. Examples include the cultivation of mammalian cells for the production of monoclonal antibodies, the growth of liver cells (hepatocytes), and for small-scale industrial waste treatment.

homogeneous 1. Substances being of the same phase. A **homogeneous catalyst** is a catalyst in the same phase as the reactants and products. *Compare* HETEROGENEOUS MIXTURE. **2.** A mathematical equation that has terms of the same degree. **3.** A differential equation in which a linear combination is set to zero.

homogenization A process in which fine particles or droplets are produced and distributed evenly throughout a continuous liquid phase. It uses a device called a homogenizer, which typically consists of a small nozzle and impact plate through which the liquid is pumped under high pressure. An example is milk in which fat globules are dispersed as fine droplets throughout the aqueous milk phase.

homologous series A series of organic compounds of the same type with each member differing from its preceding member by possessing an additional group in its molecule. Examples include alkanes, alkenes, alkynes, and alcohols. Successive members of a series are called **homologues** in which they show a gradual and regular change in physical properties.

Hooker–Raschig process *See* RASCHIG PROCESS.

hopper A wide container used for powders and particulate materials. It is funnel-shaped leading to a small discharge opening at the bottom.

horsepower (Symbol hp) A British Imperial measure of the rate of doing work. Scottish engineer James *Watt (1736–1819) used the term 'horsepower' to represent the equivalent work that his new steam engine could achieve to replace horses used in coal-mining operations. One horsepower is equivalent to 745.7 watts. The useful work that can be performed by a power device is called **brake horsepower.**

Hortonsphere A spherical pressure vessel used for the storage of compressed volatile gases in a liquid and gaseous state such as hydrocarbons and fuel gases. A trademark name registered by the Chicago Bridge and Iron Company (CBI), it is named in honour of the American bridge designer Horace Ebenezer Horton who founded the company in 1889.

(⊕) SEE WEB LINKS
• Official website of the Chicago Bridge and Iron Company.

hot gas ignition temperature The lowest temperature required for ignition of a substance by a jet of hot gas.

hot leaching process A process first used in 1860 for the extraction of potash from sylvinite ore based on the differing solubilities of sodium and potassium chloride. The ore, which is a mixture of sodium and potassium chloride, is crushed and mixed with saturated sodium chloride. It is heated to boiling in which the potassium chloride dissolves but the sodium chloride does not. On cooling, the potassium chloride crystallizes and is separated.

hot pressing A process used to create hard and brittle solid materials from a powder, reduce the porosity of metals, and increase the density of ceramic materials. It uses a very high pressure of around 1000 bar and temperature of over 1,000°C depending on the materials. Ceramics and diamond-metal composites are made using this process in which the powder particles pack to form a very dense material. An inert gas such as argon is used to prevent oxidation of the materials.

hot surface ignition temperature The lowest temperature required for ignition of a substance by a hot surface.

hot tapping A technique used to form a connection between pipes or vessels without the need for first emptying or draining the contents. The pipe or vessel can therefore remain in operation at the time of the maintenance or plant modification. It involves temporarily plugging or sealing the pipe to allow the connection to be made.

Hottel charts Graphical charts used to determine the thermal diffusion in materials in the shapes of slabs, cylinders, cubes, and spheres. They plot the *Fourier number on the x-axis and a dimensionless temperature on the y-axis, and are an extension of the

*Gurney–Lurie charts. They are named after American chemical engineer Hoyt C. Hottel (1903–98) who was professor of chemical engineering at MIT.

hot wire anemometer An instrument used to measure the velocity of a gas. It consists of a resistance wire that is electrically heated using a constant current, and works on the principle that a small diameter electric conductor gives off heat in relation to the velocity of gas passing over the wire. The observed temperature of the wire indicates the air flow using an appropriate calibration chart.

hot work A method of carrying out a task in a hazardous area using a tool or item of equipment that can provide a source of ignition, heat, or flame. Examples include welding or the use of a flame or electric arc.

HP An abbreviation for **high p**ressure. It is often used with reference to a utility or a vent line. For example, an HP air supply.

HPP *See* HIGH-PRESSURE PROCESSING.

HSE An abbreviation for the **H**ealth and **S**afety **E**xecutive. It is the authority responsible for the regulation of most of the risks to health and safety arising from work within the UK.

(()) SEE WEB LINKS
• Official website of the Health and Safety Executive in the UK.

HTU *See* HEIGHT OF A TRANSFER UNIT.

human error *See* ERROR.

human–machine interface Used in control rooms on the display screens, it presents graphically the process and includes vessels, pumps, pipes, etc. together with real-time data, graphs, and trends on temperature, pressure, flow, and levels. The process operator can control the process by adjusting set points, turning on and off pumps, opening valves, etc.

humectants A chemical material that is a wetting or moistening agent and used to provide an environment of moisture. *Compare* DESSICANTS.

humid heat The quantity of heat that is required to raise the temperature by one degree of one unit mass of dry air plus the water vapour that accompanies it. The SI units are kJ $kg^{-1} K^{-1}$.

humidification The process of adding moisture to air that results in an increase in humidity.

humidity The amount of water contained within a unit volume of dry air. The *absolute humidity is the amount of water in air expressed as the mass of water vapour per unit mass of dry air for a particular temperature and pressure condition. The *relative humidity is the moisture content of air expressed as a percentage, and corresponds to the ratio of the mass of water vapour in a given volume of air at a given temperature and pressure to the maximum quantity of water vapour that can be held in the air at those conditions.

humid volume The volume occupied by one unit mass of dry air together with the water vapour it contains expressed as:

$$V_H = \frac{RT}{p}\left(\frac{1}{29} + \frac{H}{18}\right)$$

where R is the gas constant, T is the absolute temperature, p is the total pressure, H is the absolute humidity, and 29 and 18 are the molecular weights of air and water, respectively. The SI units are $m^3\,kg^{-1}$ of dry air.

Hunter process A process used for the production of titanium metal by the reduction of titanium tetrachloride with sodium metal.

$$TiCl_4 + 4Na \rightarrow Ti + 4NaCl$$

The process takes place in batch reactors in which the reactants are heated in an inert atmosphere to 1,000°C. The salt is leached from the product using dilute hydrochloric acid. Invented by American M. A. Hunter in 1910, the process has been replaced by the *Kroll process.

Huntsman process A process used to purify steel by heating it in a graphite crucible and pouring the melt into a cast-iron mould. It is also known as the *crucible process and was developed by British clockmaker and locksmith Benjamin Huntsman (1704–76). Although not patented, the method was used until eventually superseded by the *Bessemer process.

HVAC An abbreviation for **h**eating **v**entilating and **a**ir **c**onditioning.

HX An abbreviation for **h**eat e**x**changer.

hydrate 1. The combination of certain compounds with water such as ferric chloride and copper sulphate. **2.** A compound consisting of gas molecules such as methane trapped within a cage of water molecules formed at high pressure and cool temperatures. It has the appearance and consistency of wet snow. *See* METHANE CLATHRATE.

hydrated The addition of water to a chemical to form a hydrate.

hydration *See* SOLVATION.

hydraulic fracturing *See* FRACKING.

hydraulic press A device in which forces are transferred by way of pressure in a fluid. A force F_1 is applied over an area a_1 and is transferred to a smaller area a_2 resulting in an increased applied force F_2 such that $F_2 = F_1 \times a_2 / a_1$. Extreme pressures can be obtained using pressure multipliers in which a force applied over a large area is transferred over a very small area thereby intensifying the pressure. This pressure can be intensified further by transfer onto an even smaller area.

hydraulic radius The cross section or area through which a fluid flows divided by the *wetted perimeter:

$$r_H = \frac{A}{P}$$

where A is the area and P is the wetted perimeter. It is used for *open channel flow, and for wide channels and represents the depth of the channel. Note that it is not equal to half the *mean hydraulic diameter.

hydraulic retention time The average period of time that a particular substance or component resides within a reactor or continuously fed vessel. It is also known as the **hydraulic residence time**.

hydraulics The study of the properties of liquids in motion and based on the principles of hydrodynamics and hydrostatics. The principles have been known for millennia with recorded evidence of irrigation canals and dams having been used in Egypt and Mesopotamia (now modern-day Iraq). Italian physicist Evangelista *Torricelli (1609–47) noted in 1644 that the velocity of a flowing liquid is proportional to the square root of the water head. Swiss scientist Daniel *Bernoulli (1700–82) obtained the relation between wall pressure, velocity, and elevation in 1738, using the word 'hydrodynamica' to describe the synthesis between conceptions of hydrostatics and hydraulics. Many other scientists have subsequently developed the subject area using Newtonian dynamics and differential calculus to describe the flow in pipes and channels, leading to many key concepts and the establishment of theorems including vortex motion, fluid resistance, airfoils, meteorological fronts, and statistical approaches to turbulence.

hydrocarbon Chemical compounds that consist of only hydrogen and carbon atoms. They form the main components of petroleum and natural gas and are used as fuel for energy and as the feedstock for producing many useful products such as plastics. There are three main classes based on the types of carbon–carbon bonds. Saturated hydrocarbons contain only single carbon–carbon bonds as alkanes or cycloalkanes. Unsaturated hydrocarbons contain carbon–carbon multiple bonds, such as alkenes and alkynes. Aromatic hydrocarbons are a special class of cyclic compounds and are related to benzene.

hydrochloric acid A strong and widely used acid in the chemical industry. It is largely made as a by-product from the chlorination of both aliphatic and aromatic hydrocarbons. It is also made by dissolving hydrogen chloride gas in water, or by the reaction of sulphuric acid with salt. The greatest use of the acid is in the pickling of steel. Other major uses include the production of chemicals, pharmaceuticals, and in food processing.

hydrocracking A general name given to catalytic processes that involve the hydrogenation of petroleum fractions to produce fractions with a lower molecular weight. Vaporized petroleum feed is mixed with hydrogen at high temperature and pressure and allowed to react with hydrogen in the presence of a catalyst. It includes both fixed-bed and fluidized-bed reactors. There are many variations currently used but the first hydrocracking process was the *Bergius process.

hydrocyclone A type of cyclone device used for the separation of particles suspended in a liquid. It is operated by feeding the liquid suspension tangentially into the top of the device to produce a centrifugal vortex action. This allows the more dense and coarse particles to be separated in the underflow while the smaller particles leave through the overflow. They are used in mineral ore processing, drilling mud separations, and oil–water separations on offshore operations. They have a low *capital cost, are cheap to operate, and have the ability to separate materials based on small differences in particle size.

hydrodealkylation *See* DEALKYLATION.

hydrodesulphurization A catalytic process used for the removal of sulphur from crude oil and refined petroleum products, and hydrogen sulphide from natural gas. The process involves a high temperature and pressure reaction in the presence of a nickel or alumina catalyst impregnated with molybdenum. Sulphur is required to be removed as it poisons catalysts and results in harmful sulphur dioxide in the combustion of fuels.

hydrodynamics The study of the behaviour and interaction of incompressible liquids in motion with reference to their contact with their boundaries. It is used for the study of the behaviour of water in open channel flow, and for liquids in *closed systems.

hydroelectric power Electrical power generated from the flow of water through a water turbine connected to an electrical *generator or **alternator**. The flow and pressure head of water is provided by a dam and channelled through a pipe to the water turbine below. The power generated is a product of the head, mass flow rate, and gravitational acceleration. Hydroelectric power stations produce electricity at peak times and can be used in reverse at other times to return water back into reservoirs for reuse.

hydroformylation A general name for catalytic processes involving high temperature and high pressures to extend the chain length of aliphatic compounds leading to the production of aldehydes and alkenes using carbon monoxide and hydrogen. The process was invented in the 1930s and has been modified many times since and is used in the production of fragrances and intermediates and for the production of detergents. It is also known as **oxo synthesis** or the *oxo process.

hydrogasification An exothermic process that uses hydrogen to gasify coal and produce methane. The hydrogen is provided either by using a methane steam reformer or some other source. Some of the methane is converted to carbon monoxide and hydrogen in the methane steam reformer.

hydrogenation 1. A process involving the heterogeneous reaction of adding hydrogen to molecules containing carbon such as unsaturated animal and vegetable oils. Highly unsaturated vegetable oils can be converted to solid vegetable fats by catalytically hydrogenating some or all of the double bonds in a process known as **hardening**. Margarine is produced from the hydrogenation of unsaturated vegetable oils such as soybean, peanut, or corn to produce the consistency of butter. The product can be churned with milk and artificially coloured to give the appearance or flavour of butter. **2.** Synthetic oil that is made by reacting coal with hydrogen to form synthetic hydrocarbons. *See* FISCHER–TROPSCH PROCESS.

hydroisomerization A general name given to catalytic processes that involve the use of hydrogen to isomerize aromatic hydrocarbons.

hydrometallurgy Processes that involve aqueous solutions used in the extraction and recovery of metals from ores. Leaching is used to treat metallic ores to form a solution of a salt and a metal. Dilute sulphuric acid is the most commonly used agent. The metal is then recovered from the salt solution by precipitation and/or concentration, and uses processes such as carbon adsorption, ion exchange, solvent extraction, and electrolysis. Almost all non-ferrous metals are produced using hydrometallurgy, such as gold, silver, copper, aluminium, nickel, zinc, cobalt, uranium, molybdenum, tungsten, and beryllium. *See* LIXIVIATION.

hydrometer A device used to measure the *density or *specific gravity of a liquid. The device, usually made of glass, floats in the liquid and sinks to an equilibrium depth according to *Archimedes' principle, leaving a graduated stem unimmersed from which the density or *specific gravity is directly read. It is used in the *whisky, beer, and wine industries, and can provide an indication of the concentration of alcohol after fermentation and in the distillation of spirits.

hydrophilic A substance that has an affinity with water. *Compare* HYDROPHOBIC.

hydrophobic A substance that has little or no affinity with water. *Compare* HYDROPHILIC.

hydroprocessing A general name for a process involving the refining of petrochemicals with hydrogen gas. *See* HYDROCRACKING; HYDROTREATING.

hydropyrolysis A general name given to high-temperature, high-pressure catalytic processes that convert coal into a mixture of liquid and gaseous hydrocarbon products.

hydrorefining A general name given to processes that involve the use of hydrogen for the refining of petroleum mixtures.

hydroskimming A simple form of petrochemical refinery that uses atmospheric distillation to separate crude oil, and also has the capacity for *naphtha reforming and associated treatment processes. A hydroskimming refinery can produce a surplus of desulphurized fuels and gasoline. It is therefore more complex than a *topping refinery.

hydrostatic head The *static pressure of a liquid expressed in head form representing the pressure, p, that would be exerted by a vertical column of the liquid under the influence of gravity:

$$h = \frac{p}{\rho g}$$

where ρ is the density of the liquid, g is the gravitational acceleration. Although the terms would appear to refer to water, it can be used for the static pressure of any liquid. *See* HEAD.

hydrostatic pressure The *static pressure of a liquid equivalent to the pressure that would be exerted by a vertical column of the liquid under the influence of gravity. The term can be applied to any liquid although in general most applications of hydrostatics involve water in reservoirs and dams.

hydrostatics The study of the behaviour of liquids under the action of forces and pressure when the liquid is stationary. It includes the study of liquids in storage tanks, vessels, reservoirs, and dams.

hydrostatic testing A form of pressure testing of vessels and piping or any other item of process plant equipment that is required to operate above atmospheric pressure. It involves filling the equipment with water to exert a *hydrostatic pressure. It is a safe method as water is virtually incompressible and by using colouring, leaks can also be easily detected.

hydrotreating A general name given to catalytic refining processes that involve the use of hydrogen. They are used to remove various components such as organic sulphur, nitrogen, metal, and oxygen compounds.

hygro- A prefix indicating humidity and moisture.

hygrometer An instrument used to measure the *humidity in the atmosphere. A **wet and dry bulb hygrometer** uses two thermometers to measure the *dry bulb temperature and *wet bulb temperature from which the humidity can be determined. In a *dew point hygrometer, the temperature of a polished surface is reduced to the point that moisture condenses to water. Other types of hygrometers use an organic material such as a fibre that expands and contracts due to the humidity, or use a change in electrical resistance of a *hygroscopic material to provide an indication of the humidity.

hygroscopic A material that gains or loses moisture depending on the product and the humidity and temperature of the surroundings. Biological products are hygroscopic. *See* ABSORPTION.

hypergolic Ability of a substances to ignite spontaneously when mixed with an oxidizer.

hypothesis A suggested explanation or argument to explain a phenomenon as the basis for either further verification or accepted as fact. It may be a theory or law that has the status of not being incontrovertible or universally true. There are some hypotheses in which there is no doubt such as *Avogadro's hypothesis but for which the label 'hypothesis' remains. *Compare* THEORY.

hypotonic The exertion of less osmotic pull than the medium on the other side of a *semi-permeable membrane. The material therefore has a lower concentration and loses less water during osmosis.

hypoxia A deficiency in the body when blood fails to deliver the necessary level of oxygen to tissues and organs. It can be caused by a lack of oxygen in the atmosphere due to displacement or exhaustion, particularly by the release of gases in confined spaces, or due to carbon monoxide poisoning, which inhibits the ability of haemoglobin in the blood to release the oxygen bound to it. The symptoms include headaches, tiredness, and shortness of breath, to loss of consciousness and death. Cyanosis, a blue discolouration of the skin and in particular the lips, show this in severe cases.

hypsometer An instrument used to determine elevation above sea level by measuring the change in the boiling point of water. Water and liquids generally boil at progressively lower temperatures as pressure decreases. Since atmospheric pressure decreases with altitude, the decrease in boiling point temperature is therefore an indicator of the atmospheric pressure and therefore elevation. It can also be used to calibrate thermometers.

hysteresis A lagging effect from an expected value due to a resistance such as friction. For example, in the control of processes, hysteresis is where there is a change in signal before a move in a valve.

HYSYS An abbreviation for **Hy**protech **Sys**tems, it is process-modelling software developed by AspenTech. It is used for steady-state and dynamic simulation of processes, process design, process performance monitoring, and process optimization across a wide range of industries and processes.

i A symbol used to represent $\sqrt{-1}$. It is an *imaginary number, which is distinguished from *real numbers. Engineers tend to use the symbol j instead of i. *See* COMPLEX NUMBERS.

IBC *See* INTERMEDIATE BULK CONTAINER.

ice point A freezing point of water under standard atmospheric pressure conditions in which water and ice are in equilibrium. It was once used as the *fixed point for 0°C. The kelvin and International Practical Temperature Scale has sixteen fixed points with temperatures including the triple of water.

IChemE *See* INSTITUTION OF CHEMICAL ENGINEERS.

id 1. An abbreviation for **i**nternal **d**iameter and used for specifying pipes, tubes, and some vessels of circular cross section. **2.** An abbreviation for equipment **id**entification used on flowsheets. Equipment such as a process vessel may have a specific code and number.

ideal crystal The shape of an individual crystal that has a perfectly regular lattice and has no deformations, irregularities, or imperfections.

ideal gas An idealized gas consisting of elastic and non-interacting molecules in which the effective volume occupied by the molecules is zero and obeys the **ideal gas law** $pV = nRT$ where p is the pressure, V is the volume, n is the number of moles, R is the universal gas constant, 8.314 J mol^{-1} K^{-1}, and T is the absolute temperature. Gases with a low density and far from their *critical point can be approximated as being ideal in which a gas occupies 22.4 litres per mole at 1 atmosphere and 273 K. *Compare* REAL GAS.

ideal mixture A mixture of two or more substances that form a solution in which there is no interaction between the molecules or atoms of the individual components. The total volume of the mixture is the sum of the partial volumes. *Compare* NON-IDEAL MIXTURE.

ideal solution A solution that conforms to *Raoult's law over a range of temperatures and compositions, and which shows no attractive forces between the components.

ideal stage *See* THEORETICAL STAGE.

IGPM An abbreviation for an Imperial unit of volumetric flow rate as **I**mperial **g**allons **p**er **m**inute and is used in the oil industry.

ignition The process of causing a substance to combust. Initiation of combustion is evidenced by a glow, a flame, or explosion. A substance will burn when the rate of heat gain due to oxidation exceeds the rate of heat loss. The **ignition delay** is the time to ignition from the instant reactants are mixed or exposed to a heat source. The **ignition energy** is the quantity of heat or electrical energy that must be absorbed by a substance to ignite. The **ignition temperature** is the lowest temperature at which a substance can sustain combustion in air or oxygen at a specified pressure. *See* COMBUSTION.

imaginary number The part of a *complex number that is a multiple of $\sqrt{-1}$, which is denoted as j. For example, $\sqrt{-5} = j\sqrt{5}$.

immiscibility The property of fluids to form distinct and separate phases under all relative proportions. For example, oil and water are immiscible in which oil floats on water due to its lesser density. Immiscible fluids are used in liquid–liquid extraction processes in which a solute is transferred from one phase to the other by contacting the two fluids. This often requires a large interfacial surface which can be achieved by stirrers, *pulsed columns, and in *mixer-settlers.

immobilized cell bioreactor A type of fixed bed or fluidized bed reactor in which living microbial cells or enzymes are held statically such that their movement is restricted. Depending on the characteristics of the cells or enzymes, they may be immobilized in the form of aggregates such as pellets, or adhere to porous structures such as diatomaceous earth, glass, dextran, and gelatine, or be entrapped within beads such as calcium alginate. The advantage of immobilization is that high oxygen transfer rates can be achieved and the cells can be protected from high shear forces. Cell washout can also be avoided.

impact The ultimate effect of the occurrence of an undesirable and hazardous event. It may result in loss of life, injury, property damage, destruction of equipment or process plant, economic loss, or environmental damage. The impact is evaluated in a *risk assessment.

impact pressure A measure of the pressure of a flowing gas or liquid that is brought to rest onto a surface. A *Pitot tube flow meter measures the difference in the impact pressure and the static pressure of a fluid from which the rate of flow of the fluid is deduced.

impeller A series of blades attached to a shaft used to provide radial current of a fluid when rotated. It is used in *centrifugal pumps and for mixing purposes. **Radial-flow impellers** discharge the fluid in a horizontal (radial) direction to the vessel wall. If suitable baffles are provided, this device produces strong top-to-bottom currents. If the vessel is unbaffled, then swirling and vortexing may result. **Axial-flow impellers** create a flow parallel in the direction of the axis of the shaft. These devices produce more flow per unit power supplied than radial impellers, and are generally used in flow controlled operations. As with radials, swirling and vortexing may develop without the use of baffles. *Compare* PROPELLER.

Imperial units A British non-metric system of weights and measures. The system includes the ounce, pound, stone, inch, foot, yard, mile, acre, pint, quart, gallon, etc. It has largely been replaced by *SI units for scientific and engineering applications although there is still some usage.

impingement separator A device used to separate particles or droplets in a gas stream. The gas is diverted around an obstruction such as a *baffle so that the particles or droplets hit the baffle, coalesce on the surface, and drain away without re-entrainment. *Demisters are examples of impingement separators.

implicit function A mathematical function that contains two or more variables that have not been solved explicitly. For example, in the equation $x^2 + y^2 - 2 = 0$ the variables x and y are not dependent on each other where x is an implicit function of y and y is an implicit function of x.

implosion The inward collapse of a vessel as a result of evacuation. Steam trapped inside a closed vessel that has insufficient wall strength will implode when the steam cools and condenses.

impulse 1. A quantity used to measure the total change in momentum of a body produced by a force acting on the body for a very short time. **2.** In process control, it is a disturbance to a process that occurs over a very short period of time.

incandescence The emission of light by a substance due to its high temperature. For a substance to emit white light, it is required to be heated to above 700°C. Examples include electric heaters and stoves, and the white-hot filament of a light bulb of an incandescent lamp. British chemist Humphry Davy (1778–1829) demonstrated in 1802 that a strip of metal heated to a temperature using electricity can emit light. To overcome the oxidization of the metal British chemist and physicist Joseph Wilson Swan (1828–1914) invented an electric lamp using a filament of carbonized paper in 1860 but was unable to produce a sufficient vacuum. American inventor Thomas Alva Edison (1847–1931) produced the first practical incandescent light bulb two decades later.

incineration The process of combusting materials. An **incinerator** is a type of furnace used to incinerate waste organic materials to the point that only ash remains. It is often the case that combustion of toxic wastes such as halogenated hydrocarbons, herbicides, and pesticides are best treated by incineration, otherwise they would persist within the environment over long periods of time. It uses high temperatures in the order of 1,100°C to 1,300°C with the combustion of fuel oil or natural gas in an excess of air.

inclined tube manometer An instrument used for measuring the pressure head of a gas in which one leg of the manometer is attached to a sump containing a manometric fluid and the other is a straight tube, usually made of glass, and inclined at a known angle to the horizontal. The applied differential pressure between the sump and tube gives a vertical difference between the levels. The fluid displaced from the sump moves further along the inclined tube than if the tube was vertical.

incombustible The inability of a substance to burn. *Compare* INFLAMMABLE.

incompressible fluid A fluid in which the volume or density does not change with applied pressure for a given temperature. However, truly incompressible fluids do not actually exist. Liquids are generally regarded as being incompressible to simplify calculations.

inconel A type of nickel-chromium-iron alloy used for process plant equipment, noted for its strength at high temperature and corrosion resistance. It is used for gas turbine blades, seals, and many other applications. **Incoloy** is also a nickel-chromium alloy used for process plant equipment and also noted for its strength and resistance to oxidation and carburization at high temperature. It is used in nuclear fuel reprocessing and acid production as well as for process piping, pumps, and valves.

increment (Symbol Δ) A small difference in a variable. For example, the level of a process liquid in a tank feeding to a process may change by an increment Δh over an increment of time Δt.

independent variable *See* VARIABLE.

indeterminate problem A problem that involves one or more unknown or variable quantities.

indicator 1. A substance that indicates the completion of a chemical reaction. It is often used to determine the end point in titrations, and involves a sharp colour change. **2.** A type of process controller that provides measures and controls a process variable such as temperature. On a *piping and instrumentation diagram a **TIC** is a temperature indicator controller while a **LIC** is a level indicator controller.

induction heating A means of heating an electrically conducting material by electric currents induced in the material by an alternating magnetic field. Electric heating is accomplished with an electric alternating current of high value and low magnitude voltage. It is used for melting and heat treating of certain metals as well as for welding, soldering, and in brazing. The material is heated by inducing eddy currents in the material, causing the temperature to rise.

industry An economic activity by businesses that involves the manufacture of products through the conversion of raw materials, production of goods, or supply of services. *See* CHEMICAL INDUSTRY.

inelastic collision The collision of bodies such that some of their kinetic energy is converted into internal energy and that kinetic energy is not conserved. Collisions of gas molecules may be inelastic as they cause changes in vibrational and rotational energies. In larger bodies, some of the kinetic energy is converted to vibrational energy causing a heating effect.

inert gas 1. A noble gas such as helium, neon, argon, krypton, xenon, and radon. **2.** A gas such as nitrogen or carbon dioxide, known as an *inerting agent or blanketing gas, used to protect a substance from coming into contact with air or oxygen. Blanketing gases such as carbon dioxide are used to reduce the risk of fire hazards.

inertia The tendency of a body to resist a change in the motion by an external force. A body will remain static, or travel in a straight line at a constant speed until acted upon by a force.

inerting agent An inert gas such as nitrogen or helium that can prevent the ignition and combustion of ignitable mixtures. The inert gas lowers the availability of oxygen to below the *flammability limit for combustion to occur even with an ignition source.

infinite Having no specified end, no highest or lowest value. The term *semi-infinite is sometimes used to define the geometry of an object such as in heat transfer calculations that is infinite in one or two dimensions but finite in the third. A semi-infinite cylinder has a finite radius but infinite length such that end effects can effectively be ignored. An **infinite dilution** is the concentration of a solution that corresponds to an infinite ratio of the solvent to the solute.

infinitesimal A number that is not zero but is less than any finite number. While such numbers do not exist in the conventional system of real numbers, it allows calculations to adhere to certain restrictions and requires a good understanding of limits.

inflammable A material that is combustible or has the ability to support combustion. *Compare* INCOMBUSTIBLE.

inflection A point of a curve where the gradient of the tangent changes in sign. A stationary point on a curve occurs at the point where:

$$\frac{dy}{dx} = 0$$

and can be either a maximum, a minimum, or an inflection. The point is an inflection where:

$$\frac{d^2y}{dx^2} = 0$$

influx The flow of a fluid into a process or across a system boundary. *Efflux is the flow out.

infranatant *See* SUPERNATANT.

infrared radiation A part of the *electromagnetic radiation spectrum that corresponds to wavelengths between those of microwaves and those of visible light (i.e. 10^{-6} and 10^{-3} m). While invisible to the naked eye, infrared radiation is perceived as heat. Infrared spectroscopy is an important technique in analytical chemistry since the infrared adsorption of molecules is characteristic of it and the spectrum can be used for molecular identification. Infrared radiation was discovered in 1800 by German-born British astronomer William Herschel (1738–1822) in the sun's spectrum.

Ingen-Hausz, Johannes (1730–99) A Dutch scientist and engineer who studied in four countries and made many discoveries in physics. He designed an apparatus for comparing the thermal conductivities of rods of different substances. He also noted that plants take in carbon dioxide during the day but release it out at night.

ingot A material that has been cast in moulds into a particular shape such as a bar or a block. The material is usually metal (pure or alloy) and poured in the molten state into moulds to form the ingot, which is used for further processing in order for the material to be useful as a final product. Gold ingots are used as a currency reserve. Purification of ingots can be achieved through *zone refining to remove impurities.

inherently safe A process or system that is able to operate safely without the need for external or auxiliary support. For example, a cooling water system to a process that uses heat exchangers of a sufficient duty and that are fed under gravity to ensure removal of heat is inherently safe.

inherent moisture The chemically combined and absorbed moisture of a substance; it is distinct from the surface or *free moisture.

inhibition The process of preventing or reducing a chemical reaction taking place. In enzyme kinetics, the reaction between the enzyme and substrate can be inhibited by the addition of another substance. **Competitive inhibition** is where a substance inhibits the enzyme reaction by completing with the enzyme at the active site whereas non-competitive inhibition is where a substance reduces the activity of the enzyme by binding to the enzyme whether or not it has already bound the substrate. An **inhibitor** is the substance added causing the change in rate of the chemical reaction.

initiator 1. A substance or molecule that starts a chain reaction other than a reactant. For example, peroxides are used in polymerization reactions of methyl methacrylate monomer. **2.** A small device used to start an explosive train such as a detonator. The initiator usually contains a small quantity of a sensitive explosive and receives a mechanical or electrical impulse to cause the detonation.

in-line analysis *See* OFF-LINE ANALYSIS.

in-line mixer A static device with no moving parts used in a pipeline carrying fluids that causes the fluid streamlines to cross one another thereby intimately mixing the fluid. It is used to mix fluids such as dispersing gases in liquids, assist with dissolving solids in liquids, and mixing immiscible liquids. It consists of a twisted ribbon of chemically inert and unreactive material such as metal or plastic of sufficient length to provide good mixing characteristics. It is held firmly in place. It is also known as a **static mixer**.

inoculation The injection of a small volume of liquid containing living microorganisms into a bioreactor to initiate a biochemical process such as fermentation. The microorganisms, such as yeast or bacteria, are first prepared as an inoculum by either using living cells for a previous bioreaction, or by transferring living cells from agar plates upon which they have been stored, and growing them in a shake flask. The contents of the shake flask provide the inoculum for the larger biochemical process.

input The quantity of material, energy, or power fed into a system. *Compare* OUTPUT.

insoluble A substance that is not able to dissolve in a solution.

Institution of Chemical Engineers (IChemE) A professional organization founded in 1922 representing around 38,000 (2013) members across 120 countries with its headquarters in Rugby, UK and with offices in Australia, China, Malaysia, and New Zealand. It represents benchmark standards in chemical engineering education and is licensed to award chartered chemical engineering status to qualified engineers. It also publishes research and specialist books on chemical engineering.

( ) SEE WEB LINKS
• Official website of the Institution of Chemical Engineers.

instrument air Compressed air used in chemical and process plants to operate pneumatic instruments such as pressure controllers and air-operated control valves. The air is filtered and cleaned to prevent blockage and corrosion problems and compressed to a pressure of typically 6 bar. There are strict requirements for the quality of instrument air.

insulation A material used to limit or restrict the flow of heat, sound, noise, vibration, or electricity. Thermal insulation around pipes is known as *lagging.

intalox saddle A tradename for a type of *packing material used in *packed columns. It is shaped as a saddle and perforated, and made from inert materials such as metal or plastic.

integer A whole number. It may be positive or negative and includes zero. That is {... −3, −2, −1, 0, 1, 2, 3...}.

integral action control A mode of control used to control a process in which the change in the controller output is proportional to the time integral of the deviation. The control action recognizes not only the magnitude of the deviation, but also the time during which the deviation occurs. Integral action is limited in application to processes of small capacitance and fast responses. It is therefore rarely used alone but instead is more frequently used in combination with proportional control. Integral action is the only mode of control that will bring the controlled variable back to the set point.

integration The continuous summing of the change in a function f(x) over an interval of the variable x. The result is known as the integral of f(x) with respect to x. An integral $\int_{x_1}^{x_2} f(x)dx$ can be considered as the area under the curve and the limits x_2 and x_1. Definite integrals have limits whereas indefinite integrals do not. The result of an indefinite integral has a constant of integration c. For example:

$$\int x^2 dx = \frac{x^3}{3} + c$$

There are lists of tables of standard integrals. The inverse process of integration is called *differentiation.

intensive variable A property of each phase in a mixture that is independent of the amount of material. It includes temperature, pressure, enthalpy, density, specific volume, specific heat capacity at constant volume, and mole fraction. For a multiphase mixture at equilibrium, both temperature and pressure are uniform throughout the mixture but the values for enthalpy and mole fraction are different for each phase. An **extensive variable** depends on the amount of material in the system such as volume, mass, and total energy.

intercooler A type of heat exchanger used to remove heat from a fluid between the stages in a process. It is used to cool liquids and gases in processes such as refrigeration, air conditioners, gas turbines, combustion engines, and some types of reactors, and in particular, to remove the heat of compression.

interface 1. The boundary between two phases such as two immiscible liquids, a liquid and vapour or gas, a gas and a solid, or a gas and a vapour or gas. The **interfacial area** is the area of the boundary between the two phases. The **interfacial tension** is the surface tension at the surfaces between two phases. **2.** The point of interaction or communication between a computer and some other device. A human–machine interface is the point of interaction by a process operator to control a process by receiving real-time data and making manual adjustments to pumps and valves.

interlock A safety control system that cuts in when unsafe conditions are detected, particularly in the operation of process machinery such as pumps and compressors.

intermediate bulk container (IBC) A rigid container used for storing and transporting liquids and bulk materials. It is generally cubic in shape and provides good storage geometry. It can be stacked and can vary in capacity ranging from 500 to 3,000 litres. It is called intermediate as its capacity is considered to be between that of a drum and a tank. IBCs are made of various materials and their use depends on their application. Plastic IBCs are commonly used for liquids. Steel and stainless steel IBCs are used for granular materials.

intermediate product A chemical product produced during a step in a process which is not the final product but may be used as the feedstock for the next step in the process. It is also known as a chemical intermediate.

internal diffusion The movement of a gas, vapour, or liquid without a change of state through a porous material.

internal energy (Symbol U) A conceptual state function whose absolute value can be neither measured nor calculated. It represents the total energy in a substance and includes the atoms and molecules and their intra-atomic, inter-atomic, and inter-molecular forces. It includes neither the kinetic energy nor the potential energy. It is used in energy balance equations in terms of its relative value to a reference state, whose internal energy is arbitrarily set to zero. For a *closed system, the change in internal energy ΔU is equal to the difference between the heat absorbed from the surroundings, Q, and the work done, W, by the system on its surroundings $\Delta U = Q - W$. The concept was proposed by American physicist Josiah Willard *Gibb (1839–1903).

internal reflux The partial condensation of liquid within a distillation column that descends down the column. The condensation does not take place in the condenser but on the walls of the column due to poor insulation. Alternatively, internal reflux is intentionally achieved using an internal heat exchanger that allows the more volatile component for separation to be separated more effectively from the top of the column.

International Council of Chemical Associations (ICCA) An organization that represents chemical manufacturers and producers all over the world. Its aims are to provide cooperation with major organizations such as the OECD, particularly in terms of chemical production management.

(⊕) SEE WEB LINKS
• Official website of the International Council of Chemical Associations.

interpolation A method of obtaining data that falls between two known items of data. It is used in numerical analysis in which new data can be obtained using techniques such as curve fitting.

interstitial The space between atoms, molecules, and materials. An **interstitial element** is an impurity in metals such as hydrogen, carbon, and oxygen that affects the physical properties of the metal. These atoms are small enough to fit between the crystalline lattices of the much larger metal atoms. An **interstitial compound** is where non-metal atoms or ions fit between the spaces within a metal lattice. Examples include carbides, borides, and silicides.

intumescent paint A coating applied to a surface to protect it from flame or heat. It expands on heating thereby reducing its density and creating an insulating effect. It is used in passive fire protection for pipes and other process equipment particularly in hazardous areas. It is applied as a spray. Examples include types of hydrates, which liberate their moisture content on heating, thereby maintaining a boiling point temperature of 100°C, and protecting the metal from rising temperatures.

inventory A material stock-take or form of accountancy of materials within a process plant or equipment at a given time. It can include the raw materials, products in storage, and materials within the process.

in vitro A biochemical reaction that takes place in an apparatus, and therefore distinct from the biochemical reaction taking place within the living cell (**in vivo**). In vitro literally means 'in glass'.

I/O An abbreviation for input/output. It refers to signals entering and leaving a controlled process.

iodide process *See* VAN ARKEL–DE BOER PROCESS.

ion exchange The process involving the adsorption of ions in a solution on a surface of particles or gels accompanied by the simultaneous desorption of ions of like charge. An **ion exchange resin** is a synthetic polymeric resin that is able to exchange ions on the surface of a polymeric molecular network that possesses functional groups that carry an ionic charge. The extent of the exchange is dependent on the nature of the resin and the type of exchanging ions. Ion exchange is used in *water treatment to remove calcium and magnesium ions, and replace them with soluble sodium ions. Ion exchange equipment consists of a fixed bed of the resin down through which the water for ion exchange passes. The ion exchange resin may be naturally occurring such as *zeolites, aluminosilicates, or synthetic such as cross-linked polystyrene.

ionic bond A type of chemical bond based on the electrostatic attraction between oppositely charged ions in a compound. Sodium chloride is an ionic compound in which the sodium atom loses one of its outer electrons to the chlorine atom, resulting in the formation of a sodium cation (Na^+) and a chlorine anion (Cl^-). They are mutually attracted and therefore form an ionic bond.

ionization The process of forming ions involving the absorption or emission of electrons by atoms or molecules. It can occur in several ways such as the reaction of two neutral atoms; by the combination of an already-existing ion with other particles; by the breaking of a covalent bond in such a way that each bond is associated with one of the parts such as the dissociation of water to form a hydrogen ion and a hydroxyl ion; and by the passage of energetic charged particles through gases, liquids, and solids, such as X-rays, beta particles, gamma rays, UV radiation, and electric discharges through gases.

ionizing radiation *Radiation that either directly or indirectly causes *ionization. It includes alpha, beta, gamma, and neutron radiation.

ion leakage The concentration of unwanted ions (for example, calcium ions) left in a treated liquid within an ion exchange unit.

I/P transducer *See* TRANSDUCER.

iron (Symbol Fe) A metallic element with atomic number 26. It is widely found in the element and is extracted from ore and used to make steel and other valuable alloys. *See* CAST IRON; STEEL.

irradiance (Symbol E) The radiant flux per unit area reaching a surface. The SI units are $W\ m^{-2}$. It refers to all types of *electromagnetic radiation. Illuminance refers only to visible radiation.

irradiation A low-energy form of sterilization of certain foods using ionizing radiation. Gamma rays from cobalt-60, X-rays, and electron beams are used to kill harmful moulds, bacteria, parasites, and insects. The use of irradiation also delays the ripening and the sprouting of onions and potatoes. Its use in the US was approved in 2000. In the UK, the Food Irradiation (England) Regulations 2009 lists the seven categories of food that may be irradiated (fruit, vegetables, cereals, bulbs and tubers, dried herbs, fish and shellfish, and poultry). For each the 'maximum overall average dose' that can be used is specified in units of kilograys (kGy).

irrational number A real number that cannot be represented as a fraction such as π.

irreversible reaction A chemical reaction which is considered to go forwards converting reactants to products. The reverse reaction is in comparison very slow to the point that it is non-existent. The process of combustion of a hydrocarbon fuel is an example of an irreversible reaction in which the fuel is reacting with oxygen to form carbon dioxide and water vapour.

irritant hazard A type of gaseous hazard that causes inflammation or irritation to the eyes, skin, and respiratory system.

isentropic A thermodynamic system in which the entropy remains constant in a reversible adiabatic process. An **isentropic process** is a process that takes place without any change in entropy. A friction-free adiabatic expansion and compression process is an example of an isentropic process.

isobar A condition representing constant pressure. Isobars are used on weather maps representing lines of constant atmospheric pressure. An **isobaric process** is a process that operates with constant pressure.

isochore A condition representing constant volume. It is also known as **isometric.**

isolated system The separation of the process plant from all forms of energy and material flows. *See* SYSTEM. An **isolation scheme** is used in such a way to ensure that hazardous substances are not released or that personnel are not exposed to risks during maintenance or equipment repair.

isomerism The existence of two or more substances having the same chemical composition but different arrangements of their atoms e.g. butane (C_4H_{10}) has two isomers: as a straight four-carbon chain and as a three-carbon chain with a methyl group $(-CH_3)$ in the middle. There are two types of isomerism: structural (i.e. butane) and stereoisomerism, which includes optical and geometric isomerism.

isomerization A process used in petrochemical refineries to convert linear hydrocarbon molecules to higher-octane branched molecules used for blending into gasoline or as feed to alkylating linear alkenes to heavier hydrocarbons such as xylenes.

isometric *See* ISOCHORE.

isometric drawing A three-dimensional engineering drawing of a process plant equipment or pipe run. It is characterized by an **isometric projection** in which the plane of the projection has equal angles to the three principal axes of the object being depicted. All the dimensions are the same and there is no perspective.

isomorphism The existence of identical or similar crystalline forms in different but chemically similar compounds

isopieste A curve or line on a *psychrometric chart representing equal moisture content plotted for a constant pressure.

isopleth A vertical line used in a liquid–vapour phase diagram for two substances. The line corresponds to constant composition as the pressure changes. The isopleth is perpendicular to a *tie line.

isotherm 1. A condition representing constant temperature. **2.** The relation between the moisture content of a substance and the relative humidity in the surrounding air.

isothermal efficiency A measure of the performance of a reciprocating air compressor. It is expressed as a ratio of the isothermal work input to the actual work input. An air compressor is required to raise the pressure of the air with the minimum possible work input, which occurs with isothermal compression. Isothermal compression is an ideal case since none of the work input is absorbed in raising the temperature of the compressed air, i.e. in raising its internal energy.

isothermal process A system or process in which there is no exchange of temperature with the surroundings such that the temperature remains constant. The temperature of the system or process is maintained at a constant value by removal or addition of heat at the same rate at which it is produced or removed. The change of phase of a solid to a liquid as in the process of melting, or of a liquid to a vapour as in the process of boiling, are examples of isothermal processes. *Compare* ADIABATIC PROCESS.

isotonic The exertion of the same osmotic pull by one medium as the other on the other side of a *semi-permeable membrane. This corresponds to the same concentration of particles existing on either side of the membrane.

isotope Atoms of the same element but with a different number of protons and neutrons (i.e. **nucleon number**). They have nearly identical chemical properties. For example, the two isotopes of chlorine are chlorine-35 and chlorine-37. Both have identical atomic numbers and electron arrangements, and have the same chemical properties, but the difference of two neutrons in the nucleus produces a difference of two units of *relative atomic mass. Chlorine comprises 75 per cent chlorine-35 and 25 per cent chlorine-37 by mass, giving a combined relative atomic mass of 35.45. Most elements exhibit isotopy although some do not, such as fluorine, sodium, and aluminium. Tin with an atomic number 50 has ten isotopes, ranging in mass numbers from 112 to 124.

isotropic The condition in which the same properties exist in all directions from a point or location.

iteration A repeating calculation in which the key variables are recycled in the calculation in order eventually to reach a convergence as the solution.

j A symbol used by engineers to represent the square root of –1. It is an *imaginary number which is distinguished from *real numbers. For example, the solution of $x^2 + 1 = 0$ is j and –j. Mathematicians tend to use the symbol i instead of j. *See* COMPLEX NUMBERS.

jacket 1. The outer covering or surrounding of a process vessel or pipe. The jacket is used to contain a heating or cooling medium or used as insulation to prevent heat loss or gain. Water or condensing stream is frequently used as the cooling or heating medium. **2.** The tubular steel support structure placed on the seabed to support the deck of an off-shore structure that is used for drilling and for other *topside modules used for oil and gas separation.

jack-up An offshore oil and gas platform that consists of a triangular- or rectangular-shaped structure with moveable legs that enables the platform to stand on the seabed in depths of up to 120 m. They are used mainly in drilling operations for both explo-ration and for permanent gas installations. The platform can be raised or lowered as required.

Jacobs–Dorr process A wet process used for making phosphoric acid by the reaction of phosphoric acid on phosphate rock. The rock is dissolved in phosphoric acid and calcium sulphate to form a slurry of calcium phosphate. This is then converted to phosphoric acid and calcium sulphate in the exothermic reaction:

$$Ca_3(PO_4)_2 + 3H_2SO_4 \rightarrow 3CaSO_4 + 2H_3PO_4$$

Fluorine in the rock is evolved and is required to be scrubbed from the vent gas. The process was developed in the nineteenth century and there have been a number of variations of which the Dorr–Oliver process is still used.

jarosite process A process used to extract iron from the leach liquors of *hydrometal-lurgical processes. It is named after the mineral and was first used in the processing of zinc sulphate liquors.

jaw crusher A mechanical device used to break coarse raw material such as stone and ore into small pieces. It consists of two jaws, which form a V-shape that has a fixed vertical jaw or anvil and a swinging steel jaw reciprocating on a horizontal plane. The jaws open and close up to 400 times per minute and the material is crushed as it descends under gravity through the opening and closing gap.

JET (Joint European Torus) The largest *tokamak in the world located in Culham, Eng-land and is the only operational fusion experiment that is capable of producing fusion en-ergy. It involves a partnership of several European nations and has been investigating the potential for thermonuclear power by nuclear fusion.

The origins of JET extend back to 1970 when the Council of the European Community provided the legal framework for the development of a European fusion device. After construction of the experimental nuclear reactor, which began in 1977, the first JET plasma was achieved in 1983. In 1997, 16 MW of fusion energy was produced for a total power input of 24 MW. The European Fusion Development Agreement (EFDA) was established in 1999 and has responsibility for the future use of JET and defines JET's scientific programme, allowing for the detailed study of nuclear fusion by many scientists and engineers from all over Europe. While JET is effectively a scientific experiment, ITER is a reactor-scale experiment designed to deliver ten times the amount of power that it consumes. DEMO is expected to be the first fusion nuclear power plant to provide electricity to the national power grid.

(((∰))) SEE WEB LINKS
• Official website of EFDA JET programme.

jet fire A fire that results from the ignition of a release of flammable liquid or vapour. It can be the intentional result of *flaring or accidental result such as through the rupture of a natural gas pipeline or a valve leak. The direction of the jet is dependent on the prevailing wind conditions. The hazard through the exposure to thermal radiation is dependent on the rate of release and on the heat of combustion of the fuel involved. A **jet flame** is the combustion of flammable liquid or gas emerging with significant momentum from an orifice or opening.

jet fuel A kerosene-based product of the fractional distillation of crude oil. It comprises various types of hydrocarbons for use as an aviation fuel for turbine engines. The composition of the fuel is governed by strict international regulations and usually has a freezing point of below −40ºC, which corresponds to the cold temperatures encountered at high altitudes. *Compare* AVIATION GASOLINE.

jet reactor A type of *bioreactor in which air and a fresh feed solution enter at the base in such a manner as to produce a jet of material that provides the energy necessary to agitate the contents. The air, as bubbles, rises in a tube and disengages at the top. The fermenting solution descends through the annulus for recycle. They are typically used for the industrial production of yeast from whey.

j-factor A dimensionless factor used in heat and mass transfer of fluids with turbulent flow in pipes. It is a function of Reynolds number, geometry, and boundary conditions from which the friction factor can be obtained and agrees well with convective heat transfer correlations or for determining heat transfer coefficients. It was proposed by American chemical engineer Allan P. Colburn (1904–55) and forms part of the *Chilton–Colburn analogy, which is used in heat, momentum, and mass transfer.

jig A vibrating mechanical device used to separate water-suspended particles of differing densities. The separation is based on the differences in the rate of their acceleration.

Joliot-Curie, Irène (1897–1956) A French scientist and daughter of Marie and Pierre Curie. She gained her doctorate at the Sorbonne in 1920 having first served as a nurse during the the First World War. She became Doctor of Science in 1925, having presented a thesis on the alpha rays of polonium. She is noted for her work on natural and artificial radioactivity working together with her husband **Jean Frédéric Joliot-Curie** (1900–58). She shared the Nobel Prize in Chemistry with him in 1935 in recognition of the synthesis of new radioactive elements.

Joule, James Prescott (1818–89) A British physicist who was the owner of a large brewery. Joule carried out much of his scientific work in his own private laboratory achieving a remarkably high standard of accuracy in his work. He stated a law about the heating effect of an electric current in 1840, and thereafter carried out a series of experiments to determine the mechanical equivalent of heat. His first determination of mechanical equivalent of heat was carried out with a calorimeter containing a rotating paddle. Together with Lord *Kelvin, he discovered the *Joule–Thomson effect in which there is a fall in temperature of a gas on expansion due to the attraction between molecules. The SI unit of energy is named after him.

joule (Symbol J) The SI unit of work and energy where one joule is equal to the work done by the force of one newton, in the direction of the force, a distance of one metre. That is 1 joule = 1 Nm or 1 kg m^2 s^{-2}. It is named after the British physicist James Prescott *Joule. The rate of work equal to one joule per second is a called a *watt.

Joule–Kelvin effect *See* JOULE–THOMSON EFFECT.

Joule's laws 1. In a wire of given resistance, the heat developed in a given time is proportional to the square of the current. **2.** The internal energy of a gas at constant temperature is independent of its volume, provided the gas is ideal.

Joule–Thomson effect The cooling that results when a highly compressed gas is allowed to expand adiabatically into a region of low pressure such that no work is done. The cooling effect occurs because as the molecules of the real gas separate during expansion, internal work is done in overcoming the attractive forces between them. A perfect gas, with no attractive forces between the molecules, shows no Joule–Thomson effect. The Joule–Thomson effect is more marked at lower temperatures and was used in the *Linde process for the liquefaction of air. The phenomenon was discovered by James *Joule working with William Thomson (later Lord *Kelvin). Hydrogen is anomalous to the Joule–Thomson effect, by showing a rise in temperature at ambient temperature. This effect continues down to 193 K whereupon it cools under expansion. This is called the inversion temperature. Hydrogen was liquefied by *Dewar in 1898, who cooled the gas below the inversion temperature by liquid air, and then used the principle of the Linde process. Helium, like hydrogen, is also anomalous, with an inversion temperature of 33 K.

JPL chlorinolysis process A process used for the desulphurization of coal by oxidation with chlorine. The reaction converts sulphur to sulphur monochloride. The process was developed by the Jet Propulsion Laboratory at the California Institute of Technology in the US.

Kármán *See* VON KÁRMÁN.

Kármán vortex street A phenomenon in which vortices of a moving fluid form as repeating patterns. They are caused by the unsteady separation of flow of a fluid as it passes around an object such as a wire and observed over Reynolds numbers of around 90. It is named after Hungarian-American mathematician and physicist Theodore *von Kármán (1881–1963) and is also known as the **von Kármán vortex street**.

kelvin (Symbol K) The SI unit of thermodynamic temperature. The kelvin temperature scale is equal to the Celsius degree but with zero being absolute zero (0 K). The temperature is expressed in degrees Celsius less 273.15 (i.e. °C = K-273.15). The term 'degrees Kelvin' is no longer used. The unit is named after Lord *Kelvin.

Kelvin, Lord (1824–1907) A Belfast-born Scottish scientist William Thomson, later 1st Baron Kelvin of Largs, who was the son of a gifted teacher. Both he and his brother James matriculated to Glasgow University aged 10 and 12, respectively. William Thomson was elected to the chair of natural philosophy at Glasgow in 1845 aged 22 and held the position for 53 years. His most important work was on thermodynamics, but he is most widely known for his studies of electricity applied to submarine technology. He was knighted in 1866 and made a baron in 1892.

kerosene A combustible liquid hydrocarbon fuel obtained from petroleum in the fractional distillation of oil once gasoline has been removed. It is also known as paraffin. It has a calorific value of around 43.3 kJ kg^{-1} and is used as a fuel for jet engines and domestic uses such as heating and cooking.

kettle reboiler A type of horizontal shell-and-tube type heat exchanger in which liquid to be evaporated is contained on the shell side and condensing steam or a hot liquid used to provide the heat contained on the tube side (see Fig. 24). The shell contains a relatively small bundle of hairpin tubes in a two-pass arrangement in a floating head and tube sheet, and sits in a pool of the boiling liquid, the depth being determined by an overflow weir. A flow of fresh liquid continuously enters from the bottom while vapour leaves from the top of the shell. The condensing steam in the tubes is removed through a trap. Kettle reboilers are typically used with distillation columns to boil up the bottom material and return vapour back into the column for separation.

key component Used in multicomponent distillation, key components are the two main components that are required to be identified in a mixture to be separated. The two key components are the **light key** (LK), which is identified as the lightest component in the *bottoms, and the **heavy key** (HK), which is the heaviest component in the *distillate. They are known as key components since their identification is required to unlock the problem in solving the separation where the difficulty of separation, as measured in terms of the number of trays required for a given reflux ratio, is fixed by the concentration of the key

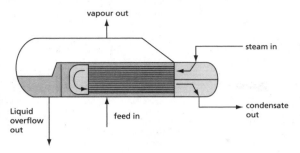

Fig. 24

components in the products. The relative volatilities of the components in a mixture are always determined with respect to the heavy key. Components that are lighter than the light key are termed LLK while those that are heavier than the heavy key are termed HHK.

K-factor A vapour–liquid equilibrium ratio of a substance:

$$K_i = \frac{y_i}{x_i}$$

where x and y are the mole fractions of the substance (i). It is used to describe the complex relationship between pressure, temperature, and equilibrium for vapour- and liquid-phase compositions, and used in multicomponent distillation calculations. For mixtures of substances of similar molecular structure, such as hydrocarbons, the K-factor is dependent on pressure and temperature, and commonly presented as *nomographs known as **K-factor charts**, or *DePriester charts. These present pressure, temperature, and K-factors for various light and heavy hydrocarbons. They are used to determine the *bubble point and *dew points of hydrocarbon mixtures. Given a multicomponent mixture of known composition and total pressure, the bubble point temperature can be found by *trial and error which satisfies the condition. The factors are either used as they stand or are converted to relative volatility values. The bubble point is found from:

$$\sum_{i=1}^{n} K_i x_i = 1.0$$

Similarly, for a given vapour composition, the dew point can be found which satisfies the condition

$$\sum_{i=1}^{n} \frac{y_i}{K_i} = 1.0$$

Sensible choices of temperature or pressure can mean that only three values are required.

kiln A type of brick-lined oven used for the hardening, burning, baking, or drying of products such as *calcining lime or firing pottery. The products may be heated in a batch kiln or continuously fed. The kiln may be fixed or stationary, or may be a rotating cylinder through which heated air is fed and used to dry and bake the product. **Kilning** is the processing of materials in a kiln.

kilo- (Symbol k) A prefix used in the metric system to denote a factor of 1,000.

kilogram (Symbol kg) The SI unit of mass, it is equal to the mass of the international platinum-iridium cylinder kept by the International Bureau of Weights and Measures at Sèvres, near Paris.

kilomole (Symbol kmol) The derived SI unit for the amount of a substance that is equal to 1,000 moles.

kilowatt-hour (Symbol kWh) A unit for electrical energy, it is equivalent to the power consumption of 1,000 watts for one hour or 3.6 MJ. It is commonly used as the commercial unit of electrical energy of electricity delivered to consumers.

kinematics The branch of mechanics that is concerned with the motion of bodies considered abstractly and without reference to force or mass. *Compare* DYNAMICS.

kinematic viscosity (Symbol v) The ratio of the *viscosity of a liquid to its density. It is a useful for quantifying the viscous properties of certain fluids such as hydrocarbon mixtures. The SI units are $m^2 s^{-1}$.

kinetic energy The energy of a body by virtue of its motion:

$$KE=\frac{1}{2}mv^2$$

where m is the mass of the body and v is its velocity. The SI units are $kg\, m^2\, s^{-2}$ or joules.

kinetic parameters The constants in a mathematical model used to describe the speed of a chemical or biochemical reaction. For example, in *Monod kinetics, they are used to describe the rate of growth of microbial cells in which the two kinetic parameters are the *maximum specific growth rate μ_{max} and the saturation constant, Ks.

kinetics 1. The science of the relationship between the motion of bodies and the forces acting upon them. **2.** The study of the rate at which chemical reactions take place under different conditions such as due to temperature, pressure, and the presence of catalysts.

kinetic theory A theory that attempts to understand the properties of liquids and gases. It is based on considering them as a large population of molecules moving freely relative to one another. The temperature of a gas is seen to be a measure of the velocity of its molecules.

Kirchhoff, Gustav Robert (1824-87) A German physicist who founded the science of spectroscopy. He discovered the laws that govern the absorption and emission of radiation and the flow of electricity in electrical networks. In 1859 he presented the law that states that the ratio of the emission and absorption powers of all materials is the same at a given temperature and a given wavelength of radiation produced. He went on to derive the concept of a perfect *black body that can absorb and emit radiation at all wavelengths.

Kirchhoff's law of radiation A law stating that the emissivity of a body is equal to its absorbance at the same temperature. The law was formulated by the German physicist Gustav *Kirchhoff (1824-87).

Kirkbride equation An equation used for the design of distillation columns to determine the ratio of the number of trays above and below the feed point. It is therefore used to determine the location of the feed tray in the column:

$$\ln\frac{N_D}{N_B} = 0.206\ln\left[\frac{B}{D}\left(\frac{x_{F(HK)}}{x_{F(LK)}}\right)\left(\frac{x_{B(LK)}}{x_{D(HK)}}\right)^2\right]$$

where N_D is the number of theoretical trays above the feed and N_B is the number below. B refers to the flow at the bottom and D at the top. x is the mole fraction and subscripts F, B, and D refer to the feed, bottom, and top, where LK and HK are the light and heavy *key components. It is named after American chemical engineer Chalmer Gatlin Kirkbride (1906–98).

Kirpichev number A dimensionless number, Ki, used in heat and mass transfer used to relate drying conditions of materials. It is expressed as the ratio of the external to internal heat or mass transfer intensity:

$$Ki_h = \frac{qL}{k-\Delta T} \text{ and } Ki_m = \frac{GL}{D\rho n}$$

where q is the heat flux, L is a characteristic length of the body, k is the thermal conductivity, ΔT is the temperature difference, G is the mass velocity, D is the diffusivity, ρ is the density, and n is the specific mass content. They are named after Russian materials scientist Victor Kirpichev (1845–1913).

k$_L$a The volumetric mass transfer coefficient seen as a measure of the transfer of a gas to a liquid in processes such as fermentation in which oxygen or air is sparged into a liquid containing microorganisms. It is the product of the mass transfer coefficient, k_L, and the interfacial area, a. It is difficult to measure the mass transfer coefficient separately but it is readily measured from gas balances in combination with the interfacial area.

kmol An abbreviation for the SI unit for *kilomole where 1 kmol is equal to a thousand moles (1,000 mol).

knocking The premature explosion of an air–fuel mixture within a piston cylinder of an internal combustion engine. It is due to the over-compression of the mixture and gives a characteristic metallic knocking sound. When the fuel and air mixture is injected into the cylinders of the engine, the mixture is combusted by the use of spark ignition. However, as the flame travels from the sparking plug towards the piston, it compresses the unburnt fuel. Ignition of the unburnt fuel occurs before the flame reaches it, resulting in shock waves that travel up and down the cylinder. The consequence is noise, a loss of power, and damage to the sparking plugs. Knocking is avoided by reducing the compression ratio, although this also results in a loss of efficiency. Antiknocking agents such as tetraethyl lead added to the fuel that increase the octane number and reduce the tendency to detonate are known to cause air pollution and environmental damage, and their use is either much reduced or banned in some countries. New fuels have been designed to have a higher octane number with a higher aromatic content. However, the presence of unburnt aromatics in the atmosphere has an adverse effect on human health.

knockout drum A vessel used to trap droplets of liquid from a gas or vapour stream from a process. They are typically used to remove liquids such as hydrocarbons that would otherwise damage a vent stack or present a hazard if released into the atmosphere. They are also known as **KO drums** or **KO pots**.

Knudsen diffusion A type of diffusion applied to gases that diffuse through the narrow pores and capillaries of a porous material in which the collisions between the gas molecules and the walls of the pores are appreciable, and consequently influence the rate of diffusion. The movement or motion of gas at low pressure is the result of individual molecular free path motion. **Knudsen flow** occurs where the mean free path is greater than the radius of the aperture by about a factor of ten. It is named after Danish physicist Martin Hans Christian Knudsen (1871–1949).

KO drum *See* KNOCKOUT DRUM.

Kohlrausch's law A law devised by German chemist Friedrich Kohlrausch (1840–1910) that states that the conductivity of a dilute solution is equal to the sum of the independent values. That is, the conductivity of a solution is equal to sum of the molar conductivity of the cations and the molar conductivity of the anions. The law is based on the independent migration of the ions and was deduced experimentally.

koji process A solid substrate fermentation process involving the fermentation of grain and soybeans by the *Aspergillus* fungus. The cooked substrate is inoculated with a culture of the filamentous fungi and grown as a shallow layer in which amylase and protease enzymes break down the substrate to form koji, which is a proteolytic enzyme used for treating flour proteins. It is also the basis of other similar fermentations such as the production of citric acid, and rice fermentations including the rice wine saké.

Kolmogorov, Andrey Nikolaevich (1903–87) A Russian mathematician noted for major contributions in various scientific fields including classical mechanics, probability theory, and fluid mechanics. He studied at the Moscow State University and the Chemistry Technological Institute, graduating in 1929. He became a professor at Moscow University two years later, and went on to become the world's foremost expert in probability theory. In 1935 he became the first chairman of probability theory at the Moscow State University. He was elected to the USSR Academy of Sciences in 1939. He held a number of positions at Moscow State University including head of several departments and served as the dean of the faculty of mechanics and mathematics.

Kolmogorov eddies The smallest size of eddy that can be formed in a fluid. The size of an eddy is dependent on the viscosity of the fluid. In a stirred vessel, the Kolmogorov eddy size also decreases with increasing stirrer speed. Large-sized eddies are unstable and eventually break up to form smaller eddies with the kinetic energy of the initial large eddies being dissipated into the smaller eddies stemming from it. These smaller eddies undergo the same process, giving rise to even smaller eddies. Thus the energy is dissipated to the point that the viscosity of the fluid can effectively dissipate the kinetic energy into internal energy. The theory was first proposed in 1941 by Russian mathematician Andrey Nikolaevich *Kolmogorov (1903–87).

Kopp's rule A generalized rule used to determine the heat capacity of solids. The heat capacity of a solid compound is approximately equal to the sum of the heat capacities of the constituent elements. The rule is used only where experimental values are not available and applies only over a limited temperature range.

Kossovitch number A dimensionless number, Ko, equal to the ratio of the heat energy required for a change in phase to the heat required for heating or cooling:

$$Ko = \frac{\lambda}{c_p \Delta \theta}$$

where λ is the latent heat (of fusion or vaporization), c_p is the specific heat of the body, and $\Delta \theta$ the difference between the phase change temperature point of the material and the temperature of the heating or cooling medium.

Kozeny–Carman equation *See* CARMAN–KOZENY EQUATION.

kraft process A process used for the conversion of wood into wood pulp that was invented by Carl Dahl in 1879. The process is also known as the **sulfate process** and consists of cooking wood in a basic solution of sodium hydroxide, sodium sulphide, and sodium carbonate. The process involves the hydrolysis of lignin to acids and alcohols. Both sodium hydroxide and sodium hydrosulphide (NaSH) are formed by the reaction:

$$Na_2S + H_2O \rightarrow NaOH + NaSH$$

which are active in breaking down the bonds between the lignin and cellulose. The pulping takes place in a continuous digester. Spent cooking liquor known as black liquor contains all the chemicals used in the digestion, and is recovered and concentrated by *multiple effect evaporation before being burnt in a recovery furnace. A fused salt melt known as green liquor is collected at the bottom, which is treated with burned lime to convert sodium carbonate to sodium hydroxide:

$$CaO + H_2O \rightarrow Ca(OH)_2$$
$$Na_2CO_3 \rightarrow Ca(OH)_2 \rightarrow CaCO_3 + 2NaOH$$

The calcium carbonate precipitate is removed by filtration. *Calcination is then used to produce calcium oxide in a lime *kiln:

$$CaCO_3 \rightarrow CaO + CO_2$$

After treatment with calcium oxide the filtrate serves as fresh cooking liquid for the digester. With the small loss of sodium salts in the digestion process, sodium sulphate is added to the recovery furnace as the make-up chemical, hence the alternative name for the process:

$$Na_2SO_4 + 2C \rightarrow Na_2S + 2CO_2$$

Kremser equation An equation used in the calculation of the number of theoretical stages required in the design of an absorption column used for absorption processes such as the drying of natural gas. It is based on the condition that the pressure divided by the product of *Henry's law constant and the ratio of moles liquid to moles vapour is a constant.

Kroll process A process used in the production of certain metals such as titanium by reduction of its chloride with magnesium metal. It was named after William J. Kroll, who invented the process in 1937, replacing the earlier *Hunter process. The process involves reducing refined rutile from its ore at 1,000°C within a *fluidized bed with chlorine gas to produce titanium tetrachloride ($TiCl_4$). This is then reduced using an excess of liquid magnesium at around 800°C in the reaction:

$$2Mg + TiCl_4 \rightarrow 2MgCl_2 + Ti$$

The titanium is then purified by *leaching and *vacuum distillation. The process is expensive to operate, but the value of titanium metal is far greater than stainless steel and has numerous applications, particularly in the nuclear industry. The process is also used for the commercial production of tantalum, zirconium, and niobium.

k

laboratory A building, room, or facility used for carrying out scientific or engineering research experiments or for practical teaching purposes. Routine and specialist analytical testing, *lab-scale process testing, and validation are also carried out in laboratories.

lab-scale The small-scale testing of processes, chemical reactions, mixing, heating, cooling, separation, and other unit operations used in the validation of the design of full-scale process plant or the testing of existing process plant parameters. It involves small quantities of materials with relatively inexpensive equipment tested over short timescales.

lagging Insulation material used to surround process pipes and process vessels to reduce heat transfer. The properties of the insulation material are a low thermal conductivity and effectiveness at minimizing heat transfer, cost, and durability. It should also be corrosion-proof, water and weather-proof. The correct thickness of lagging should be determined based on economics. There is an optimum thickness of lagging for pipes and it is a balance in cost between thickness and surface area which increases with increasing thickness.

lagoon A wide, shallow body of water used for settling and separating fine particles from process liquids. It can also be used for collecting process water and other collected water prior to discharge. The water can also be used as emergency water for fire fighting.

lag phase The initial period of time in a bioreactor after inoculation of microorganims into the nutrient medium in which there is no growth. It is the period of time of apparent inactivity in which the microorganims adjust to their new environment.

Lagrange, Comte Joseph Louis (1736–1813) An Italian-born French mathematician and astronomer noted for his work in mechanics, harmonics, and in the calculus of variations. He also established the theory of differential equations. He succeeded Swiss mathematician and physicist Leonhard Euler (1707–83) as the director of mathematics at the Prussian Academy of Sciences in Berlin, during which time he published his work in *Mécanique Analytique* (1788), that covered every area of pure mathematics.

Lagrange fluid dynamics The study of fluid mechanics in which the movement of small elements in the moving fluid are studied. It is named after Joseph Louis Lagrange (1736–1813). *Compare* EULERIAN FLUID DYNAMICS.

lag time 1. The apparent delay in growth of a microorganism within a bioreactor after inoculation as the living cells adapt and establish themselves to their environment and multiply to appreciate levels. *See* LAG PHASE. **2.** The delay in response of a system or process that is being controlled to a step change disturbance.

laminar flow The streamline flow of a fluid in which a fluid flows without fluctuations or turbulence. The velocities of fluid molecules are in the direction of flow with only minor movement across the streamlines caused by molecular diffusion. The existence was first demonstrated in 1883 by Osborne *Reynolds, who injected a trace of coloured fluid into a flow of water in a glass pipe. At low flow rates the coloured fluid was observed to remain as discrete filaments along the tube axis, indicating flow in parallel streams. At increased flow rates, oscillations were observed in the filaments, which eventually broke up and dispersed across the tube. There appeared to be a critical point for a particular tube and fluid above which the oscillations occurred. By varying the various parameters Reynolds found that his results could be correlated into terms of a dimensionless number called the *Reynolds number, Re, as:

$$Re = \frac{\rho v d}{\mu}$$

where ρ is the density of the fluid, v is the velocity of the fluid, d is the diameter of the pipe, and μ is the viscosity of the fluid. The critical value of Re for the break-up of laminar flow in pipes of circular cross section is about 2,000.

Langmuir, Irving (1881–1957) An American chemist who taught chemistry before becoming a research chemist and later research director for General Electric Company's research laboratory at Schenectady, a post he held for 41 years. He proposed a theory of atomic structure and discovered how to make atomic hydrogen, and invented an atomic hydrogen blowpipe giving an extremely hot flame for welding metals. He developed a high-vacuum pump and electric discharge tubes for radio. He measured the sizes of virus molecules, which he obtained in layers one molecule thick. He is also noted for his work on surface chemistry and the molecular orientation on surfaces, the theory of adsorption catalysis, and the understanding of plasmas. He was awarded a Nobel Prize in Chemistry in 1932.

Langmuir adsorption isotherm An equation expressing the equilibrium fraction, f, of an adsorbent homogeneous surface which is covered with a single layer of adsorbed and non-interacting molecules as:

$$f = \frac{ap}{1+ap}$$

where a is an empirical adsorption constant known as the **Langmuir adsorption constant** and p is the equilibrium partial pressure of the adsorbate in the gas phase. Adsorption isotherms are used in the design of gas–solid processes such as solid catalyzed gas phase reactions as in the synthesis of ammonia. It is named after American chemist Irving *Langmuir (1881–1957).

Laplace, Pierre Simon (1749–1827) A French professor of mathematics noted for his contribution to mathematics. Laplace was the son of a farm labourer who owed his education to some rich neighbours. He solved many problems concerning the motion of the solar system. He also worked on surface tension and, with Antoine *Lavoisier (1743–94), did work on the measurement of the heat produced during chemical changes. In 1822 he persuaded the French Academy to carry out experiments to measure the velocity of sound over a distance of eleven miles by firing cannons. The result confirmed his belief that when sound waves pass, the air cools as it expands and heats up as it contracts.

Laplace transform A method of obtaining a solution to a differential equation where the unknown integration constants are obtained using straightforward algebra. It is commonly used in process control applications since differential equations do not readily enable the relationship between the input and output to be discerned. The Laplace transformation therefore allows a simpler algebraic calculation to be performed. The Laplace transform of a function $f(t)$ is therefore multiplied by e^{-st} and the product integrated between zero and infinity. It is denoted by $\mathcal{L}\{f(t)\}$ as:

$$F(s)=\mathcal{L}\{f(t)\}=\int_0^\infty e^{-st}f(t)dt$$

where s is a variable whose values are chosen such that the semi-infinite integral converges (i.e. the integration is between 0 and $+\infty$ and is therefore one-sided). For the Laplace transform to exist, the integrand $e^{-st}f(t)$ must converge to zero as t approaches infinity.
As an example, the Laplace transform of a unit step function is:

$$\int_0^\infty 1e^{-st}dt=\left[\frac{e^{-st}}{-s}\right]_0^\infty=\frac{1}{s}$$

If $F(s)$ is the Laplace transform of $f(t)$ then $f(t)$ is the inverse Laplace transform of $F(s)$. Thus:

$$f(t)=\mathcal{L}^{-1}\{F(s)\}$$

There is no simple definition of the inverse transform and the solution is found in reverse. Tables of Laplace transforms and their inverse transforms are used.

latent heat (Symbol λ or L) Literally meaning 'hidden heat', it is the quantity of heat absorbed or released when a substance changes its physical phase at constant temperature and pressure. The **latent heat of fusion** is the energy needed to be removed to solidify a liquid, whereas the **latent heat of vaporization** is the energy required to be absorbed to change a liquid to a gas. The *specific latent heat is the heat energy absorbed or released per unit mass at constant temperature, while the *molar latent heat is measured per mole of substance. The SI units are J kg^{-1} or J mol^{-1}, respectively.

lattice The regular and organized arrangement of atoms, ions, or molecules in a stationary structure, such as in a crystal. The arrangement is usually three-dimensional such as in a diamond. However, two-dimensional crystal structures exist such as graphite which has a layer lattice in which the atoms of carbon are chemically bonded in planes, with relatively weak forces between the planes. The **lattice energy** is the energy that is released when one mole of a solid ionic compound is formed from its constituent gaseous ions. It is alternatively defined as the energy required to separate completely the constituent ions of a crystal from each other to an infinite distance. The lattice energy is effectively a measure of the stability of a crystal lattice.

Lavoisier, Antoine Laurent (1743–94) A French chemist noted for determining the components of water as being oxygen and hydrogen, and establishing that sulphur is not a compound but an element. He also helped to formulate the metric system. Through very careful experimentation he was the first to show that although matter can change its state in a chemical reaction, the total mass remains the same at the end as at the beginning of every chemical change. He was the first to state the conservation of mass. During the French Revolution, he was accused of various crimes and guillotined.

law A description of a principle of nature that covers all circumstances by the wording of the law. There is no ambiguity or exceptions. Eponymous laws are named after their discoverers such as *Charles's law while others are known by their subject matter such as the *law of conservation of mass.

law of chemical equilibrium *See* EQUILIBRIUM CONSTANT.

law of conservation A law that states that the total quantity of something remains unchanged within a system even though there may be changes taking place within the system such as chemical reactions, changes of states, and other physical, chemical, and biochemical changes. The conservation of mass is a law that states that the total amount of material within a system such as a chemical process remains unchanged. That is, the total amount within the system boundary does not increase nor decline. It forms the basis of a material balance for a process. The conservation of energy is a law that states that the total energy within a system remains unchanged. The Bernoulli theorem is used to show the total energy in a fluid is conserved and takes into consideration pressure-volume, kinetic, and static energies. The total enthalpy of chemical components remains unchanged in spite of any chemical reactions. It forms the basis of an energy balance for a process.

law of universal attraction A law proposed by Sir Isaac *Newton (1642–1727) stating that any two bodies attract each other in proportion to the product of their masses and inversely to the square of their distances apart. To do this, he had to define the meaning of mass, momentum, force, inertia, and to state his three *laws of motion. These ideas appeared to show that the entire universe obeyed simple mechanical laws and remained undisputed until Einstein.

laws of motion Three laws proposed by Sir Isaac *Newton (1642–1727) that state:

First law: Every body continues in its state of rest or of uniform motion in a straight line unless acted on by an impressed force.

Second law: The rate of change of momentum of a body is proportional to the impressed force and takes place in the direction of the force.

Third law: To every action there is a reaction, which is equal and opposite to the action.

layer of protection analysis (LOPA) A form of semi-quantitative *risk analysis used for a process or system that analyses individually the different forms or layers of protection; it is used to reduce or mitigate process risk in terms of reducing the likelihood of the occurrence of an undesirable event or reducing its severity. LOPA is used to determine and understand the *safety integrity level (SIL) and is a widely used technique in the oil and gas industry. It uses various hazard analysis techniques and evaluates the frequency of potential incidents as well as the probability of failure of the layers of protection.

LCA *See* LIFE CYCLE ANALYSIS.

LC$_{50}$ The *lethal concentration of an airborne or waterborne chemical substance that causes the death of half (50 per cent) of a group of exposed test animals in a given time, which is usually four hours. It is used as a measure of the toxicity of a substance and is expressed in micrograms or milligrams per litre, or parts per million of air or water. The lower the value, the more toxic the substance.

LD$_{50}$ The *lethal dose of a chemical substance or other agent that is expected to produce death in 50 per cent of the group of test animals such as mice or rats exposed to one dose. It is often quoted in milligrams per kilogram of body weight. A substance is considered to be highly toxic if the LD$_{50}$ is less than 50 mg per kg of body weight.

leachate The solution that leaves a *leaching process and consists of the solvent containing the leached material as the solute. For example, in the decaffeination of coffee beans using either water or supercritical carbon dioxide, the leachate is the solvent carrying the dissolved caffeine.

leaching A separation process involving the extraction of soluble components of a solid by percolating a solvent through it. The solid carrier material is inert and contains the solute for extraction. The solvent is added to selectively dissolve the solute. The overflow or *leachate consists of the solvent and dissolved solute, while the underflow consists of a slurry with a similar composition to the overflow and also contains the solid carrier. In an ideal leaching equilibrium stage, all the solute is dissolved by the solvent and none of the carrier is dissolved. The efficiency of the leaching process is dependent on the type of equipment used and the properties of the materials used. Examples include the separation of metal from ore using acid, the extraction of caffeine from coffee beans or tea leaves using hot water, and the extraction of sugar from sugar beets using hot water.

lead chamber process A process once used for the large-scale manufacture of sulphuric acid by the oxidation of sulphur dioxide with air using a potassium nitrate catalyst in water. This involved burning iron pyrites, FeS_2, or some other source of sulphur such as zinc sulphide, ZnS, to produce sulphur dioxide which was then oxidized in large lead chambers to sulphuric acid by the action of air, oxides of nitrogen, and water:

$$4FeS_2 + 11O_2 \rightarrow 2Fe_2O_3 + 8SO_3$$
$$SO_3 + H_2O + NO_2 \rightarrow H_2SO_4 + NO$$
$$2NO + O_2 \rightarrow 2NO_2$$

The process was carried out in expensive lead chambers but produced only dilute acid. The process was invented by English physician and industrialist John Roebuck (1718–94) in 1746 and was later replaced by the more economic *contact process in 1876.

leading edge In fluid mechanics, the edge of an aerofoil that first encounters a fluid stream such as air.

least squares method A statistical method of deriving an average value from a set of approximate or inaccurate values by introducing errors as unknown variables and then minimizing the sum of their squares. The method was devised by German mathematician Karl Friedrich Gauss (1777–1855) as a way of best fitting a straight line through a set of plotted data points. *See* REGRESSION ANALYSIS.

Leblanc, Nicolas (1742–1806) A French chemist and physician who invented a process for the manufacture of soda ash (sodium carbonate) from salt (sodium chloride). He began his studies in medicine in Paris to train as a physician. When the French Academy of Sciences offered a prize for the conversion of inexpensive salt to the more highly valued soda ash in 1775, he invented a two-step process. He was awarded the prize for the process that used sea salt and sulphuric acid as the raw materials. The process was successfully developed but was confiscated by the French revolutionary government, which refused to pay him the prize money he had rightfully earned. Napoleon, however, later returned the plant to him in 1802 although not the prize. Unfortunately, Leblanc was not able to afford to run the process and he committed suicide in 1806.

Leblanc process A process used in the nineteenth century for the manufacture of sodium carbonate from salt and superseded an earlier process of manufacturing soda from the ash of seaweed. The essential features of the Leblanc process were the conversion of sodium chloride to sodium sulphate by the action of heated sulphuric acid:

$$NaCl + H_2SO_4 = Na_2HSO_4 + HCl$$
$$Na_2HSO_4 + NaCl = Na_2SO_4 + HCl$$

The sodium sulphite was then reduced to sulphide and converted to carbonate by double decomposition with calcium carbonate in the form of limestone:

$$Na_2SO_4 + 4C = Na_2S + 4CO$$
$$Na_2S + CaCO_3 = CaS + Na_2CO_3$$

The salt cake was mixed with limestone and powdered coal and heated in a rotary furnace. The product was known as black ash. The sodium carbonate was extracted with water and crystallized from the calcium sulphide by-product. Invented in 1783 by the French chemist Nicolas *Leblanc (1742–1806), this was the first synthetic production of sodium carbonate. The Leblanc process was superseded by the cheaper *Solvay process and is the only method still used for making sodium carbonate.

Le Chatelier, Henry Louis (1850–1936) A French professor of chemistry who investigated the chemistry of silicates, cement, ceramics, and steels. In 1887 he developed the platinum/platinum-rhodium thermocouple, which made possible the accurate measurement of very high temperatures. He worked on electrical conductivity and on specific heats of gases and discovered the effects of temperature and pressure on a system in equilibrium. *See* LE CHATELIER'S PRINCIPLE.

Le Chatelier's principle A law that states that if a system in physical or chemical equilibrium is subjected to a change in temperature, pressure, or concentration, then the system will automatically alter itself so as to reduce the effects of the change. It can be shown that this law must be true if energy is neither created nor destroyed.

Lee–Kesler equation of state An extended form of the *Benedict-Webb–Rubin equation of state used to describe the vapour–liquid equilibrium data of various substances. It expresses the *compressibility factor, z, as a linear function of an *acentric factor.

Lee's disc A device for determining the thermal conductivity of poor conductors in which a thin, cylindrical slice of the substance under study is sandwiched between two copper discs. A heating coil or steam chest is placed on one of these discs and the temperatures of the two copper discs are measured, from which the thermal conductivity of the substance can be determined from the heating profile across the thickness.

Leidenfrost effect A phenomenon in which a liquid in near contact with a surface at a much higher temperature than the liquid's boiling point produces an insulating vapour layer which prevents the liquid from boiling rapidly. It is named after German doctor of medicine and theologian Johann Gottlob Leidenfrost (1715–94).

Lessing ring A type of packing material used in packed columns and towers that provides a high surface-to-volume ratio. It consists of a hollow cylinder with an internal structure such as a bar or cross arrangement that provides strength and additional surface area (see Fig. 25). They are made from various materials such as plastic, glass, and ceramic, and are available in various sizes. *See* PACKING.

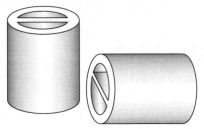

Fig. 25

lethal dose A measure of the toxicity of a chemical substance or ionizing radiation that leads to death. It is usually expressed as the amount of chemical or ionizing radiation per unit body weight. The median lethal dose denoted as *LD_{50} represents the dose that results in the death of half (50 per cent) of the exposed animals within a population within a defined period of time such as within a two-week period. The subscript represents the statistically expected level of mortality. For example, LD_{99} represents a lethal dose leading to the death of 99 per cent of those exposed. The **lethal concentration** is the concentration of an airborne chemical substance over a four-hour period that causes the death of exposed test animals. Where half of test animals in a population die, the lethal concentration is denoted as *LC_{50}.

level alarm An alarm alerting an operator that the level of a substance within a vessel has deviated significantly from the required *set point sufficient to be of concern. A *low-level alarm signifies that a vessel may be nearly empty whereas a high-level alarm may indicate that a vessel is at capacity or near to overflowing.

level control A method of controlling the level of substance such as a liquid contained within a vessel. The level can be controlled either by adjusting a valve on the inlet to the vessel or, if there is a constant flow in, by adjusting a valve on the outlet. A simple form of level control consists of a ballcock in which the position of a float on the surface of the liquid activates the opening and closing of a flow valve.

level gauge An instrument used to indicate the level of a liquid or solid in a tank or process vessel. A simple level gauge consists of a vertical glass or plastic tube attached to the bottom of the tank allowing the liquid to rise to the level in the tank. A *pneumercator is a device that is used to determine the level of a liquid in a tank and consists of a vertical tube through which a gas is gently passed as bubbles through the liquid. The level in the tank is determined from the supplied pressure to overcome hydrostatic pressure.

lever rule A rule that enables the relative amount of two phases in a mixture that are in equilibrium to be found by the construction of a *phase diagram. The lever rule states that $n_1 l_1 = n_2 l_2$ where n_1 and n_2 are the amounts of phase 1 and 2, and l_1 and l_2 are the distances along the horizontal *tie lines of the phase diagram. The lever rule takes its name from the mechanical lever rule relating the moments of two masses about a pivot.

Lewis–Matheson A stage-by-stage calculation used for solving multicomponent distillation problems. The calculation operates from both ends of the column and works towards the middle. The calculations are required to 'mesh' somewhere in the column, which is usually at a feed stage. For more than one feed a choice of mesh points is made for each component and, if the components vary widely in volatility, the same mesh point cannot be used for all components if serious numerical difficulties are to be avoided. The procedure is subject to large truncation-error build-up where components differ widely in volatility. Arbitrary procedures are also needed for non-distributed components whose concentration in one of the product streams is smaller than the smallest number carried by the computer. The concentrations for these components do not naturally take on non-zero values at the proper point as the calculations proceed through the column.

Lewis number A dimensionless number, Le, used in heat and mass transfer that characterizes a particular substance as the ratio of the *thermal diffusivity, α, to the molecular diffusivity, D:

$$Le = \frac{\alpha}{D}$$

It is named after American chemical engineer Warren K. Lewis (1882–1975). *See* LYKOV NUMBER.

Lewis–Randall rule A thermodynamic rule stating that the fugacity of the species in an ideal solution is proportional to the mole fraction of each species in the liquid phase. It is named after American chemists Gilbert Newton Lewis (1875–1946) and Merle Randall (1888–1950).

LFL *See* LOWER FLAMMABLE LIMIT.

life cycle analysis (LCA) A systematic set of procedures for compiling and examining the inputs and outputs of materials and energy consumed within a process, and the associated environmental impacts directly attributable to the functioning of a product or service throughout its life cycle. It takes a cradle-to-grave approach, starting from the origin of the raw materials from natural resources such as oil wells or extraction from ores, and follows them through transformation into useful products, the use by the consumer, recycling where possible, and eventual disposal. Within each step of the life cycle, waste is created. It is therefore more useful than just concentrating on minimizing waste since problems often lie elsewhere within the life cycle.

The components of a life cycle analysis include the **life cycle inventory**. These involve a complete resource requirement to be identified in terms of materials and energy. The **life cycle impact assessment** characterizes and assesses the effects of the environmental emissions. The **life cycle improvement analysis** is used to quantify the life cycle inventory and import, and is used to assess possible environmental improvements that can be made.

light ends Hydrocarbon fractions in oil refining that have a low boiling point and are easily evaporated. This corresponds to butane or lighter components.

light key *See* KEY COMPONENTS.

light phase The liquid with the lower density in a liquid–liquid extraction separation process and sits above the *heavy phase, which has the higher density.

lignin An organic substance found in the woody parts of plants and is associated with cellulose. It is the raw material used in papermaking such as the sulphite process. In the *kraft process it is cooked with a solution containing sodium hydroxide and sodium sulphide.

limiting factor An environmental variable or a nutrient restriction on the growth of a microorganism in a biotechnological process. While limiting factors may be seen as a form of deficiency, they are useful in bioreactors for the growth of microorganisms where growth may be intentionally limited such that the level of a particular nutrient exceeds the limits of tolerance for the microorganism.

limiting oxygen index The lowest oxygen concentration in an oxygen–nitrogen mixture at which a substance will continue to burn by itself.

limiting reactant A reactant consumed in a chemical reaction that is present in the smallest relative amount to the other reactants, which are not in their stoichiometric proportion. The *excess reactants remain unreacted. For example, in the oxidation of o-xylene to phthalic anhydride:

$$C_8H_{10} + 3O_2 \rightarrow C_8H_4O_3 + 3H_2O$$

three moles of oxygen are required in the reaction for each mole of o-xylene. However, in practice an excess amount of oxygen is added and the o-xylene is therefore the limiting reactant.

limiting substrate A nutrient used in the growth of microorganims in a concentration that restricts or limits the rate of growth. It is intentionally used to control biotechnological processes in which all other essential nutrients are added in excess such that the limiting substrate determines the rate of growth and can lead to the controlled production of a desired bioproduct. In fermentation processes, sugar is used as the limiting substrate to ensure that the yeast cells convert the sugar to alcohol. Where another nutrient is the limiting substrate, other undesirable bioproducts may result or there may be a poor growth of cells.

Linde process 1. An early process used for the *liquefaction of air. It involves first purifying the air by passage over soda lime to remove carbon dioxide, and then drying it. It is then pressurized to around 200 atmospheres, and cooled back to ambient temperature and passed through closely spiralled thin copper tubing surrounded by thicker copper tubing. The air is allowed to expand to about 20 atmospheres and cooled by the *Joule–Thomson effect. The cooled air passes through the annulus between the two tubes thereby cooling further incoming compressed air. The process was developed by German engineer Carl von Linde (1842–1934) and later improved by George *Claude (1870–1960) who made the expansion of the gas doing useful work in an expansion engine. **2.** A process used for the removal of hydrogen sulphide and mercaptans from petroleum fractions by reaction with oxygen in the presence of a metal amino acid chelate in an aqueous solution containing an amine.

linear equation An equation with two variables that gives a straight line when plotted on a graph of the form $y = mx + c$ where m is the gradient and c is the intercept on the y-axis.

linear programming A mathematical technique used to provide an optimal solution to a set of *linear equations. The technique uses a model of a process, which consists of a set of equations and also an *objective function. The objective function may typically represent the economics of the process. The set of linear equations are known as constraint equations and define a region of feasibility that has an infinite number of solutions. The objective function is used to evaluate the optimal solution from these. Linear programming is widely used for process optimization, product supply and distribution, project management, and general resource allocation.

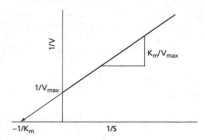

Fig. 26

Lineweaver–Burk plot A double-reciprocal plot used to determine the two constants featured in simple enzyme kinetic equations such as *Michaelis–Menten kinetics, *Monod kinetics, and in similar adsorption isotherm models such as the *Langmuir adsorption isotherm. The constants are determined from the intercept with the y-axis and the gradient (see Fig. 26). It was devised and published in 1934 by American chemist Hans Lineweaver (1907–2009) and American biochemist Dean Burk (1904–1988).

lining An inner surface applied to process equipment to prevent or reduce the occurrence of corrosion, erosion, or the damaging effects of temperature. The surface may be a coating or a shield such as glass, lead, or in the case of furnaces, heat-resistant brick.

Linton–Sherwood correlation A dimensionless equation used to determine the mass transfer in gas absorption:

$$Sh = 0.023\,Re^{0.83}\,Sc^{0.33}$$

It was developed in 1950 based on experimental data from wetted wall columns and is valid for *Reynolds numbers between 2,000 and 70,000 and *Schmidt numbers between 0.6 and 2,500.

liquation A separation and purification process used for metals from ores. It involves heating to a temperature at which the lower melting metals liquefy and can then be separated by draining. It was once used to separate lead containing silver from copper in the fifteenth century. The lead melts at 327°C whereas copper melts at 1,084°C allowing the silver-rich lead to melt and freely drain away. To avoid oxidation, the liquation furnace operates with a reducing atmosphere.

liquefaction The process of making or becoming liquid. Gases can be liquefied through refrigeration, pressurization, and the *Joule–Thomson effect. *Andrews' work on carbon dioxide showed that to liquefy any gas, it is necessary to cool it below its critical temperature, and then to apply sufficient pressure to induce liquefaction. The liquefied gases occupy a smaller volume than gases for storage, and *multicomponent mixtures can also be readily separated by the process of *distillation. Solid particles can also be made into a fluid-like mass through mechanical vibration.

liquefied natural gas (LNG) Natural gas, largely composed of methane, which has been refrigerated to a liquid. The boiling point of methane at atmospheric pressure is

-161°C. Because of its low critical point, it must be cooled to below this temperature before it can be liquefied by pressure. Once liquefied, it can be stored in well-insulated vessels and can be conveniently transported. It is used as an engine fuel.

liquefied petroleum gas (LPG) A mixture of gaseous petroleum products normally stored and transported as liquid under pressure. The main constituents are propane and butane. LPG has an ignition temperature of 450°C and calorific value or 45.3 MJ kg^{-1}. It is largely used as an engine fuel and has a wide range of other industrial and domestic uses.

liquefied refinery gas A group of gases including ethane, propane, butane, isobutane, and their various derivatives that are produced in petroleum refineries by fractionation. Also known as *still gas, it is kept in the liquid state through compression and/or refrigeration. It is used as a refinery fuel or petrochemical feedstock.

liquid A state of matter between a gas and a solid in which the molecules are free to move with respect to each other, but are held by cohesive forces such that they maintain a definite volume but not a fixed shape. There is still no simple comprehensive theory of the liquid state even though liquids have been studied for many years. *Compare* SOLID; GAS.

liquid distributor A device used to distribute liquid over the top of packing material contained within a packed column. It consists of channels located above the packing through which the liquid flows. The liquid either overflows over small notches and weirs cut in the side of the distributor channel or drains through orifices in the bottom.

liquid–liquid extraction *See* SOLVENT EXTRACTION.

liquidus A boundary line or curve of a *phase diagram between liquid and liquid/solid at equilibrium. Above the line, the substance is liquid.

liquor 1. Liquid as the product of a process used as a wash. **2.** Water used in brewing as the fermented or distilled product. **3.** Water in which food has been boiled. **4.** A solution of a pharmaceutical dissolved in water.

litre (Symbol l) An *SI unit of volume that occupies one cubic decimetre ($1\,dm^3 = 10^{-3}\,m^3$). It was previously defined as the volume of one kilogram of pure water under standard conditions of 4°C and standard pressure, which is equivalent to $1.000\,028\,dm^3$.

Little, Arthur Dehon (1863–1935) An American chemist and chemical engineer who founded a major consulting company. He was instrumental in founding chemical engineering at MIT and is credited with coining the term *unit operations. A graduate of chemistry from MIT, he formed a partnership with chemist Roger B. Griffin who was later to die in a laboratory accident in 1893. Working by himself and then closely with MIT and, in particular, with William Hultz Walker, he formed another partnership, Little and Walker, in 1900. After dissolving the company five years later he continued again on his own and established Arthur D. Little in 1909, which is today a leading international management consultancy that covers many industrial sectors. Arthur D. Little taught papermaking at MIT from 1893 to 1916. He was president of the American Chemical Society (1912–14), president of the *American Institute of Chemical Engineers (1919), and president of the Society of Chemical Engineering (1928–29).

lixiviation The process of separating mixtures by dissolving soluble constituents in water. The **lixiviant** is the liquid medium that is used in *hydrometallurgy to extract selectively a desired metal from the ore or mineral and is used to enable quick and complete leaching. The metal can then be recovered from it in a concentrated form after leaching.

LNG *See* LIQUEFIED NATURAL GAS.

lock and key mechanism A mechanism proposed in 1890 by Emil Fischer (1852–1919) to explain the binding between the active site of an enzyme and a substrate molecule. The active site is seen as the fixed structure (lock) and exactly matches the structure of the substrate (key). An example is the interaction of an enzyme and substrate in which the lock and key mechanism produces an enzyme–substrate complex. The substrate is then converted to products, which no longer fit the active site, releasing the product and liberating the enzyme. Recent X-ray diffraction studies have shown that the active site of an enzyme is more flexible than this simple mechanism would suggest.

lockout-tagout (LOTO) A safety procedure used to disable machinery and process equipment being serviced by disengaging or isolating it from its power source. This is to ensure that it cannot be restarted until associated maintence is completed. It uses a system of locks and tags in which tags on the locked equipment are used to indicate that the equipment cannot be restarted. The procedure involves notifying employees, shutting down and isolating the equipment, locking and tagging, the release of any stored energy, and verification that the equipment has been isolated.

logarithm Any real number can be written as another number raised to a power in the form $y = x^n$ where n is the the logarithm to the base x of y, i.e. $n = \log_x y$. If base 10 is used, the logarithms are called *common logarithms. **Natural logarithms** or (Naperian logarithms) are written to the base $e = 2.718\ 28...$ and written as either $\log_e$ or ln and named after Scottish mathematician John Napier (1550–1617). Logarithms contain two parts: the characteristic is the integer and the mantissa is the decimal. For example, the logarithm to the base e of 10 is 2.302 where 2 is the characteristic and 0.302 is the mantissa. Note that for any base, the logarithm of 1 is zero, the logarithm of 0 is not defined, the logarithm of a number greater than 1 is positive, the logarithm of a number between 0 and 1 is negative, and the logarithm of a negative number cannot be evaluated as a real number. In the past, tables were constructed called Tables of Logarithms. Nowadays, electronic calculators have superseded the use of these tables.

logarithmic mean temperature difference A temperature driving force applied to the heat transfer between fluids with constant heat capacities. It is applied to fluids that transfer heat in countercurrent directions and is determined from the difference in the temperature difference in fluid temperatures at either end of the heat exchanger divided by the natural logarithm of the ratio of the same two temperature differences:

$$\Delta T_{lm} = \frac{(T_3 - T_2) - (T_4 - T_1)}{\ln\left(\frac{(T_3 - T_2)}{(T_4 - T_1)}\right)}$$

It should not be used where the *overall heat transfer coefficient varies appreciably, with *multipass heat exchangers, nor where heat is generated such as on one side of the heat transfer surface as in an exothermic reaction in a water-cooled reactor (see Fig. 27).

logarithmic scale A scale of measurement in which an increase or decrease of one unit represents a tenfold increase or decrease of that measurement: pH measurement is an example of a logarithmic scale. Data that follows a logarithmic relationship can be plotted

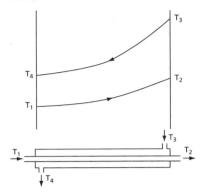

Fig. 27 Log mean temperature difference

on special graph paper with a logarithmic scale. Alternatively, the relationship can be linearized. For example, the Fanning friction factor for turbulent flow in smooth pipes is related to *Reynolds number as $f = a\,\mathrm{Re}^{-b}$. This can be represented as a linear relationship on a plot of $\ln f$ versus $\ln Re$. This is a straight line, therefore $\ln f = \ln a - b\,\ln\mathrm{Re}$ which enables the gradient b to be determined.

log phase A period of time in which microbial cells grow at a constant maximum rate, unrestricted in terms of nutrients or substrate such that there is a repetitive doubling of cells. After inoculation of a bioreactor with living cells, there is an initial *lag phase in which the cells adjust to their new environment. There is then an *acceleration phase until the cells begin to grow at their maximum rate. The increase in population is then expressed as:

$$\frac{dN}{dt} = kN$$

Integrating from an initial population N_o at time $t = 0$, gives $t = \dfrac{1}{k}\ln\!\left(\dfrac{N}{N_o}\right)$. It can also be expressed as $N = N_o e^{kt}$ and alternatively is known as the *exponential phase.

loop reactor A type of chemical reactor that has an external loop from the main body of the reactor through which a portion of the reacting material flows. Used for liquid-phase reactions, it operates as a form of gas lift in which gas is admitted into the bottom and rises as bubbles. After disengagement of the bubbles, the reacting materials descend into the bottom of the reactor via the external loop. This type of reactor does not require mechanical mixing to achieve good gas–liquid dispersions. There are, however, difficulties in controlling the rate of mass transfer and the consumption of energy.

LOPA *See* LAYER OF PROTECTION ANALYSIS.

loss The amount of material, energy, power, or pressure lost during a process. An increase in loss is usually synonymous with a reduction in efficiency.

loss prevention A comprehensive and systematic approach to the study of safety, the prevention of accidents, and the minimization of their effects on people, processes, and equipment. Derived from an insurance term for the financial loss incurred due to an accident, it considers the cost of replacing damaged plant, third-party claims, loss of production and sales. The whole system is examined in detail in order to ensure process efficiency and safety can be achieved within reasonable economic limitations. Within process design, safety and loss prevention considers the identification and assessment of the hazards, their control such as containment of flammable and toxic materials, the control of the process through the provision of control systems, alarms, trips, and operating practices, and the limitation of damage and injury through pressure relief, plant layout, and provision of fire-fighting equipment.

lost time incident (LTI) A work-related incident that involves an injury or illness where an employee of an organization or a contractor is unable to return to work as a result of their injury.

LOTO *See* LOCKOUT-TAGOUT.

low and vacuum pressure safety valve (LVPSV) An automatic system that relieves the pressure on a gas in a vessel or pipeline. It is used for small, negative or positive pressure differences near to atmospheric pressure.

lower flammable limit (LFL) The lowest concentration of a flammable vapour or gas mixed with an oxidant such that it will propagate flame at a specified temperature and pressure. *Compare* UPPER LIMIT OF FLAMMABILITY. *See* FLAMMABILITY LIMITS.

low-level alarm *See* LEVEL ALARM.

low pressure safety valve (LPSV) An automatic valve system that relieves pressure on a gas. It is used for small differences between the pressure in a vessel and atmospheric pressure.

LP An abbreviation for low pressure. It is often used with reference to a utility or a vent line, for example, an LP air supply.

LPG *See* LIQUEFIED PETROLEUM GAS.

LPSV *See* LOW PRESSURE SAFETY VALVE.

LTI *See* LOST TIME INCIDENT.

lubrication The use of substances to reduce the frictional forces between two surfaces. Lubrication is required in the smooth running of machinery such as pumps, gears, and bearings to prevent wearing and overheating. Most lubricants are based on hydrocarbons. Additives are often added to control the viscosity and prevent oxidation. Greases are used to seal against moisture and dirt and other particulates and are often used on vertical surfaces. A **lubricating oil** is a fraction of crude oil distillation at around 300°C to 370°C that is used for motor oil and other lubricants designed to reduce the friction in machinery. Impurities are removed that affect the viscosity before adding additives that enable the lubes to be used in various circumstances such as at low temperatures.

lumen (Symbol lm) An *SI derived unit of luminous flux equal to the flux emitted in a solid angle of one steradian by a point source having a uniform intensity of one candela.

Lurgi process A coal gasification process that was originally developed in Germany. It is used for the gasification of poor-quality coal, and uses high temperature and pressure in the presence of steam and oxygen. The gas produced is mainly hydrogen and carbon dioxide, which is then mixed with steam and passed over a catalyst to convert the carbon monoxide and steam into carbon dioxide and hydrogen by the *water-gas shift reaction.

lute A U-shaped loop in a pipe or tube in which a liquid is trapped and used to provide a seal for gas flow. For example, the pipe from a gravity-fed reflux from the condenser of a distillation column contains a lute. The distillation column has a higher pressure than the condenser and the lute prevents vapour entering from the wrong direction. The depth of the lute is determined from hydrostatics and is greater than the maximum vapour pressure difference between the column and the condenser outlet.

lux (Symbol lx) An *SI derived unit of illumination equal to a luminous flux of one *lumen per square metre.

LVPSV *See* LOW AND VACUUM PRESSURE SAFETY VALVE.

Lykov number (Luikov number) A dimensionless number, Lu, that combines heat and mass transfer as the ratio of mass diffusivity, D, to thermal diffusivity, α:

$$Lu = \frac{D}{\alpha}$$

See LEWIS NUMBER.

lyophilization The removal of moisture from a material such as a biological substance in which the properties are preserved by freezing under vacuum. It is used at temperatures of around $-80°C$, followed by drying in a vacuum. The moisture is removed by *sublimation. It is used for the long-term preservation of microbial cultures and protein samples. *See* FREEZE DRYING.

lyophobic A term used to mean lacking an affinity for a solvent. Where water is the solvent, the term hydrophobic is used. The opposite is **lyophilic**, or in the case of water being the solvent, hydrophilic.

lysis The process of rupturing the cell wall and membranes (if present) of microorganisms by physical, chemical, or biochemical means to release the contents of the cells. Presses and liquid shear such as *bead mills provide physical disruption to the cells, whereas chemical agents, including detergents such as Triton X, remove the lipid molecules thereby causing disruption of cell membranes. The enzyme lysozyme, obtained from hen egg white, dissolves the cell wall causing disruption. It is, however, expensive and rarely used for large-scale extraction processes. The processes are often used in combination.

M

Mach number The ratio of the speed of a body or a fluid to the speed of sound in that body or fluid at the same temperature and pressure conditions. Where the speed exceeds the speed of sound (i.e. Mach 1), the speed is supersonic. *Choked flow of air in a pipe or through an aperture occurs at Mach 1. It is named after Austrian physicist and philosopher Ernst Mach (1838–1916).

macro- A prefix meaning large. A **macromolecule** is a high molecular weight polymeric molecule such as a protein, carbohydrate, or nucleic acid. Their diameter is typically between 10^{-8} and 10^{-6} m. Natural and synthetic polymers such as cotton and plastics are also examples of macromolecules. A **macroporous resin** is a type of *ion exchange resin that has an open pore structure capable of absorbing large ions. The rate of absorption of ions and resistance to osmotic shock is better than for gel-type resins, although their overall capacity is lower. Something that is **macroscopic** is large enough to be visible to the naked eye.

magnex process A process used to remove minerals from heated pulverized coal by making it magnetic using gaseous iron pentacarbonyl ($Fe(CO_5)$). This deposits a layer of magnetic material on the minerals but not the coal itself. Magnetic separation is then used to remove the minerals.

Maillard reaction Also known as non-enzymatic browning, it is the chemical reaction between proteins and reducing sugars during heating which produces colours ranging from brown to red. It is responsible for the colour and flavour in many processed foods. It is named after French chemist Louis Camille Maillard (1878–1936).

maintenance The routine action of ensuring that a process facility or item of process equipment is fully serviceable and operable at its original or designed capacity, requirement, and efficiency. It includes a programme of inspection, testing, and servicing. **Scheduled maintenance** is planned and recurring work designed to ensure that a facility or process is serviced before the inconvenience of a breakdown, whereas **unscheduled maintenance** can be a costly inconvenience.

make-up The addition of a process material or substance to a process to allow for its gradual loss. The loss, which may or may not be intended, may be due to leakage, through purging, evaporation, or accompanying a product as an impurity. Additional solvent make-up is added to extractive distillation processes, which lose solvent as an impurity to the product. Cooling towers that cool process water for reuse require make-up water for the water lost as vapour.

malleability The ability of a material such as gold to be hammered into thin sheets.

manifold A pipe or chamber which directs a gas or liquid into a series of smaller pipes or chambers.

Manning formula An empirical equation used in open channel flow under the influence of gravity of the form:

$$C = \frac{m^{\frac{1}{6}}}{n}$$

where C is the Manning roughness coefficient and is used in the *Chézy formula, m is the *mean hydraulic depth, and n is a dimensional roughness factor, the magnitude of which depends on the type of surface. It is also known as the **Gauckler–Manning formula** and was first developed by French engineer Philippe G. Gauckler in 1867 and further developed by Irish engineer Robert Manning (1816–97) in 1890.

manometer An instrument used to determine the pressure difference between two points, usually by measuring the height difference of liquid in two vertical legs. There are many variations of instrument, and all consist of a vertical leg up which a liquid moves in proportion to the pressure applied.

A **gas-filled manometer** is used to measure the pressure of process gases and consists of a sealed container at some pressure and the other end is attached to a process via a U-tube containing a liquid. The pressure of the process is determined from the levels of manometer liquid in the U-tube and the pressure in the sealed vessel.

A **differential manometer** is used to measure the pressure of a process fluid and measures the difference in pressure between two points in a process, such as the pressure drop across a heat exchanger. An **inverted manometer** is essentially an inverted differential manometer in which the process fluid is used to measure its own pressure. A head of trapped gas (usually air) is used in the device.

Single leg manometers replace the U-tube containing a manometer liquid, with a large *sump containing the liquid that extends up one leg when a differential pressure is applied. An **inclined leg manometer** is a variation of the single leg manometer but with the protruding leg from the sump inclined at some angle. This provides a magnification of the movement of the liquid along the leg to an applied pressure and is particularly useful for measuring small pressures.

manual control A type of *feedback control used to control a process in which a process operator makes the judgement to adjust a *process variable. The operator receives a signal, usually from a gauge monitoring the temperature, pressure, flow, or level, and manually adjusts a valve. Opening or closing the valve to alter the flow of cooling water, steam, gases, etc., results in changes to the process conditions. The process operator continually monitors the gauges making more adjustments until the required process conditions are met.

manway An access port into major process vessels such as reactors, columns, or tanks. It is normally sealed with a lid or covered with an O-ring on a flange and locked with nuts. They are sufficiently wide to enable a person to enter, wearing appropriate *personal protective equipment, to carry out internal inspections, cleaning, and *maintenance, etc.

Margules' activity coefficient model A simple thermodynamic model used to describe the excess *Gibbs free energy of a liquid mixture. It uses activity coefficients that are a measure of the deviation from ideality of solubility of a compound in a liquid. *See* RAOULT'S LAW. In the case of a binary mixture, the excess Gibbs free energy is expressed as a power series of the mole fraction in which the constants are regressed with experimental data. The activity coefficients are found by differentiation of the equation. Unlike other models such as UNIQAC and NRTL, the model has the ability to predict a wide range of reliable thermodynamic values for the coefficients. It is named after Austrian scientist Max Margules (1856–1920) who developed it in 1895.

Marshall and Swift index A method of determining the installed equipment cost of a chemical plant. It is based on average cost data from nearly 50 industries involving both process industry and more general industry equipment costs. The cost at any particular time is based on the original cost at some earlier specified time multiplied by a ratio of cost indices for the present to the earlier time of interest. The base year is taken as 1926 with a value of 100. Until April 2012, the index was published in the monthly issues of the magazine *Chemical Engineering*.

mass (Symbol m) A measure of the quantity of material. It is defined as the resistance or inertia of a body to acceleration. Newton's laws of motion state that if two bodies of equal mass, m, each acquire an acceleration, a, then $m_1 a_1 = m_2 a_2$. That is, the mass of one body can be compared to the other. A standard mass is therefore used to compare all other masses. This is a one-kilogram cylinder of platinum-iridium alloy called the international standard of mass.

mass balance The analysis of a process in which the total mass of the chemical reactants is correlated with the total mass of the products according to the *law of conservation of mass. Where there is an inequality, there is either an unaccounted loss or accumulation within the process. A mass balance calculation is used in the design stage of a process. It is also used during process operation as a form of inventory or accountancy. It is also known as a *material balance.

mass fire A large-scale fire involving many buildings or structures. A forest fire is a mass fire.

mass flow rate The flow of materials measured in terms of its mass per unit time. The mass flow rate of a fluid within a pipeline is the product of its density, ρ, mean velocity, v, and cross-sectional area, a, of the pipe. $m = \rho a v$. The *SI units are kg s^{-1}.

mass flux The rate of mass of a fluid flowing per unit area. It is used to describe the flow of materials through pipes and process equipment such as membranes and filters. In a two-phase system such as a liquid and a gas flowing through a pipe or through an absorption column, the mass fluxes of the different phases are considered and expressed separately. The *SI units are kg m^{-2} s^{-1}.

mass fraction The proportion of a substance in a mixture expressed as a fraction of the overall mass of the mixture.

mass loading (Symbol G and L for gas and liquid, respectively) A measure of the mass flow rate per unit area, and used for the flow of fluids within packed columns to ensure that good mass transfer takes place between a liquid and gas, and that the internal hydraulics of the column provide the required operability. The *SI units are kg m^{-2} s^{-1}.

mass number The number of neutrons and protons in the atomic nucleus of an atom (i.e. **nucleons**). For example, uranium-238 comprises 92 protons and 146 neutrons.

mass transfer The movement of the mass of a component in a system containing two or more components from a region of high concentration to that of a lower concentration. It forms the basis of many chemical and biological processes such as crystallization, adsorption, liquid–liquid extraction, etc. The movement of mass is either by random molecular motion or by convective forces where mass is transferred from a surface into a moving fluid. The *mass transfer coefficient is a measure of the diffusion rate that relates the mass transfer rate, mass transfer area, and concentration gradient as the *driving force.

mass transfer coefficient (Symbol k_c) A quantity that characterizes the extent of the movement of mass across a boundary. It is defined as the ratio of the mass flux to the difference between mass fractions on either side of the boundary:

$$k_c = \frac{n}{a\Delta C_A}$$

where n is the mass transfer rate, a is the effective mass transfer area, and ΔC_A is the driving force concentration difference. It is used to quantify the mass transfer between phases and in mixtures, and is used to design and size separation equipment. Mass transfer coefficients can be obtained from theoretical equations, correlations, and material properties, depending on the material or process being studied. For example, the mass transfer coefficient can be correlated using the Sherwood number (Sh) with the Reynolds number (Re), Grashof (Gr) number, and the Schmidt number (Sc) for various flow regimes such as:

$$Sh = a\,Re^b\,Sc^c$$

or for natural convection:

$$Sh = a'\,Gr^{b'}\,Sc^{c'}$$

The coefficients and exponents are determined experimentally.

material Any substance or matter (element, compound, or mixture) in any physical state (gas, liquid, or solid). In a process, material may undergo change in its form but the total mass remains constant. Material and mass are therefore not synonymous.

material balance The exact accounting of all *material that enters, leaves, accumulates, or is depleted within a given period of operation across an imaginary boundary of a process or part of a process. It is in effect an expression of the *law of conservation of mass in accounting terms. Where the conditions of a process are steady and unvarying with time, the material input flow across the boundary equals the output flow. Depending on whether chemical or physical change occurs, material balances may be conveniently calculated and expressed in either mass or molar terms.

material safety data sheet (MSDS) The documented information associated with a supplied chemical for its safe use and handling. The information includes physical data such as melting points, boiling points, flash points, and also includes information of its potential hazards and health effects such as toxicity. It is intended to protect those intending to use or handle the chemical, and includes the necessary *personal protective equipment to be used as well as first-aid information. It is also known as a **product safety data sheet**.

mathematical model A mathematic description of a process or system used to understand or predict its behaviour. It can also be used to aid understanding of the influences on a process as well as to control the process or system itself. A model may take the form of a set of differential equations and describe both steady-state behaviour and dynamic behaviour, and may be combined with experimental data for validation purposes or for leading to new theories and hypothesises. *Unstructured models, such as *Monod kinetics, are used to describe the time variation of a component such as accumulation of biomass or depletion of a limiting substrate in which the environmental changes are ignored. Although mathematically simple, they are frequently useful for design, control, and optimization purposes. Structured models are more complex in nature, and endeavour to describe and predict the influence of the environment on the internal mechanisms of a living organism.

matrix 1. A solid substance within which another substance is embedded such as a metal that constitutes the major part of an alloy. **2.** A rock material within which a mineral is embedded. **3.** In algebra, a rectangular array of elements set out in rows and columns used to facilitate the solution to mathematical problems. They can be used to present the coefficients of simultaneous linear equations in which each row corresponds to one equation.

matte An impure mixture of metal sulphides such as copper and iron.

maturation An aging process used in the production of *whisky following the process of distillation in which the distilled spirit is placed in wooden casks for long periods of time. There is a legal requirement to mature Scotch whisky for a minimum of three years and it is often much longer. During this time flavours from the casks that have been previously used in the wine, sherry, cognac, and port industries are imparted, and include lignin and vanillin from the wood. The aging process also imparts colour. Around 2 per cent is lost each year through evaporation known as the *angel's share.

maximum allowable working pressure The maximum pressure at which process equipment can be operated safely at a designated temperature. It applies to vessels and pipes.

maximum drying rate The maximum rate at which *moisture can be removed from a substance.

maximum specific growth rate (Symbol μ_{max}) The maximum possible rate of growth of a culture of microorganisms in a bioreactor. It is dependent on the temperature, pH, dissolved oxygen, and presence of nutrients. It has dimensions of reciprocal time and commonly has the unit of h^{-1}. *See* MONOD KINETICS.

Maxwell, James Clerk (1831–79) A Scottish physicist and mathematician noted for his pioneering work on the kinetic theory of gases and electromagnetism. He was known affectionately at school as 'daftie' and at fifteen he had a mathematical paper read to the Royal Society of Edinburgh. He retired in 1865 as professor at Aberdeen and London, but in 1871 he was recalled to be the first professor of experimental physics at Cambridge where he also prepared plans for the Cavendish Laboratory that was named after the Hon. Henry *Cavendish (1731–1810). He also made major contributions to astronomy, colour, and colour-blindness.

Maxwell–Boltzmann distribution A mathematical function used in the *kinetic theory of gases. It is derived on the basis of *statistical mechanics and gives the distribution of particle velocities in a gas at a particular temperature.

MBOED A unit of volumetric flow in the oil industry as **m**illion **b**arrels of **o**il **e**quivalent per **d**ay. The barrel of oil equivalent (**BOE**) is a unit of energy that is released from the combustion of one *barrel of crude oil. It is used for financial purposes to combine both oil and gas into a single measure.

McCabe–Thiele A rigorous graphical method used in the analysis of the separation of two heterogeneous liquids by distillation. It is used to determine the number of stages required to bring about a required separation, and is based on a stage-wise approach requiring vapour liquid equilibrium data between the two liquids presented in terms of the more volatile component. The diagram (see Fig. 28) presents the vapour liquid equilibrium data based on the more volatile component as the mole fraction in the vapour phase (y) and the mole fraction in the liquid phase (x). Molar balances over the stripping and rectifying sections of the distillation column provide *operating lines. The step-wise lines represent the vapour liquid equilibrium on each theoretical tray or equilibrium stage and provide an indication of the total number of trays or stages required for a separation. The intercept of the stripping and rectification sections meets the *q-line and represents the condition of the feed to the distillation column. The method assumes a constant molar overflow requiring constant molar heats of vaporization, that for every mole of liquid vaporized a mole of vapour is condensed, that there is no heat of mixing or solution, and that there is no heat transfer to or from the distillation column. The method was developed by American chemical engineers Warren L. McCabe (1899–1982) and Ernest W. Thiele (1895–1993) in 1925, but is now largely obsolete due to the use of computers able to carry out rapid and detailed calculations. Its simple graphical representation makes it a useful teaching tool. *Compare* PONCHON–SAVARIT.

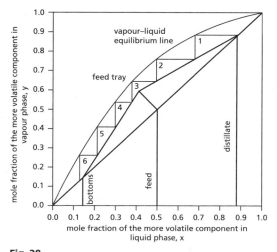

Fig. 28

mean A representative value of a set of values. The arithmetic mean or average value of a number of values is obtained by summing the values in the set and dividing by the number of values, n. The *geometric mean is obtained by the $1/n^{th}$ power of the product of the values. The harmonic mean is obtained from:

$$H = \frac{n}{\dfrac{1}{x_1} + \dfrac{1}{x_2} + \ldots \dfrac{1}{x_n}}$$

mean free path The mean distance between successive elastic collisions of molecules in a gas. According to the *kinetic theory of gases, the mean free path is inversely proportional to the pressure of the gas.

mean hydraulic depth The ratio of the flow area to the *wetted perimeter in an open channel used to transport a liquid. It is used in the *Manning formula to determine the *Chézy coefficient needed for determining the rate of flow.

mean hydraulic diameter A parameter used in the calculation of fluids flowing in pipes, enclosed channels, and ducts in which the cross-sectional area is not necessarily circular. It is expressed as the ratio of four times the flow area to the *wetted perimeter.

mean velocity The average velocity of a fluid in a pipe based on all the local velocities in the cross section. A simple form of calculation is based on the total flow divided by the cross-sectional area. For laminar flow in circular pipes, the mean or average velocity is half the maximum velocity, which occurs along the axis.

mechanical engineering A branch of engineering that is concerned with the study and production of mechanical devices, such as machines, tools, and vehicles that are capable of carrying out specific tasks.

mechanics The study of the behaviour of bodies under the action of forces. *Statics is the study of forces but without a change in momentum, whereas the study of dynamics concerns changes in momentum. The study of *kinematics deals with the motion of bodies without reference to the forces that affect the motion. Newtonian mechanics is concerned with the bodies in the solid state, whereas *fluid mechanics is concerned with the study of forces on fluids.

mechanism 1. A description of the steps that are involved when a chemical reaction takes place. It includes the reactants and describes the steps that result in products as well as any intermediates. **2.** The workings of a mechanical device.

median A type of statistical average equal to or exceeded by half the values in a statistical distribution. The medium has the advantage that it is not influenced by extreme values and can be calculated from incomplete data. However, it cannot be used in any other form of mathematical processing. It is useful in quantifying data such as the statistical distribution of particle sizes.

medium The aqueous environment containing a dissolved substrate and all the other necessary nutrients in which microorganisms are cultured such as in the process of fermentation. All microorganims require water and a source of carbon, nitrogen, minerals, salts, trace elements, and vitamins. Oxygen is also required for some types of microorganisms. A *defined medium consists of pure compounds in which the composition and concentration of all the chemicals is known. Common media used to culture microorganims as colonies on petri dishes include agar and gelatine.

mega- (Symbol M) A prefix used to denote a million, for example one megajoule (1 MJ) is equal to a million joules.

melt crystallization A crystallization process used to separate components of a liquid mixture by cooling until a particular component is crystallized as a solid from the liquid phase within a *crystallizer. It involves either a gradual deposition of a crystalline layer on a chilled surface of a flowing or static melt, or the fast generation of discrete crystals dispersed in the crystallizer.

melting point The temperature at which a substance changes state from a solid to a liquid. The transition occurs at a constant temperature and requires the absorption of heat known as the *latent heat of fusion. The melting point of ice at standard atmospheric pressure is 0°C. The melting point temperature is influenced by pressure.

membrane distillation A form of distillation process in which a substance is distilled by way of a temperature difference across a *semi-permeable membrane. The pores of the membrane are filled with the substance being distilled, and it is bounded by two liquid phases. *Compare* PERVAPORATION.

membrane filtration A form of separation process in which particles are separated from a fluid using a semi-permeable polymeric membrane in which the pore size and construction of membrane permits certain particle types and sizes to pass and retains others. Examples of membrane processes include *dialysis, *microfiltration, *reverse osmosis, and *ultrafiltration.

membrane permeation flux The volume of fluid that flows through a *semi-permeable membrane per unit area per unit time. For vapours and gases, the flow is dependent on temperature and pressure. The SI units are $m^3 \, m^{-2} \, s^{-1}$.

membrane separation A separation process that involves the use of a *semi-permeable membrane that permits the diffusion of one or more selective components to pass. The components able to pass through the membrane are known as the *permeate, while the components unable to pass are known as the *retentate. The membrane is highly porous and the size of the pores can be controlled during its manufacture using an organic polymer evaporating solvent. For example, the pore size selectivity can be used to control the separation of nitrogen molecules from air. The membranes are manufactured as sheets or as hollow fibres and packed into bundles called **modules**.

Mendeleev, Dmitri Ivanovich (1834–1907) A Russian chemist noted for the formulation of the periodic table of elements, which was published in 1869. He was appointed professor of chemistry at St Petersburg in 1866.

meniscus The curved free surface of a liquid that is separated from a gas in a capillary tube. It can be concave or convex, and is the result of *surface tension.

Merox process A catalytic process used in petroleum refineries to remove mercaptans from LPG, propane, butane, naphtha, kerosene, and *jet fuel by oxidation to form hydrocarbon disulphides. An abbreviation of **mer**captan **ox**idation, the process uses an aqueous solution of sodium hydroxide and a water-soluble catalyst to remove the foul-smelling mercaptans.

mesh A way of designating the number of openings in a screen or sieve per linear inch. A mesh of ten has ten openings per inch (100 per square inch). A standard wire size is used so that the area of opening is standardized. The screen or sieve is used to separate solid particles of a particular size.

MESH equations A set of equations used for calculating the material, energy, and thermodynamic profiles within a distillation column. They combine material balances, equilibrium conditions, summation conditions, and heat balances, which are required to be solved simultaneously. Due to the complexity of the computations, computer simulation software uses algorithms based on *Newton-Raphson methods to converge onto solutions. Once the calculations have been solved rigorously for one stage, the next stage is computed. When the traverse of the entire column is completed, a comparison is made of the computed and assumed overall material balance. Another trial is used where there is an unacceptable difference.

metabolism The biochemical processes that take place within a living organism and result in the breakdown of large or complex organic molecules into smaller and more simple molecules (catabolism). Metabolism is also accompanied by the release of energy, or the building up of large and complex molecules from small and simple molecules (anabolism), together with the storage of energy. Metabolic processes are usually regulated by enzymes.

metal One of a number of chemical elements such as nickel, iron, and copper that are characterized as being lustrous ductile solids at room temperature. They form basic oxides, form positive ions, and are good conductors of heat and electricity. An *alloy, such as steel, contains one or more of these chemical elements.

metal fatigue A cumulative effect of repeated applications of applied stress culminating in the failure of a metal. The applied stress is applied on a cyclic basis and does not exceed the ultimate tensile strength. The number of cycles that lead to failure is dependent on factors such as the extent of the applied stress and strain as well as temperature and corrosion.

metallic bond A covalent bond between atoms in metals and alloys, in which the valence electrons are free to move through the crystal lattice structure.

metalline Resembling or relating to metals.

metallography A branch of metallurgy that is concerned with the composition and structures of metals and alloys.

metalloid A non-metallic element such as silicon that possesses some of the properties of a *metal.

metallurgy The study of the metals. It includes the processes of extraction, refining, alloying, and fabrication of metals as well the study of their properties and structure.

metal spraying A process that involves spraying one metal onto another in the molten state.

metastable A form of temporary or indefinite stability of chemical and physical equilibria. Metastable phases of matter include the melting of solids, freezing of liquids, condensing of vapours, sublimation of solids, super-cooling and super-heating of liquids. Pure liquids can remain in the liquid state below their freezing point until initiated to crystallize by either applying vibration or seeding.

metering The addition of a small quantity of a fluid or solid particles at a constant rate to a process stream used for accurate determination of flow. Positive displacement metering pumps are generally used for dosing or the dispensing of fluids. They operate by trapping a fixed volume of fluid and moving this fluid via gears, pistons, diaphragms, vanes, or other devices such as *peristaltic pumps. They are not particularly sensitive to changes in discharge and suction conditions, and permit flow regulation by adjusting speed and displacement.

methanation A petrochemical process in which carbon monoxide and carbon dioxide are converted to methane using hydrogen gas:

$$CO + 3H_2 \rightarrow CH_4 + H_2O$$
$$CO_2 + 4H_2 \rightarrow CH_4 + 2H_2O$$

The reactions are exothermic and carried out in a fixed-bed nickel-based catalytic reactor at a temperature between 370°C and 430°C.

methane clathrate (methane hydrate) A solid ice substance in which methane is entrapped within the crystal structure of water. It occurs naturally at the bottom of the ocean floor and in deep sedimentary structures at low ambient temperature and high pressure. It is responsible for oil and gas sub-sea pipeline blockages. Its formation is controlled by the use of additives such as methanol. It is also known as gas hydrate, methane ice, and fire ice.

method A defined or systematic way of doing something such as carrying out a chemical analysis.

metre (Symbol m) The base *SI unit of length, it is precisely defined as length of the path of light in vacuum during a time interval of $1/2.99\ 792\ 458 \times 10^8$ second. It was defined by the General Conference on Weights and Measures in 1983 and replaces earlier definitions.

metric system A decimal system of units based on the metre, gram, and second for length, mass, and time, respectively. It was originally devised by the French Academy in 1791 that included French mathematicians J. L. *Lagrange and P. S. *Laplace. The system replaced a base twelve system. The SI system is used widely in scientific work and is based on the metric system and uses the kilogram as the SI unit of mass.

metrology The study of weights and measures, and the units of measurement.

Michaelis–Menten kinetics A general two-parameter enzyme reaction model in which the reaction velocity, V, is given by:

$$V = \frac{V_{max} S}{K_M + S}$$

where S is the substrate concentration, V_{max} is the maximum reaction velocity, and K_M is the Michaelis constant. For simple enzyme reactions, the two parameters, V and K_M, can be readily obtained from experimental data by linearizing the equation in a double reciprocal plot of $1/V$ against $1/S$ such that the intercept on the y-axis is $1/V_{max}$ and the slope is K_M/V_{max}. It is named after German chemist Leonor Michaelis (1875–1949) and Canadian scientist Maud Menten (1879–1960) who proposed the reaction in 1913.

MIChemE Post-nominal letters used after someone's name to indicate that they are a *chartered chemical engineer and a member of the *Institution of Chemical Engineers.

micro- (Symbol μ) A prefix used to denote 10^{-6} of something; for example, one micrometre is equal to 10^{-6} m.

microfiltration A *membrane filtration separation process used to separate fine particles and molecules. The membrane consists of a polymeric microporous material with a pore size of between 0.1 and 10 micrometres. The filtration separation uses an applied pressure across the membrane as the driving force. It is typically used for the separation of proteins from yeast cells following the process of fermentation, clarification of liquid beverages, separating cream from whey in milk, and the sterilization of liquids by filtering out microorganisms.

micron The former name for the unit of length equal to 10^{-6} m and is now called the **micrometre**.

microorganism A microscopic living organism that is too small to be seen with the naked eye. Microorganisms are simple life forms and usually unicellular. Commonly used microorganisms in industrial processes include bacteria, fungi, and protozoa, and are noted for their rapid rate of reproduction and ability to utilize waste products as substrates such as agricultural waste. They can be cultivated in batch and continuous processes, and on a large scale with high yields. Certain microorganisms can be genetically manipulated through *recombinant DNA technology and other techniques to produce bioproducts such as viral antigens without the need for alternative hazardous processes. They usually produce non-toxic waste products.

microwave *Electromagnetic radiation that has a wavelength between 10^{-3} m and 0.03 m. Microwaves are used in the drying of materials as well as heating and domestic cooking of foods.

Midgley, Thomas (1889–1944) An American mechanical engineer who developed the petrol anti-knock additive tetraethyl lead and contributed to the production of Freon refrigerants. The son of an inventor, he qualified with a degree in mechanical engineering in 1911. While working for General Motors, he developed the lead additive for which he won an award from the American Chemical Society in recognition of his achievements. However, his health suffered from the effects of working with lead. He later worked on the development of a new type of chemically inert refrigerant to replace earlier refrigerants that were toxic, flammable, and explosive such as ammonia. Dichlorofluoromethane was the first CFC to be developed called Freon and others followed for which he received various awards of recognition. He was president of the American Chemical Society in 1944.

milli- (Symbol m) A prefix used to denote 10^{-3} such as in millimetre (1 mm $= 10^{-3}$ m) or millibar (1 mbar $= 10^{-3}$ bar). Standard atmospheric pressure corresponds to 1,013 mbar.

milling 1. A mechanical process used to grind grain such as wheat into flour or meal. **2.** A mechanical operation involving cutting, shaping, finishing, or working products manufactured in a mill.

mineral A solid substance that occurs in nature characterized by having a definite chemical composition or being made up of several chemical compositions with a distinctive molecular structure. Most minerals are inorganic, although minerals such as coal are organic in origin. Ore is a mineral or aggregate that contains minerals that may be valuable and can be extracted, or contain undesired minerals called *gangue.

mineral acid A generic name for organic acids such as nitric, sulphuric, phosphoric, and hydrochloric acids.

mineral dressing A series of unit operations used in the mineral processing industries to prepare minerals and ores for extraction. These include crushing, grinding, particle size distribution, and various forms of separation such as flotation.

mineral oil A generic name for diesel fuels obtained from petrochemical feed stocks.

minimal medium A growth medium for microorganisms in which only the minimum number of nutrients is supplied for the growth of a particular microorganism. *See* MEDIUM.

minimum reflux ratio The smallest possible reflux ratio in the operation of a fractional distillation column that can produce top and bottom products of specified compositions. It corresponds to a distillation column with an infinite number of theoretical stages and is therefore a hypothetical quantity.

mist flow A type of two-phase gas–liquid flow in which a gas flows with a very high velocity carrying a fine dispersion of the liquid as droplets.

mitigation The act of reducing the severity of the consequence of an undesirable event.

mitosis The normal process of microbial cell division in which the paired chromosomes duplicate at the beginning of the process, and each of the daughter cells formed has pairs consisting of one original and one new and replicated chromosome. Microorganisms are used as biological catalysts in biotechnological processes such as fermentation.

mixer 1. A device used to intimately combine materials. There are various types, the choice of which is dependent on the application. A commonly used mixer consists of a rotating shaft upon which blades are attached. These are either flat or pitched at an angle to provide radial or axial mixing, or a combination of both. Other designs include *static mixers, which are ribbons of metal held within a pipe causing flowing liquids to alter their movement and mix. **2.** A unit operation used in flowsheets to represent process input streams that combine together to provide a single output. It may be a simple vessel with a stirrer. A mixer does not allow a chemical reaction to occur, nor change of phase to take place.

mixer-settler A device used in *solvent extraction processes. It consists of a vessel in which two immiscible liquids are dispersed within one another using a *mixer within one section or compartment, and another partitioned section where the two liquids are permitted sufficient time to separate into two layers. Their operation constitutes one theoretical stage. They are therefore often arranged in series and operate with *countercurrent flow.

mixing The intimate contact of two or more components and/or phases used to achieve a desired product quality. Mixing is also used to promote mass transfer or enhance the rate of a chemical reaction. *Blending is a type of mixing that involves particulates or powders. Mixing can involve dispersing one phase through another such as sparging gases through liquids, or the use of rotating agitators, impellors, and propellers. *Fluidized beds provide excellent mixing to enhance both heat and mass transfer in which a fluid, such as a gas, is driven up through a bed of particles, such as a catalyst.

mixing rule A way of calculating the combined effects of mixtures using pure component parameters. Mixing rules are used for calculating the effects of two or more components for viscosity and density, and used in thermodynamic models such as cubic equations of state.

mixture The combination of substances in a process without chemical reaction. It can involve more than one phase such as a solid and a liquid, or an emulsion of liquids. A homogenous mixture has a complete dispersion of components in a single phase, whereas a heterogeneous mixture has distinct phases. Air is a mixture of mainly oxygen and nitrogen.

mmHg A unit of pressure used in *manometers and *barometers in which mercury is used as the manometric fluid. *Standard atmospheric pressure (101 325 Pa) is equal to 760 mmHg. A pressure of 1 mmHg is equal to 133 322 Pa.

MMSCFPD An abbreviation for **m**illions of **s**tandard **c**ubic **f**eet **p**er **d**ay, which is a volumetric measure of gas flow.

mode 1. A type of statistical average. In a discrete set of data, it is a single value. If the data is continuous, the mode is the point of greatest clustering of occurrences. It is used when the most typical value is required. **2.** A type of operation of a controller used to control a process that can either be manual, automatic, or remote. In automatic mode, the controller calculates the output value as the difference between the set point and the measured *process variable. In manual mode, the process operator sets the output, while in remote control, the controller obtains the set point from another controller.

model A mathematical representation of a system or process using equations that predict its behaviour. See MATHEMATICAL MODEL.

moderator A material, usually graphite, used to slow down fast-moving neutrons in a nuclear reactor in order to allow them to strike fissile material such as uranium-235 and cause a controlled nuclear chain reaction. The rate of neutrons absorption is controlled by lowering the moderator into the reactor. Other moderators include water and *heavy water (deuterium oxide).

Moh's scale of hardness A decimal scale of scratch hardness for materials based on ten minerals arranged in order of hardness with 1 being the softest and 10 the most resistant. These are: 1 talc; 2 gypsum; 3 calcite; 4 fluorite; 5 apatite; 6 orthoclase; 7 quartz; 8 topaz; 9 corundum; and 10 diamond. The scale is linear between 1 to 9 although diamond is about ten times harder than corundum. There is a modification to this scale in which the first six minerals are the same, but continues with 7 pure silica glass; 8 quartz; 9 topaz; 10 garnet; 11 fused zircon; 12 corundum; 13 silicon carbide; 14 boron carbide; and 15 diamond. The scale was devised in 1812 by German mineralogist Friedrich Mohs (1773–1839), and is still used today.

moisture A liquid, usually in the form of water, that is dispersed through a gas or vapour, or in a solid contained within pores.

moisture content The amount of liquid, usually in the form of water, that is contained within a substance. The relationship is expressed either on a **dry basis** as the amount of water per unit mass of substance, or on a **wet basis** as the amount of water per unit mass of substance with water. Both are expressed as a percentage for which the relation between the two is:

$$x_d = \frac{x_w}{100 - x_w}$$

where x_d is the moisture content on a dry basis and x_w is the moisture content on a wet basis. The moisture content measured on a wet basis can exceed 100 per cent.

molar 1. The concentration of one mole of a substance dissolved in one litre of solvent. **2.** An amount of a substance. For example, the *molar latent heat is the quantity of heat absorbed or released when an amount of substance changes its physical phase at constant temperature and pressure per mole for which the SI units are J mol^{-1}.

molar density The number of moles of a substance within the volume occupied by the substance. The SI units are mol m^{-3}.

molar flow rate The rate of flow of material in a process stream expressed in terms of moles of material. The material is independent of phase and may be expressed as either one component, some, or all of the components in the process stream. The molar flow rate is often presented in a *stream table.

molar flux The rate of flow of a fluid expressed as the number of moles per unit area per unit time. It is used for describing the flow of materials in terms of the transport of molecules or particles rather than bulk mass. The SI units are mol m^{-2} s^{-1}.

molar heat capacity (Symbol c$_m$) The amount of heat required to raise the temperature of one mole of a substance through 1 K. The SI units are J mol^{-1} K^{-1}.

molarity A measure of the concentration of a substance expressed as moles of solute dissolved in one litre of solvent. For example, a 1 molar solution of sodium chloride contains 58.44 g of NaCl (since its molecular mass is 58.44 g) per litre of solution.

molar latent heat The amount of heat absorbed or released when a substance changes its physical phase at constant temperature and pressure per mole. The SI units are J mol^{-1}.

molar volume The volume occupied by one mole of a substance. It is equal to the product of the *specific volume and molar mass. The SI units are m^3 mol^{-1}.

mole (Symbol mol) The amount of a substance that has many elementary entities as there are atoms in 0.012 kilograms of the isotope carbon-12, and corresponds to the amount of material in *Avogadro's constant or 6.023x10^{23}. The number of moles of a particular species or element is its mass divided by its molar mass.

mole balance A calculation carried out to determine the total amount of substances that flow in and out of a process or system volume that is defined by a boundary. The mole balance does not take into consideration the physical form of the substances, only the number of moles of each substance.

molecular diameter A diameter of a molecule in which the molecule is assumed to be spherical. It is usually expressed in *angström unit (10^{-10} m) and is multiplied by a factor depending on the element or compound.

molecular diffusion A form of mass transfer in which molecules of a component in a system involving two or more components are transported from a region of high concentration to a lower concentration by random molecular motion. It is independent of any convective forces that may be present. *See* MASS TRANSFER.

molecular distillation A vacuum distillation used to separate and purify substances that would otherwise be adversely affected by distillation at a higher temperature. Using pressures of below 0.01 torr, the gaseous phase no longer exerts a significant pressure on the substance being evaporated. The rate of evaporation is therefore not dependent on the pressure since mass transport is governed by molecular dynamics and not fluid dynamics. Molecular distillation is used for the purification of oils, the manufacture of lubricants from petroleum, and the purification of vitamins.

molecular formula A way of expressing a chemical compound using symbols for the atoms. *Compare* EMPIRICAL FORMULA.

molecularity The number of molecules or ions involved in a chemical reaction. Unlike the order of a reaction, the molecularity must be a whole number. It cannot be fractional or zero.

molecular modelling A computational technique based on theoretical knowledge of molecules to understand their behaviour. The theoretical models typically involve treating atoms as individual particles but can also include the modelling of the electrons of each atom. The interactions of atoms are described as spring-like interactions. The Lennard-Jones potential is often used to describe *van der Waals' forces. While simple calculations can be performed by hand, the complexity of the interactions between molecules requires powerful computers to undertake the necessary number of computations. Molecular modelling is used in the design of pharmaceutical drugs, material science, enzyme catalysis, and other areas of study requiring the understanding of the behaviour of complex molecular systems.

molecular sieve An absorbent and inert material such as aluminosilicate or *zeolite used to remove moisture from gases or organic liquids. The absorbent has a high porosity of uniform size in the order of 4 to 5 angströms and adsorptivity at low water vapour pressures.

molecular weight *See* RELATIVE MOLECULAR MASS.

molecule The smallest part of an element or chemical compound that can exist separately and retain its properties. A molecule of helium consists of a single atom, whereas a molecule of water consists of an aggregation of oxygen and hydrogen held together by valence forces.

mole fraction The ratio of the number of moles of a component in a mixture to the total number of moles in the mixture expressed as a fraction. The symbol x is used for the mole fraction of liquids and y is used for vapour. The sum of the mole fractions in the liquid or in the gas phase is equal to unity.

Mollier chart The graphical representation of the thermodynamic properties of a pure substance with enthalpy on the y-axis and entropy on the x-axis. Other properties are also included on the chart (see Fig. 29), including pressure and temperature, and the critical point. They are used to visualize thermodynamic cycles such as in power plants, refrigeration, and air conditioning. It is named after German physicist Richard Mollier (1863–1935).

moment of momentum *See* ANGULAR MOMENTUM.

momentum The quantity given by the mass of a body multiplied by its velocity. Used in the study of dynamics, its quantity is conserved under certain circumstances, and its rate of change gives the amount of force acting on the body.

momentum balance A balance in which the sum of all the forces acting on a moving fluid in one direction is equal to the difference between the momentum leaving with the

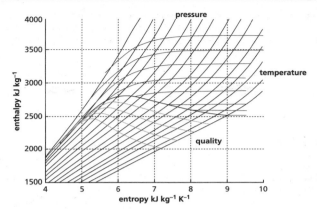

Fig. 29

fluid per unit time and that brought in per unit time by the fluid. The momentum flow rate of a fluid stream is the product of the mass flow rate and velocity.

monatomic molecule A molecule that consists of one atom such as argon or helium. It is distinct from diatomic molecules that consist of two atoms, or polyatomic molecules than comprise many atoms.

Mond, Ludwig (1838–1909) A German-born British chemist noted for the industrialization of several major chemical processes. After studying chemistry under Robert Bunsen (1811–99) at Heidelberg, he worked at various chemical works applying his knowledge to industry. Moving to Britain in 1862, he developed a process to remove sulphur from the waste products from the *Leblanc process. He joined a partnership with John *Brunner in 1873 and established a factory in Cheshire to manufacture soda by the new ammonia process invented by Ernest *Solvay (1838–1922). Mond also developed several other processes including the Mond process for purifying nickel.

Mond nickel process A process invented in 1889 by Ludwig *Mond (1838–1909) for extracting and purifying nickel. The process was developed after Mond noticed that nickel valves in a chemical manufacturing process became corroded by carbon monoxide. He discovered a new compound called nickel carbonyl $Ni(CO)_4$ and that pure nickel could be obtained by reaction with carbon monoxide at 50°C, and that the new compound decomposes on further heating to around 200°C.

Monel A type of corrosion-resistant nickel-copper alloy that combines high strength with high ductility. It is used for process equipment in salt and seawater applications.

monochromatic Meaning one colour, it is the property of some sources of electromagnetic radiation to emit waves of one frequency (or wavelength). A laser is an example.

Monod, Jacques Lucien (1910–76) A French biochemist noted for his work on the study of molecular biology. He studied first in Paris before working at Columbia University having gained a Rockerfeller Fellowship. He gained his PhD from the Sorbonne for his work on bacterial growth. He joined the French Resistance during the Second

World War after which he joined the Pasteur Institut in Paris, becoming its head in 1954 and director in 1971. He was professor of molecular biology at the College de France from 1967. The Monod equation describing the kinetics for molecular growth is named after him.

Monod kinetics An *unstructured model used to describe the correlation of substrate concentration with microbial growth kinetics. The model is based on enzymatic *Michaelis–Menten kinetics:

$$\mu_{(S)} = \mu_{MAX} \frac{S}{Ks + S}$$

where μ_{MAX} is the maximum specific growth rate of the microorganism, S is the substrate concentration, and Ks is the saturation constant and equivalent to the Michaelis–Menten constants. Ks varies with the type of substrate while the maximum specific growth rate is dependent on process conditions such as temperature, pH, and dissolved oxygen. When homogeneous conditions are achieved in the bioreactor and the microorganisms are freely suspended, a relatively low substrate is required to maintain maximal growth. At high substrate concentrations, substrates can inhibit growth for which the Monod kinetic model can be extended accordingly.

monolayer A single layer of molecules that are absorbed onto a support, surface, or fluid interface. The layer has a depth of one molecule. *See* LANGMUIR ADSORPTION ISOTHERM.

monomer A simple molecule or a compound that joins with others to form a larger molecule such as a dimer, trimer, or a polymer in a repeating form. An example is vinyl chloride monomer, which can be polymerized to form polyvinyl chloride (PVC), used for pipes and tubing.

Mono pump A type of *positive displacement pump that consists of a rotating helical worm contained within a casing or stator. A self-priming pump, it is particularly suitable for fluids with high viscosities and solid content such as wastewater, sludge, and slurries. It is widely used in the chemical, oil, and gas industries, as well as food, paper, mining, and mineral processing industries. It can produce high-delivery heads such that *pressure relief valves are required on the delivery line. They were first developed by Mono in 1935.

(⊕) SEE WEB LINKS
• Official website of Mono Company.

monosepsis A biochemical process such as fermentation that involves the culture of a single type or strain of microorganism. A process bioreactor or fermenter is first fully sterilized before being inoculated with a small volume of the required living and viable microorganism. The process is then operated in a monoseptic state by controlling the ports into and out of the bioreactor by filtration and the use of septic seals to avoid microbial contamination by other microorganisms.

Monte Carlo simulation A computer-based iterative statistical method that uses sets of random values from a set of ranges or probability distributions to determine a quantitative outcome in complex process simulations. The outcome of the iterations is to build up a distribution of the possible outcomes of the simulation as a frequency distribution or in graphical form. It was originally developed as a way of simulating random neutron diffusion in fissionable material by workers at the Los Alamos National Laboratory in the 1940s under the secret code name Monte Carlo, which is famous for its casino.

Montejus An egg-shaped closed tank used to transport a liquid or slurry by using compressed air. By applying a pressure into the surface of the liquid in the tank, the liquid is forced out by displacement through a dip leg that extends into the liquid. Since the tank does not involve complex and expensive pumps other than a supply of compressed air, it is useful for transporting corrosive liquids such as very concentrated acids. It is also known as an *acid egg on account of its shape and application.

Moody plot, chart, diagram A dimensionless representation of friction factor with Reynolds number for a fluid flowing in a pipe. Presented on log-log scales, the diagram includes laminar, transition, and turbulent flow regimes. It also includes the effects of pipe relative roughness as a dimensionless ratio of absolute roughness with internal pipe diameter. The plot was developed in 1942 by American engineer and professor of hydraulics at Princeton, Louis Ferry Moody (1880–1953).

more volatile component The component in a heterogeneous mixture that has the highest vapour pressure. It also has the lowest boiling point temperature. In the separation of a mixture of volatile liquids by distillation, the more volatile component rises to the top of the distillation column for separation while the least volatile component is retained at the bottom of the column.

Morton, Frank (1906–99) A British chemical engineer after whom a sports day is named. After gaining a degree in chemistry and a PhD in Manchester College of Science and Technology (later UMIST and then Manchester University), he joined the then newly formed department of chemical engineering at the University of Birmingham, before moving to take up head of the department of chemical engineering at the Manchester College of Science and Technology in 1956. He was president of the Institution of Chemical Engineers (1963–64). A very active sportsman, a sports day for UK chemical engineering students and a medal 'for excellence in chemical engineering education' are named after him.

mothballing A procedure used to close a process facility for an extended period of time. The facility is shut down and left in a preserved state such that it no longer operates, but may at some future time be successfully operated again if required. All process materials are removed and the process equipment cleaned and shut down thereby preserving the equipment and preventing the effects of corrosion. Mothballing is generally used when a process is no longer economic due to loss of demand for its product. Some nuclear facilities have been mothballed due to political reasons. It is named after mothballs that were once commonly used to deter moths from eating clothes thereby preserving them.

mother liquor A term used in crystallization that refers to a highly saturated solution of crystallizable solids from which crystals are nucleated. Once crystallization has taken place, the mother liquor is separated from the crystals by filtration or centrifugation and any residual mother liquor remaining is rapidly washed away to prevent the crystals from redissolving.

motor A machine that converts electrical or chemical energy into useful work. Pumps used to transport fluids are driven by electric motors, which convert electrical energy into mechanical energy. A simple form of electric motor uses electric current to power a set of electromagnets on a rotor in the magnetic field of a permanent magnet. An internal combustion engine is also a motor in which fuel is combusted with oxygen in the form of air and converted to useful mechanical energy.

motor gasoline See PETROLEUM.

MOV An abbreviation for **motor-operated valve**. See VALVE.

MSDS See MATERIAL SAFETY DATA SHEET.

MTG An abbreviation for any process used for the conversion of **methanol to gasoline**.

MUF An abbreviation for **material unaccounted for**. Used in the accountancy of nuclear material during reprocessing, it is the difference between the actual *inventory and expected amount of nuclear material.

muffle furnace A type of electrically heated laboratory furnace used for drying substances or carrying out high-temperature reactions using controlled atmospheres such as the pyrolysis of organic materials. The furnace is insulated and typically operates at temperatures up to 1,200°C.

Müller–Kühne process A process used to produce sulphuric acid from calcium sulphate. It uses calcium sulphide to produce lime (CaO) and sulphur dioxide in the so-called **Müller–Kühne reaction**, which takes place in a kiln operated at 1,150°C:

$$CaS + 3CaSO_4 \rightarrow 4CaO + 4SO_2$$

It is named after German chemist inventors W. S. Müller and Hans Kühne (1880–1969) who developed the process in the First World War (1914–18) when Germany was unable to obtain imported supplies of iron pyrites used for making sulphuric acid.

multicomponent A mixture having more than two components. The components in a process stream or item of plant equipment may be of the same phase, such as all being gases. *Crude oil is an example of a multicomponent mixture.

multicomponent distillation The separation of mixtures containing more than two components in significant amounts. If all of the components are present in small amounts except for two, the mixture can be treated as a binary and the same design principles apply as for binary. Complexity arises from the number of components present. For a given pressure and a particular component composition in the liquid phase, there no longer exists a unique vapour composition and temperature. These depend on the amount and type of the other components present. For hydrocarbons, *K-factors can be used. It is also possible to use vapour pressure data for ideal systems. Calculation methods often depend essentially on an assumed temperature profile that can be used to link vapour and liquid phase compositions for a given column pressure. Checks are made to ensure that the sum of the mole fractions in the liquid and vapour phases are equal unity. The process is repeated until the differences in compositions or flows at product take-off or at the feed point are below an acceptable figure. This assumes equimolar flows in any section.

Together with knowledge of feed condition and reflux ratio, the flow in the column profile can be calculated. Short-cut methods are available for preliminary investigations. The usual ones are the methods of *Fenske, *Underwood, *Erbar–Maddox, and *Smith–Brinkley. Two of the earliest rigorous methods were developed by *Lewis–Matheson and *Thiele–Geddes.

To perform multicomponent distillation calculations requires the introduction of the concept of *key components. These are the components between which the split between components is required and must be present in appreciable amounts in the feed, and effectively all of one will appear in the distillate (light key) and all of the other in the bottoms (heavy key). More correctly, the light key is the lightest component in the bottoms whose composition is to be specified. Some simplified calculations assume that all LLK (lighter than light key) components are absent from the bottoms and all HHK (heavier than heavy key) from the distillate. The key components effectively unlock the problem.

multipass The flow of material or energy through a system on more than one occasion. Multipass operations are used to improve process efficiencies. **A multipass heat exchanger** is a heat exchanger device in which one or both fluids pass more than once. They are used when the length of tube for heat transfer is too great to make a single pass operation practical. The heat transfer medium passes through either the *shell side or *tube side transferring heat, and is then returned back for further heat exchange. The temperature profiles are more complex than for single pass operations since some parts of the equipment will be cocurrent and other parts will be countercurrent. For the number of tube passes used, correction factor charts are used to adjust the log mean temperature difference calculations.

multiphase Having more than one phase. Material in a process stream or plant equipment may be in combinations of solid, liquid, and gas or vapour phases, including liquid-liquid. It is not dependent on the number of components present. A single component may have more than one phase present such as liquid and its vapour flowing through a pipeline in the form of bubbles (see Fig. 30 for examples of horizontal and vertical pipe flow). The majority of industrial chemical processes are concerned with multiphase flow systems such as power generation, refrigeration, and distillation, and depend on multiphase evaporation and condensation cycles. Desalination, steel-making, paper manufacturing, and food processing all contain critical steps that depend on the nature of multiphase behaviour.

multiple effect A number of process units, such as evaporators, operated in series to perform a particular duty. In the case of multiple-effect evaporators, each successive stage has a greater vacuum and operates at correspondingly lower temperature. This enables the heat in the vapour leaving each successive effect to be used to heat the next. There are several arrangements that are commonly used, including feed forward in which the product and vapour both move on to subsequent effects. In backward feed operations, the vapour feeds in the reverse direction starting in the last effect. This requires pumping to overcome the higher pressures. The heating of the feed in the last effect, which has the lowest temperature, requires a lower temperature difference for evaporation. In parallel feed operation, the vapour feeds from one effect to another while fresh feed is fed to each effect independently.

multistage 1. A process that involves more than one step. The flow from one stage forms the feed to the next. *See* MULTIPLE EFFECT. **2.** Process equipment such as pumps or compressors in which the flow of material from one stage feeds into the next. It is used to generate high pressure. A multistage compressor may operate with the stages on a common rotating shaft.

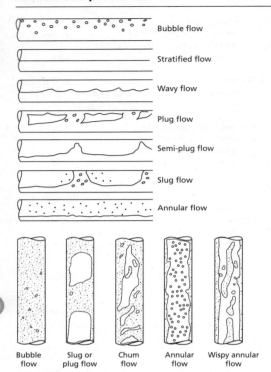

Bubble flow

Stratified flow

Wavy flow

Plug flow

Semi-plug flow

Slug flow

Annular flow

| Bubble flow | Slug or plug flow | Chum flow | Annular flow | Wispy annular flow |

Fig. 30 Multiphase flow

multivariable process control *See* STATISTICAL PROCESS CONTROL.

Murphree plate efficiency A measure of the closeness to equilibrium on a plate or stage within a fractional distillation column. The vapour and liquid on an ideal stage are in equilibrium. However, in practice, this may not be the case. The Murphree plate efficiency is therefore expressed as the ratio of the increase in mole fraction of vapour of a volatile component passing through a plate in a column to the same increase when the vapour is in equilibrium. In effect, more stages are therefore required to bring about a desired separation. For a binary distillation, it is presented as:

$$\eta = \frac{y_n - y_{n-1}}{y^* - y_{n-1}}$$

where the subscripts n and $n-1$ refer to the outlet and inlet vapour streams to a stage, and y^* is the equilibrium vapour concentration. It was proposed in 1925 by Eger V. Murphree (1892–1962).

muster area A designated location where process plant personnel will gather in the event of an emergency. The muster area is required to be clearly defined and located with direct access by all personnel. On offshore installations, the muster area has direct access to survival craft and other life-saving equipment. The area is of a sufficient size to enable all gathered personnel to don *personal protective equipment.

nano- (Symbol n) A prefix used to denote a scale of 10^{-9}. For example, a distance of 10^{-9} m or one nanometre (nm) is typically the scale used in the study of molecular structures of macromolecules.

nanomaterials A material in which one, two, or three external dimensions range in size from 1 to 100 nm. Nano-sized objects can be classified as **nanoplates** in which one external dimension is at the nanoscale; **nanofibres,** which have two external dimensions at the nanoscale; and **nanoparticles,** which have all three external dimensions at the nanoscale. A **nanotube** is a hollow nanofibre while a **nanorod** is a solid **nanofibre**.

nanotechnology The study of materials of the nanometre scale ranging from 1 to 100 nm. Nanotechnology involves the creation and/or manipulation of *nanomaterials, either by scaling up from single groups of atoms, or by refining or reducing bulk materials.

naphtha A hydrocarbon distillate mixture obtained from the processing of crude oil in a petroleum refinery. It has a boiling point range of between 150°C to 180°C. Naphtha is used as the feedstock to produce aromatics, by reforming and extracting ethylene and light gases using high-temperature catalytic cracking.

naphthalene A white crystalline solid with a melting point of 80.5°C and boiling point of 218°C, once used as mothballs and widely used as the starting material for plasticers, lubricants, resins, and dyes. It was once produced from coal tar as a *by-product of coke production for the steel industry and is now produced as a by-product from petroleum refining as a more pure product. Naphthalene is produced by the process of **hydrodealkylation** from catalytic reformer bottoms or from recycled materials from cracking operations in the presence of hydrogen using chromium and aluminium oxide catalysts at high temperature.

Natta, Giulio (1903–79) An Italian chemical engineer noted for his work on polymerization and the development of commercially important polymerization processes. After graduating from the Polytechnic of Milan, he began his career with the study of solids using X-rays and electron diffraction, which he extended to catalysts using the same methods. He was awarded the Nobel Prize in Chemistry in 1963, together with the German chemist Karl Waldemar *Ziegler (1898–1973).

natural convection A process by which thermal energy as heat is transported unassisted within a fluid by the movement of the fluid itself. The motion of the fluid is caused by natural means such as the buoyancy effect that manifests itself as the rise of warmer fluid and the fall of the cooler fluid. In contrast, *forced convection is the transport of heat within a fluid by external means such as a pump or a fan.

natural gas An odourless and flammable gas consisting largely of methane. It is found in its natural state in particular geologic formations as a product of the decomposition of organic matter. Natural gas is used as a fuel for domestic heating. Mercaptans and other *stenching agents are added to enable consumers to identify the presence of the gas in the event of leakage.

natural logarithm A *logarithm that is to the base e (e = 2.718). It is abbreviated to ln or log$_e$. All logarithmic functions are the inverse of a power function.

Navier, Claude-Louis (1785–1836) A French engineer and physicist who studied at the École Polytechnique and then continued his studies at the École Nationale des Ponts et Chaussées (1804–06). He was admitted to the French Academy of Science in 1824 and became professor at the École Nationale des Ponts et Chaussées in 1830 before taking the position of professor of calculus and mechanics at the École Polytechnique. He directed the construction of bridges at Choisy, Asnières, and Argenteuil, and is noted for his work in fluid mechanics for which the *Navier–Stokes equations are best known.

Navier–Stokes equations A set of mathematical expressions used to study the motion of fluids. They are expressed in terms of velocity gradients for a *Newtonian fluid with constant density and gradient. Using Cartesian or rectangular coordinates, the equations represent inertia of the left-hand side and body force, pressure, and viscous terms of the right-hand side:

$$\rho\left(\frac{\partial v_x}{\partial t}+v_x\frac{\partial v_x}{\partial x}+v_y\frac{\partial v_y}{\partial y}+v_z\frac{\partial v_z}{\partial z}\right)=\rho g_x-\frac{\partial p}{\partial x}+\mu\left(\frac{\partial^2 v_x}{\partial x^2}+\frac{\partial^2 v_x}{\partial y^2}+\frac{\partial^2 v_x}{\partial z^2}\right)$$

$$\rho\left(\frac{\partial v_y}{\partial t}+v_x\frac{\partial v_y}{\partial x}+v_y\frac{\partial v_y}{\partial y}+v_z\frac{\partial v_z}{\partial z}\right)=\rho g_y-\frac{\partial p}{\partial y}+\mu\left(\frac{\partial^2 v_y}{\partial x^2}+\frac{\partial^2 v_y}{\partial y^2}+\frac{\partial^2 v_y}{\partial z^2}\right)$$

$$\rho\left(\frac{\partial v_z}{\partial t}+v_x\frac{\partial v_z}{\partial x}+v_y\frac{\partial v_y}{\partial y}+v_z\frac{\partial v_z}{\partial z}\right)=\rho g_z-\frac{\partial p}{\partial z}+\mu\left(\frac{\partial^2 v_z}{\partial x^2}+\frac{\partial^2 v_z}{\partial y^2}+\frac{\partial^2 v_z}{\partial z^2}\right)$$

The equations can also be written using cylindrical and spherical coordinates. The solutions to the equations are called velocity fields or flow fields. The equations were developed by Claude-Louis *Navier (1785–1836) in 1822, and developed further by George *Stokes (1819–1903), and find many applications including the study of the flow of fluids in pipes and over surfaces.

nb An abbreviation for *nominal **b**ore used for specifying pipe and tube sizes.

negative pressure A pressure below atmospheric pressure expressed as a gauge pressure. At sea level using *standard atmospheric pressure, the lowest possible pressure is −101325 Nm^{-2} or −760 mmHg. Negative pressures are used in processes where air is required to be drawn into the process as a way of preventing the release of potentially harmful substances from contaminating personnel or the environment. Fume cupboards operate with a negative pressure.

Nelson–Farrar cost index A method of determining the construction cost of a petroleum refinery or a process unit. It is used to compare operating costs over time and involves normalizing cost during that time span and includes changes in productivity costs such as labour and utilities, etc. The purchase cost at any particular time is based on the original cost at some earlier specified time multiplied by a ratio of cost indices for the present to the earlier time of interest. Established in 1926, the original index was given a value of 100 in

1946. The Nelson–Farrar cost index is published in the first issue of each month of *Oil and Gas Journal,* and not considered to be suitable for refineries and process units that are older than five years.

(⊕) SEE WEB LINKS
• Official website of *Oil and Gas Journal* containing the Nelson–Farrar index.

net positive suction head (NPSH) The total head (or pressure) at the suction nozzle side of a centrifugal pump that is required to avoid the potentially destructive phenomenon of *cavitation. To avoid vapour bubble formation at the eye of the impeller, the pressure head must be greater than the head corresponding to the vapour pressure of the liquid. The **available net positive suction head** is a function of the particular system and layout of pipework, whereas the **required net positive suction head** is a function of the pump. The available net positive suction head is required to exceed the required net positive suction head for cavitation to be avoided. Details of the required NPSH for a particular pump are supplied by the pump manufacturer.

neutral spirit A commercially produced alcohol (ethanol) obtained through the process of *fermentation of any cereal such as maize and purified through distillation. It is usually produced continuously. It is used for the manufacture of vodka and gin.

neutron scattering An analytical technique used to examine the properties of materials at the atomic scale. It involves emitting neutrons from a radioactive source and directed into a substance being tested. The atomic nuclei of the substance are bombarded by neutrons. When a neutron collides with heavy nuclei of the substance, less energy is transferred from the neutron than if it hits light nuclei. The neutrons scattered back to the vicinity of the source are measured by a sensitive radiation counter. Neutron scattering is widely used in the study of materials in many fields of science and engineering such as pharmaceuticals, healthcare, nanotechnology, and IT.

Newton, Sir Isaac (1642-1727) An English scientist and politician who made a major contribution to the understanding of many aspects of science and engineering. Although not overly successful at school, he was taken away from school to work on his mother's farm. His uncle, however, noticed his talent in mathematics and had him sent back to school, and eventually on to Trinity College, Cambridge, of which he became Fellow in 1667. He became professor of physics two years later, was a Member of Parliament from 1689, and was Master of the Mint in 1699. He was president of the Royal Society from 1703 until his death. His most productive years were 1665–66, during which he produced a wealth of ideas. He discovered the binomial theorem and the beginnings of differential and integral calculus. He started research on light to show that sunlight was made up of the seven colours of the rainbow, and began ideas of gravity to account for the path of the moon. He proposed a law that any two bodies attract each other in proportion to the product of their masses and inversely to the square of their distances apart. To do this, he had to define the meaning of mass, momentum, force, inertia, and to state his three *laws of motion. He also established other laws such as *Newton's law of cooling.

newton (Symbol N) The *SI unit of force defined as the force that when applied to a mass of one kilogram, produces in that mass an acceleration of one metre per second per second.

Newtonian fluid A classification of fluids in which viscosity is independent of shear stress and time. It is named after Sir Isaac *Newton (1642-1727) who first proposed that an applied shear stress, τ, is proportional to the deformation of the fluid or velocity gradient (or shear rate), γ:

$$\mu = \frac{\tau}{\gamma}$$

Examples of Newtonian fluids include water, mercury, treacle, tar, mineral oils, glycerol, sucrose solutions, standard calibration oils (e.g. octane), milk, fruit juices, and honey. Fluids that do not fall into this classification are known as *non-Newtonian fluids.

Newton number *See* POWER NUMBER.

Newton–Raphson An iterative mathematical method used to find the solution to a mathematic function $y = f(x)$. It involves using an initial approximate value (x_o) of a root, and from the gradient of the function a better or improved approximation (x_1) can then be obtained using:

$$x_1 = x_o - \frac{f(x_o)}{f'(x_o)}$$

By using the new value, the method is then repeated in order to converge to a solution.

Newton's law of cooling An empirical law proposed by Sir Isaac *Newton (1642–1727) that states that for small ranges of temperature, the rate of loss of heat by a body is proportional to the mean difference of temperature between the body and its surroundings. It is valid only for considerable differences in temperature between the body and the surroundings where the heat loss is by forced convection or conduction.

Newton's laws of motion *See* LAWS OF MOTION.

NGL An abbreviation for natural gas liquids that consist predominantly of a mixture of liquefied ethane, propane, normal butane, isobutane, and pentane. With a high calorific value, they are used as petrochemical feedstocks, as domestic heating fuels, refinery blending, and for drying agricultural crops amongst other applications.

Nichols plot A type of frequency response diagram used for analysing the frequency response of a controlled system to a disturbance signal. It involves plotting the magnitude and phase-angle measurements with frequency as a parameter. It is similar to the *Nyquist plot and uses Cartesian coordinates in which the real and imaginary parts are plotted on the x and y axes.

nitrification The biological conversion of ammonia to nitrate ions by the action of bacteria. The bacteria *nitrosomonas* is able to convert ammonia to nitrate and *nitrobacter* bacteria converts nitrites to nitrates. Nitrification is an important natural part of the nitrogen cycle in soil.

Nobel, Alfred Bernhard (1833–96) A Swedish industrial chemist and inventor of dynamite. One of eight children to poor parents, he showed an early interest in explosives. His father was an inventor and having moved to St Petersburg made his fortune making explosives. Now that it was affordable, Alfred first received private tuition and studied chemistry before moving to Paris and then the US. The Nobel family produced armaments for the Crimean War (1853–56) but thereafter filed for bankruptcy. His brother Ludwig improved the business, while Alfred improved his inventions, including dynamite, and ways in which it could be stabilized. He invented gelignite in 1875. An explosion at his factory in Stockholm killed, amongst others, his younger brother Emil. When his brother Ludwig died in 1888, a French newspaper carried an obituary of Alfred in error condemning him for his invention of dynamite. Having amassed a vast fortune in his lifetime, Alfred left most of his wealth in a trust to fund awards that are known as Nobel Prizes. It is thought that the erroneous obituary convinced him of the need to leave a better legacy after his death.

noble gases A group of six monatomic gaseous elements of group 18 in the *periodic table that comprise helium, neon, argon, krypton, xenon, and radon. They differ from other gases in that they have a full set of electrons in their outer shell making them very stable. They were once known as inert gases. However, some of the noble gases can take part in chemical reactions. Apart from helium, all are found in trace amounts in the atmosphere.

noble metal A metal that is generally non-reactive to acids and atmospheric oxidizing conditions. Examples include gold and platinum.

no flux surface A surface that reflects all the incident thermal radiation. There is no net interchange of radiant heat at the surface.

nominal bore *See* NOMINAL PIPE SIZE.

nominal pipe size (NPS) A defined set of standard pipe sizes used in the US. Pipes are specified with two non-dimensional numbers for diameter based on the Imperial units of inches, and a schedule describing wall thickness. From the NPS and *schedule number of a pipe, the pipe outside diameter and wall thickness can be obtained from reference tables. These are based on ASME standards. For example, NPS 14 Sch 40 has an outside diameter of 14 inches and a wall thickness of 0.437 inches. The European equivalent to NPS is DN (nominal diameter, *diamètre nominal*, or *Durchmesser nach Norm*), in which sizes are measured in millimetres. The term **nominal bore (nb)** is also used and is equivalent to NPS.

nomograph (nomogram) A two-dimensional diagram that uses a parallel coordinate system in which scaled vertical axes are used to evaluate the dependence of two properties such as temperature and pressure on a third property such as the *K-factor. The procedure is to draw a straight line between the left- and right-hand scales, intercepting with a third scale between them. They were once widely used for quick calculations before the advent of the calculator and computer. Nomography was invented in 1884 by French engineer Philbert Maurice d'Ocagne (1862–1938).

non-combustibility The property of a material to withstand high temperature without ignition.

non-competitive inhibition *See* ENZYME INHIBITION.

non-destructive testing A procedure in which materials are tested without damage or destruction. It is used to test and inspect process equipment such as vessels and pipelines for corrosion, damage, material deposits, and blockage without losing integrity. Methods typically used include magnetic particle inspection, ultrasound, X-rays, and gamma rays. It can also be used to determine the presence of fissile material within sealed process equipment and plant.

non-ferrous metal A metal or alloy that does not contain iron. Examples include copper, tin, lead, zinc, and aluminium, and their alloys such as brass, which is an alloy of copper and zinc.

non-Fickian *See* FICK'S FIRST LAW OF DIFFUSION.

non-flammable A substance that is not liable to ignite or burn when exposed to flame. *Compare* INFLAMMABLE.

non-ideal flow The flow of fluids that deviates from idealized flow. Idealized flow patterns in chemical reactors include plug flow and mixed flow. However, in reality, there may be appreciable deviation from ideality causing lower process performance. The causes of the deviation may be channelling, by-passing, or stagnant regions. Understanding non-ideal flow is important for *scale-up since the flow varies as the size of equipment increases. It is therefore difficult to predict how the flow changes as size increases. An important measure is the *residence time distribution.

non-ideal mixture A mixture of two or more substances in which there is an interaction between the molecules or atoms of the individual components. *Raoult's law applies to ideal mixtures in which the forces between the particles in the mixture are the same as those in the pure liquids. Mixtures that exhibit a positive deviation from Raoult's law have vapour pressures greater than that of an ideal mixture (see Fig. 31). This is caused by the intermolecular forces between the molecules being less than for the pure liquids. Heat is absorbed when the liquids mix. That is, the enthalpy change of mixing is endothermic. An example is ethanol and water, which has a maximum vapour pressure for a mixture containing 95.6 per cent of ethanol by mass. Mixtures that exhibit a negative deviation from Raoult's law have vapour pressures that are less than those of an ideal mixture. This is due to stronger intermolecular forces in the mixture than for the pure liquids. Heat is therefore evolved on mixing. An example is the mixing of nitric acid and water, which react to form hydroxonium ions and nitrate ions.

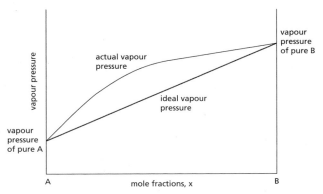

Fig. 31 Non-ideal mixture (positive deviation)

non-Newtonian fluids Fluids that do not exhibit Newtonian behaviour. That is, the rate of shear is not directly proportional to the shear stress over all values of shear stress.

Instead, the viscosity depends on shear stress and/or time. Non-Newtonian fluids are classified as being time-independent, time-dependent, and viscoelastic. They exhibit characteristics where the apparent viscosity either increases as the rate of shear increases, such as polymer melts, paper pulp, wall paper paste, printing inks, tomato purée, mustard, rubber solutions, protein concentrations, and are known as pseudoplastic, or decreases as the rate of shear increases. Examples of the latter are comparatively rare but include TiO_2 suspensions, cornflour/sugar suspensions, cement aggregates, starch solutions, and certain honeys.

With time-dependent fluids, the relation between shear stress and shear rate depends on the time and flow history of the fluid. They can be classified as either thixotropic or antithixotropic (or rheopectic). For thixotropic fluids, the shear stress will decrease with time for a fixed value of shear rate. A simple explanation is that as the liquid is sheared, the structure breaks down. If a cyclic experiment is carried out, a hysteresis loop is formed. Examples include greases, printing inks, jelly, paints, and drilling muds. For antithixotropic (rheopectic) fluids, the shear stress will increase with time for a fixed value of shear rate. Examples include clay suspensions and gypsum suspensions.

Viscoelastic fluids possess the properties of both viscosity and elasticity. Unlike purely viscous fluids where the flow is irreversible, viscoelastic fluids recover part of their deformation. Examples include polymeric solutions, partially hydrolyzed polymer melts such as polyacrylamide, thick soups, crème fraîche, ice cream, and some melted products such as cheese.

non-polar compound A chemical compound in which the molecule has no permanent dipole moment. Depending on the relative electro-negativities of the two atoms sharing electrons, there may be partial transfer of electron density from one atom to the other. When the electro-negativities are not equal, electrons are not shared equally and partial ionic charges develop. The greater the difference, the greater the ionic bond. Bonds that are partly ionic are known as polar covalent bonds. Non-polar covalent bonds have an equal share of the bond electrons and arise when the electro-negativities of the two atoms are equal. There are many non-polar substances. Some are completely non-polar while others are considered to be non-polar since they lack any significant polarity. Completely non-polar compounds include nitrogen, oxygen, and chlorine gases, diatomic molecules such as Br_2, I_2, and F_2, acetylene, and carbon tetrachloride. They are all perfectly symmetrical in which the dipole moments of any polar bonds is completely annulled by equal and opposite dipole moments from the other bonds. Hydrocarbons such as propane, butane, pentane, hexane, cyclohexane, octane, as well as fats and oils are mostly non-polar.

non-random two-liquid (NRTL) A thermodynamic model used in phase equilibria calculations that correlates the activity coefficients of a component in a liquid mixture with its mole fraction. It is based on the hypothesis that the concentration around a molecule is different from the bulk concentration caused by the difference in interaction energy of the central molecule with the molecules of its own kind and that of the molecules of the other kind. The energy difference also introduces a non-randomness at the local molecular level. The model belongs to the local-composition type which includes the Wilson, *UNIQUAC, and *UNIFAC models.

non-renewable energy *See* RENEWABLE ENERGY.

non-return valve A type of valve used in a pipeline to ensure that the flow of a fluid is in one direction only and therefore prevent the inadvertent change of direction of flow. Also known as a *check valve or *one-way valve, there are various types and designs used for a wide variety of applications. They all consist of a mechanical mechanism that prevents the reversing of flow such as using a spring-loaded ball that seats and blocks flow in the event of a reversal of pressure. Flexible diaphragms are also used which flex and open in one flow direction but seal in the other. Others consist of a hinged disc that opens freely with flow but seals shut when the flow is reversed.

normal distribution A statistical probability density function that is represented as a symmetrically bell-shaped graph in which data is symmetrical with a mean, μ, and standard deviation, σ. It has a near-zero start and rises in a smooth-bell shaped peak of probability in the centre and descends to near-zero at the end. It was also known as the **Gaussian distribution**, named after German mathematician Karl Friedrich Gauss (1777–1855) and was first derived by French mathematician Abraham de Moivre (1667–1754). The probability density function is:

$$\Pr(x) = \frac{1}{\sigma\sqrt{2\pi}} e^{\frac{-(x-\mu)^2}{2\sigma^2}}$$

normal operating conditions The expected range of conditions used to operate a process or equipment and under which operating influences are usually stated.

normal solution A solution having a concentration of one gram equivalent of the solute per litre (1 dm³). It is denoted by N. In comparison, the *molarity of a solution is the number of moles of the solution dissolved per litre of the solution and is denoted by M.

no-slip boundary condition A boundary condition assumption used in fluid mechanics that proposes that the velocity of the fluid layer in contact with a surface has the same velocity as the surface. *See* BOUNDARY SLIP.

notch A rectangular or triangular cut in a weir over which a liquid flows. They are used in liquid distributors such as within absorption columns and cooling towers.

NOx A general term used for nitrogen oxide gases that include nitric oxide (NO), and nitrogen dioxide (NO_2), and other oxides of nitrogen. These highly reactive gases have a major role in the formation of ozone. While many NOx gases are colourless and odourless, nitrogen dioxide has a reddish colour. They are formed during the combustion of fossil fuels such as coal as well as the combustion of fuels in motor vehicles.

nozzle A device used in spraying or atomizing fluids to form droplets with a high surface area. For example, within a spray dryer used to dry heat-labile liquids such as milk, the nozzle produces a fine spray enabling rapid liquid evaporation to take place.

nozzle meter A fluid flow measuring device that consists of a tapered nozzle arrangement within a pipeline. The device is designed to increase the velocity of the fluid with a corresponding decrease in pressure. A measurement of the difference between the upstream and nozzle pressure gives an indication of flow rate. The *coefficient of discharge for high flows is around 0.8 and is therefore better than an *orifice plate meter although it is comparatively more expensive to fabricate.

NPSH *See* NET POSITIVE SUCTION HEAD.

NPV An abbreviation for **n**et **p**resent **v**alue, representing the cost of a process or product calculated in the present day currency. The premise is that the investment of money today can earn interest and so has a greater present value than money received at some point in the future. The NPV is therefore determined from the sum of the present values of each individual cash flow starting with the start of the process or product. The annual discounted cash flow is obtained for each year's rate of interest. The greater the NPV, the more economically attractive the process. A negative NPV indicates an economically unviable process. The NPV is dependent on the choice of interest rate used in the calculation.

NRTL *See* NON-RANDOM TWO-LIQUID.

n.t.p. Normal temperature and pressure of a gas and used for standard conditions of thermodynamic calculations and tabulations. It is defined as 20°C and 1 atmosphere. *Compare* s.t.p.

NTU *See* NUMBER OF TRANSFER UNITS.

nuclear decommissioning The shutdown and permanent removal from service of nuclear installations such as reactors and reprocessing plants at the end of their commercial life. There are several stages to taking such an installation permanently out of service. *Mothballing involves shutdown and washout to attain a low level of residual radioactivity. The plant may be re-opened at a later date. Removal of nuclear material, decontamination and sealing equipment, dismantling and removing process and ancillary equipment leaves the facility inoperable. The final stage involves removal of major plant such as the reactor, demolishing the buildings, and returning to a green-field site.

nuclear energy The energy associated with nuclear reactions. The amount of energy released is considerably greater than that associated with chemical reactions. For example, the amount of energy released from radium, which loses an alpha-particle to form radon, is 4.2×10^{11} joules per mole of radium. The possibilities of harnessing nuclear energy changes has been known since the time when *Rutherford brought about the first transmutation of one element to another. This was accomplished by the action of alpha-particles on nitrogen gas to produce the isotope of oxygen-17.

nuclear fission A nuclear reaction or a radioactive decay process. In a nuclear reaction, which forms the basis of a nuclear reactor, the exothermic reaction involves the collision of a neutron in which an atomic nucleus is divided into smaller parts together with the release of energy. Nuclear fission was discovered in 1939 by Austrian-born Swedish physicist Lise Meitner (1878–1968), German physicist Otto Hahn (1879–1968), and German chemist Friedrich Wilhelm (Fritz) Strassmann (1902–80). They discovered that the nucleus of uranium-235 could absorb a neutron and then break into two roughly equal parts with mass numbers lying between 72 and 162 and with neutrons also being emitted. During this fission there was also a loss of mass which is converted into energy according to the equation $E = mc^2$ where m is the mass and c is the velocity of light.

nuclear fuel A substance capable of nuclear fission for the production of nuclear energy. It can refer to the substance itself or a mixture of substances and assembly of rods. By slowing down fast-moving neutrons that strike the nuclear fuel using a moderator such as graphite, other neutrons are emitted and form a chain reaction generating heat. The most common nuclear fuels are uranium-235 and plutonium-239. The nuclear reaction is contained within a *nuclear reactor.

nuclear fuel reprocessing A process developed to separate and recover fissionable plutonium from irradiated nuclear fuel. Originally developed to extract plutonium for nuclear weapons, nuclear reprocessing is used to recover plutonium and uranium for reuse as fuel in commercial nuclear power stations. Spent uranium fuel contains both fissionable uranium-235 and plutonium-239, and has substantially higher fuel content than natural uranium. It is necessary to separate the uranium and plutonium from the neutron-absorbing fission products, followed by separation of the plutonium from the uranium, and can include enrichment of the uranium. The reprocessing involves a solvent extraction process and based on the fact that both uranium and plutonium nitrate form complexes with tri-butyl phosphate (TBP). The process involves chopping up the spent fuel, which is then charged to dissolvers containing nitric acid. The uranium and plutonium nitrates are extracted from the solution by a solution of TBP in kerosene. Other fission products are retained in the aqueous phase, and then concentrated by evaporation. The uranium and plutonium complexes are backwashed from the kerosene phase into the nitric acid solution. The plutonium is reduced from a valency of 4 to 3 with ferrous sulphate since the valent state of 3 does not form complexes with TBP and can therefore be separated from the uranium. The plutonium is then oxidized back to the 4 state.

There are various variations on this separation. The *purex solvent extraction process is currently used to reprocess spent nuclear fuel to separate plutonium, uranium, and fission products.

nuclear fusion *See* FUSION.

nuclear meltdown A term loosely used to describe the overheating of a nuclear reactor. A nuclear core melt occurs when the heat that is generated in a nuclear reactor is greater than the ability of the cooling system to remove it, resulting in the nuclear fuel reaching its melting point. A nuclear meltdown may be due to the loss or reduction in the rate of flow of coolant. *See* FUKUSHIMA DAIICHI.

nuclear power (atomic power) Electrical or motive power that is produced by a *nuclear reactor.

nuclear power plant A thermal power station used to generate electricity through the generation of heat from the controlled process of nuclear fission. The heat from the nuclear fission contained within a nuclear reactor is used to generate steam and converted to electricity using a turbine and alternator.

nuclear reactor A process in which nuclear energy is produced from a controlled nuclear fission reaction. The reaction takes place in the core and involves fissile materials known as nuclear fuel for which the common nuclear fuels are uranium-235 and plutonium-239. By slowing down fast-moving neutrons using a moderator such as graphite, the neutrons that strike the nuclear fuel results in the emission of other neutrons to form a chain reaction generating heat which is converted into electrical energy through steam turbines. The world's first commercial nuclear power station at Calder Hall in the UK came into operation in 1956. It consisted of four Magnox reactors that were originally capable of generating 60 MW of electrical and thermal power. Named after an abbreviation for the magnesium oxide non-oxidizing cladding of the unenriched (i.e. natural) uranium fuel rods, the reactors were designed as pressurized carbon dioxide-cooled, graphite-moderated reactors. The reactors used boron-steel control rods. The power station was closed in 2003.

There are nearly 500 nuclear reactors in operations around the world and the most widely used is the pressurized water reactor (PWR). These nuclear reactors use water pressurized to around 160 bar to dissipate heat to achieve a high temperature and avoid boiling. The heat is transferred to a secondary system in a steam generator.

nuclear waste *See* RADIOACTIVE WASTE.

nucleon A proton or a neutron in the nucleus of an atom. The **nucleon number** (*mass number) is the number of nucleons in the atomic nucleus of a nuclide.

nucleus 1. The central core of an atom that comprises neutrons and protons. It has a positive charge and is determined by the number of protons. The nucleus is surrounded by electrons of negative charge. Atoms are neutral so the number of electrons around the nucleus must equal the number of protons in the nucleus. The number of neutrons contribute to the atomic mass. It is possible for two or more atoms to exist possessing the same number of protons, the same number of electrons, and the same chemical properties, but different numbers of neutrons, and therefore they have different atomic mass. For example, chlorine-35 and chlorine-37 have the same number of protons (17) and electrons (17), but unequal numbers of neutrons (18 and 20). *See* ISOTOPE. **2.** The spherical part of a living cell that is bounded by a membrane and contains the chromosomes and other essential molecules that determine the growth of the living cell.

nuclide An atomic species in which all the atoms have the same atomic number and mass number, such as carbon-14. An isotope refers to a series of different atoms that have the same atomic number but different neutron numbers, such as uranium-235 and uranium-238.

null hypothesis Used in statistical probability theory to test the validity of statistical data, it assumes that events occur on a purely random basis.

number of transfer units (NTU) 1. A measure of the difficulty of the separation of two components in a liquid–liquid separation. In an equimolar counter diffusion process in which the mole fraction of the more volatile component in a mixture increases from y_1 to y_2, the NTU is determined from the integral:

$$NTU = \int_{y_1}^{y_2} \frac{dy}{y_i - y}$$

where y_i is the mole fraction of the more volatile component at the interface. The number of transfer units is also used in the cooling tower calculations in which the calculation is based on the differential of the enthalpy of the water. **2.** A parameter used to quantify the performance of a heat exchanger expressed as the ratio of the product of heat transfer area, A, and heat transfer coefficient, U, to the minimum heat capacity rate, mc_p:

$$NTU = \frac{UA}{(mc_p)_{min}}$$

numerator The top part of a mathematical fraction. For example, in the fraction ½, 1 is the numerator and 2 is the denominator. The numerator is the dividend.

numerical integration A mathematical procedure used to calculate the approximate value of an integral. It is often used when a function is known only as a set of variables for corresponding values of a variable and not as a general formula that can be integrated. *Simpson's rule and the *trapezium rule are examples of numerical integration. These are used to estimate the area under a curve and involve dividing the area into vertical columns of equal width with each column representing two values for x for which f(x) is known.

Nusselt number A dimensionless number, Nu, used in heat transfer calculations characterizing the relation between the convective heat transfer of the boundary layer of a fluid and its thermal conductivity:

$$Nu = \frac{hd}{k}$$

where h is the surface heat transfer coefficient, d is the thickness of the fluid film, and k is the thermal conductivity. It is named after German engineer Ernst Kraft Wilhelm Nusselt (1882–1957). *Compare* BIOT NUMBER.

Nyquist plot A type of frequency response diagram used for analysing the behaviour of a controlled system to a disturbance signal particularly in terms of assessing the stability with feedback. It involves plotting the magnitude and phase angle on a chart with frequency as a parameter. Using polar coordinates, the gain of the transfer function is plotted on the radial coordinate with the phase angle plotted as the angular coordinate. It is named after Swedish engineer Harry Nyquist (1889–1976). *Compare* NICHOLS PLOT.

n

objective function A mathematical equation used to determine the optimal solution to a mathematical problem. Used in *linear programming which consists of a set of *linear equations, the objective function may typically represent the economics of a chemical process. Linear programming is widely used for process optimization in which the objective function may be maximized to represent the greatest economic return, or minimized to reduce resource wastage.

obligate anaerobe A bacterium that requires no oxygen for reproduction. The presence of minute traces of oxygen are sufficient to inhibit or kill it. *See* FACULTATIVE ANAEROBE.

occupational health The management of personal safety. It deals with the exposure of harmful and hazardous materials, their effect on humans, and seeks ways of reducing risk by comprehensive risk assessments. Exposure can be reduced by containment, distance, reducing the time of exposure, and through the use of *personal protective equipment.

octane number, rating A number indicating the resistance of fuels to spontaneous ignition, known as knocking, in spark-ignition engines. The higher the octane number, the greater is the fuel's resistance to spontaneous ignition. The rating is derived from comparison between a fuel and a benchmark blend of iso-octane (C_8H_{18}) with normal heptane (C_7H_{16}). Higher-octane ratings are more suitable for higher-performance engines whose compression ratios are more likely to make the fuel detonate. *Compare* CETANE NUMBER.

od An abbreviation for **o**utside **d**iameter and used for specifying pipes, tubes, and some vessels of circular cross section.

odourizing The addition of a substance to another to give a distinctive and strong odour. Pungent-smelling substances such as diethylsulphide and mercaptans are added to odourless natural gas and LPG for the purposes of safety so that leakage can be readily detected. It is also known as *stenching.

off-gas A gas that is produced as a *by-product of a chemical or biochemical process, it is often treated before being discharged into the environment. The treatment of off-gases includes the use of filtering, washing columns, venturi washers, and wet filtering. For hot gases, combustion may be used together in pre- or post-filtering and dust retention.

off-line analysis The measurement technique that involves withdrawing a sample from the process for its determination. It is used in monitoring a process in which continuous measurement is either not possible or not required. For example, solid, liquid or gas samples can be taken and analyzed using various analytical techniques such as gas or liquid chromatography. These procedures require the discrete processing of samples and it may take many minutes to return a result. **In-line analysis**, on the other hand, involves immediate and direct sensing and reading of process parameters. This form of continuous process monitoring has the advantage of eliminating errors caused by delays. Examples of in-line analysis probes include thermocouples, pressure gauges, pH probes, and dissolved oxygen probes.

offset Used in process control, it is the deviation from a controlled *set point for a process or system at steady state.

off-sites Part of a chemical process or petrochemical refinery that supports the actual process or refinery and includes tanks, utilities such as power and steam generation, waste effluent treatment, and flares, etc.

oil One of various liquid, viscid, unctuous, usually inflammable, chemically neutral substances that is lighter than and insoluble in water, but soluble in alcohol and ether and classified as non-volatile. Natural plant oils comprise terpenes and simple esters such as essential oils. Animal oils are glycerides of fatty acids. Mineral oils are mixtures of hydrocarbons. Oils have many uses and include fuels, lubricants, soap constituents, varnishes, etc., and are used as the feedstock for the production of many other products.

oil and gas Refers to the industry associated with the recovery of liquid and gaseous hydrocarbons from underground deposits as reservoirs found both onshore and offshore around the world. A collection of localized deposits is known as an oil field or gas field. When they are drilled, they are known as oil and gas wells. Oil is mainly used as fuel for transportation purposes, whereas gas is primarily used as fuel for domestic and industrial purposes, and for converting into other chemicals such as plastic. Oil is widely transported in ships. Gas is transported in underground, sub-sea, or overland pipelines covering large distances. Oil from offshore installations is also brought onshore by sub-sea pipelines. *See* FPSO.

oil gasification A general name for processes that convert liquid petroleum fractions into gaseous fuels such as through the *cracking of heaving petroleum fractions and reaction of *naphtha with steam in the presence of a *catalyst.

oil refinery An industrial process plant where *crude oil is converted into useful products such as naphtha, diesel fuel, kerosene, and LPG. Also known as a *petroleum refinery, the process involves the separation of the crude oil into fractions in the process of fractional distillation. By boiling the crude oil, the light or more volatile components with the lowest boiling point rise towards the top of the column, whereas the heavy fractions with the highest boiling points remain at the bottom. The heavy bottom fractions are then thermally cracked to form more useful light products. All the fractions are then processed further in other parts of the oil refinery, which may typically feature vacuum distillation used to distill the bottoms; *hydrotreating, which is used to remove sulphur from naphtha; *catalytic reforming; *fluid catalytic cracking; *hydrocracking; *visbreaking; *isomerization; *steam reforming; *alkylation; hydrodesulphurization; and the *Claus process used to convert hydrogen sulphide into sulphur, solvent dewaxing, and *water treatment.

oil shale An oil-bearing fine-grained carbonaceous sedimentary rock containing an organic matter called kerogen. It is generally uneconomic to extract oil from oil shale unless the cost of extraction falls below the cost of petroleum. A number of recovery methods, such as *fracking, are currently being pioneered in response to threats of declining oil reserves.

Oldshue–Rushton column A type of column used for *liquid–liquid extraction that has fitted agitators, horizontal rings with a central opening, and vertical *baffles attached to the inside of the column wall. The rings divide the column into a series of mixing zones and the vertical baffles enhance mixing. In the centre of each mixing zone is a flat-bladed *Rushton turbine. All the turbines are mounted on a common shaft along the axis of the column. The light phase is fed into the bottom of the column and the heavy phase into the top. These are respectively removed from the top and bottom of the column once extraction is complete.

olefins A former name for the class of unsaturated hydrocarbons now known as *alkenes.

oleum A fuming liquid of disulphuric acid ($H_2S_2O_7$) produced during the *contact process and used in the sulphonation of organic compounds.

Olin–Raschig process *See* RASCHIG PROCESS.

one-dimensional flow A simple approach used to determine the flow of a fluid that has a single coordinate x and a velocity in that direction. The flow in a pipe or tube is considered to be one-dimensional.

one-pot synthesis A method of synthesizing organic compounds in a single vessel rather than using a series of vessels or stages.

one-way valve A type of valve that permits the flow of a fluid in one direction only with no opportunity of reverse flow. *See* CHECK VALVE.

on/off control A basic mode of control involving the opening and closing of a valve in response to a measured variable such as temperature. It is a type of *proportional control.

OPEC *See* ORGANIZATION OF THE PETROLEUM EXPORTING COUNTRIES.

open channel A conduit carrying a liquid with a *free surface. Open channels are used for transporting large volumes of water at low velocities. They are used to transport water from rivers to process plants. The rate of flow is dependent on the slope of the channel, surface roughness, and dimensions. The maximum rate of flow is achieved in an open channel with a trapezoidal cross section. *See* CHANNEL.

open hearth process An early method used for the manufacture of steel involving the heating together of scrap, pig iron, and hot metal in a refractory-lined shallow open furnace heated by the burning of *producer gas in air. This causes excess carbon and other impurities to be burnt out. The method was developed to overcome the very high temperatures that are required. Developed in the nineteenth century, it was gradually replaced during the twentieth century with basic oxygen and electric arc furnaces due to its comparatively slow operation.

open loop control The manual control of a process or system in which information about a controlled variable is not used to adjust any of the system inputs to compensate for variations in any of the measured process variables. There is therefore no automatic feedback used to adjust the process or system. The term is often used to indicate that uncontrolled process dynamics are being studied.

open system A process in which both material and energy are transferred across a defined system boundary. A *continuous process is an example of an open system. *Compare* CLOSED SYSTEM.

operating conditions The conditions of temperature and pressure to which a process or item of equipment is subjected. The *normal operating conditions are the expected range of operating conditions used to operate the process or equipment, and under which process operating influences are usually stated.

operating line A line used in the graphical determination of the number of theoretical plates or stages in a *multistage process such as distillation. The equation of the operating line is based on the mass balance of the more volatile component and represents the actual vapour–liquid relationship. This is in contrast to the actual equilibrium relationship between the components. *See* MCCABE–THIELE.

operating manual, instructions Written information used for the general operation of a process or item of process plant equipment. It includes details of the normal *operating conditions and procedures, start-up, shutdown, *emergency shutdown procedures, sampling, *maintenance, repair, and supervision.

operator 1. A trained person who has responsibility for carrying out the day-to-day operation of a process or plant. **2.** An industrial company that is a member of a joint venture and appointed to carry out all the activities and operations for a particular process plant, or for activities such as exploration and drilling for oil. **3.** A mathematical entity used to perform a specific operation such as ∫ (the integral operator) or Δ (the differential operator).

opex An abbreviation for **op**erational **ex**penditure, it is the on-going cost for operating a process, business, or system. It includes the cost of materials, energy, maintenance, personnel, support services, and utilities. *Compare* CAPEX.

optical density The measure of the reduction in intensity of incident radiation that passes through an absorbent medium. It is largely replaced by absorbance and used to measure the density of materials and media such as the microbial cell concentration in a bioreactor. The optical density or absorbance is often directly correlated with *cell dry weight.

optimization A procedure or set of procedures used to find the best compromise or optimal solution to a problem between conflicting requirements. Solved mathematically, it can be used to maximize or minimize a mathematical functional within defined constraints. For example, it can be used to maximize the yield from a chemical reaction or the revenue generated from a process while minimizing the consumption of energy, production of a *by-product, or waste.

optimum reflux ratio The reflux ratio used in the operation of a distillation column that corresponds to the minimum cost that combines both operating and fixed costs.

order of magnitude A designation used to describe a difference between two numbers with one being ten times greater than that of the other.

order of reaction The power to which the concentration of a component in a chemical reaction is raised. It provides an indication of the mechanism of the reaction and is determined experimentally. Most reactions are usually first or second order. In the reaction $A + B \rightarrow products$ the rate may be related to the concentration of component A by $-r_A = k[A]^X$. If X = 1 then the reaction is first order with respect to A and k is the velocity or rate constant. If X = 2 then the reaction is second order with respect to A. If the rate is related to the concentration of component B, then the order of reaction with respect to B is Y. The overall rate equation is $-r_A = k[A]^X[B]^Y$ and the overall order of the reaction is X+Y. The order of reaction can be fractional. If the rate of a reaction is independent of the concentration of a particular reactant, then the reaction is zero-order with respect to that reactant. The order of reaction cannot be deduced from a balanced chemical reaction equation.

ordinate The vertical or y-coordinate in a two-dimensional Cartesian coordinate system such as a *chart or graph. *Compare* ABSCISSA.

ore A naturally occurring mineral aggregate from which metal or other valuable constituents can be usefully extracted. Metals may be present in their native form such as gold but most are usually in the form of oxides, sulphides, sulphates, silicates, etc.

ore dressing Another name for *mineral dressing and involves the extraction of minerals from ores using crushing, grinding, particle size distribution, and various forms of separation such as flotation. *See* BENEFICIATION.

ore flotation A process used to extract potash from an ore known as sylvinite. The ore, which is a mix of sodium and potassium chloride, is crushed and treated using a cationic detergent or a fatty amine. Air is bubbled through a vessel known as a collector, in which, depending on the design, the bubbles attach to either the sodium or potassium chloride crystals, which float to the top and are separated.

organic chemistry A branch of chemistry concerned with all aspects of compounds of carbon.

organic compound A compound that contains carbon. However, there are some carbon-containing compounds that are inorganic such as carbides and carbonates as well as some oxides of carbon such as carbon monoxide and carbon dioxide, as well as graphite and diamond that are inorganic. The term organic is historical and dates from an age when it was believed that compounds had a connection with living organisms and could be synthesized through alchemy. Organic compounds include all living materials, polymers, rubbers, carbohydrates, fats and oils, vitamins, proteins, peptides, and nucleic acids.

organic synthesis A branch of chemical synthesis that is concerned with the formation and methodology of preparation of *organic compounds.

Organization of the Petroleum Exporting Countries (OPEC) An international intergovernmental organization of member countries that produce oil and gas, which regulates the supply of petroleum for the purpose of ensuring stabilization of world demand and use. It also aims to ensure a fair economic return to producers and for countries investing in the industry. OPEC was formed in 1960 by the five founding member countries of Iran, Iraq, Kuwait, Saudi Arabia, and Venezuela. Others have subsequently joined and the membership is currently twelve member countries. The headquarters is based in Vienna, Austria.

• Official website of OPEC.

orifice A small opening through which a fluid passes.

orifice plate meter A device used to measure the flow rate of a fluid. It consists of a flat plate in which an orifice is drilled. The plate is fitted across the face of flow of a fluid in a pipe

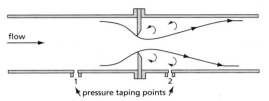

flow

1   2
↖ pressure taping points ↗

Fig. 32

between two flanges (see Fig. 32). The shape of the orifice is usually circular and concentrically aligned with the axis of the pipe although other designs are used and may be offset to below the centreline to enable the passage of fluids carrying suspended solids. The flow of fluid through the orifice results in an increase in velocity and corresponding decrease in pressure. A measurement of the pressure drop provides a measure of the rate of flow. This type of flow meter does not allow for the recovery of energy in the fluid for which there is a high permanent energy loss. This is reflected in the *coefficient of discharge of around 0.6 for high flows.

Orsat analysis A measurement of the oxygen, carbon dioxide, and carbon monoxide in a mixture of gases, usually from the exhaust of combustion processes such as boilers, furnaces, fired heaters, and combustion engines. Named after its inventor H. Orsat in 1873, it involves absorption of the gases onto materials contained in pipette tubes. The method has been largely replaced by other techniques.

osmosis The movement by diffusion of a fluid or solvent from a low to a more concentrated solution through a *semi-permeable membrane. The movement causes the concentrations on either side of the membrane to equalize. The **osmotic pressure** is the equilibrium pressure difference between the fluids of differing composition on either side of the membrane. It is a function of the solute concentration and is independent of the number of ions or molecules in solution. *Compare* REVERSE OSMOSIS.

osmotic shock A method used to extract intracellular materials such as proteins from the cells of yeast and bacteria that have been harvested from a bioreactor. It involves transferring water across the outer membrane of the living cells to build up the **osmotic pressure** inside them, resulting in their eventual disruption. Although it is less energy-intensive than the use of *bead mills for cell disruption, it is relatively inefficient and therefore largely confined to use in laboratory practices.

Ostwald, Friedrich Wilhelm (1853–1932) A Latvian-born German physical chemist noted for his work on chemical reactions and electrolytes. Having studied chemistry and physics in Riga, he was appointed professor of chemistry at Riga in 1881. Six years later he accepted the then only chair in physical chemistry in Germany at the University of Leipzig. He retired in 1906 having been appointed as the first German exchange professor to Harvard. He was awarded the Nobel Prize for Chemistry in 1909 for his work on catalysis, chemical equilibria, and reaction velocities.

Ostwald–de Waele equation A simplified power law relationship used to describe non-Newtonian fluids as $\tau = a\gamma^n$. Depending on the value of the power index, n, the fluid can be classified as being pseudoplastic (n < 1), Newtonian (n = 1), or dilatant (n > 1). It is named after German chemist Friedrich Wilhelm Ostwald (1853–1932) and British chemist Armand de Waele (1887–1966).

Ostwald process A catalytic process used for the production of nitric acid by the oxidation of ammonia with air. The first step involves mixing air as a supply of oxygen and ammonia over a catalyst at a temperature of 700°C:

$$4NH_3 + 5O_2 = 6H_2O + 4NO$$

The gases are then cooled in two towers and the oxidation is completed as:

$$2NO + O_2 = 2NO_2$$
$$3NO_2 + H_2O = 2HNO_3 + NO$$

The NO is in part reoxidized to form more nitric acid in successive repetitions of the process. It is named after German chemist Friedrich Wilhelm *Ostwald (1853–1932).

Othmer, Donald Frederick (1904–95) An American professor of chemical engineering who was responsible for cofounding and the editorship of the *Kirk–Othmer Encyclopedia of Chemical Technology*. A graduate of chemical engineering from the University of Nebraska in 1924, he gained his masters and PhD from the University of Michigan in 1927. He joined the department of chemical engineering at the Polytechnic Institute of Brooklyn and became its head in 1937. He became distinguished professor in 1961 and professor emeritus after his retirement, and continued teaching up until his death.

OUR *See* OXYGEN UPTAKE RATE.

output The quantity of material, energy, or power out of a system. *Compare* INPUT.

output signal The output from a controller of a controlled process. In a manually controlled process, the process operator sets the output, whereas in an automatically controlled process, the controller computes the output based on a calculation using the *error signal. The error is the difference between the set point and measured *process variable.

overall heat transfer coefficient (Symbol U) Used in heat transfer calculations, it is a measure of the overall transfer of heat through convective and conductive barriers. It is expressed as the *heat flux divided by the difference between the temperature in the bulk of two fluids. Since many types of heat transfer equipment use pipes and tubes, it is necessary to specify the area of heat transfer, where:

$$U_i = \cfrac{1}{\cfrac{1}{h_i} + \cfrac{x}{k} + \cfrac{1}{h_o}}$$

where h_i and h_o are the inside and outside surface heat transfer coefficients, x is the tube wall thickness, and k is the thermal conductivity of the wall. If the outside area is specified, the overall heat transfer coefficient is based on the outside area and is written as U_o, while if the inside area is specified, then the coefficient is denoted by U_i. The SI units are $W\,m^{-2}\,K^{-1}$.

overall mass transfer coefficient (Symbols K_G and K_L) Used in mass transfer calculations, it is the mass flux of one fluid to another across an interface divided by the difference between the concentrations of the diffusing component in the bulk of the two fluids. The subscripts refer to the overall mass transfer coefficients in the gas and liquid phases, respectively. The overall mass transfer coefficient is defined as either the liquid- or gas-side since the interfacial composition cannot be determined. The bulk driving force is used.

That is, the composition in the gas phase is assumed to be in equilibrium with the bulk concentration in the liquid phase, or the liquid composition is in equilibrium with the composition of the bulk gas phase. The SI units are $m^2 s^{-1}$. The mass transfer coefficients are dependent on the geometry of the contacting equipment as well as fluid behaviour. Many correlations have been established to determine the coefficients. An example is the use of the *Sherwood number for describing the gas phase in terms of the *Reynolds number and *Schmidt number.

overall plate efficiency The ratio of the actual number of plates or trays to the theoretical number that are required in a *distillation column. The actual number of plates or trays that are required is greater than the theoretical number since vapour–liquid equilibrium is not always reached on each plate or tray. *See* MURPHREE PLATE EFFICIENCY.

overburden The material in mining processes that lies above the material that is of economic interest. An example is the rock that lies over an ore or a seam of coal. It is also known as waste or spoil. *Compare* TAILINGS.

overflow Used in liquid–liquid extraction and leaching processes, it is the flow of less dense material or particulate-carrying liquid, and moves from one stage to another. The underflow is the heavier phase and moves in the opposite direction in countercurrent extraction processes.

overhead product The product removed from the top of a distillation column as either liquid, distillate, or vapour. It is also known as the *top product.

overpressure 1. The pressure within a process vessel or some other equipment, which exceeds its expected or intended design pressure. **2.** A pulse of pressure that is above atmospheric pressure in the form of a blast wave as the result of an *explosion. The peak positive overpressure is the maximum overpressure that is generated in the explosion. The side-on and reflected overpressures are the pressures experienced by a body placed in the path of the blast wave that offers no obstruction and diffracts the wave, respectively.

overshoot A measure of how far a controlled signal responds in a system to a disturbance rising beyond the final steady-state value. The overshoot is the difference between the height of the first peak and the final steady-state value.

oxidative coupling A general name given to heterogeneous catalytic processes used for the direct conversion of natural gas, which comprises mainly methane, to ethane, ethylene, and other longer hydrocarbons that have a higher added value, with ethylene being the world's largest commodity chemical. For the production of ethylene, the gas phase reaction $2CH_4 + O_2 \rightarrow C_2H_4 + 2H_2O$ is very exothermic and is therefore carried out at a high temperature. The methane is converted first to ethane and then goes through a process of dehydrogenation to form ethylene.

oxo process A high-temperature, high-pressure process used for the production of aldehydes and alkenes from the catalytic reaction between alkanes, carbon monoxide, and hydrogen. The process was invented in Germany in 1938 by O. Roelen, although modified many times since, and is used in the production of fragrances and intermediates for the production of detergents. The name is derived from the German *oxierung* meaning ketonization. It is also known as **oxo synthesis** or **hydroformylation.**

oxychlorination process A process used to convert ethylene to ethylene dichloride (1,2-dichloromethane). It involves reacting ethylene, air, and hydrogen chloride in the presence of a catalyst of cupric chloride on potassium chloride. In the production of vinyl chloride monomer, the ethylene dichloride is cracked and the hydrogen chloride recycled.

oxydesulphurization process 1. A general name for processes that remove sulphur from coal using oxygen. These use pulverized coal that is heated to a high temperature and pressure in a *fluidized bed. **2.** A process used to remove sulphur from carbon dioxide or natural gas (methane).

oxygen uptake rate (OUR) The rate of change of the dissolved oxygen of a nutrient medium used for cultivating microbial cells within a *bioreactor.

oxyhydrochlorination process A process developed in the 1920s and used to convert methane into chloromethane that can then be converted in petroleum fuels. It involves mixing the methane with oxygen and hydrochloric acid in the presence of a catalyst:

$$2CH_4 + O_2 + 2HCl \rightarrow 2CH_3Cl + 2H_2O$$

The chloromethane is then converted to petroleum fuel using a *zeolite catalyst. Hydrogen chloride is produced and recycled back to the process.

ozone An unstable gas, O_3. It is formed naturally in the atmosphere and also by an electric discharge in oxygen. It has a bluish colour and distinctive odour and is used in the purification of air and water.

ozonolysis The chemical reaction of alkenes with *ozone to form a group of unstable compounds called ozonides. It was once used to investigate the structure of alkenes by hydrolyzing the ozonide to give aldehydes or ketones.

packed bed A vessel that is filled with a solid *packing material for the purpose of enhancing a chemical reaction or a separation by improving the contact area available for the reaction or separation. The packing material is usually inert and resistant to the process materials and is typically made of ceramic, plastic, glass, or metal. Structured packing consists of corrugated sheets that allow the two phases such as a liquid and a gas to intimately contact one another through complicated channels. Unstructured or random packing consists of small objects that have been designed to provide a high contact area for a minimum volume. *Raschig rings, *Lessing rings, and *Berl saddles are some of the most commonly encountered types of packing material and are loaded into vessels such as distillation columns, adsorption and absorption towers. Catalytic particles also form packed beds to promote chemical reactions and are loaded into vessels or tubes depending on the design. Adsorption materials such as *zeolites are loaded into vessels to form packed beds.

packed bed reactor (PBR) A tubular reactor consisting of one or more tubes within which a solid catalyst is packed to promote a chemical reaction. They are mainly used in heterogeneous gas-phase reactions and operated continuously.

packed column A cylindrical column used to bring about the intimate contact between a rising gas or vapour and a descending liquid typically used for a gas–liquid separation such as stripping. The vessel is packed with small objects that produce a large contact area relative to their size such as *Raschig rings or *Berl saddles, and are used in *distillation columns and adsorption and absorption towers. The liquid descends through the packing through a tortuous route and is in intimate contact with the rising gas or vapour for the purpose of raising the mass transfer area within a limited volume. The amount of packing is determined from calculations involving the number of theoretical stages required determined from vapour–liquid equilibria and the *height equivalent to a theoretical plate (HETP).

packing The small objects that are used in *packed columns and *packed beds. They provide a high surface area per unit volume allowing intimate contact between a rising vapour or gas with a descending liquid. There are many types commonly used, such as saddles and rings, and they are made from plastics, glass, metals, and ceramics that are inert and unreactive to the substances in which they are in contact. They may be arranged in packed columns either in a structured form or randomly, and sit on a support plate designed to take their bulk weight. They are required to allow good passage of liquid and vapour and offer a low pressure drop across the entire packing. The **packing fraction** is the volume taken by the number of particles within a given space or volume.

paddle A type of stirrer used for mixing in vessels to ensure good homogeneity of materials. It consists of flat plates attached to the rotating shaft. The paddle gives *radial flow patterns for which the diameter of the paddle is typically less than half the diameter of the vessel itself.

Pall ring A type of packing material used in a *packed column and consists of a perforated cylinder with a height equal to its diameter. Pall rings are sometimes used in distillation columns to provide an intimate contact between the liquid and vapour. Metal Pall rings have strips that bend inwards, while plastic versions have an internal cross structure.

P&ID *See* PIPING AND INSTRUMENTATION DIAGRAM.

paraffin A former name for the class of saturated hydrocarbons known as *alkanes.

parallel-disc rheometer An instrument used to obtain the rheological properties of fluids. It consists of a fixed flat surface with another rotating surface held at a fixed elevation above, with a sample of the fluid sandwiched between them (see Fig. 33). The rotational speed and distance defines the shear rate. The torque to resist the motion defines the characteristic shear rate. The apparent viscosity is calculated from the ratio of the shear stress to shear rate. The surface can be heated or cooled to determine the rheological properties as a function of temperature.

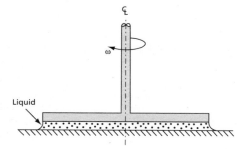

Fig. 33

parallel flow *See* COCURRENT FLOW.

parameter A *variable whose characteristic is not considered and may therefore be taken as being constant.

Parkes process A metallurgical process for the removal of silver from lead using zinc in *liquid–liquid extraction. Zinc is immiscible with lead and its addition to lead containing silver as a contaminant results in silver dissolving in the zinc. The zinc-silver solution is then separated and the zinc removed from the silver by distillation. The process was patented in 1850 by British metallurgist Alexander Parkes (1813–90).

partial condenser A condenser-type heat exchanger used in a distillation process in which only part of the vapour from the top of the distillation column condenses and returns as reflux. The rest of the vapour is used as the top product. A partial condenser is also known as a *dephlegmator. *Compare* TOTAL CONDENSER.

partial derivative The derivative of a function of two or more variables with respect to one of the variables and with the others remaining constant. For example, for the function $z = f(x, y) = x^3 y^2 + x^2 y + y^3 + x$ then the partial derivative of x with respect to z, is:

$$\frac{\partial z}{\partial x} = 3x^2 y^2 + 2xy + 1$$

where y is unchanged; and;

$$\frac{\partial z}{\partial y} = 2x^3 y + x^2 + 3y^2$$

where x is fixed, y is the independent variable, and x is the dependent variable. There are many other examples that use various mathematical rules in their solution such as the chain rule and the quotient rule. They are widely used in chemical engineering, such as in applications involving the movement of fluids or solving energy balances within a space that does not have any defined boundary conditions.

partial fractions A set of simple fractions from which a more complicated fraction can be resolved.

partially miscible liquids Liquids that can dissolve within one another within certain concentration limits to form a single phase, but remain as separate liquids outside those limits. An example is n-butanol and water, in which the liquids separate to a 98 mol per cent water-rich phase and a 56 mol per cent n-butanol-rich phase. Partially miscible liquids are used in the partitioning of a third phase in *liquid–liquid extraction.

partial pressure The pressure exerted by a gas or vapour in a mixture at constant temperature that the gas would exert if it alone occupied the whole volume actually occupied by the mixture. The ratio of partial pressures in a mixture of gases is therefore the same as the ratio of volumes. The partial pressure of a component within a mixture of liquids is the product of the vapour pressure of the pure component and its mole fraction in the liquid phase. The partial pressure, p, of a component, i, of a gas mixture is directly proportional to the total pressure, p, where for an ideal gas:

$$p_i = y_i p$$

where y_i is the mole fraction in the vapour phase.

partial product The mathematical result that is obtained by multiplying a number with one digit of a multiplier.

partial volume The volume occupied by a component in a gas mixture if it alone were present at the total pressure and temperature of the mixture. *See* AMAGAT'S LAW.

particle size distribution The classification of particles in a mixture by size based on their diameter or some other physical characteristic such as weight. The particle size distribution is the broad range of particle sizes in a mixture of particles, expressed as the quantity of the particles whose sizes fall between two measureable characteristics such as the weight percentage of the mixture between two defined diameters. The particle size distribution can also be expressed as a cumulative quantity such as the total weight expressed as a percentage of the mixture above a certain size.

particulate fluidization *See* FLUIDIZATION.

particulates Very small solid bodies often having an irregular shape. They may be a single substance or a mixture of chemical substances often as the product emanating from a process waste stream. They include dirt, dust, smoke, soot, fumes, aerosols, mists, and sprays. They are sufficiently small to be able to penetrate the lower reaches of the lungs.

Dust is classified as particulates that have an aerodynamic diameter of less than 75 micrometres (75 μm), whereas smoke has an aerodynamic diameter of less 15 μm. Inhalable particulates are generally less than 15 μm while respirable dust is less than 5 μm. Cement dust ranges from 1 to 100 μm, while fly ash is approximately 10 μm. Fine particulates have aerodynamic diameters of less than 2.5 μm and abbreviated as $PM_{2.5}$. *See* AIR POLLUTION.

parting A very old process used for separating gold, silver, and platinum from each other by dissolving them in nitric acid. These days, electrochemical processes are used to separate them.

partition coefficient The equilibrium distribution of a substance between two immiscible or *partially miscible liquids. For example, oxalic acid is soluble in both water and diethyl ether, which are themselves immiscible in one another. The partition coefficient is used to determine the effectiveness of a *liquid–liquid extraction process.

Pascal, Blaise (1623–62) A French mathematician, physicist, and thinker on religion and philosophy. With help from his brother-in-law, he arranged an experiment on the Puy de Dome, a mountain in the Auvergne, and was able to show that the height of mercury in a barometer decreases with elevation. He also discovered that liquids transmit pressure in all directions. To show how columns of liquid balance one another, he designed 'Pascal's vases', which are unusually shaped vessels but the pressure at the bottom of each when filled to the same depth is the same. The unit of pressure is named after him. He suffered ill health throughout most of his life and died at the early age of 39.

pascal (Symbol Pa) The *SI unit of pressure equal to one newton per square metre. It is named after Blaise *Pascal (1623–62). Standard atmospheric pressure is equal to 101 325 Pa.

Pascal's pressure law A law that states that the pressure applied to a fluid is transmitted equally in all directions. It is also known as **Pascal's principle.**

pass The flow of material or energy through a system. In a heat exchanger, the heat transfer medium passes through either the *shell side or *tube side transferring heat. The heat transfer medium can be made to return back through the heat exchanger for further heat exchange. For systems where the medium passes on numerous occasions, the term *multipass is used. *See* PASS PARTITION PLATE.

passive The appearance of being chemically unreactive and is usually the result of the formation of a protective layer that thereby prevents further reaction. For example, aluminium metal reacts with oxygen in air to form a layer of aluminium oxide covering the metal thereby preventing further oxidation to the metal beneath.

pass partition plate A horizontal divider used within the fixed or floating head cover of a shell and tube heat exchanger. It is used to direct the heat transfer fluid from one end of the heat exchanger to the other. Each traverse is called a *pass. The number of passes on the shell side is usually quoted first such as a 2:4 exchanger, which has two *shell side passes and four *tube side passes.

Pasteur, Louis (1822–95) A French chemist and microbiologist who studied the fermentation of wine and introduced the thermal food preservation process of *pasteurization. He later studied diseases and developed vaccines for cholera, anthrax, and rabies. He was director of the Pasteur Institut in Paris from 1888 until his death.

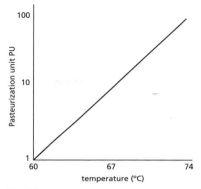

Fig. 34

pasteurization A thermal process used to extend the shelf-life of foods such as fruit juices, beer, milk, and other dairy products by destroying or inactivating harmful and pathogenic microorganisms that might be potentially present. Pasteurization is usually achieved at a specified temperature and time combination. For example, milk can be pasteurized by heating to 72°C and holding the temperature for 15 seconds. There are other temperature-time combinations that will achieve the same effect such as the Holder process, which involves heating between 63°C to 66°C and holding for 30 minutes. The rapid heating and cooling can be achieved using *plate heat exchangers. The process is named after French chemist and biologist Louis *Pasteur (1822–95).

pasteurization unit A sterilizing effect achieved when heating a food product at 60°C for one minute (see Fig. 34). It is used in the pasteurization of foods such as dairy products, beer, and fruit juices in which harmful bacteria can be reduced ten-fold by an increase in temperature of 7°C and is calculated from:

$$PU = t10^{\frac{T-60}{7}}$$

where T is the temperature and t is the holding time. The number of pasteurization units required for a particular product depends on the product, specific type of bacteria, packaging, and expected shelf-life.

patent An official document granted by a government to an inventor of a novel product or process. It assures the legal protection and sole rights to manufacture, use, and sell the product or process within a limited period. A **patentee** is a person, group, or company that has been granted a patent.

pathogen Any harmful disease-causing microorganism or virus. Pathogenic bacteria in foods such as *Clostridium botulinum*, which can produce harmful toxins can be eliminated through processes such as *canning, which involves sterilization by heat.

Pattinson process A process formerly used for the extraction of silver from lead. Developed in 1833 in the UK by H. L. Pattinson, it involved a repeated melting and cooling process that allowed crystals of lead to form and be skimmed off to yield a silver concentrate. The process was superseded by the *Parkes process and is still used for the recovery of bismuth.

PBR An abbreviation for a *packed bed reactor or *pulsed baffle reactor.

Péclet number A dimensionless number, Pe, used in the study of the thermal transport phenomena of flowing fluids. For the diffusion of heat, it is expressed as the ratio of the molecular to convective heat transfer of a substance:

$$Pe = \frac{vl}{\alpha}$$

where v is the velocity of the substance, l is the distance travelled, and α is the thermal diffusivity. For the diffusion of mass, it is expressed as the ratio of the molecular and mass diffusion of a substance:

$$Pe = \frac{vl}{D}$$

where D is the mass diffusion coefficient. It is named after French physicist Jean Claude Eugène Péclet (1793–1857).

Pekilo process A biochemical process developed in Finland for the production of single-cell protein using a fungal biomass. It uses various forms of carbohydrates as the substrate including spent sulphite liquors from the wood and paper industries as well as waste liquors from the potato and sugar industries.

Peltier effect A thermoelectric cooling effect used in small-scale refrigeration units. It involves a solid-state heat pump that transfers heat from one side of the device to the other through the consumption of electricity. The thermoelectric cooling effect is created between the junction of two different types of conducting materials. The device can also be used for heating. It is named after French physicist Jean Charles Athanase Peltier (1785–1845) who discovered the effect in 1834.

Peng–Robinson (PR) equation of state An equation of state used to predict the behaviour of real gases based on the *van der Waals' equation of state. It describes the variation of molar gas volume and pressure with temperature for many substances in a cubic equation as:

$$p = \frac{RT}{V-b} - \frac{a\alpha(T)}{(V-b)(V+b)}$$

The constants a and b are determined from critical point data. The factor α is related to reduced temperature as:

$$\alpha = \left[1 + (0.37464 + 1.54226\omega - 0.26992\omega^2)(1 - T_r^{0.5})\right]^2$$

where ω is the acentric factor. Developed in 1976, it is also useful for predicting liquid densities and is widely used in the petrochemical industry. It was further modified by Stryjek and Vera in 1986 to improve the accuracy of the model. The **Peng–Robinson-Stryjek-Vera**

(PRSV) equations of state feature an adjustable pure component parameter and a modified polynomial fit for the acentric factor, and can be used for polar substances.

perfect gas A theoretical gas that differs from a real gas that consists of elastic molecules in which the effective volume occupied by the molecules is zero and the gas obeys the *ideal gas law: $pV = nRT$ where p is the pressure, V is the volume occupied by the gas, n is the number of molecules, R is the *universal gas constant, and T is the absolute temperature. It applies to a gas with a low density in which the intermolecular forces between the molecules are neglected. Perfect and ideal gases are often used interchangeably although perfect gases assume a constant specific heat at constant volume.

periodic table The tabular arrangement of all the chemical elements in order of their atomic numbers. The table consists of vertical columns containing groups of elements with similar properties, demonstrating the periodic law. The elements in each group have the same number of electrons in their outer orbitals and therefore share the same valency, thereby accounting for their similar chemical properties. The horizontal rows are periods. The table was originally devised and published by Russian chemist Dmitri Ivanovich Mendeleev (1834–1907) in 1869 and was based on relative atomic masses.

peristaltic pump A positive displacement pump that consists of a flexible tube which contains the process fluid to be transported. The tube is compressed and squeezed between rollers on a rotating wheel thereby moving the fluid along the tubing (see Fig. 35). The roller pressure traps portions of the fluid and carries them forward as discrete volumes. The flow is, however, not pulsation-free. The rate of flow is adjusted by changes in the rotational speed of the wheel. By maintaining the fluid in the sealed tube such that it has no contact with the moving parts of the pump, it is therefore often used in medical applications for fluids such as blood, and for metering applications.

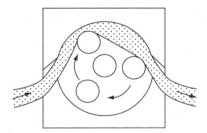

Fig. 35

permanent gas A gas that was once thought to be impossible to liquefy, such as oxygen and nitrogen. It is now considered to be a gas that is not able to be liquefied by pressure alone at normal room temperature. That is, the gas has a *critical temperature below normal room temperature.

permeability The ability of a gas or liquid to flow through a *semi-permeable membrane per unit driving force per unit membrane thickness. It is also known as the **permeability coefficient** and is experimentally determined for a particular separation. The permeability is dependent on the type of permeate, temperature, membrane pore size, flow area, and thickness. For gas separation, the driving force is the pressure difference across the membrane. For the permeability of oxygen, the barrer is a non-SI unit in which one barrer is equal to 3.348×10^{-19} kmol m m^{-2} s^{-1} Pa^{-1}. It is named after Richard Barrer (1910–96).

permeate The liquid, gas, or vapour that passes through a *semi-permeable membrane in a separation process. *Hollow fibre membranes can be used to separate selectively gases such as nitrogen from oxygen in air in which the smaller nitrogen molecules diffuse through the pores of the membrane as the permeate. In membrane modules used to separate activated sludge, water molecules pass through the membrane as permeate. The material remaining is called the *retentate.

permit to work A formal written and documented procedure that authorizes certain people to carry out specific work within a specified time frame. It is used to control certain types of work that are potentially hazardous and forms an essential part of safe systems of work for many maintenance activities and allows work to begin only after safe procedures have been defined. It provides a clear record that all foreseeable hazards have been considered. Where the proposed work is assessed as having a high risk, strict controls are required detailing the work and how it will be done. It may require a declaration from those involved in a shift handover of the procedures or extensions to the work that is required to be done. Before process equipment or machinery is returned to service, it may also require a declaration from the permit originator that it is ready for normal operational use.

Perry, Robert H. (1924–78) The second editor of *Perry's Chemical Engineers' Handbook*, which was first edited by his father, John H. Perry and published in 1934. He served as chairman of the Department of Chemical Engineering at the University of Oklahoma and was programme director for graduate research at the National Science Research Foundation. He also taught at the University of Rochester and University of Delaware and was advisor to the United Nations and other international organizations.

persistent organic pollutant (POP) A toxic substance that is harmful to the environment, due to being resistant to biodegradation. POPs enter into the food chain causing health issues and concerns. The *Stockholm Convention was a major international conference convened in 2001 to address POPs and identify the twelve substances of greatest concern, including DDT, dioxin, and polychlorinated biphenyls. Other substances have subsequently been added to the list.

personal protective equipment (PPE) Specialist equipment worn or held by a person to reduce or minimize the exposure or contact with injurious workplace hazards. The equipment is used as a barrier to reduce the risk of injury. PPE includes equipment such as safety footwear, hard hats, ear protection, high-visibility waistcoats, goggles, life jackets, respirators, and safety harnesses. In the UK, the principal legislation governing these is the Personal Protective Equipment at Work Regulations 2002 based on the European Council (EC) Directive 89/656/EEC, which requires similar basic laws throughout the European Union on the use of PPE in the workplace.

(())) SEE WEB LINKS
• Official website of the UK Health and Safety Executive.

PERT *See* PROGRAMME EVALUATION AND REVIEW TECHNIQUE.

pervaporation A membrane separation process used to separate volatile substances from dilute solutions in which the membrane provides a selective barrier. Being independent of vapour–liquid equilibria, pervaporation operates by permeation of a substance through the membrane followed by its evaporation into the vapour phase. The separation is based on a difference in transport rates of individual components in the liquid *retentate and vapour *permeate on either side of the membrane. The upstream side of the membrane is typically at ambient pressure and the downstream side is under vacuum to allow the evaporation of the selective component after permeation through the membrane. The driving force for the separation is the difference in the partial pressures of the components on either side of the membrane. Pervaporation is used by many industries for purification and separation and is popular due to its simplicity, low energy consumption, and low temperature and pressure operation. It is effective for separating diluting solutions containing small amounts of a component to be removed. Hydrophilic membranes are used for dehydration of alcohols containing small amounts of water, while hydrophobic membranes are used for the removal of organic substances from aqueous solutions. It is a less aggressive separation process than distillation and is therefore used for the separation of ethanol from yeast fermentations, removal of water from esterification reactions, organic solvents from industrial wastewater, and for separating hydrophobic flavour compounds from aqueous solutions.

petrochemicals Organic chemicals produced from petroleum or natural gas.

petroleum (motor gasoline) A liquid mixture of light hydrocarbons used as a fuel in internal combustion engines. It is distilled from crude oil between 35°C and 215°C and can include oxygenates to reduce the amount of carbon monoxide formed during combustion, as well as octane enhancers. It can also be mixed with anhydrous ethanol such as bioethanol. The word 'petroleum' is derived from the Latin *petra* meaning 'rock or stone' and *oleum* meaning 'oil'.

petroleum coke A black solid residue used as a feedstock in coke ovens used in the steel industry for heating and chemical production. Obtained by the process of *cracking and carbonizing petroleum-derived feedstocks, tars, and pitches, it has a high carbon content of up to 95 per cent and a low ash content. However, it has a high sulphur content, which can result in environmental issues.

petroleum ether A mixture of volatile alkanes comprising mainly pentane and hexane. It has a boiling point of between 30°C and 70°C. It is widely used as a solvent and in the extraction of edible oils. It is also an anaesthetic.

petroleum feedstock Chemicals derived from petroleum and natural gas such as naphtha that are used as the raw materials to produce other chemicals, plastics, and synthetic rubbers.

petroleum refinery *See* OIL REFINERY.

petroleum reservoir A naturally occurring subsurface pool of hydrocarbons trapped within rock formations in the Earth's crust. The hydrocarbons may exist as crude oil or natural gas, and are the result of the high pressure and temperature decomposition of aquatic organisms that lived millions of years ago. Petroleum reservoirs are broadly classified as being oil or gas reservoirs and quantified in terms of the composition of the hydrocarbon mixture, their initial temperature and pressure, and surface production temperature and pressure. *See* WELL.

PFD 1. An abbreviation for *process flow diagram. **2.** An abbreviation for probability of failure on demand, which is the likelihood that a process, system, or item of process plant will fail to operate in the required and expected manner on demand. *See* PROBABILITY.

PFR *See* PLUG FLOW REACTOR.

pH A measure of the acidity or alkalinity of a liquid based on the negative logarithm of the hydrogen ion concentration:

$$pH = \log_{10} \frac{1}{\left[H^+\right]}$$

The scale ranges from 1.0 (highly acidic) to 14.0 (highly alkaline) with neutral solutions such as deionized water with a pH of 7.0. The scale was developed in 1909 by Danish chemist Søren Peder Lauritz Sørensen (1868–1939).

phase A state of matter being either solid, liquid, or gas. *Single-phase flow is the flow of a substance or mixtures of substances of the same phase. *Multiphase flow involves a substance or a mixture of substance of different phases. A refrigerant in the evaporator in a *refrigeration cycle is an example of multiphase flow involving a liquid and a vapour.

phase diagram A diagram representing the relationship of solid, liquid, and gaseous phases over a range of temperatures and pressures (see Fig. 36 for a typical example). Phase diagrams for pure substances of pressure and temperature show the existence of triple and critical points. Phase diagrams for binary mixtures showing liquid and gaseous phases at constant temperature illustrate the variation of pressure, and where *tie lines join the two phases in equilibrium. Phase diagrams for binary mixtures such as metals with temperature show the existence of eutectic points. The *liquidus is the line or curve between liquid and liquid/solid at equilibrium, and the *solidus is the line or curve between solid and liquid/solid at equilibrium.

phase rule *See* GIBBS' PHASE RULE.

phase separation The separation of two or more distinct phases. An example is the separation of immiscible liquids such as oil and water, which settle to form two distinct layers where the oil, being of a lower density, floats on the water. The oil and water can then be readily separated using overflow weirs. Phase separation is used to separate crude oil, natural gas, water, and sand in separators on offshore oil platforms.

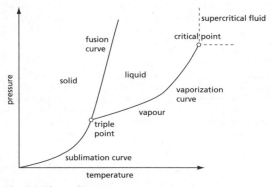

Fig. 36 Phase diagram

phase transition The change from one phase of a substance into another such as a solid to a liquid, a liquid to a gas, etc. The boiling of water to steam is a phase transition, as is the freezing of water to ice.

Phillips process 1. A liquid-phase catalytic process used for polymerizing linear olefins (alkenes) such as ethylene into linear thermoplastic polymers such as high density polyethylene. It uses a Phillips catalyst, which is based on a chromium (VI) oxide on silica. *Compare* ZIEGLER-NATTA CATALYST. **2.** A fractional crystallization process used for freeze-concentrating beer and fruit juices. **3.** A two-stage process used for the dehydrogenation of butane to form butadiene.

photochemistry The study of chemical reactions that are initiated or accelerated by exposure to visible light, ultraviolet radiation, or infrared radiation. The basic laws of photochemistry state that light must be absorbed by a compound for a chemical reaction to take place, known as the first or Grotthus–Draper law. The second or Stark–Einstein law states that for each photon of light absorbed by a chemical system, one molecule is activated for the subsequent reaction. It was derived by Albert *Einstein during his development of the quantum theory of light. The Bunsen–Roscoe law of reciprocity states that a photochemical effect is directly proportional to the total energy dose, irrespective of the time required to deliver the dose.

photolytic reaction A chemical reaction that is the result of exposure to visible light or ultraviolet radiation. The reactions often involve free radicals. Photolysis is an important reaction in the photosynthesis of plants, in which energy from sunlight is absorbed by chlorophyll to produce gaseous oxygen, electrons, and hydrogen ions.

physical chemistry A branch of chemistry that is concerned with the effects of chemical structure on the physical properties of substances.

physical explosion An explosion that does not involve any form of combustion or chemical reaction. The most common is the rupture and bursting of a vessel of which a *boiling liquid expanding vapour explosion (BLEVE) is an example, as is the rupture of a gas cylinder with the rapid release of energy.

pi (Symbol π) A transcendental number with a value of 3.141 592 . . . It is the ratio of the circumference of a circle to its diameter.

pickling The removal of scale such as oxides from substances by immersion in a liquid containing sulphuric acid or hydrochloric acid. It is usually used between hot- and cold-rolling in the processing of sheet steel.

PID control The modes of control used to control processes or part of a process. The three basic modes of control are *proportional control, *integral control, and *derivative control. Derivative control is always used in combination with proportional control or both proportional and integral control. Integral control is generally used in combination with proportional or with both proportional and derivative control. PID control is also known as **three-term control**.

Pidgeon process A process used for the production of magnesium developed by Canadian scientist Lloyd Montgomery Pidgeon (1903–99) in 1941. It involves the high temperature reaction of silicon and magnesia to form silica and magnesium:

$$Si + 2MgO \leftrightarrow SiO_2 + 2Mg$$

The magnesium vapour produced is removed by distillation and recovered as magnesium crystals.

piezometer An instrument used to determine the static pressure of a liquid such as in a pipe. It consists of a vertical tube in which the vertical elevation of the liquid is a measure of the pressure. **Piezoelectric pressure sensors** are used to measure the pressure and convert it to an electrical signal. There are various types commonly used including pneumatic, strain gauges, and vibrating wire sensors. The data is then captured on a data logger.

pig A device used to clean or clear away the inside of a pipeline, particularly those used offshore carrying natural gas and crude oil which are prone to the build-up of deposits such as waxes and hydrates. Pigs are made of rubber or polyurethane and the basic design consists of two plates held apart by a short rod. They can also incorporate various sensing and recording equipment. They are launched into the pipeline and move under the effect of an applied pressure; they emit a squealing sound as their blades scrape along the pipeline and are recovered from a receiving trap, which are loops in the pipeline that can be isolated by shut-off valves. **Smart pigs** or intelligent pigs are inspection devices used to measure the condition of the pipe including metal loss, restrictions, and pipe deformation.

pig iron Another name for cast iron.

pilot-operated safety relief valve (POSRV) An automatic valve system that relieves pressure in a vessel or pipeline by the remote command from a pilot. *See* SAFETY VALVE.

pilot plant A small-scale process unit based on laboratory-scale research or findings. It is used to evaluate the feasibility and potential for full-scale process design and operation. A **pilot scale experiment** is a small-scale preliminary study used to check the feasibility or to improve the design of a process or piece of research. They are often carried out before committing to large-scale process construction and operation in order to avoid wasting time and money. They can also be used to determine the optimum conditions required in a full-scale process or to gain a valuable insight into a process under carefully controlled conditions.

pinch analysis A technique for minimizing energy usage in a process. It is based on calculating the minimum energy consumption by optimizing the heat recovery, energy supply, and process operating conditions. It uses process data represented as energy flows, or streams, as a function of heat load against temperature. These data are combined for all the hot and cold streams in the process to give two composite curves—the hot streams releasing heat and the cold streams requiring heat. The point of closest approach between the hot and cold composite curves is called the **pinch point** and corresponds to the point where the design is most constrained. Using this point, the energy targets can be achieved using heat exchange to recover heat between the hot and cold streams in two separate systems, with one for temperatures above the pinch temperature and one for temperatures below pinch temperatures. First developed by chemical engineer Bodo Linnhoff in 1977 at the University of Leeds, it is also known as *process integration, heat integration, and **pinch technology**.

pinch point 1. Used in multistage separation calculations, it represents the point of no further enrichment between stages. On the *McCabe–Thiele diagram, the point is located where the operating line touches the equilibrium curve. **2.** The point in a *pinch analysis that corresponds to the point where the hot and cold streams in an integrated process are most constrained (see Fig. 37).

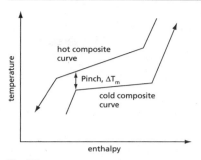

Fig. 37

pipe An enclosed conduit used to transport fluids. These are circular in cross section and available in widely varying sizes, wall thicknesses, and materials. Pipes are specified in terms of their diameter and wall thickness. The wall thickness is indicated by the *schedule number. Pipes are commonly made from metals and alloys, glass, and various plastics. Low-carbon steel pipes are most commonly used in process plants. PVC pipes are typically used for water and gas lines. Lengths of pipes are joined together by *flanges, welding, or screw fittings.

pipe chase An enclosed space used to conceal pipes. Electrical wires and cables may also run through the pipe chase.

pipeline A long section of large-bore pipe used to transport fluids. Natural gas and crude oil from offshore reservoirs are transported over long distances across the seabed up to platforms for separation and treatment, and then transported sub-sea through pipelines for onshore processing. Pipelines are the principal means of transporting hydrocarbon feedstocks such as ethylene over long distances through buried pipelines between petrochemical refineries. Natural gas is transported in similar pipelines over long distances for distribution. Process water including seawater is also transported in great quantities. Pumps and compressors are used to transport such fluids in large volumes at the required pressure. Single-phase flow is easier to transport and meter than multiphase fluid flow. In wet gas lines, large volumes of liquid are required to be routinely swept out using a *pig. *Methane clathrate, waxes, and other deposits can also complicate flow.

pipestill *See* FRACTIONAL DISTILLATION.

pipetrack The organized routing of pipework above ground to transport raw materials, products, and utilities such as water and steam to and from process equipment. The pipetrack is supported on structures such as trestles and gantries with sufficient overhead clearance. Being overhead, leaks are easy to detect and are harmlessly dispersed into the atmosphere. *Compare* TRENCHED PIPING.

pipework A collective term used for all the piping in a process plant irrespective of its purpose and size. A **pipe run** is the route taken by a length of pipe in a process.

piping and instrumentation diagram (P&ID) A schematic representation of the interconnecting pipelines and control systems for a process or part of a process (see Fig. 38 for an example). Using a standard set of symbols for process equipment and controllers, it includes the layout of branches, reducers, valves, equipment, instrumentation, and control interlocks. They also include process equipment names and numbers; process piping including sizes and identification; valves and their identification; flow directions, instrumentation, and designations; vents, drains, samplings lines, and flush lines. P&IDs are used to operate the process system as well as being used in plant maintenance and process modifications. At the design stage, they are useful in carrying out safety and operations investigations such as *HAZOP.

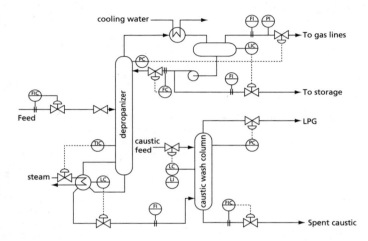

Fig. 38 Piping and instrumentation diagram

piston A movable device used in a cylindrical chamber to transmit a force either onto a fluid in the chamber, such as in a reciprocating pump, or by the expansion of a fluid in the chamber, such as in an internal combustion engine (see Fig. 39). The piston is attached to a connecting rod and crank to transmit or provide power. A **piston pump** is a type of self-priming positive displacement pump used to transfer and meter fluids using a piston that sweeps into a chamber containing the fluid and displaces it. Check valves are used to ensure that the flow is in the correct direction. The rate of flow is dependent on the frequency of the stroke of the piston and the swept volume. They are capable of generating very high

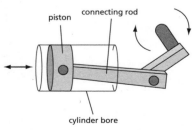

Fig. 39

pressures of up to 500 bar, and therefore have a pressure relief valve installed on the delivery side. The sealing of the piston and the check valves are the weak points and they are usually only used for clean fluids of low viscosity, due to the need for quick action by the check valves and the possibility of scoring of the piston.

pitch 1. The distance between the centre-line of tubes used to carry fluids for heat transfer in a shell-and-tube heat exchanger. Square and triangular pitches are commonly used (see Fig. 40). **2.** A generic term used for a flammable semi-solid tar-like substance produced naturally or by the distillation of heavy or long-chained hydrocarbons. It is also known as bitumen or asphalt and used to surface roads.

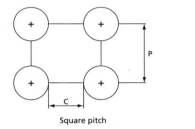
Square pitch

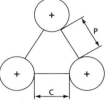
Triangular pitch

Fig. 40

Pitot tube An instrument used to measure the velocity of a flowing fluid by measuring the difference between the impact pressure and static pressure in the fluid. The device normally consists of two concentric tubes arranged in parallel; one with a face directed towards the flow to measure the impact pressure, the other face perpendicular to the flow to measure the static pressure. By taking a number of readings at various points in the cross section of a pipe or duct, known as a **Pitot traverse**, the overall rate of flow can be determined. As with all flow measurement devices, Pitot tubes should ideally be located away from disturbances such as bends. The device was devised by Italian-born French engineer Henri de Pitot (1695–1771).

P/I transducer *See* TRANSDUCER.

pitting A type of corrosion of a metal surface resulting in cavities. It is also one of the damaging effects of long-term *cavitation in centrifugal pumps that occurs on the surface of impellers.

pK$_a$ value A measure of the strength of an acid expressed as the negative base-10 logarithm of the acid dissociation constant, K$_a$:

$$pK_a = \log_{10} \frac{1}{\left[K_a \right]}$$

The dissociation constant can be calculated from the dissociation reaction. For example, for the reaction $HB_{aq} \leftrightarrow H^+_{aq} + B^-_{aq}$::

$$K_a = \frac{\left[H^+ \right]\left[B^- \right]}{\left[HB \right]}$$

The pK value is often used to compare the strengths of different acids.

plait point A point on a triangular diagram used to represent ternary liquid systems where two conjugate phases are mutually soluble and where there is no *tie line. *See* TRIANGULAR DIAGRAM.

plant Major equipment and machinery used in industrial processes. A *process plant is the entire industrial process or factory in which raw materials are converted to products through chemical, physical, or biochemical transformation.

plant layout study An analysis of the different possible physical configurations for an industrial *process plant. Due to the complexity of modern plants and manufacturing facilities that involve complex operations, the study typically involves the physical space and proximity of vessels and equipment, materials handling, piping and auxiliary equipment, utilities and services, communications systems, emergency systems, structural and architectural considerations, and general site work.

plasma 1. A state of matter resulting from the ionization of gases in which the number of positive and negative ions is approximately the same. In a *thermonuclear reactor, a very high temperature is maintained by retaining the plasma away from the walls using strong electromagnetic fields. It was first identified in 1879 and the nature of the matter was then identified by British scientist Sir Joseph John Thomson (1856–1940) in 1897. Irving *Langmuir first used the term plasma in 1928. **2.** The colourless part of blood in which corpuscles are suspended.

plasmid A small extra-chromosomal piece of DNA that is independent of the main chromosomes in the nucleus of a living cell. The DNA can be modified to contain the genetic code for drug resistance and can be transmitted between bacteria and yeast of the same or different species. They are used in *genetic engineering. *See* RECOMBINANT DNA TECHNOLOGY.

plastic A high molecular weight polymer that can be shaped at some stage in its manufacture by the application of heat and pressure to give a product that is stable at normal temperature. A thermoplastic is a substance that can be softened by the application of heat so that it can be shaped and moulded indefinitely. A thermosetting plastic is not able to be softened by reheating due to a chemical change such as the formation of cross-links between chains of the polymerized molecules. PVC (polyvinyl chloride) is a thermoplastic that is often used for piping systems and is stronger and more rigid than most other thermoplastic materials.

plasticity The ability of a body to retain a deformation in shape when a particular loading has been applied and then withdrawn. Fluids that exhibit plasticity require an applied shear stress to exceed the *yield stress before flow can occur. The greater the yield stress, the greater the plasticity.

plate A flat perforated horizontal sheet of metal used in distillation and absorption columns designed to provide an intimate contact between a rising vapour or gas and a liquid to allow vapour–liquid equilibrium to be achieved. The plate lies across the entire cross section of the column and has perforations or small openings that allow the vapour or gas to pass yet allow the liquid to remain on the plate. A variety of designs of plates are commonly used and include sieve plates, *valve trays, and bubble caps. A *distillation column may typically have many plates. *See* PASS PARTITION PLATE.

plate and fin heat exchanger A compact type of heat exchanger with a particularly high heat-transfer area that consists of cross flow channels between parallel plates in which the parting sheets form the primary heat-transfer surfaces and corrugated fins between the parting sheets. The corrugations are sealed to provide an enclosed and pressure-retaining unit.

plate and frame filter press An apparatus used for the batch separation of solid particles from a liquid slurry. It consists of cloths stretched across a metal plate with a rippled surface and sealed in a metal frame. A number of plates, cloths, and frames are firmly sandwiched together to ensure no leakage from the sides. The liquid slurry to be separated is pumped into each compartment formed by the frames through channels onto one side of the cloth. The cloth allows the liquid to pass through while retaining the solids or **filtrate,** which builds up as a cake. This corresponds to a build-up of pressure and a slow reduction in the rate of flow. The filtration process is halted once a sizable cake has built up. The plate and frames are then decoupled and the cake manually removed from the cloths. The cleaned cloths are then reused.

plate efficiency The measure of the performance of plate used in an *absorption tower or *distillation column. The plate is usually perforated to allow vapour or gas to rise up and make contact with liquid. Ideally, a vapour–liquid equilibrium is reached. However, this is not always the case. Due to these contact inefficiencies, more plates are required to achieve a desired separation. The plate efficiency is usually expressed as a percentage of the ideal or theoretical separation. *See* MURPHREE PLATE EFFICIENCY.

plate heat exchanger A compact and versatile type of heat exchanger consisting of rows of compacted and sealed corrugated metal plates between which liquid flows, transferring heat through the surface of the plates (see Fig. 41). The hot and cold liquids are

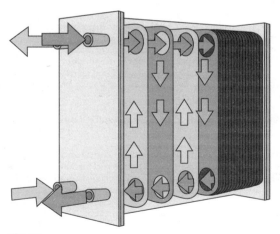

Fig. 41

channelled between alternating plates. The corrugation effect causes turbulence of the heat-transfer liquids thereby enhancing heat transfer. The size and area of the plate heat exchanger can be increased or decreased by adding or removing plates.

platforming process A catalytic reforming process used to convert straight-chain aliphatic hydrocarbons into aromatic hydrocarbons and hydrogen using a platinum catalyst.

The name is derived from **plat**inum-re**forming**. While platinum was the first catalyst to be developed for catalytic reforming, other catalysts and related processes have subsequently been developed over the past 60 years. Continuous catalytic regeneration or *CCR platforming is a widely used continuous version of the process.

platner process A process developed in the nineteenth century to extract gold from ore by a chlorination reaction. The gold chloride is then extracted using water and reduced by ferrous sulphate to gold metal:

$$AuCl_3 + 3FeSO_4 \rightarrow Au + Fe_2(SO_4)_3 + FeCl_3$$

PLC An abbreviation for **p**rogrammable **l**ogic **c**ontroller. It is a fast-acting computer-based monitoring and control system used to control complex processes in which the control actions are based on process data. It uses a computer to continuously monitor the various input signals and logically manipulate the necessary outputs to maintain control.

plenum A large gas chamber usually connected to ducts and maintained under either positive or negative pressure. It is used in air conditioning.

plug flow An intermittent two-phase gas–liquid flow regime found in vertical pipes or tubes often referred to as *slug flow characterized by bullet-shaped bubbles, which tend in size towards the diameter of the pipe. In the continuous liquid phase, the higher-velocity gas bubbles coalesce to form rising plugs or slugs. A **Taylor bubble** is the liquid film around the plug that may move downwards at low velocity. The liquid between Taylor bubbles often contains a dispersion of smaller bubbles.

plug flow reactor (PFR) A type of idealized tubular reactor that features no radial or axial mixing. All the components that flow through the reactor therefore possess the same residence time. A PFR may consist of either one long reactor or many short reactors as a tube bundle. The reactants flow through the length of the reactor during which the chemical reaction rate changes. They are usually used for gas-phase reactions requiring high temperatures. Used for both exothermic and endothermic reactions, heat transfer is effective through the tube walls.

plume The continuous release of a gas as a cloud. In contrast, a puff is the instantaneous release of a gas cloud.

plunger pump A type of reciprocating pump that uses a plunger instead of a piston within a cylinder to displace a liquid. They are used as metering pumps and for liquids that may be abrasive since the plunger moves into the cylinder and is clear of the walls. Check valves on the suction and delivery side of the pump ensure that the flow is in the correct direction.

PM An abbreviation for **p**articulate **m**atter. Often presented with a subscript, for example PM_{10}, which refers to particulates that have an aerodynamic diameter of less than 10 micrometres (10 μm). These pose respiratory health concerns as they can be inhaled and can penetrate the lower reaches of the lungs, and therefore accumulate within the respiratory system. $PM_{2.5}$ are particulates of less than 2.5 micrometers and are referred to as fine particles, which can present severe respiratory problems leading to respiratory, pulmonary, and cardiovascular diseases. *See* PARTICULATES.

pneumatic control valve A valve used to control the flow of process material through a pipe by controlling the flow area through the valve by way of an applied air pressure signal.

The operation of a control valve involves an air supply that positions its movable part (i.e. plug, ball, or vane) relative to the stationary seat of the valve. A valve actuator accurately locates the valve plug in a position determined by the pneumatic control signal and operates to move the valve to either fully open or fully closed positions. The actuators may be either piston or diaphragm types. Air-to-open valves require air to open and therefore automatically close in the event of fail closure. They are therefore used on fuel lines to furnaces. Air-to-close valves fail to open on a loss of air pressure and are used on air lines into fuel burners. In general, fail-to-open and fail-to-close valves operate when the supplied air pressure drops below a minimum value.

pneumatic conveying The transportation of granular free-flowing solids that are suspended in a fast-moving flow of air or gas. The particles move freely and can be readily transported to and from hoppers and silos through pipes and ducts. Pneumatic conveyors are typically used for the transportation of grains and catalyst particles.

pneumatic mixing A type of mixing that involves sparging gas into a liquid to cause turbulence. Often used for biological reactions in bioreactors, bubbles of gas (usually oxygen or filtered air) rise in the liquid generating turbulence. Bubble size and number is dependent on the type of sparger and can influence the rate of mass transfer.

pneumatics The study and production of devices that rely on air pressure for their operation.

pneumercator An instrument used to determine the depth and volume of liquid in a vessel, tank, or reservoir. It consists of a vertical leg down which a compressed gas is gently discharged in the form of bubbles. The pressure of the applied gas is measured from which the hydrostatic depth is determined. Using two such legs, it is possible to determine the density of the liquid, and in the case of immiscible liquids forming two layers, the location of the interface.

pneumoconiosis An occupational disease of the lungs caused by the inhalation of mineral and organic dust particles. Depending on the type of dust, the disease can also be called black lung arising from coal dust, asbestosis from asbestos fibres, siderosis from iron dust, silicosis from silica, and silicosiderosis from a combination of silica and iron dusts.

poise (Symbol P) A c.g.s. unit used for the viscosity of fluids. It is the tangential force expressed in dynes per square centimetre that is required to maintain the difference in velocity of two parallel plates that sandwich the fluid by a distance of one centimetre at a velocity of one centimetre per second. The centipoise (cP) is more commonly used in which one cP is equal to 10^{-3} Pa s.

Poiseuille, Jean Louis Marie (1797–1869) A French physician and physiologist noted for his work on fluids. He had a major interest in the flow of blood through the body. Using narrow glass capillaries, he made detailed studies of flow. He experimentally derived a formula and published his work in 1840, crediting the work of German engineer Gotthilf Hagen who also independently derived the formula.

Poiseuille's equation A relationship used to determine the rate of flow of a fluid with laminar flow through a horizontal cylindrical tube or pipe. Named after French physician and physiologist Jean Louis Marie *Poiseuille(1797–1869), it is given by

$$Q = \frac{\pi}{8\mu} \frac{\Delta p}{L} R^4$$

where μ is the viscosity of the fluid, $\Delta p/L$ is the pressure drop along the tube, and R is the internal radius of the tube.

poison 1. A chemical substance that results in the reduction in effectiveness of a catalyst as the result of it being contaminated by a reactant, a product of the reaction, or some other extraneous material. The poison accumulates on the surface of the catalyst, reducing the effectiveness by reducing active sites for the reaction to occur. **2.** Another name for a *toxic substance.

Poisson's ratio The ratio of the lateral strain to the longitudinal strain in a material held under tension. When a sample of material is stretched (or squeezed), there is a contraction (or extension) in the direction perpendicular to the applied load. Poisson's ratio is the ratio between these two quantities. The value lies between −1.0 and 0.5. It was introduced by French mathematician and physicist Siméon-Denis Poisson (1781–1840).

polarimeter An instrument that measures the polarization of any form of *electromagnetic radiation.

polishing The process of removing the traces of contaminants from a liquid to produce a very clear product, such as in polishing filtration used for beer, or polishing *ion exchange.

pollutant A chemical substance that when released into the environment gives rise to harmful and damaging effects on living organisms. The substance can be either a toxic substance that is harmful to the environment by being resistant to biodegradation such as pesticides, or can already be present in the environment but is added in excessive amounts such as nitrogen into the soil that accumulates in lakes and rivers. *See* PERSISTENT ORGANIC POLLUTANT.

pollution A substance whose uncontrolled release can cause damage to human and animal life, plants, trees and other vegetation, and the environment in general. Airborne pollutants may be in the form of gases, mists, vapours, clouds, dust, smoke, soot, and fumes. Waterborne pollutants may contaminate land and water courses such as rivers. Non-biodegradable pollutants include heavy metals, certain pesticides, many types of plastics, and chlorinated hydrocarbons. Biodegradable pollutants such as certain plastics and sewage can be broken down by microorganisms over a period of time and rendered harmless. Airborne and waterborne radioactive pollutants have a lasting effect on the environment and can enter the food chain. *See* AIR POLLUTION; WATER POLLUTION.

polymerization A process that involves the chemical reaction of simple molecules called *monomers to combine to form longer and more complex molecules called **polymers**. These are large macromolecules made up of many repeating units derived from a small simple number of simple molecules. The polymers are formed into sheets, chains, or three-dimensional structures, and held together by covalent bonds. Functional groups are attracted to each other by intermolecular forces (*van der Waals' forces) and in some cases ionic and hydrogen bonds. There are many types of polymer that exist naturally such as proteins, rubber, and polysaccharides. Many others are produced synthetically such as polyethylene and polypropylene. Synthetic polymers have many applications such as textiles, plastics, rubber, coatings, and adhesives. The feedstock for most polymers is ethylene and is essential in the production of vinyl chloride and styrene, which are used for plastics. Propylene and butadiene are also used in high quantities and are by-products from the manufacture of ethylene. Addition polymers have identical monomer subunits linked to form a polymer that has the same empirical formula as the monomer. Condensation polymers have monomers joined during a condensation reaction with the elimination of water during the reaction, to form a polymer with a different empirical formula to that of the monomer. Copolymers are composed of two or more different types of monomers. Naturally forming polymers include polysaccharides and proteins. Synthesized polymers include polyvinyl chloride and polyester.

polytropic gas A gas that can be represented by $pV^n = k$ to describe its compression and expansion. The special cases are $n = 1$ (isothermal); $n = k$ (isentropic); $n = 0$ (pressure constant); $n = \infty$ (volume constant).

Ponchon–Savarit A rigorous graphical method used in the analysis of the separation of two heterogeneous liquids by distillation. It is used to determine the number of stages required to bring about a required separation and is based on a stage-wise approach using enthalpy and composition of each theoretical tray or equilibrium stage. The graphical technique is based on the vapour–liquid equilibrium data for the more volatile component in the feed. The method was developed by M. Ponchon and P. Savarit independently between 1921 and 1922.

pool boiling A type of liquid boiling that occurs on a submerged surface. The surface in a stagnant pool of liquid is heated to a temperature above that of the boiling point of the liquid. The boiling that occurs produces vapour whose motion relative to the surrounding liquid is due to the buoyancy effects of the vapour with the bulk of the liquid being at rest.

pool fire The combustion of flammable liquid that is evaporating from the base of the fire. The liquid, such as a pool of vaporizing hydrocarbon fuel, has no or little initial momentum. Well-ventilated open fires are fuel-controlled whereas fires within enclosures may become ventilation-controlled. Pool fires represent a significant element of risk associated with major accidents on offshore oil and gas installations that may have large liquid hydrocarbon inventories.

POP *See* PERSISTENT ORGANIC POLLUTANT.

porosity The proportion of the volume of a porous body that is not occupied by the body itself. It is usually given as a fraction, percentage, or decimal. The porosity of a *packed bed is the ratio of the volume of voids to the total volume of the bed. The value is dependent on the shape and size distribution of the particles, the ratio of the particle size to bed diameter, and the method used for filling the bed. It is usually determined by measuring how much water is used to fill the voids compared with the total volume of the bed. *Compare* VOIDAGE.

port An access point to a process vessel and used for instrument probes and transfer lines, etc.

positive displacement pump A classification of pump type in which fluids are transported by its displacement from one place to another for flow through pipelines, conduits, ducts, and channels. Examples include *reciprocating pumps, which involve the displacement of a fluid from a chamber by a piston or plunger, and *rotary pumps, which involve rotating gears or lobes with the fluid being transported between the teeth. There are many variations such as the *diaphragm pump, which involves a diaphragm flexed to and fro, and the *Mono pump, which involves a rotating helical worm.

POSRV *See* PILOT-OPERATED SAFETY RELIEF VALVE.

potential energy The capacity of a body to do work due to its elevation above a reference point. It is the product of weight and height.

pot still Used for batch distillation, it consists of a boiler with a condenser attached.

pound A unit of mass in *f.p.s. engineering units. One pound (lb_m) is equal to 0.453 592 kilograms.

poundal A unit of force in the f.p.s. system of units and equal to the force needed to accelerate a unit of mass of one *pound by one foot per second.

Pourbaix diagram A diagram used to illustrate the pH dependence of the oxidation-reduction behaviour for compounds of a given element (see Fig. 42).

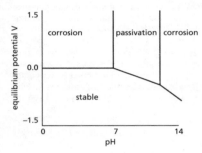

Fig. 42

pour point A rough indicator of the relative paraffinicity and aromaticity of a crude oil. Given as a temperature in degrees, either Fahrenheit or Celsius, a low pour point indicates a low paraffin content and a greater content of aromatics.

powder A mass of fine, dry, particulate matter. Powders are obtained through the process of pulverizing, milling, crushing, and grinding of ores and spray drying. *Compare* DUST.

power The work done or energy transferred per unit time. In SI units, it is measured in watts (that is, joules per second).

power consumption The amount of energy per unit time used by a process or process equipment.

powerforming A catalytic reforming process used to produce aromatics. It uses a platinum catalyst and reforming straight-chain hydrocarbons in the C_6 to C_8 range from *naphtha or gasoline fractions into compounds containing benzene rings. *See* BTX.

power generation The generation of electricity such as from the combustion of natural gas and coal, or nuclear reaction that raises steam to drive turbines linked to generators to produce electricity. Gas turbines are used to generate electricity directly by the combustion of fuel gas in the turbine. The electricity is used as a utility to support process plant operations or sold to the national network.

power input The work or energy per unit time that is applied to a process or item of equipment such as to a pump, an agitator, or stirrer. It is measured using a wattmeter.

power number A dimensionless number, P_o, used to represent the power required in mixing or agitation, and is also used in centrifugal pump design and sizing. It is also known as the *Newton number and given by:

$$P_o = \frac{P}{D^3 N^5 \rho}$$

where P is the power to the shaft, D is the impeller diameter, N is the rotational speed, and ρ is the density of the liquid.

PPE *See* PERSONAL PROTECTIVE EQUIPMENT.

ppm An abbreviation for **p**arts **p**er **m**illion. That is, the number of parts of something per million parts in total. It is often used as a measure of the level of impurities in solids, liquids, and gases.

Prandtl, Ludwig (1875–1953) A German pioneer of aerodynamics; after gaining his PhD at the University of Munich, he was appointed professor of mechanics at the University of Hanover, and was professor of applied mechanics from 1904 to 1953 at the University of Göttingen. He taught and carried out research in fluid mechanics and made a significant contribution to the understanding of boundary-layer theory, which led to the understanding of skin friction and drag on aircraft wings.

Prandtl number A dimensionless number, Pr, representing the ratio of the momentum of diffusivity to thermal diffusivity in fluid convection:

$$Pr = \frac{c_p \mu}{k} = \frac{Pe}{Re}$$

where c_p is the specific heat, μ is the viscosity, and k is the thermal conduction, Pe is the *Péclet number, and Re is the *Reynolds number. It is named after German scientist Ludwig *Prandtl (1875–1953).

Prandtl's one-seventh power law A relationship used to determine the velocity distribution for turbulent flow in pipes carrying fluids given as:

$$v = v_m \left(\frac{y}{r_o} \right)^{\frac{1}{7}}$$

where v is the local velocity, at a distance y from the wall of a pipe of radius r_o, and v_m is the maximum velocity at the centre line of the pipe. It was formulated by German scientist Ludwig Prandtl (1875–1953).

precipitation The formation of a suspension of solid particles, known as a **precipitate,** as the result of a chemical reaction or a physical change such as through the reduction in the solubility of the dissolved material in the solvent. Solid materials can be formed in a solution by the addition of compound, which causes a reaction in the solution that converts the material to be separated into an insoluble state. The solid material can then be separated by centrifugation or filtration. Liquid materials can also be precipitated as a condensate from a gas. Precipitation is used to separate metals from aqueous solutions, and also be used to separate impurities and contaminants from biochemical processes. A precipitator is a vessel used to carry out the process of precipitation.

precision A measure of the exactness of a measured quantity. Used in statistics, the precision may be increased by increasing the sample size. On a calculator, it is the number of significant figures to which a reading is taken: the greater the number of significant figures, the greater the precision. *Compare* ACCURACY.

pressing A mechanical process used to extract liquid from a liquid slurry or liquid-bearing solid, and involves applying a controlled pressure in order to free the liquid and retain the solid material. Juice is extracted from grapes as a first stage in making wine by pressing. Likewise, pressing is used to extract oil from nuts and seeds.

pressure The force applied over a given area. Instrument gauges used to measure the pressure of fluids are either expressed as *absolute pressure, which is measured above a vacuum, and *gauge pressure, which is the pressure measured above atmospheric pressure, which is variable. The SI units are $N\,m^{-2}$ or pascals. Some gauges are calibrated in the Imperial units of psi (pounds force per square inch). Gauges used to measure a vacuum are expressed in *torr.

pressure drop The decrease in pressure between two points in a system caused by frictional losses of a moving fluid in a pipe or duct, or by some other resistance such as across a filter, packed bed, or catalyst, or due to the effects of hydrostatic head such as across the liquid on the tray of a *distillation column.

pressure drop multiplier (Symbol φ^2) A parameter used in two-phase gas–liquid frictional pressure drop calculations where the overall pressure drop along a length of pipe is due to contributions from the flowing gas and liquid. That is:

$$\frac{dp_f}{dz} = \varphi_g^2 \left(\frac{dp_g}{dz}\right)_g = \varphi_L^2 \left(\frac{dp_L}{dz}\right)_L$$

where φ_g^2 and φ_L^2 are the pressure drop multipliers for the liquid and gas phases in which the parameter X^2 is defined as:

$$X^2 = \frac{\left(\frac{dp_L}{dz}\right)_L}{\left(\frac{dp_g}{dz}\right)_g} = \frac{\varphi_g^2}{\varphi_L^2}$$

Correlations have been developed to determine relationships for the multipliers for combinations of laminar and turbulent gas and liquid phases.

pressure gauge An instrument used to determine the pressure of a fluid. There are various types commonly used, such as *manometers that consist of a column of liquid whose vertical elevation is dependent on the applied pressure; the expanding element type such as the *Bourdon gauge; and the electrical transducer type that use strain gauges.

pressure head The equivalent height of a column of liquid that can produce a given applied pressure:

$$h = \frac{p}{\rho g}$$

where p is the pressure, ρ the density, and g the gravitational acceleration. A barometer is used to measure atmospheric pressure from the elevation of a fluid such as mercury in a sealed vertical glass tube. The pressure head of standard atmospheric pressure corresponds to 760 mm Hg.

pressure relief valve (PRV) A valve used on a process vessel or pipe designed to operate as a safety device and activate to discharge the pressurized gas or liquid whenever it has reached a set point pressure. The valve will automatically close again once the pressure has fallen below the set point. In comparison, a **pressure safety valve** (PSV) has a level that can be manually activated in the event of an emergency.

pressure-swing adsorption (PSA) A process used to selectively adsorb one or more components in a gas mixture under pressure onto a porous solid surface and then to release them again under reduced pressure. It is used in the separation of gases from mixtures such as the absorption of nitrogen from air onto a *zeolite. The air leaving the process is therefore richer in oxygen. The zeolite is regenerated by reducing the pressure. Other commonly used absorbents include activated carbon, silica gel, and alumina.

pressure-temperature diagram A phase diagram representing thermodynamic data for a multicomponent mixture illustrating the existence of liquid and gas phases, and the coexistent liquid and gas phases (see Fig. 43). They are typically used to classify oil hydrocarbon systems and describe the phase behaviour found in oil and gas reservoirs.

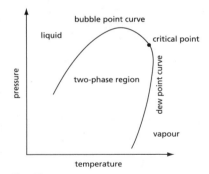

Fig. 43

pressure vessel A closed container designed to hold gases or liquids at a pressure greater than ambient pressure. Pressure vessels are used in a wide variety of applications including compressed air receivers, distillation columns, autoclaves, and as storage vessels for liquefied gases such as propane, butane, chlorine, ammonia, and LPG. Pressure vessels are broadly divided into simple vessels and those that have more complex features. Simple vessels have a cylindrical body with dished ends and no supports or sections; they are classified based on the product of pressure and volume: Class I: 3,000 to 10,000 bar litres; Class II: 200 to 3,000 bar litres; Class III 50 to 200 bar litres. More complex pressure and higher-pressure vessels follow accepted international design codes such as the Pressure Equipment Directive legislation and EN13444 design code in the European Union, the ASME Boiler and Pressure Code Section VIII in the US, and BS PD5500 in the UK.

pressure-volume diagram A diagram used to illustrate the variation of volume and pressure for a substance (see Fig. 44). Spanning a wide range of volumes and pressures, isotherms (lines of constant temperature) are used to illustrate features such as liquid phase, gas phase, and critical point. Developed in the eighteenth century and once known as indicator diagrams, they were used to analyse the behaviour and efficiency of steam engines.

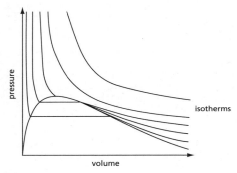

isotherms

Fig. 44

preventative maintenance The scheduled inspection and testing of process equipment and machinery inspection aimed at avoiding premature and costly failure. It involves cleaning, minor component replacement, and lubrication as a means of extending the life of the equipment. It is intended to prevent emergency and unscheduled repairs and downtime by detecting problems before they arise. *See* MAINTENANCE.

prevention The act of not allowing something to happen. *See* LOSS PREVENTION.

prilling A process used to make solid pellets from a molten material or solution. The liquid is poured under gravity into an upward flow of air. It is used in the manufacture of fertilizers such as ammonium nitrate.

primary heating The use of heat to maintain heavy grades of oil in storage tanks at a controlled viscosity so that it can be pumped. The heating is achieved using steam or hot-water coils in the tank, or by electric heating elements. The minimum temperature required depends on the grade of oil. **Secondary heating** is the temperature required for the oil to be efficiently atomized.

prime mover Any machine or device used to transfer energy, power, or motion to another device. Examples include water and steam turbines used to turn a generator for producing electricity, or an electrical motor to drive a *compressor.

priming 1. The filling of a centrifugal-type pump with the process liquid to be pumped by displacing residual air or vapour. It is carried out manually or, in some cases, remotely by activated control valves. Centrifugal pumps are not self-priming. **2.** The entrainment of boiler water in steam produced in a boiler. It may be caused by excessively high water levels or foaming caused by excessive salts in the water.

probability The statistical likelihood of an event or a sequence of events occurring during a defined interval of time, or the chances of a success or failure of an event. It is used in *quality control, *risk assessment, and also to determine the reliability of process equipment and operations (*see* FMEA). The probability of something happening or not happening is expressed as a dimensionless number ranging from 0 and 1. For example, if there are m possible outcomes and n ways an event can occur, then there is a probability of n/m chances of an occurrence. If the probability of success of a pump operating is S, then the probability of pump failure is $F = 1 - S$. For example, if a process operates with three pumps, A, B, and C, each with equal chances of success in operating, then the possible combination of success and failure for the pumps is:

Pump

A	B	C	Probability
S	S	S	S^3
S	S	F	S^2F
S	F	S	S^2F
S	F	F	SF^2
F	S	S	S^2F
F	F	S	SF^2
F	S	F	SF^2
F	F	F	F^3

The sum of the probabilities is:

$$S^3 + 3S^2F + 3SF^2 + F^3 = (S+F)^3$$

If there is an equal chance of each pump functioning being 1 in 10, say, then the probability of at least one pump being functional is:

$$S^3 + 3S^2F + 3SF^2 = 0.9^3 + 3 \times 0.9^2 \times 0.1 + 3 \times 0.9 \times 0.1^2 = 0.999$$

whereas the chance of at least two pumps being functional is:

$$S^3 + 3S^2F = 0.9^3 + 3 \times 0.9^2 \times 0.1 = 0.972.$$

The **probability of failure on demand** is therefore the likelihood that a process, system, or item of process plant will fail to operate in the required and expected manner on demand.

problem-solving A set of mental techniques used to solve complex problems. There are many techniques commonly used including: brainstorming, which is used to present solutions and developed to reach an optimum; root cause analysis, which aims to eliminate the problem; *trial and error, which involves systematically testing possible solutions; and lateral thinking, which seeks solutions either indirectly or creatively.

process 1. Refers to the changes from one equilibrium state, or steady state, to another that take place within a *system. For example, in a heating process, heat is transferred to the system from the surroundings. **2.** A term used in process control to describe the collective functions performed in and by the process in which the process variables such as temperature, pressure, flow, and level are controlled. **3.** A shorthand for *chemical process, which is a collective description of the way that useful chemical products are manufactured from raw materials through chemical, physical, or biochemical means.

process control The adjustment of process variables such as pressure, flow rate, level, and temperature in a process plant at an intended or desired value. Processes are designed to operate safely and in a designed and intended manner to achieve product quality, energy efficiency, and waste minimization according to *flowsheet conditions. However, all processes are dynamic and subject to disturbances such that they tend to deviate from the designed conditions. They therefore need to be controlled and returned back to the desired conditions of operation.

Processes may be controlled automatically or manually. Manual operation involves process operators who make adjustments to temperatures, pressures, flow, and levels. Automatic control is used to control complex processes, or for remote and hazardous processes, or where a high level of product quality is required. Computers are routinely used to operate and control processes.

A control system consists of several components including the process itself, a controller, and a valve. A good understanding is required to control the process. However, this may not always be possible since many processes are multivariable or non-linear in behaviour, or not well understood such as in the case of biological processes.

*Feedback control and *feedforward control are two forms of automatic process control. The former involves comparing a controlled signal with the desired value of a process variable and automatically making adjustments to minimize the difference. Feedforward control involves controlling a process in which the disturbance is detected before it enters the system. The controller calculates the required counteracting disturbance. Process disturbances are measured and compensated for without waiting for a change in the controlled variable to indicate that a disturbance has occurred. It is useful where the final controlled variable cannot be measured.

process costing A way of determining the cost of production in a process in which all the costs are obtained and the average unit costs of production then determined. The main product, coproducts, and by-products are each separately distinguished.

process design The design of industrial processes that use physical, chemical, or biochemical transformations for the production of useful products. It is used for the design of new processes, plant modifications, and revamps. It starts with conceptual and feasibility studies, and includes detailed material and energy balances, the production of *block flow diagrams (BFDs), *process flow diagrams (PFDs), *engineering line diagrams (ELDs), and *piping and instrumentation diagrams (P&IDs). It also includes the production of reports and documents for plant construction, commissioning, start-up, operation, and shutdown. The reports and documents are used by vendors, regulatory bodies, operators, and other engineering disciplines.

process dynamics The behaviour of a system or process with time. All processes are inherently unstable and dynamic as their properties vary with time. It is therefore necessary to control the system or process to disturbances, and to restore the controlled variable back to its design or desirable value. This is done by using mathematical rate equations to describe the process such as material, energy, and momentum equations. Dynamic analysis is then used to determine how the associated process variables change with time and how their behaviour changes in response to disturbances.

process economics An evaluation of a process in terms of all the costs that are involved. It considers the cost of raw materials and how they are processed, as well as the costs associated with waste processing such as recycling or disposal. It also includes the optimization of a process to best utilize materials and energy. The fixed costs of a process are not dependent on the rate of production whereas the variable costs are and must be met by the revenue generated by sales. Taxes are deducted to leave the net profit.

process engineering A branch of engineering that encompasses petrochemical, mineral processing, advanced material, food, pharmaceutical, and biotechnological industries. It focuses on the design, operation, control, and optimization of chemical, physical, and biological processes. It is also known as **process systems engineering**, which is a specialist area of research in US, Europe, Japan, Korea, and China.

process flow diagram (PFD) A schematic representation of a process or part of a process that converts raw materials to products through the various unit operations. It typically uses a symbolic representation for the major items of equipment such as storage vessels, reactors, and separators, process piping to and from the equipment, as well as by-pass and recirculation lines, and the principal flow routes. Key temperatures and pressures corresponding to normal operation are included, as well as equipment ratings, minimum and maximum operational values. Material flows and compositions are included. It may also include important aspects of control and pumping, as well as any interaction with other process equipment or flows. The design duties or sizes of all the major equipment are also featured, which can collectively provide a comprehensive representation of the process. It is also known as a **system flow diagram**. *Compare* FLOWSHEET; ENGINEERING LINE DIAGRAM.

process integration 1. A holistic approach used in process design that considers the process as a whole with the interactions between unit operations in comparison with the optimization of unit operations separately and independently. It is also known as *process synthesis. **2.** A technique used to minimize the energy consumption and heat recovery in a process. It is also known as **process heat integration** and *pinch analysis.

process intensification An approach to engineering design, manufacture, and operation of processes that aims to substantially improve process performance through energy efficiency, cost-effectiveness, reduction in waste, improvement in purification steps, reduction of equipment size, increase in safety, and operational simplicity. It involves a wide range of innovative mixing, reactor, and separation technologies that can result in dramatic improvements in process performance. Involving an integrative approach that considers overall process objectives rather than the separate performance of individual unit operations, process intensification can enable a process to achieve its maximal performance leading to the development of cheaper, smaller, cleaner, safer, and sustainable technologies.

process plant A collective name for an industrial facility used to convert raw materials into useful products. It includes all the process equipment such as mixers, reactors, and separation units, all the associated pipework and pumps, heat exchangers, and utilities such as steam and cooling water. It is often used interchangeably with *chemical plant, although process plants may not always involve chemical transformation or the production of chemicals as in the case of power stations.

process reaction-curve method A widely used empirical procedure used to tune and control a process using *PID control with optimum controller settings. Developed by J. G. Ziegler (1898–1973) and N. B. Nichols (1903–79) in 1942, it is based on using results from open-loop tests. The settings from the tuned controller result in an underdamped transient response with a *decay ratio of one-quarter.

process safety A comprehensive management system that focuses on the management and control of potential major hazards that arise from process operations. It aims at reducing risk to a level that is as low as is reasonably practicable by the prevention of fires, explosions, and accidental or unintended chemical releases that can cause harm to human life and to the environment. It includes the prevention of leaks, spills, equipment failure, over- and under-pressurization, over-temperatures, corrosion, and metal fatigue. It covers a range of tools and techniques required to ensure safe operation of plant and machinery to ensure the safety of personnel, the environment, and others, through detailed design and engineering of facilities, maintenance of equipment, use of effective alarms and control points, procedures, and training. It also includes risk assessments, *layers of protection analysis, and use of *permit to work authorizations.

process simulation The use of computers to model and predict the operational and thermodynamic behaviour of a process. Sophisticated commercial software packages are used to simulate and model batch, continuous, steady-state, and dynamic processes. They require combined material and energy balances, the properties of the materials being processed, and sometimes combine the use of experimental data with mathematical descriptions of the process being simulated. Most software packages feature optimization capabilities involving the use of complex cost models and detailed process equipment size models.

process synthesis The conceptual design of a process that identifies the best process *flowsheet structure, such as the conversion of raw materials into a product. This requires the consideration of many alternative designs. Due to the complexity of most processes, the flowsheet is divided into smaller parts and each considered in turn, then choices and decisions made. Various techniques are used to arrive at the best flowsheet such as those based on total cost, which needs to be minimized, the use of graphical methods, the use of *heuristics, and various other forms of minimization such as the use of *process integration.

process upset A sudden, gradual, or unintended change in the operational behaviour of a process. It may be due to process equipment failure or malfunction, operator intervention, a surge or fall in pressure, flow, level, concentration, etc.

process validation The documented evidence that a procedure, process, or change has been fully evaluated before its implementation that it, which can provide a high degree of assurance of meeting pre-determined specifications and quality. First used by the Food and Drugs Administration in the US in the 1970s to improve the quality of pharmaceuticals, it is widely used in many industries and, in particular, the pharmaceutical and allied medical industries. It is used alongside other regulatory requirements such as Good Manufacturing Practice. Before the validation process begins, it is necessary to ensure that the system is properly qualified, which includes a design, installation, operational, performance, and component qualification. These define the function, operation, and specification of the equipment, process and product.

process variable A dynamic feature of a process or system that is required to be controlled to ensure that it operates according to design requirements and does not deviate so as to be unsafe or result in undesirable consequences. The commonly measured process variables include temperature, pressure, flow, level, and concentration.

process waste *See* WASTE.

producer gas A mixture of carbon monoxide and nitrogen used in a *gasification process and formed by the partial combustion of coal, coke, or anthracite in a blast of air. The exothermic reaction produces carbon monoxide:

$$2C + O_2 \rightarrow 2CO$$

Nitrogen in the air remains unchanged in the process. Adding steam to the gasification process results in a gas mixture also containing hydrogen. Producer gas is used as a *fuel gas in furnaces and the generation of power in gas turbines. A **producer process** is a generic name for processes that convert solid fuels into gaseous fuels.

product A chemical substance formed as the output from a process or unit operation that has undergone chemical, physical, or biological change.

production platform An offshore structure that features all the necessary equipment to maintain an oil or gas field in production. It has facilities for temporarily storing the output of several wells.

production separator A horizontal cylindrical vessel used on offshore platforms for the separation of gas and water from several crude oil wells. The oil enters the separator in which the reduction of pressure causes the release of dissolved gases that are removed from the top of the vessel. The water and oil separate by virtue of being immiscible and having different densities. The oil and condensate is separated from the water by overflowing a weir.

product recovery The extraction and purification of valuable chemicals, including biochemicals such as therapeutic proteins produced in biochemical reactions. It is also used for the recovery and recycling of waste materials.

programme evaluation and review technique (PERT) A project planning technique used to help plan and manage complex projects. In its simplest form, it is known as *critical path analysis or network analysis. It is marketed as computer software under various trade names and has five identifiable, sequential steps. These involve identifying and listing all the individual activities, establishing the dependencies for the activities, creating a network of how they fit together, completing the critical path analysis that introduces a timescale to the project, and finally, producing a Gantt chart giving a visual representation of progress.

project management An activity concerned with the overall planning and coordination of a project from its conception to its completion. It is aimed at meeting the defined requirements and ensuring completion on time, within budget, and to defined quality standards.

project network techniques A group of management tools aimed at planning, analysing, and managing projects. They consider the logical interrelationships of all the project activities and are concerned with time, resources, costs, and other influencing factors such as uncertainty. They have three identifiable phases of planning, scheduling, and control.

*Critical path analysis and *programme evaluation and review techniques are examples of project network techniques.

proof 1. A rigorously defined, logical, and complete demonstration of the correctness of a statement, formula, law, or *theorem. It involves a set of basic assumptions known as axioms or premises that are used to derive and lead to a conclusion to show that the statement, formula, law, or theorem has been proved. **2.** A former measure of the amount of alcohol in *whisky defined as the most dilute spirit that would ignite gunpowder: 100 per cent proof corresponds to 57.15 per cent ABV (alcohol by volume).

propellant 1. A chemical substance used as a fuel that burns in a controlled manner and is used to propel projectiles such as rockets. **2.** A volatile substance that is used to produce a spray in an aerosol can. It can be liquefied by pressure and dissolved into the working substance. On release of pressure, the liquefied propellant vaporizes producing the spray. Chlorofluorocarbons (CFCs) were once widely used but have been discontinued due to their harmful effects on the ozone in the atmosphere; they have now been largely replaced by hydrocarbons such as pentane or mixtures of hydrocarbons.

propeller An agitation device that consists of blades attached to a rotating shaft giving axial flow. They are used for the agitation of low-viscosity liquids. **Propeller fans** or *blowers have two or more propeller blades on a shaft contained in a cylindrical casing and used for transporting gases.

property diagram A graphical representation of the thermodynamic properties of a fluid. Any two properties can be plotted and from the *two-property rule, the other state properties can be determined once two have been fixed. The most commonly encountered property diagrams are the *pressure-volume diagram, the *temperature-entropy diagram, the pressure-enthalpy diagram, and the entropy-enthalpy diagram.

proportional (Symbol $\propto$) A mathematical relationship between two quantities that vary in a constant ratio. For example, the viscosity of a *Newtonian fluid is the ratio of the shear stress to shear rate over all values, $\tau \propto \gamma$. That is, the shear stress, τ, is directly proportional to the shear rate, γ. The **proportionality constant** is the constant that links the two quantities. In this case, it is the viscosity, μ. If the quantities are inversely proportional, then the product is a constant. For example, *Boyle's law states that the volume of a fixed mass of gas is inversely proportional to its pressure at constant temperature.

proportional band Used in the control of a process, the controller output is proportional to the error or a change in a *process variable. The proportional band is the fractional or percentage change in input that is required to produce a 100 per cent change in controller output. The proportional band is therefore a value of 100 divided by the gain of the process. For example, a 100 per cent controller output may be a signal that fully opens a valve, while a 0 per cent controller output fully closes it.

proportional control A mode of feedback control of a process in which the output from the controller is proportional to the error in the signal. This is the difference between desired (set point) and measured values. *See* PID.

prototype An experimental version of a machine or process, in which the initial design can be tested and improved based on testing.

proximate analysis A type of compositional analysis of fuels in terms of moisture content, volatile matter, ash, and fixed carbon content. *Compare* ULTIMATE ANALYSIS.

PRV *See* PRESSURE RELIEF VALVE.

PSA *See* PRESSURE-SWING ADSORPTION.

PSD An abbreviation for **p**rocess **s**hut**d**own. *See* SHUTDOWN.

PSDS An abbreviation for **p**roduct **s**afety **d**ata **s**heet. *See* MSDS.

pseudocomponent Used in computations and computer simulations of complex processes such as hydrocarbon processes and petroleum refining in which many non-polar molecules may be present in a mixture. Rather than representing the mixture by all the components, it is easier to group them by some useful property thereby reducing the number of components. The properties of these groupings can be represented by an average boiling point, specific gravity, and molecular weight. The converse is referred to as **real components**.

pseudocritical properties Empirical values used for the critical properties in multi-component chemical systems. They are based on mixing rules of pure components and often have very different properties from their true critical points, and have no real physical significance. For example, the **pseudocritical viscosity** is a viscosity parameter used in an empirical relationship to determine the viscosity of liquids as a function of temperature. Tabulated data are available for hydrocarbons, ethers, ketones, aldehydes, acetates, alcohols, and organic acids.

pseudo-order The order of a chemical reaction that appears to be less than the actual order as the result of the experimental conditions used. It occurs when one reactant is present in excess such that the reaction may appear to be due to the large amount of the reactant present.

pseudoplastic A time-independent *non-Newtonian fluid, which is shear thinning. The apparent viscosity expressing the ratio of shear stress, τ, to shear rate, γ, decreases with increasing shear rate: $\tau = k\gamma^n$ where k is a proportionality constant and n has a value between 0 and 1. Examples include polymer melts, paper pulp, wallpaper paste, printing inks, tomato purée, mustard, rubber solutions, and protein concentrations.

psi An Imperial unit of pressure expressed in pounds force per square inch. Some pressure gauges are given in **psig**, which represents the gauge pressure and is distinguished from **psia** or absolute pressure in pounds per square inch. Standard atmospheric pressure corresponds to a gauge pressure of 14.7 psi.

PSV An abbreviation for **p**ressure **s**afety **v**alve, which is a type of spring-loaded pressure relief valve. It is activated by the static pressure upstream of the valve and characterized by its ability to open rapidly.

psychrometer An instrument used to determine the humidity of air. It comprises both dry and *wet bulb temperature measurements. The difference in the two thermometer readings is used to determine the humidity. A whirling psychrometer is used to provide a consistent reading with greater accuracy and involves swinging the instruments to ensure a high air velocity over the bulbs of the thermometers.

psychrometric chart A graphical representation of the thermodynamic properties of water and air with dry bulb temperature on the y-axis and moisture content on the x-axis

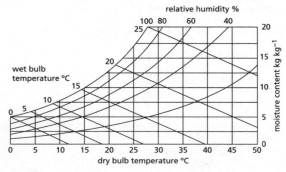

Fig. 45

(see Fig. 45). Other characteristic curves and lines include *wet bulb temperature and *relative humidity (shown here), and *specific volume and enthalpy of humid air (not shown here). The charts are usually presented at atmospheric pressure and are used for the design of air-conditioning systems and water-cooling towers.

psychrometry The study and measurement of the humidity of air.

puddling process A process once used for the production of wrought iron from pig iron. It was based on the partial decarburization of the pig iron in a reverberatory furnace, which is a long, low structure built of fire bricks and fuelled by coal. By ensuring the smoke and flame remained above the iron, the carbon content could be controlled.

pulsed baffle reactor (PBR) A tubular vessel used to carry out a chemical reaction such as polymerization in which the reactants are gently oscillated by either pulsing the feed by way of a reciprocating pump or by bellows of the entire contents on the reactor. The reactor contains either ring-shaped or plate-shaped baffles separated along the length of the reactor that encourage fluid turbulence and mixing, and promote reaction.

pulsed column A column used in *liquid–liquid extraction in which two immiscible liquids enter the column in the same direction. A reciprocating-type pump causes sharp pulsations in the direction of the flow, causing the liquids to disperse in one another, allowing intimate mixing of the liquids, and encouraging mass transfer. The dispersion is encouraged in columns that may contain perforated distribution plates, packing, or horizontal discs to prevent coalescence. Pulsed columns have found many applications, particularly in the nuclear-reprocessing industry for the separation of radioactive waste materials.

pulverizer A device used for grinding solid fuels such as coal as finely as possible with a minimum of power required to achieve a given *particle size distribution. Pulverized fuels are used in power stations for electricity generation and also for boilers for raising steam. The small particle size enables the fuel to be carried into the furnace with a current of air and allows for rapid and controlled combustion. Pulverizers are classified by the speed of operation and include various forms of mill such as the *ball mill.

pump A mechanical device used to transport a fluid from one place or level to another by imparting energy to the fluid. The three broad groupings are reciprocating, rotary, and centrifugal-type pumps. The most commonly used pump is the centrifugal type, which has a rotating impeller used to increase the velocity of the fluid and where part of the energy is converted to pressure energy. Rotary and reciprocating pumps are *positive displacement pumps in which portions of fluid are moved in the pump between the teeth of gears, and by the action of a piston in a cylinder, respectively. There are many variations of these types and each has a particular application and suitability for a fluid in terms of its properties, required flow rate, and delivery pressure. Pumps that do not conform to these groupings include acid eggs, air-lift pumps, and steam ejectors. All involve the use of energy to transport the fluid.

pump priming Used for the start-up and successful operation of centrifugal pumps in which the casing housing the *impeller is first filled or primed with liquid before operation begins. Since the density of a liquid is many times greater than that of a gas, vapour, or air, the suction pressure is otherwise insufficient to draw in more liquid. Depending on the type of pump, priming can be achieved either manually or by drawing liquid in using a vacuum pump. Valves can be used to prevent drainage and ensure that the pump does not require priming once the pump stops. Alternatively, the pump can be configured in such a way that it always maintains a reservoir of liquid at the suction side.

pump selection chart A diagram supplied by manufacturers of pumps and used to identify a pump for a particular duty (see Fig. 46). The rate of flow is presented on the x-axis and the delivered head or pressure on the y-axis. The performances of a number of pumps are typically presented spanning an acceptable range of efficiencies. Each area corresponds to a name or code, which is a combination of case number, impeller size, and speed.

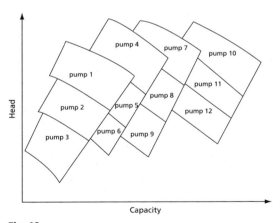

Fig. 46

purex process A process used to recover spent nuclear fuel from nuclear reactors by solvent extraction for recycling and fabrication into new fuels. An abbreviation for **p**lutonium and **u**ranium **r**ecovery by **ex**traction, it is a hydrometallurgical process that was developed in the late 1940s in which spent nuclear fuel is initially dissolved in refluxing nitric acid. Solvent extraction is then used to extract the uranium and plutonium from the fission product containing nitric acid solution into a kerosene diluent containing the solvent tri-butyl phosphate (TBP). Final separation of the uranium from the plutonium involves manipulation of the plutonium oxidation state. It superseded the earlier butex process, which used solvent extraction of spent nuclear fuel by solvent extraction from a nitric acid solution using diethylene-glycol dibutyl ether as the solvent. *See* NUCLEAR REPROCESSING.

purge The controlled removal of a small amount of material from a process as a sidestream that prevents an accumulation of undesired materials. It is also known as a *bleed.

purification The process of removing impurities. The **purity** is the extent to which a substance is free from extraneous or contaminating material. A substance is considered to be pure when the level of impurity can no longer be detected. The purity is typically expressed as a percentage of the substance that is free from the extraneous material.

pyro- A prefix meaning fire, burning, heat, etc.

pyrolysis The irreversible chemical decomposition or transformation due to temperature, and without the reaction with oxygen. It is used to produce coke from coal for use in making steel using temperatures of up to 2,000°C. Coke can also be made by pyrolysis using other hydrocarbon substances in petroleum refining.

pyrometer An instrument used to measure high temperatures such as in furnaces. Since the level of thermal radiation varies in processes, there are various types of pyrometers commonly used. These include thermocouple, optical, and radiation types. Optical pyrometers measure the infrared wavelength of heat using a lens whereas a radiation pyrometer measures the radiation wavelength without being near the hot object.

pyrophoricity A material that is capable of auto-ignition upon contact with ambient air. Certain gases, liquids, and solids are capable of auto-ignition such as sodium, potassium, finely divided uranium, and iron sulphide. Pyrophoric materials can often react with water, and therefore are required to be stored and transported under careful conditions.

Pythagoras' theorem A theorem that states that in a right-angled triangle, the square of the hypotenuse is equal to the sum of the squares on the other two sides. It is named after the Greek mathematician and philosopher Pythagoras (*c.*570 BC–*c.*495 BC), who provided a mathematical proof.

q-line A representation of the condition of the feed to a binary distillation process used in the *McCabe–Thiele graphical method (see Fig. 47). The line represents the condition of the feed in which q is the mole fraction of the more volatile component in the feed. The slope of the q-line is $q / (q - 1)$ and can be used to indicate the condition of the feed. Where the feed is a saturated liquid then q = 1 and the slope of the q-line is vertical. If the feed is saturated vapour then q = 0 and the slope of the q-line is horizontal.

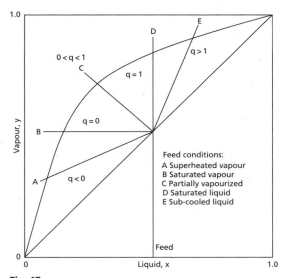

Fig. 47

QSL process A smelting process used to extract lead from ores. It involves feeding pellets of lead sulphide ore concentrate into a bath of molten slag in a rotating cylindrical furnace. By feeding oxygen below the slag on the surface, sulphur dioxide is formed along with heat. The lead oxide is reduced to metal by the addition of powdered coal. The process is named after its inventors P. E. Queneau, R. Schumann, and is now operated by Lurgi GmbH.

quadratic equation A type of polynomial equation in which the highest power of the unknown variable is two, such as x^2. It has the general form:

$$ax^2 + bx + c = 0$$

where a, b, and c are constants. The two roots of the equation can be obtained using the formula:

$$x = \frac{-b \pm \sqrt{b^2 - 4ac}}{2a}$$

quadruple-effect A multiple-effect unit, such as an evaporator, in which there are four stages. *See* MULTIPLE EFFECT.

qualitative analysis The identification of the part or parts present in a sample without consideration to the relative proportions of each.

quality 1. The features and characteristics of a product set against defined standards. The quality of a product such as a material produced in a manufacturing process or a detailed engineering design is measured in terms of a compliance with predetermined standards and specifications such as composition, level of impurities, functionality, etc. Deviation from these standards leads to poor quality. **2.** The percentage of saturation of a vapour. A fully saturated vapour has a quality of 100 per cent while a totally dry vapour has a quality of 0 per cent.

quality assurance (QA) A set of procedures or planned system of activities used to ensure that the quality control programme for a manufactured product from a process or a service is properly implemented. Planned and systematic actions are taken to provide adequate confidence that the product or service satisfies specified requirements. *Compare* QUALITY CONTROL.

quality control (QC) A set of procedures or planned system of activities used to ensure that a manufactured product from a process or a service meets a defined set of standards or criteria. For example, a product may be required to contain a maximum level of impurity, achieve a certain density specification, or have a certain viscosity. The standards may be set by the customer or by a regulatory body, as is the case with pharmaceutical manufacture. Not all the product manufactured is required to be tested but instead representative random samples are taken and tested. Where unacceptable statistical deviations are found, remedial action is required to correct the process in order to achieve the prescribed quality control standards.

quantitative analysis The determination of the amount or proportion of one or more constituents in a sample.

quantity of heat (Symbol Q) The total amount of heat in a body, process, or reaction. The SI unit is joules. The *calorie is also still used, which is the amount of heat required to raise the temperature of one gram of water by 1°C. The definition is not very precise since the specific heat capacity of water varies with temperature. In the *f.p.s. engineering unit system, the *British thermal unit is defined as the heat required to raise the temperature of one pound of water through one degree Fahrenheit. The therm is 10^5 Btu.

quantum mechanics A branch of mechanics that is based on the *quantum theory used for interpreting and understanding the behaviour of elementary particles, atoms, and molecules, which do not obey Newtonian mechanics.

quantum state The state of a system that is characterized by a set of quantum numbers and represented by an eigenfunction. Each state has an energy that has a value, which is precise within the limits imposed by the uncertainty principle but which may be changed by applying a field force.

quantum statistics The statistics concerned with the distribution of the number of identical elementary particles, atoms, ions, or molecules amongst possible quantum states.

quantum theory A theory concerning the behaviour of physical systems first developed by German physicist Max Planck (1858–1947) in 1900. The theory is based on the idea that such systems possess certain properties, such as energy and angular momentum, in discrete amounts or quanta.

quarl The refractory throat surrounding the burner port of a furnace. It is designed to direct air into the flame from the combustion of fuel and the radiation from the hot refractory. It also assists with efficient combustion of the fuel.

quart A British Imperial volumetric unit of measure equal to two pints and is a quarter of the volume of a gallon. It is equal to 1.136 522 litres. The Winchester quart is no longer used and is approximately equal to two quarts (2.25 litres). Some laboratory chemicals are supplied in Winchester quart bottles that have a volume of 2.5 litres.

quarter damping The ratio of successive peaks or troughs of the signal in an underdamped controlled system above or below the final steady-state value equal to 4:1 (see Fig. 48). That is, the peaks are progressively smaller and equal to one quarter of the height or depth of the previous peak or trough. This corresponds to an optimal control of a process. *See* DECAY RATIO.

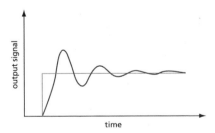

Fig. 48

quasi-steady state A system that is considered to be operating in a steady state although it may actually be an unsteady state process in which a *process variable varies with time. An example is a vessel containing a viscous liquid that discharges through a small orifice. A steady-state material balance is conveniently used to describe the rate of discharge and change in capacity in the vessel although the progressive fall in level is dependent on the characteristics of the orifice. The validity of a quasi-steady-state mathematic model can be assessed by setting the time derivative to zero.

Queeny, John Francis (1859–1933) An American industrialist who founded the Monsanto Company in 1901. It was the first American company to produce saccharine as an artificial sweetener at a time when the only commercial sources were from German chemical companies. He commercialized many processes including a butter substitute and hydrogenated vegetable fats. In the 1920s he developed a process to manufacture sulphuric acid using a vanadium catalyst as an alternative to the more expensive platinum catalyst that was used at the time. He named his company after his wife, whose maiden name was Olga Mendez Monsanto.

quench column A tall cylindrical vessel used to bring about the rapid cooling of a liquid or gas. Cooled liquid is sprayed down the column and intimately mixed with the hot rising vapour.

Quentin process A process using a solution of magnesium chloride that regenerates ion exchange resin used in the sugar refining industry. It is named after its inventor G. Quentin who developed it in 1957.

quick-acting valve A manually operated valve used to rapidly shut off the flow of a fluid such as fuel to a burner.

quick freezing A process used to rapidly freeze foodstuffs in which the temperature zone of maximum crystallization takes place as quickly as possible. After stabilization, the food is held at a temperature of $-18°C$ or lower. The purpose of the rapid freezing process is to avoid slow ice-crystal growth that results in damage to delicate cell walls causing a loss of mechanical structure and quality of the foodstuff on subsequent thawing. To achieve quick freezing, either liquid nitrogen or liquid carbon dioxide is used as a spray into a chamber containing the food to be quick frozen. *See* FLASH FREEZING.

quiescence A non-dividing state of microbial cells. In batch fermentations, viable yeast cells reach a state of inactivity or dormancy and no longer proceed through the cell cycle once all the limiting substrate has been exhausted.

Quinan, Kenneth Bingham (1878–1948) An American chemical engineer who designed and built explosives factories. With no formal chemical engineering qualifications, he began working for his uncle in an explosives factory in California. He followed his uncle to South Africa when offered the opportunity to build an explosives factory for the De Beers mining company. He built a factory that manufactured glycerine used in explosives just as the First World War was beginning. He was then commissioned to design, construct, and operate munitions and explosives factories in the UK. He received recognition with awards including the Croix de Guerre from the French government. He was the first vice president of the *Institution of Chemical Engineers.

quotient The mathematical result of dividing one quantity by another. For example, 13 divided by 3 gives a quotient of 4 remainder 1.

rad 1. A unit of radioactivity defined as the absorbed dose of *ionizing radiation when 1 g of material absorbs 0.01 J kg⁻¹ of energy. **2.** The symbol for *radian which is the plane angle between two radii of a circle that form an arc on the circumference that is equal in length to the radius.

radial In the direction of the radius. For example, **radial flow** is the flow of fluid in the direction of the radius such as from the effects of a *Rushton turbine type impeller, whereas a propeller provides **axial flow**.

radial-flow fan A mechanical device used to move air or a gas and comprises a power-driven shaft with a **radial impeller** or vanes. The air or gas enters near the axis of rotation and moves the air along the radius using the centrifugal force or rotation.

radial-flow fixed-bed reactor A type of chemical reactor that has a bed of catalyst with very small pressure drop across it. The catalyst is held in a toroidal basket placed in a cylindrical shell. The reaction gases enter either through the centre core or within the annulus and the products leave on the other side of the bed depending on the application. Unlike axial flow reactors, the design is more complicated, involving the flow distribution along the length of the bed, *voidage, catalyst settling, bed expansion, and contraction. It is used in the large-scale synthesis of ammonia, ethylbenzene dehydrogenation, and *catalytic reforming.

radial velocity The speed of a particle or fluid away from a central point.

radian (Symbol rad) The plane angle between two radii of a circle that form an arc on the circumference that is equal in length to the radius.

radiation 1. The transfer of energy by electromagnetic waves through a transmitting medium. The radiation may be absorbed, transmitted, or reflected by a receiving body although only the absorbed radiation is converted to heat. **2.** Emission of particles, especially alpha or beta particles from a radioactive source or neutrons from a nuclear reactor.

radioactive A substance such as an element that exhibits radioactivity.

radioactive decay The process in which unstable atomic nuclei spontaneously lose some of their excess energy by disintegrating into more stable nuclei. This is accompanied by the emission of alpha particles, beta particles, or gamma rays. Several of the heavier radioactive elements decay through a series of unstable radioisotopes before reaching a stable end-product. It is possible, in some cases, to induce artificial radioactivity through bombardment of the nuclei with particles such as neutrons.

radioactive fallout *See* FALLOUT.

radioactive isotope *See* RADIOISOTOPE.

radioactive series A sequence of radioactive nuclides in which each member is formed by the *radioactive decay of the nuclide proceeding it. The series ends with a stable nuclide. For example, in the thorium decay series, each step involves the loss of either an alpha particle or a beta particle, or occasionally both, until eventually a stable isotope of lead is formed.

radioactive waste (nuclear waste) The solid, liquid, or gaseous substances and materials that contain radionuclides that remain after the mining of radioactive ores, the reprocessing of nuclear fuels, the operation of nuclear reactors, the manufacture or decommissioning of nuclear weapons, and the waste from hospitals and research laboratories. The radionuclides may be highly radioactive and have very long half-lives presenting a danger to all living organisms. Their disposal is therefore highly regulated. Radioactive waste is classified as being either high-level, intermediate-level, or low-level waste. High-level waste includes spent nuclear fuel from nuclear power stations and is first required to be cooled for very long periods of time before being disposed of. Intermediate-level waste arises from the reprocessing of nuclear fuels and includes sludges, liquid, and the equipment used in the reprocessing. The waste is solidified, mixed with concrete, and packed into steel drums. It is stored in deep geologically stable mines. Low-level waste arises from materials used in the everyday activities of nuclear reprocessing, hospitals, and research laboratories, and includes liquids and solids. It is packed into steel drums and disposed of in special concrete-lined landfill sites. In the UK, these sites are under the authority of the UK Nuclear Decommissioning Authority that manages the site at Drigg near the UK reprocessing plant at Sellafield in Cumbria.

 SEE WEB LINKS
• Official website of the UK Nuclear Decommissioning Authority.

radioactivity A qualitative term used to describe the phenomenon resulting from the spontaneous disintegration of atomic nuclei usually with the emission of penetrating radiation or particles. The phenomenon was first noticed by Henri *Becquerel in 1896 in uranium salts in which rays emitted by these salts were found to affect a photographic plate. Marie and Pierre *Curie detected more intense radioactivity in a new element which they called polonium, and later isolated another radioactive element, which they named radium.

radiobiology A branch of biology concerned with the effects of radioactive substances on living organisms and the use of radioactive substances as *tracers to study the biological processes of living cells.

radiocarbon dating A technique used to determine the age of organic materials such as wood and based on the content of the isotope carbon-14 acquired from the atmosphere when they were formed. The *half-life of the radioisotope is 5,730 years and decays to the isotope nitrogen-14. Measurement of the amount of carbon remaining in the material therefore gives an estimate of its age.

radiochemistry A branch of chemistry concerned with radioactive elements, their compounds, and ionization. It includes the use of radionuclides in chemical reactions. *See* RADIOLYSIS.

radiogenic Produced or resulting from radioactive decay.

radiography A process that produces an image on a sensitive plate or film by X-rays or other ionizing radiation. As well as being used in medicine and surgery, it is used as a *non-destructive testing technique to identify cracks and faults in metal such as welds in vessels and pipes. It is used in the nuclear industry particularly to check the quality of welds. The advantage of this method is that the source of radiation can be very small.

radioisotope (radioactive isotope) An isotope of an element that is radioactive. Every element has at least one isotope, although for many they can only be obtained by bombarding the elements with high-energy particles or in nuclear reactions.

radiology The study and use of X-rays, radioactive materials, and other ionizing radiations for medical purpose especially for the diagnosis and treatment of cancer and associated diseases.

radiolysis The use of ionizing radiation in chemical reactions, it involves the use of alpha particles, electrons, X-rays, and gamma radiations emitted from radioactive materials.

radionuclide A nuclide that is radioactive.

radiosity Used to calculate the thermal radiation heat transfer exchange between surfaces, it is the sum of the emissive power and the portion of the radiation that is reflected by a surface. By conveniently combining the radiation being emitted by and reflected from a surface, it is useful in determining the net thermal energy exchange between multiple surfaces.

radiotherapy The use of ionizing radiation such as X-rays and radioactive isotopes in medical treatment.

raffinate The residual or waste liquid stream leaving a liquid–liquid solvent extraction process after the extraction has taken place with an immiscible liquid to remove solutes from the original liquid. The word is derived from the French *raffiner* meaning to refine. *See* SOLVENT EXTRACTION.

ramp response The behaviour of a system or process to a change resulting from a sudden increase in the rate of change from zero to some finite value. It includes the combined transient and steady-state behaviour.

Ramsay, Sir William (1852–1916) A British scientist noted for the discovery of radon, which at the time was the last of the noble gases to be discovered. He worked under German chemist Robert Bunsen (1811–99) before taking up professorships at Bristol (1880–87) and then London (1887–1912). He worked with *Rayleigh on the gases in air. Together they discovered argon in 1894. He worked with Morris Travers (1872–1961) and discovered neon, krypton, and xenon, discovering radon in 1904. He was awarded the Nobel Prize for Chemistry in 1904.

random error *See* ERROR.

random packing Used in packed columns, it is the arrangement of packing materials such as saddles and rings that have no structured order. The orientation of packing materials that are charged into columns in bulk typically lie randomly on support plates. The packing is used to intimately contact a gas and a liquid to bring about effective mass transfer by providing a high surface area per unit volume.

Raney nickel A porous solid catalyst made from an activated alloy of nickel and aluminium. The nickel is the catalytic metal with the aluminium as the structural support. It was developed by American mechanical engineer Murray Raney (1885–1966) in 1926 for the hydrogenation of vegetable oil and is now used in hydrogenation reactions in various forms of organic synthesis. It is widely used as an industrial catalyst for the conversion of olefins and acetylenes to paraffins, nitriles, and nitro compounds to amines, and benzene to cyclohexane amongst others.

range The extent over which a quantity is measured such as temperature, pressure, and level. It is defined by stating both its lower and upper range values.

Rankine cycle An ideal reversible thermodynamic cycle used in steam power plants (see Fig. 49) that more closely approximates to the cycle of a real steam engine than the *Carnot cycle and converts heat into mechanical work. It involves water being introduced under pressure into a boiler and evaporation taking place, followed by expansion of the vapour without the loss of heat, ending in condensation. The cycle therefore consists of four stages: i) steam passes from the boiler to the cylinder at constant pressure; ii) the steam expands adiabatically to the condenser pressure; iii) heat is given to the condenser at constant temperature; iv) condensation is completed and the condensate is returned to the boiler. In the Rankine cycle, the work done is equivalent to the total heat in the steam at the end of the adiabatic expansion subtracted from the total heat in the steam at the beginning of the expansion. The heat supplied is equal to the sensible heat in the condensed steam subtracted from the total heat.

The Rankine cycle is used to describe the way steam-operated heat engines that are found in thermal power generating plants generate power for which the heat sources are nuclear fission, or the combustion of coal, oil, or gas. It is named after Scottish civil engineer and physicist William John Macquorn Rankine (1820–72).

Rankine scale A thermodynamic temperature scale based on the absolute zero of the Fahrenheit temperature system such that −459.67°F is equal to 0°R. It was proposed by Scottish engineer William Macquorn Rankine (1820–72) in 1859.

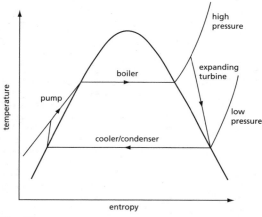

Fig. 49

Raoult, François-Marie (1830–1901) A French chemist who studied the behaviour of solutions and their physical properties. He is best known for his work on solutions, in particular the depression of freezing points and the depression of a solvent's vapour pressure due to a solute that was shown to be proportional to the solute's molecular weight. Both became ways of determining molecular weights of organic substances. *Raoult's law is named after him.

Raoult's law A law that states that the partial pressure exerted by each component in an ideal liquid mixture is the product of the vapour pressure of each component and its mole fraction at the temperature of the liquid expressed as:

$$P_A = x_A p_A$$

where P_A is the partial pressure, x_A is the mole fraction of component A, and p_A is the vapour pressure. Formulated by French chemist François-Marie Raoult (1830–1901) in 1882.

rare earth elements The oxides of a group of elements in the *periodic table known as lanthanoids. Unlike their name would suggest, they are not rare but are widely found. They are extracted for use in electronic equipment such as mobile telephones.

rarefaction A process of making less dense or the state of being less dense. A reduction in the pressure of a fluid therefore reduces its density.

rare gas Another name for a noble gas and includes helium, neon, argon, krypton, xenon, and radon.

Raschig, Friedrich (Fritz) August (1863–1928) A German chemist who, after gaining his PhD in 1884 from the University of Berlin, started working for BASF but began his own chemical company in 1891. He patented many processes, mainly based on phenols. He also made many process improvements such as to distillation. He also developed a small ring, known as a *Raschig ring, that is used in *distillation columns and packed columns to improve extraction and separation performance.

Raschig process 1. A process used for the production of chlorobenzene. It uses a gas-phase catalytic reaction between benzene vapour, hydrogen chloride, and oxygen in the form of air at a temperature of 230°C in the presence of a copper chloride catalyst:

$$2C_6H_6 + 2HCl + O_2 \rightarrow 2H_2O + 2C_6H_5Cl$$

The chlorobenzene produced is then used for the production of phenol:

$$C_6H_5Cl + H_2O \rightarrow C_6H_5OH + HCl$$

The reaction takes place at 430°C in the presence of a silicon catalyst. The process is named after German chemist Fritz *Raschig (1863–1928). The **Hooker–Raschig process** is a variation and uses a different type of catalyst. **2.** A two-step process used for the production of hydrazine using the oxidation of ammonia with sodium hypochlorite in the presence of gelatine:

$$NH_3 + NaOCl \rightarrow NaOH + H_2NCl$$
$$H_2NCl + NH_3 \rightarrow N_2H_4 + HCl$$

The **Olin–Raschig process** is a refinement of the process; developed by the Olin Corporation, it involves mixing a solution of sodium hypochlorite with ammonia at 5°C to form chloramine and sodium hydroxide, and then quickly adding it to anhydrous ammonia under pressure and a temperature of 130°C to produce the hydrazine, water, and sodium chlorine.

Raschig ring A type of packing material used in packed columns and towers (see Fig. 50). It consists of a hollow cylinder whose length is equal to its diameter. Raschig rings are made from various materials such as plastic, glass, and ceramic. Raschig rings made from borosilicate glass are sometimes used in vessels containing nuclear materials such as uranyl nitrate to act as neutron absorbers to prevent the possibility of a *criticality incident. They are named after German chemist Fritz *Raschig (1863–1928).

Fig. 50

rate constant A constant associated with chemical reactions and is a measure of the rate at which the reaction takes place. The SI unit is dependent on the order of the reaction and is expressed in terms of one of the reacting species. The *Arrhenius equation is often used to relate the rate constant for the reaction to reaction temperature.

rate-determining step Most chemical reactions involve several stages. If one of these stages is slower than the others then the overall rate of the reaction will be determined by the rate of the slowest stage. In catalytic reactions, for example, the overall process involves diffusion, adsorption, reaction, and desorption. The rate-limiting step is the step that consumes most of the driving force.

rational number The ratio of two integers presented as a quotient or fraction, for example 3/2. They can be both positive and negative. A number that is not rational is irrational.

raw data Data that has not been organized or processed in any way.

raw materials The starting materials used in a process that may be characterized on the basis of their chemical and physical form. For example, crude oil is the starting material in the petroleum industry and is converted into many useful and valuable products. Coal and natural gas are also the starting materials for many other useful products. Inorganic starting materials include metal ores, salt, air, water, and sulphur.

Rayleigh, Lord (1842–1919) A British scientist born as John William Strutt. He studied mathematics at Cambridge and was appointed professor of experimental physics and head of the Cavendish laboratory at Cambridge in 1879. He succeeded to the barony on the death of his father in 1873. He was appointed professor of natural philosophy at the Royal Institution until 1905, was president of the Royal Society between 1905 and 1908, and was awarded the Nobel Prize for Physics in 1904. His contribution to science covered many aspects of physics including optics, vibrating systems, sound, wave theory, colour vision, electrodynamics, electromagnetism, light scattering, flow of liquids, hydrodynamics, density of gases, viscosity, capillarity, elasticity, and photography. He published extensively and contributed to the *Encyclopaedia Britannica*.

Rayleigh equation An equation that relates the quantity to the concentration of the more volatile component in *batch distillation. The equation is formulated from an unsteady-state material balance in which vapour leaving the still is in equilibrium with the liquid. As the vapour is richer in the more volatile component, the composition of the liquid and vapour are not constant:

$$\ln\frac{L_1}{L_2} = \int_{x_2}^{x_1} \frac{dx}{y-x}$$

where L_1 and L_2 are the initial and final number of moles in the still, x_1 and x_2 are the initial and final mole fractions of the more volatile liquid, and y is the mole fraction in the vapour phase. The integral can be integrated graphically or numerically using equilibrium data. It is named after Lord *Rayleigh (1842–1919) who first derived it.

Rayleigh–Jeans law A formula giving the intensity of *black-body radiation at long wavelengths for a body at a particular temperature. It is an approximation to Planck's full formula for the black-body intensity based on quantum concepts.

Rayleigh's dimensional analysis method A method used in *dimensional analysis that expresses a functional relationship for the variables involved in an observable phenomenon. It involves expressing all the independent variables that influence a dependent variable in the form:

$$y = fa^x b^y c^z ... z^n$$

where f is a dimensionless constant. Grouping the exponents, x, y, z ...n, in their fundamental units and solving the set of equations simultaneously leads to the formation of dimensionless groups by grouping the variables with each exponent. The number of dimensionless groups formed is equal to the *fundamental units subtracted from the total number of variables. It is named after Lord *Rayleigh (1842–1919).

reaction The transformation of a substance by physical, chemical, or biochemical means into other substances with the release or consumption of energy. Single reactions are rare in practice but instead are generally complex resulting in several products, intermediates, and undesirable products. In parallel reactions, the concentration level of desired product is key to the control of the product distribution. For reactions in series, the mixing of the different compositions is the key to the formation of an intermediate. Series-parallel reactions are seen from the point of view of the constituents. The thermodynamic feasibility of a chemical reaction can be determined from the free energy while the kinetics of the reaction determines the rate at which the reaction takes place, and therefore determines the size of the reactor required to contain the reaction. The choice of reactor, such as *plug flow reactor or *continuous stirred-tank reactor, is dependent on the heating and cooling requirements of the reaction. Heats of reaction can be determined from heats of combustion and heats of formation of the individual components in the reaction.

reaction rate (Symbol r) The number of moles of a species, i, undergoing chemical change per unit volume of reactor, V, per unit time:

$$r_i = \frac{-1}{V}\frac{dn_i}{dt}$$

It may be expressed in terms of the rate of decrease in concentration of a reactant or the rate of increase in concentration of a product. The rate of reaction may be affected by the concentration of reactants, pressure, temperature, physical form of the reactants, and the presence of catalysts. Convention uses a negative sign to represent the disappearance of reactants and a positive sign to represent product formation.

reactive distillation A distillation process that involves a chemical reaction taking place within the *distillation column. As with conventional distillation, the reactants and the products are separated by virtue of differing boiling points and are therefore dependent on phase equilibria properties as well as hydraulic behaviour. Reactive distillation can also involve the presence of a catalyst and is known as **catalytic distillation**. Reactive distillation is used for esterification and etherification amongst other processes where the reaction takes place in the liquid phase. The benefits of reactive distillation include the elimination of product recovery along with the separation and recycling of unconverted reactants.

reactor A containment within which a controlled chemical, biochemical, or nuclear reaction takes place. There are various forms of chemical reactors whose design and application is dependent on the form of the chemical reaction. They may be large vessels or pipes, operated continuously or as a batch. The design of the reactor used for contacting of the reactants is important in determining the yield of desired product. In general, batch reactors are used where chemical conversion is dependent on reaction time only, whereas tubular reactors are used where chemical conversion is dependent on the position. *Continuous stirred-tank reactors are used where neither time nor position is critical. Batch reactors require a charge of reactants, whereas continuous reactors operate at steady state with a continuous flow of fresh reactants and the removal of products. The reactants and products may be liquid or gaseous. Catalysts may be required within the reactor or the presence of inert materials such as in *fluidized bed reactors. Continuous stirred-tank reactors are commonly used in which liquid reactants are continuously charged to the reactor, which is equipped with an agitator to ensure homogeneity of the reactants. There is a continuous overflow to maintain a constant volume within the reactor. A *plug flow reactor also operates continuously and consists of tubes that may be packed with a catalyst and used for gas- and liquid-phase reactions. They assume idealized plug flow of materials flowing through them and are used for both exothermic and endothermic reactions that require rapid heat transfer from or to the reaction. Fluidized bed reactors involve reactants to be injected under a bed of solid particles such as a catalyst and are effective at promoting heat and mass transfer as well as reaction rates. There are many other types of reactor such as *pulsed baffle reactors and semi-batch reactors, which operate with both continuous and batch feeds and outputs.

real gas A gas that does not behave as an ideal gas. This is, its behaviour deviates from ideality. The molecules of a real gas are finite in size and an attractive force exists between them. A number of *equations of state have been established such as the *van der Waals' equation of state that allows for the size of the molecules and the attractive forces. The ideal gas law can also be used to mathematically describe the behaviour of a real gas or gas mixture and includes a *fugacity to represent a pseudo-pressure in place of partial pressure.

real number A number that can be expressed using a decimal point.

real-time control The control of processes in which a computer receives data signals from a process plant and returns control signals within a defined time constraint. The data signals are rapidly processed and can use past trends, experimental data, and databases, as well as permitting operator intervention. The output-controlled signals are used to make adjustments to valve positions, etc. and the controlled process variables displayed on screen in a control room.

reboiler A type of heat exchanger associated with *distillation columns for boiling the liquid from the bottom of the column to totally or partially vaporize it and be returned to the column to drive the separation. Reboilers usually consist of a horizontal cylindrical body containing hairpin tubes for heating with a characteristic vapour dome above the boiling liquid. Vertical *thermo-syphon reboilers are also used. Condensing steam, waste heat, or a hot liquid from the process is usually used as the heating medium. *See* KETTLE REBOILER.

reciprocating pump, compressor A positive displacement device consisting of a piston and cylinder arrangement. The piston is driven by a crankshaft though a connecting rod such that the piston enters and leaves the cylinder expelling and drawing in liquid for transportation in a cyclic fashion. A tight tolerance is required between the piston and cylinder to ensure no leakage. Check valves are used to ensure that the flow is in the desired direction. Multiple cylinders operated out of phase are used to provide a more continuous flow. Variations include the *plunger pump and the *diaphragm pump.

recombinant DNA technology A biotechnological process used to produce many useful and valuable products with medical, healthcare, agricultural, and veterinary applications. It involves taking genes from the cells of a living organism and transferring them to the cells of another organism. The genes transferred contain the genetic code for the expression of a required biochemical product, such as a protein with therapeutic properties, or provide resistance to an antibiotic or other substance. Recombinant human insulin, used for the treatment of diabetes, has virtually replaced insulin derived from pigs and cows, and is produced by bacteria containing human genes. Human growth hormones, hepatitis B vaccine, blood-clotting agents known as Factor VIII, anti-cancer drugs, and vaccines against scours, a toxic diarrhoea in pigs, are all produced by recombinant DNA technology. Crops developed through recombinant DNA technology that are resistance to herbicides, insects, low moisture, and other environmental conditions include rice, maize, canola (oil-seed rape), and cotton. The biotechnology was pioneered by Stanley Cohen from Stanford School of Medicine and Herbert Boyer from the California School of Medicine, San Francisco in 1973.

recrystallization 1. A process used to purify a substance or to obtain more regular crystals of a purified substance by repeated crystallization. **2.** A solid-state process that involves the reformation of the lattice structure of a crystal. It requires a rapid diffusion to the lattice and is carried out at an elevated temperature but below its melting point temperature. The homologous temperature is the absolute temperature as a fraction of the melting point also in absolute temperature. Recrystallization typically takes place at a homologous temperature of 0.6.

rectification The process of purifying liquids by way of *distillation.

rectification section The part of a *distillation column above the *feed point where the more volatile component in both the liquid and vapour increases. The part of a distillation column below the feed point is known as the *stripping section.

Rectisol process The tradename for an *acid gas removal process. It uses methanol as the solvent to remove hydrogen sulphide and carbon dioxide from gas streams.

recuperator A type of heat exchanger used to heat process fluids using hot waste fluids. The hot and cold fluids pass through separate channels as either single or multipass.

recycle 1. The return of materials or energy to a previous stage in a cyclic process. It is used to increase yields, to enrich a product, to conserve heat, and to improve operations. **2.** The conversion of waste or used products to reusable material as a way of preventing the waste of potentially useful materials, reducing the consumption of fresh *raw materials, reducing energy usage, and reducing pollution. Recycling is the third component of the 'Reduce, Reuse, Recycle' waste hierarchy.

recycle ratio The ratio of recirculated reactant for reprocessing to the same reactant entering the process as fresh feed.

Redlich–Kwong (RK) equation of state A cubic equation of state based on the *van der Waals' equation of state used to describe the behaviour of real gases:

$$p = \frac{RT}{V - b} - \frac{a}{T^{0.5} V(V + b)}$$

The constants a and b are determined from the critical point data of the gas. The compressibility factor at the critical point is equal to one-third. The equation was developed by Austrian chemist Otto Redlich (1896–1978) and Joseph Neng Shun Kwong in 1949. The equation was further developed and modified by Pitzer, Wilson, and Soave to improve the predictability of gas behaviour. *See* SOAVE–REDLICH–KWONG (SRK) EQUATION OF STATE.

redox reaction *See* REDUCTION.

reduced pressure (Symbol p_r) The ratio of the pressure of a component to its critical pressure, p_c:

$$p_r = \frac{p}{p_c}$$

reduced temperature (Symbol T_r) The ratio of the temperature of a component to its critical temperature, T_c:

$$T_r = \frac{T}{T_c}$$

reduction Any chemical reaction in which oxygen is removed from a substance, hydrogen is added, or in which an atom or group of atoms gain electrons. Reduction is accompanied by oxidation in many reactions. Such combined oxidation and reduction reactions are called **redox reactions**.

redundant Superfluous to requirements such as an item of process plant that is no longer required or in operation. *Compare* MOTHBALLING.

refine To make or become free from impurities. Metals, oils, and sugar are all refined.

refinery A factory for the purification of some crude or unprocessed material such as crude oil and sugar. Petroleum refining processes begin with distillation by boiling the crude oil into separate fractions or *cuts. These cuts are then converted to changing the size and structure of the molecules through *cracking, *reforming, and other conversion processes, followed by various treatment and separation processes to remove undesirable components and improve product quality. Refineries are complex and highly integrated facilities.

refining 1. A process used to purify an impure material but that does not involve chemical change. The refining of metals involves either hydrometallurgical or pyrometallurgical processes. The *cupellation process involves the extraction of silver from lead by melting the lead in a vessel known as a cupel and blowing air over the surface to oxidize the lead and leave pure silver. **2.** The refining of petroleum products involving the conversion of crude oil into usual petroleum products such as naphtha, diesel fuel, kerosene, and LPG. It is carried out in petroleum refineries.

reflectivity The portion of radiant energy falling on a surface that is reflected back. *Compare* ABSORPTIVITY; TRANSMISSIVITY.

reflux The boiling, condensing, and subsequent return to the boiler of a volatile liquid. It is used to enhance the separation of liquids of differing volatilities as in the process of distillation. In a *distillation column used to separate volatile liquids by virtue of their differing boiling points, using more reflux decreases the number of plates required for the separation. Conversely, using less reflux increases the number of plates required. The *reflux ratio is the ratio of *distillate taken from the distillation process to the reflux returned as liquid for further separation. Where no distillate is taken, the process operates on **total reflux**. The *minimum reflux ratio corresponds to an infinite number of trays to bring about a separation. Ideally, the reflux ratio should be optimal for which the combined fixed costs and the cost of operation should be least.

reflux drum A vessel used with *distillation columns to receive condensed reflux liquid from the top of the column to be returned back to the column in order to provide cooling and condensation of the rising vapour. Level control is used to control the amount of condensed liquid that leaves the drum and returns to the column.

reflux ratio Used in the control of the purity of top products from a distillation column, it is the ratio of the molar flow of liquid, L, returned to the top of the column to the amount of top product or distillate, D, removed.

$$R = \frac{L}{D}$$

reforming A process used to rearrange the molecular structure of hydrocarbons to alter their properties by *cracking or by *catalytic conversion. In *petroleum refining, reforming is used to produce hydrocarbons for use in gasoline. *Steam reforming is the conversion of methane from natural gas into hydrogen. It is used in ammonia production, which is produced from desulphurized and scrubbed natural gas that is mixed with steam and passed over nickel catalyst packed in tubes at a high temperature of around 900°C.

$$CH_4 + H_2O \rightarrow CO + 3H_2$$
$$CH_4 + 2H_2O \rightarrow CO_2 + 4H_2$$

As both reactions are endothermic, heat is supplied to the reformer.

refractory A non-metallic material used to line furnaces, kilns, ovens, and some high-temperature reactors, which can maintain its mechanic strength at high temperature. They are made from chemically inert materials such as fire clay, silicon oxide, and aluminium oxide, and are able to withstand very high temperatures without cracking or expansion. Some refractory materials are also required to be resistant to alkaline and acidic environments.

refrigerant A low boiling liquid used as the working fluid in the process of *refrigeration that is capable of phase change at low temperatures. CFCs were once popular due to their excellent thermodynamic properties but have been internationally banned due to their ozone-depleting potential. Modern refrigerants do not contain harmful chlorine. Ammonia is the refrigerant used in absorption refrigeration.

refrigeration The process of cooling or freezing substances such as foods or process liquids and maintaining them at a temperature below that of their surroundings. The most commonly used refrigeration plants use a vapour compression refrigeration cycle. This consists of a compressor, evaporator, expansion valve, and condenser connected in series. Heat is absorbed by a refrigerant causing it to boil. The vapour is then compressed to an elevated pressure, and then condensed to a liquid releasing the absorbed heat. The liquid is then returned to the low-pressure evaporator through a throttle valve for reuse (see Fig. 51). The ideal cycle can be presented on a temperature-entropy diagram or a pressure-volume diagram. Within the refrigeration process, the *coefficient of performance is the ratio of the refrigeration effect to the work input. The *vapour absorption cycle is also used in industrial refrigeration systems and involves a refrigerant such as ammonia. Small-scale refrigeration can be achieved using the *Peltier effect.

refuel A process that receives a supply of fresh fuel such as a coal, coke, hydrocarbon liquid or gas, or nuclear fuel within a nuclear reactor.

regenerate The process of restoring or of being restored to its original chemical or physical state. For example, *ion exchange units consist of a fixed bed of ion exchange resins used to exchange ions within a solution, and are used in water treatment to remove calcium and magnesium ions replacing them with soluble sodium ions. The resins are regenerated for reuse by applying pressure and temperature to return them to their original state.

regenerator A type of heat exchanger in which hot and cold fluids pass alternately through the same chamber. The process is unsteady in which the chamber is first heated and then cooled by the cold fluid.

Regnault, Henri Victor (1810–78) A French scientist who was born in poverty and rose to become an outstanding chemist and physicist. He discovered carbon tetrachloride and measured the chemical composition of the atmosphere around the world. He made a series

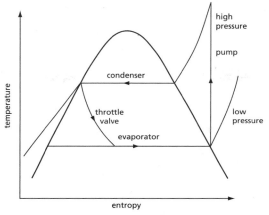

Fig. 51

of measurement of the specific heats of many solids, liquids, and gases, of the expansion of gases, and of vapour pressures. He developed the use of an accurate air thermometer. He also designed an efficient hygrometer for measuring *dew point and invented an apparatus for measuring the coefficient of expansion of mercury.

Regnault's method A method of measuring the density of a gas involving weighing an evacuated glass bulb of known volume and then admitting the gas at a known pressure and temperature. The method is named after French chemist and physicist Henri Victor *Regnault (1810–78).

regression analysis (least squares method) The statistical analysis and measure of the association between a dependent variable (y) and an independent variable (x) that involves calculating the best straight line for a linear equation between x and y as $y = a + bx$ where a and b are constants. While x can be measured accurately, there is a statistical random error associated with y. The least squares method aims to minimize the sum of the square of the difference between all (n) values of y on the line at x:

$$an + b\Sigma x = \Sigma y$$
$$a\Sigma x + b\Sigma x^2 = \Sigma xy$$

The **regression line** is the calculated line of best fit.

regulator 1. A device such as a valve used to control the flow of fluid, pressure, temperature, voltage, etc. in a process. **2.** A person or organization that has responsibility for regulating the activities of an industry such as the release of emissions into the atmosphere. **3.** A gene within the DNA of living organisms responsible for controlling the synthesis of a product from another gene. *See* RECOMBINANT DNA TECHNOLOGY.

Reid vapour pressure The vapour pressure of a fluid measured at 100°F (38°C) in a bomb where the initial liquid volume is about a quarter of the air volume. It is applied to the vapour pressure of hydrocarbons.

reimbursable A form of engineering contract in which the contractor is reimbursed for the time taken to carry out and complete a project. *Compare* TURN-KEY.

rejection coefficient A measure of the amount of a solute in a feed solution fed to a *semi-permeable membrane that does not permeate the membrane in a separation process such as *ultrafiltration:

$$R = 1 - \frac{C_P}{C_R}$$

where R is the solute rejection coefficient, C_P is the concentration of the solute in permeate, and C_R is the concentration of solute in the retentate. For a given membrane with a pore size distribution there is a relationship between the rejection coefficient and the solute molecular weight.

relative atomic mass The mass of an atom relative to one-twelfth the mass of an atom of carbon-12. It is dependent on the number of protons and neutrons in the atoms. The electrons have negligible mass.

relative density The ratio of the density of a substance to a reference substance. A substance with a relative density less than one is less dense than the reference; if greater than one then it is denser than the reference. If the relative density of the substance and reference substance is exactly the same, then the densities are equal; i.e. equal volumes of the two substances have the same mass. The temperature and pressure must be specified for both the sample and the reference. *Compare* SPECIFIC DENSITY.

relative humidity The moisture content of dry air expressed as a percentage. This corresponds to the ratio of the mass of water vapour in a given volume of air at a given temperature and pressure, to the maximum quantity of water vapour that can be held in the air at those conditions.

relative molecular mass (molecular weight) The mass of a molecule relative to one-twelfth of the mass of an atom of carbon-12. It is determined by adding the relative atomic masses of the atoms that comprise the molecule.

relative roughness (Symbol ε) Ratio of the absolute pipe wall roughness to the inside diameter of the pipe in consistent units. Being dimensionless, the magnitude of the surface roughness is relative only to the inside pipe diameter as a form of shape factor. *See* ABSOLUTE ROUGHNESS.

relative volatility (Symbol α) The ratio of the vapour pressure of one liquid component to another in a heterogeneous mixture and is a measure of their separability. For a binary mixture, the relative volatility can be expressed in terms of the mole fraction of the more volatile component in the liquid and vapour phases, x and y, as:

$$\alpha = \frac{y(1-x)}{x(1-y)}$$

The greater the value of the relative volatility, the greater the degree of separation. If $y = x$ then no separation is possible.

release agent A substance that is applied to surfaces to prevent them from sticking together. They are used in many industrial processes such as food, plastics, polymers, and glass. They include waxes, silicones, and glycerides, amongst others. For example, zinc stearate is used to prevent polymerized methyl methacrylate from sticking to steel moulds used in the production of household appliances.

reliability The statistical chance of a system, process, item of equipment, performing its intended purpose or function without failure or within required levels of performance.

relief valve (RV) An automatic type of safety valve that is activated by the static pressure in a liquid-filled vessel. It opens in proportion to increasing pressure. *See* SAFETY VALVE.

renaturation The reconstruction of a denatured protein or nucleic acid to its original conformation and function. The process of renaturation may be aided by reversing the process conditions that causes *denaturation, but in practice may be very slow or may not actually occur at all.

renewable energy Energy derived from sources that do not deplete the Earth's finite and non-renewable mineral resources. Examples include geothermal energy, tidal power, wave power, wind power, solar energy, biomass, biofuels, and hydroelectric power. Many others are currently being investigated due to concerns that fossil fuel reserves are running out, as well as the effects of carbon dioxide emissions from their combustion in terms of climate change. The use of renewable energy is not new. Hydropower uses kinetic energy in water flow to drive a water turbine, which in turn drives an electric generator. The first hydropower plant was open at Niagara Falls in North America in 1899. Tidal power is generated by passing water through a two-way turbine and was first used in the 1900s. Wind turbines generate electricity by using the kinetic energy of the wind to drive a set of turbine blades and thus a generator. There is no atmospheric pollution and no fuel costs. However, there is a visual impact and some noise. Wind turbines were once a common site across Europe for milling grain since the fourteenth century.

repeatability The closeness of agreement of a consecutively measured statistical data sample.

Reppe process A general name for catalytic processes used to produce vinyl compounds from acetylene. They use high pressure in the presence of metal acelylide catalysts. They are named after German chemist Walter Reppe (1892–1969) who developed the safe industrial use of acetylene.

reproducibility The closeness of agreement amongst repeated measurements of data in a statistical sample.

residence time The duration of a substance within a process or item of equipment. The residence time is used to give a measure of chemical reaction or required equipment size.

residence time distribution (RTD) A concept used to characterize the macro-mixing in a continuous flow system such as a reactor. It is expressed as the fraction of a material remaining in the system as a function of time given either as a cumulative distribution or as the derivative of the cumulative curve. The RTD can be determined by adding either a step increase in a tracer substance or an instantaneous pulse of a tracer into the flow system, and measuring the response at the other end or output. The RTD information is useful to determine the performance of a reactor throughput.

residual moisture The moisture that remains within a substance following a normal drying process to remove the moisture.

residue 1. The material remaining after boiling off the volatile liquid in a mixture containing solids or high boiling-point liquids. **2.** The distillation of all but the heaviest components from liquid materials such as crude oil.

resistance 1. A force, such as friction, that opposes the direction of motion having the effect of slowing a body such as fluid flowing in a pipe. **2.** The ability of the human body to withstand the harmful effects of chemical pollutants, viruses, microorganisms, and physical conditions such as extremes of temperature and pressure.

resistance in parallel A system arranged so that a flow is divided through each part such that the resistance is the same in each part. In electric circuits, resistors are connected such that the current divides between them. For pipes in parallel, the flow of fluid divides into each pipe such that the pressure drop resistance of each pipe is the same:

$$\Delta p_A = \Delta p_B = \Delta p_C = ...$$

resistance in series A system arranged in sequence such that the flow through each part is the same. In electric circuits in which the circuit elements are arranged in sequence, the same current flows through each in turn where the total resistance is equal to the sum of the individual resistances. In heat transfer, the temperature drop across a wall comprising several materials is the sum of the temperature drops for each material:

$$\Delta T = \Delta T_A + \Delta T_B + ...$$

For pipe flow, the total pressure drop across several pipes linked in series is the sum of the pressure drops for each pipe:

$$\Delta p = \Delta p_A + \Delta p_B + ...$$

resistance temperature detector An instrument used to measure temperature through changes in electrical resistance. It consists of an element of fine coiled wire made from copper, nickel, or platinum wrapped around a glass core. The wire has a known change of resistance with temperature. It is noted for its accuracy and reliability across a wide range of temperatures.

resolution 1. The smallest increment of change that can be detected by an instrument or other measurement system. **2.** The process of separating something into its constituent parts.

respiration The process of oxygen exchange in living organisms from the surroundings, releasing energy and carbon dioxide. The **respiration quotient** is the ratio of the production rate of carbon dioxide to the oxygen uptake by a culture of microorganisms within a bioreactor. It is a measure of the performance of the bioreactor in terms of the ability of the culture to use a carbon-limiting substrate and is often correlated with the product yield. It is also used to measure the performance of the human body as the ratio of carbon dioxide liberated from the lungs to the oxygen consumed.

response The behaviour of a process to a disturbance such as a change in flow, temperature, pressure, etc. It is measured using an instrument and the information used by a controller to return the process back to its desired value. A **step response** is the behaviour of the controlled process to adjust from one steady-state condition to another. A **dynamic response** is the behaviour of the controlled process with respect to time. The **response time** is the time for a process or instrument to adjust to a disturbance and return to a new steady-state value. It is expressed as the time required to fall within a specified percentage of the change final steady-state value. This is often taken as 5 per cent or occasionally 3 per cent.

retentate The material (liquid, gas, or vapour) that does not pass through a semi-permeable membrane in a separation process such as *ultrafiltration. In membrane modules used to separate activated sludge, for example, water molecules pass through the membrane as *permeate while the material remaining is the retentate.

retention time The time at which a certain amount of a substance has been removed or eluted from a process or process pipe. In a gas or liquid chromatograph, which is used to separate mixtures of unknown components by injection into a packed column and carried by a solvent or gas to a detector, the time for a substance to appear is measured relative to an internal standard component so that an accurate identification on the components can be made against known standards.

revamp An informal term used for the maintenance, repair, redesign, and return to operation of process plant or an item of process equipment such as a pump.

reverse-acting controller A type of controller used to control a process in which the output signal from the controller decreases with increasing measured (input) value. It has a positive gain. *Compare* DIRECT-ACTING CONTROLLER.

reverse flow plate A type of plate used in distillation columns that consists of a downcomer and a separation or baffle running perpendicular to the downcomer weir across most of the plate, causing liquid entering the plate to flow along one side and around the other.

reverse osmosis A process in which liquid flows through a *semi-permeable membrane used to separate two solutions. The most concentrated solution is a pressure above that of the osmotic pressure. This causes the liquid to flow from the more concentrated solution to the less concentrated solution. A common use of reverse osmosis is water purification from seawater.

reversible A phenomenon or action that can take place in two directions. It is applied to mechanical, thermal, chemical, and biological processes. For example, the *Carnot cycle is a reversible thermodynamic cycle. A **reversible process** is a process in which the *process variables that define the equilibrium thermodynamics can be made to change in such a way that the process can be made to return to its original thermodynamic equilibrium. All natural processes are irreversible. A **reversible reaction** is a chemical or physical phenomenon that can take place in two directions. Strictly, all chemical reactions are reversible. For many chemical reactions, the equilibrium conversion is effectively complete such as the combustion of fuel and therefore considered to be irreversible. For example, the reversible reaction:

$$aA + bB \underset{k_2}{\overset{k_1}{\longleftrightarrow}} cC + dD$$

where k_1 and k_2 are the forwards and reverse reaction constants. The rate equation for component A assuming a constant volume reactor is therefore:

$$\frac{dC_A}{dt} = -k_1 C_A^a C_B^b + k_2 C_C^c C_D^d$$

At equilibrium:

$$\frac{dC_A}{dt} = 0$$

such that the equilibrium constant, K, for the reaction is:

$$K = \frac{k_1}{k_2} = \frac{C_C^c C_D^d}{C_A^a C_B^b}$$

Reynolds, Osborne (1842–1912) An English engineer born in Belfast. A graduate of Cambridge, he was appointed as the first professor of engineering at the University of Manchester in 1868, where he remained for 37 years. He studied a wide range of engineering problems and his most notable work was in the field of hydrodynamics. He studied cavitation in hydraulic machines, designed the first multistage centrifugal pump, and patented the use of guide vanes in centrifugal pumps, which are still used today. The *Reynolds number is named after him. He was elected a member of the Royal Society in 1877.

Reynolds number A dimensionless number, Re, expressing the ratio of inertial to viscous forces in a flowing fluid, and can be used to determine the flow regime. For a fluid in a pipe of circular cross section:

$$\text{Re} = \frac{\rho v d}{\mu}$$

where ρ is the density, v is the mean velocity, d is the diameter of a pipe, and μ is the viscosity. Where the value for circular pipes falls below 2,000, the flow is *laminar flow or streamline. For Reynolds numbers above 4,000, the flow is turbulent.

Various forms of Reynolds number are used, such as the flow through an annulus in which case a *mean hydraulic diameter is used, or the flow of fluid in the vicinity of an impeller of diameter, D, and rotational speed, N, given as:

$$\text{Re} = \frac{\rho N D^2}{\mu}$$

It is named after British engineer Osborne *Reynolds (1842–1912).

rheogram A graphical plot of rheological parameters for fluids such as shear stress with shear rate. They are used in the study of the behaviour of fluids.

rheology The study of the deformation and flow of fluids and includes the fundamental parameters of *elasticity, *plasticity, and *viscosity. The state of a fluid is characterized by three principal types of behaviour depending on the nature of flow:

1 Time-independent fluids. Examples include *Newtonian fluids in which the ratio of shear stress to shear rate is a constant, such as water, oils, and honey; *pseudoplastic fluids in which the apparent viscosity decreases with increasing shear rate, such as

polymer melts, paper pulp, wallpaper paste, printing inks, tomato purée, mustard, and rubber solutions; and dilatants in which the apparent viscosity increases with increasing shear rate. These are more rare and include titanium dioxide suspensions, cornflour/sugar suspensions, cement aggregates, starch solutions, and certain honeys.

2 Time-dependent fluids. For these fluids, the relation between shear stress and shear rate depends on the time and flow history of the fluid. They are classified as being either thixotropic or antithixotropic (or rheopectic). For thixotropic fluids, the shear stress decreases with time for a given shear stress. This is due to a structural breakdown of the fluid. If a cyclic experiment is carried out, a hysteresis loop is formed. Examples include greases, printing inks, jelly, paints, and drilling muds. In antithixotropic fluids the shear stress increases with time for a given shear stress. Examples include clay suspensions and gypsum suspensions.

3 Viscoelastic fluids. Fluids that possess the properties of both viscosity and elasticity. Unlike purely viscous fluids where the flow is irreversible, viscoelastic fluids recover part of their deformation. The rheological measurement of these fluids is understandably very difficult. Examples include polymeric solutions, partially hydrolyzed polymer melts such as polyacrylamide, thick soups, crème fraîche, ice cream, and some melted products such as cheese.

rheometer An instrument used to obtain the rheological properties of fluids. It involves a small quantity of the fluid being entrapped between two geometric surfaces with one surface static and the other in motion at constant speed. This defines the shear rate. The torque to resist the motion defines the characteristic shear rate. A limitation with the rotational speed is the need to avoid secondary flow patterns that give false readings. *Parallel-disc, *cone and plate rheometer, and *Couette are all forms of rheometer.

rheometry The study of the flow of fluids and fluid deformation.

rheopexy A non-Newtonian property of certain fluids that shows a time-dependent change in viscosity setting more rapidly when they are stirred or shaken. **Rheopectic fluids** are relatively rare and include gypsum suspensions, Bentonite clay suspensions, certain lubricants, and printing inks. They are also known as *antithixotropic fluids. *Compare* THIXOTROPIC FLUIDS.

Richardson, John Francis (1920–2011) A British chemical engineer best known as a coauthor and editor for a series of six textbooks entitled *Chemical Engineering*, which were first published in 1954. After gaining a BSc and PhD in chemical engineering at Imperial College, London, 'Jack' joined the academic staff there before being appointed as head of the department of chemical engineering at the University of Swansea in 1960 where he remained until his retirement in 1987. He was president of the Institution of Chemical Engineers (1975–76). He was awarded an OBE in 1981 for services to industry based on his various government involvements.

rig An apparatus constructed specifically for obtaining or validating process plant data. It is often a small-scale version or a specific part of a process plant. Offshore structures or platforms for oil and gas recovery are loosely termed as rigs.

rigorous multicomponent distillation methods Computational procedures that enable the rigorous solution of equilibrium-stage multicomponent distillation problems to be effected rapidly to an accuracy depending only on equilibrium and enthalpy data. There are four types of restricting equations in the equilibrium stage model.

1 The equilibrium relation:

$$y_{i,n} = K_{i,n} x_{i,n}$$

2 Componential mass balance for stage n:

$$L_{i,n} + V_{i,n} - L_{i,n+1} - V_{i,n+1} - f_{i,n} = 0$$

where $f_{i,n}$ represents the feed or side stream.

3 Energy balance for stage n:

$$L_n h_n + V_n h_n - L_{n+1} h_{n+1} - V_{n-1} h_{n-1} - f_n h_{f,n} - q_n = 0$$

where q_n represents the heat input on stage n.

4 Inherent restrictions on mole fractions:

$$\sum x_{i,n} = 1.0$$
$$\sum y_{i,n} = 1.0$$

These equations are insufficient in number to define a unique operation of a column and fix the concentrations, temperature, pressure, and flow-rate variable in the column. For a column with a single feed, a total condenser, and a partial reboiler, the number of design variables is C+2N+9 for C components with N-1 trays.

riser 1. The vertical tube under a bubble cap plate used in a *distillation column for vapour to rise. **2.** Used in offshore oil and gas processing, it is a vertical pipe rising from the seabed to a production platform carrying oil, gas, or a mixture of the two.

rise time The time taken for the response of a signal in a controlled process to reach its first peak.

rising film evaporator A type of evaporator used to concentrate solutions. It consists of a vertical tube heat exchanger in which the liquid to be evaporated is fed to the bottom of the evaporator and which forms a thin film on the surface. The evaporator is operated under vacuum and the overall flow of the liquid is upward. It is also known as a **climbing film evaporator**.

risk The statistical probability or likelihood of an undesired event occurring within a defined period or in specified circumstances and the severity or impact it could have if it were to happen. It is therefore a measure of the potential for causing human injury, loss of life, economic loss, or environmental impact. The likelihood may be expressed either as a frequency of specified events occurring in a given time, or as a probability of a specified event occurring. Acute risk is where the consequence occurs quickly, such as the effect of an electrical storm. Chronic risk is where the consequence builds up slowly over time although a single dosage may be small.

risk analysis The quantitative analysis of the risk associated with the likelihood of undesirable events occurring. The analysis is based on engineering evaluations and mathematical techniques that combine the statistical probability of occurrence with the consequences.

risk assessment The evaluation of the likelihood of undesired events and the likelihood of harm or damage being caused, together with the judgements made concerning the significance of the results. A risk assessment considers the elimination, substitution, reduction, and adequate control of risks, through the identification of hazards, impact, and likelihood of events occurring. The decisions are taken based on relative ranking of risk reduction strategies or through the use of targets levels that are required to be met.

risk management A management procedure used to identify, assess, control, and reduce risks. This is through the reduction of the likelihood of an event occurring or the effect of loss due to an event occurring. The risk may be associated with human injury, loss of life, damage to process plant, loss of production, or financial loss. It is proactive and not reactive in its approach and achieved through applying engineering principles.

rivet A pin that fixes together two sheets of material by insertion through a hole in the two sheets. The pin is usually metal and the head is expanded by striking it flat.

roasting A process involving the burning of a finely ground sulphide-containing ore with air to produce sulphur dioxide, and used prior to smelting. The roasting process is also used to remove any chemically bound moisture and the ore is converted to an oxide. A fluidized bed is used for the roasting of pyrite ore.

rock An aggregate of minerals that makes up part of the Earth's surface. Sand, mud, and clay are unconsolidated forms, while granite, limestone, and coal are consolidated forms. It may contain minerals of economic value that require extraction and purification.

roentgen (Symbol r) A unit of measurement for ionizing electromagnetic radiation from X-rays and gamma rays, defined as the charge in air of 2.58×10^{-4} coulombs on all ions of one sign when all the electrons of both signs liberated are completely stopped by a mass of one kilogram of air. It is named after German physicist Wilhelm Konrad Röentgen (1845–1923) who discovered X-rays. A lethal dose to humans corresponds to an exposure of 200 roentgens over a five-hour period. The unit is no longer used.

Romankov number A dimensionless number, Ro. Used in drying, it is the ratio of the dry bulb temperature to the absolute temperature of the product:

$$Ro = \frac{T_{db}}{T_p}$$

root The solution to an algebraic equation involving real or complex numbers.

root locus A method of analysis used in process control to determine the stability of a closed-loop control system.

root-mean-square (RMS) A statistical measure defined as the square root of the mean of the squares of a sample:

$$RMS = \frac{\sqrt{x_1^2 + x_2^2 + x_3^2 + \ldots x_N^2}}{N}$$

Roots blower A type of low-gauge pressure compressor for the transport of large volumes of gas. It consists of two rotating lobe-type rotors that have a small clearance between the rotors and the casing. It is named after its American inventor Francis Roots (1824–89) and his brother Philander (1813–79), who founded the Roots Blower Company.

Rotameter A registered name for a type of *variable area flow meter used to measure the rate of flow of fluids. It consists of a tapered tube and contains a float. The elevation of the float in the tube gives a measure of the rate of flow and is read from a calibrated scale on the tube.

rotary drum filter A type of continuous filter used to remove bulk liquid from a liquid suspension such as a precipitate. The drum operates at reduced pressure and sits semi-submerged in the liquid. As the drum slowly rotates, a filter cake is formed on the drum, lifted from the liquid, and removed using a blade.

rotary kiln A type of furnace used for *roasting various sulphide ores prior to metal extraction. The kiln consists of a slowly rotating cylinder inclined slightly to the horizontal with a refractory lining. Material is fed into the rotating kiln allowing it to cascade down. Hot air from a burner which provides the heat is also fed into the kiln either cocurrently but more usually countercurrently. At the end, the product is collected and any entrained product with the hot gases separated by a cyclone. They are used in the manufacture of cement and pelletizing of iron ore, titanium dioxide, and alumina.

rotary pump A type of positive displacement pump consisting of gears with meshing teeth (see Fig. 52). The rate of flow is determined by the number and size of the spaces between the teeth and their rotational speed. A close tolerance is required between the teeth to prevent leakage, although unlike the reciprocating type pump, check valves are not required. They require liquids with good lubricating properties and are unsuitable for fluids with abrasive properties.

rotary vacuum filter A type of continuous filter used to remove bulk liquid from a suspension of particles such as a precipitate to produce a wet filter cake (see Fig. 53). It consists of a perforated circular plate that slowly rotates. Material is fed onto a point on the plate that

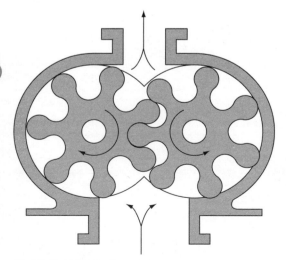

Fig. 52 Gear-type rotary pump

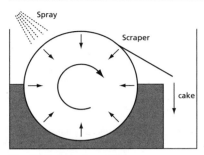

Fig. 53 Rotary drum filter

rotates over a plenum operating at a reduced pressure drawing the liquid through the plate. The remaining filter cake is continuously scraped from the surface.

rotating disc atomizer A device used to atomize a liquid in a gas. It consists of a disc that rotates at a very high speed. Liquid is fed onto the disc where the high-speed shearing effect reduces the liquid to a fine spray. It is typically used in *spray dryers to reduce the liquid containing a heat-labile solid to a fine spray for rapid low-temperature evaporation such as in spray-drying milk.

rotor Rotating part of a pump, turbine, or generator, etc.

Royal Australian Chemical Institute (RACI) The qualifying body in Australia for professional chemists and a learned society promoting the science and practice of chemistry in all its branches.

(⊕) SEE WEB LINKS
• Official website of the Royal Australian Chemical Institute.

rules of thumb A set of suggested values used by engineers that are considered to be reasonable and based on experience. While based on the application of fundamentals, they do not replace them but instead assist with solving problems. They are typically used to judge the reasonableness of answers, quickly assess assumptions, guide understanding of complex systems and situations, and provide rapid order of magnitude estimates.

runaway reaction A chemical reaction in which control has been lost and it continues to accelerate to a point that either it exhausts its reactants or the vessel containing the reaction over-pressurizes and loses containment. This is often in the form of an explosion with catastrophic consequences. **Thermal runaway** exothermic reactions may occur due to insufficient cooling or loss of cooling, a loss of or excessive mixing, excessive reactant or catalyst, the loss of an inhibitor, or external fire. Where a runaway is likely, pressure relief systems are used. The *Seveso and Bhopal accidents are examples of runaway reactions with disastrous consequences.

rupture disc *See* BURSTING DISC.

Rushton, John Henry (1905–85) An American chemical engineer noted for his work on mixing. He obtained his first degree in chemical engineering from the University of

Pennsylvania in 1926 and worked as an engineer before becoming an academic. He was professor and head of chemical engineering at the University of Virginia. In 1946 he moved to the Illinois Institute of Technology as professor and director of the Department of Chemical Engineering. He joined Purdue University in 1955 and remained there until 1971. He was considered a world expert on design and application of large-scale mixing, mass transfer equipment, and on process design, and was president of *AIChE in 1957.

Rushton turbine A type of impeller used for gas dispersion such as in biochemical reactors and consists of a flat disc with six vertical flat blades mounted on the circumference (see Fig. 54). It therefore provides radial-flow mixing. It is named after American chemical engineer John Henry *Rushton (1905–85), who was noted for his work on mixing. There have been many subsequent modifications and improvements to the basic design.

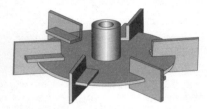

Fig. 54 Rushton impeller

rusting An electrochemical corrosion process involving the oxidation of iron to form a hydrated iron oxide that occurs in the presence of both water and oxygen. It is very damaging to process equipment and support structures. Rust has a characteristic red colour and is permeable to both water and air. Rust prevention involves coatings that can exclude oxygen from the iron surface. Certain paints can provide the necessary seal. *Galvanization is an effective process involving coating the iron with zinc. The use of stainless steels involves an *alloy with chromium that forms a passivation layer of chromium oxide on the surface. *Compare* CORROSION.

Rutherford, Lord Ernest (1871–1937) A New Zealand-born British scientist who became Baron Rutherford of Nelson in 1931. He studied in New Zealand, gaining a double First in physics and mathematics, before going to Cambridge to study under Sir J. J. Thomson in the Cavendish Laboratory. He became professor of physics at the age 27 at McGill University, Canada and studied radioactivity. Working with British chemist Frederick Soddy (1877–1956), he discovered alpha particles, beta particles, and gamma radiation. He moved to Manchester University in 1907 and, working with H. Geiger and E. Marsden, he developed an experiment that led to the discovery of the atomic nucleus in 1911. In 1919 he succeeded in transforming nitrogen into an isotope of oxygen by bombarding it with alpha particles. He was awarded the Nobel Prize for Chemistry in 1908 for his work on radioactivity.

 SEE WEB LINKS
• Official website of McGill University and its museum to Rutherford.

rutherford A unit of radioactivity corresponding to 10^6 disintegrations per second. The curie is another unit of radioactivity corresponding to 3.7×10^{10} disintegrations where 37 rutherford = 1 mCi.

RV *See* RELIEF VALVE.

Sabatier, Paul FRS (1854–1941) A French chemist noted for his work on catalysts used in organic chemical reactions. He worked on the industrial applications of hydrogenation and developed the use of nickel to facilitate the addition of hydrogen to carbon compounds. Working at the University of Toulouse, he rose to become dean of the faculty of science in 1905. He was awarded the Nobel Prize for Chemistry in 1912 together with French chemist François Auguste Victor Grignard (1871–1935). The Paul Sabatier University in Toulouse is named in his honour and the *Sabatier process is named after him.

Sabatier process A catalytic process used for the production of methane involving the reaction of carbon dioxide and hydrogen:

$$CO_2 + 4H_2 \rightarrow CH_4 + 2H_2O$$

The process uses high temperature and pressure in the presence of a nickel catalyst. More efficient catalysts have subsequently been developed such as ruthenium on aluminium oxide. The process was originally developed by French chemist Paul Sabatier (1854–1941).

Sabatier–Senderens process A process for the hydrogenation of an organic compound such as an unsaturated fat and used to produce margarine. It uses hydrogen gas and a nickel catalyst at around 150ºC. The process is named after 1912 Nobel Prize winner Paul *Sabatier (1854–1941) and Jean-Baptiste Senderens (1856–1937).

Sachse process A process used to convert methane to form acetylene (ethyne):

$$CH_4 \rightarrow C_2H_2 + 3H_2$$

The high temperature reaction takes place at around 1,500°C. The high temperature is obtained by burning part of the methane in air.

sacrificial protection *See* CATHODIC PROTECTION.

safety *See* PROCESS SAFETY.

safety audit A critical examination of all aspects of safety for a process or a defined part of an installation. The audit covers process design, management, safety culture, training, operating procedures, emergency plans, personnel protection, and accident reporting, and is carried out by qualified personnel. The report presents necessary actions for improvement to safety practices.

safety factor The ratio of the breaking stress of a material to the calculated maximum stress. Pressure vessels are designed with a safety factor to provide an allowance that the maximum stress is never exceeded.

safety integrity level (SIL) A safety management technique used to measure the performance of a process or an activity in terms of risk reduction which is numerically rated with the values of 1 to 4. SIL 1 presents a high probability of failure while SIL 4 presents a low probability of failure. The SIL is determined using methodologies such as *layers of protection analysis (LOPA) and risk matrices. SILs are useful for allowing risk of failure to be appreciated without having to understand fully all of the technical details.

safety management system (SMS) A systematic and comprehensive approach to identifying hazards and controlling risks while maintaining assurance that risk controls are effective. It provides goal-setting, planning, and measuring performance. An effective SMS defines how a business manages risk, identifies workplace risk and implements suitable controls, effective communications, and a process to identify and correct non-conformities. It acts as a framework to allow an organization to meet its legal obligations but is itself not a legal requirement.

safety relief valve (SRV) A type of automatic valve that opens to relieve pressure from both gas and liquid applications. It is used on process equipment and pipelines.

safety report A document that presents the justification for the safety of an installation. A **safety evaluation** is a report used in the assessment of process equipment and pipelines.

safety valve A generic name for a device used for the automatic release of a substance from a vessel or system when the pressure or temperature exceeds preset limits. Safety valves were first used on steam boilers during the Industrial Revolution in which boilers were prone to accidental explosion. Examples of safety valves include **pressure relief valves (PRV)**, **pressure safety valves (PSV)**, and **relief valves (RV)**. PSVs differ from PRVs by having a manual lever to open in the event of an emergency. Vacuum safety valves are used to prevent a vessel from collapsing when emptying or when a vacuum may accidentally form due to a trapped vapour that then condenses.

salt A compound form when one or more hydrogen atoms of an acid are replaced by a metal atom or by an electropositive ion such as ammonium. Most salts are crystalline ionic compounds. Soluble salts dissolve into ions in solution.

sample A small part or portion of a larger quantity of a product used to evaluate product quality. The sample is intended to be a representative part of the larger quantity. Statistical methods can be used to determine statistical variability of the sample from which judgements can be made such as for quality control purposes. Data of *process variables such as temperature, pressure, flow, and level are used in the computer control of processes. A **sample interval** is the time between the gathering of information or data. In some cases, the time interval is very small such that the flow of information is virtually instantaneous whereas in others, the time interval is much greater.

sand filtration A process used in *water treatment involving the removal of suspended particles of organic matter by *filtration through beds of sand to produce water of the required *quality. Sand filtration is used prior to chlorination to produce a safe supply of drinking water. Sand is the general name for particles of rock composed largely of quartz that have diameters in the range of 0.06 to 2.00 mm.

saponification A process used for the manufacture of soap. The process involves the hydrolysis of triglycerides using a strong base such as caustic soda to form a sodium salt of a carboxylate and glycerol. The triglycerides are esters of fatty acids obtained from vegetable and animal fats. The process of producing common soap involves blowing steam into large pans containing a fat such as glyceryl tristearate and a solution of sodium hydroxide to form sodium stearate and glycerol:

$$C_3H_5(O \cdot OC \cdot C_{17}H_{35})_3 + 3NaOH = C_3H_5(OH)_3 + 3C_{17}H_{35} \cdot CO \cdot ONa$$

The soap is separated as a curd by adding strong brine. The lower aqueous layer is run off and processed to give glycerol. The soap is given further washes, left to reach the right condition for mixing with perfume and colouring before being run into frames for cooling. If the fat is hydrolyzed using potassium hydroxide, the resulting solution sets on cooling and yields a jelly known as soft soap containing water and glycerol.

SARA An abbreviation for **s**aturates, **a**romatics, **r**esins, and **a**sphaltenes, which are the four solubility classes of hydrocarbon fractions of *crude oil. The saturates are generally iso- and cyclo-paraffins whereas aromatics, resins, and asphaltenes form a continuum of molecules with increasing molecular weight, aromaticity, and heteroatom contents. Asphaltenes can also contain metals such as nickel and vanadium. A **SARA analysis** is a method used for the characterization of heavy oils based on fractionation.

saturated 1. A molecule that has only single bonds (i.e. no double or triple bonds) such as *octane (C_8H_{18}) which is a straight-chain hydrocarbon saturated with hydrogen. **2.** A solution containing the maximum possible amount of a solute at a given temperature. The solution is in equilibrium with the undissolved solute. A solution that contains more than the equilibrium amount is known as *supersaturation. A solution that contain less than the equilibrium amount is *unsaturated.

saturated solution A solution that is in equilibrium with an undissolved solute at a particular temperature. The solubility of a substance is defined with reference to a saturated solution of it at a stated temperature in a standard amount of solvent. For example, the solubility of a substance in water at a given temperature is the mass of the substance in grams that will saturate 100 grams of water at that temperature.

saturated vapour pressure The pressure exerted by a vapour that exists in equilibrium with its liquid.

saturation temperature The boiling point of a liquid. The liquid is deemed to be saturated with the energy required to bring about the phase change. The boiling point is the temperature at which the vapour pressure of the liquid is equal to the surrounding pressure of the liquid. The **saturation pressure** at the saturation temperature corresponds to the pressure at which the liquid boils.

Sauter mean drop diameter A method used to characterize the average drop size in a population of droplets in immiscible liquid–liquid dispersions. It is related to the volume fraction of the dispersed phase, Φ and the interfacial area, a. It is also known as the volume-to-area average drop diameter:

$$d_{SM} = \frac{6\Phi}{a} = \frac{\sum_{i=1}^{n} n_i d_i^3}{\sum_{i=1}^{n} n_i d_i^2}$$

where n is the number of droplets and d is the diameter.

Saybolt viscometer A type of instrument used to determine the viscosity of petroleum oils. It is based on the time in seconds for a given volume of oil to pass through an aperture at a controlled temperature and collect in a container with a volume of 60 millilitres. The Saybolt universal second is the unit used as a measure of the *kinematic viscosity.

SCADA An abbreviation for **s**upervisory **c**ontrol **a**nd **d**ata **a**cquisition, it is a computer software package used to control and monitor process plant or equipment. It displays process information, logs and stores data, and shows all the associated process control alarms. It also allows operator control of a process from a PC.

scalar quantity A mathematical representation of a quantity with magnitude and no direction. Examples include temperature, pressure, mass, speed, and concentration.

scaled distance A technique used to calculate the damage caused by an explosion at various distances from the epicentre. Also used in blasting operations, it is the actual distance from the blast to the point of concern divided by the square root of the total charge weight of explosive.

scale-up The translation of a process design from the laboratory or experimental scale to the larger pilot plant scale, or commercial or industrial scale. Scale-up is an important part of commercializing a process in which it is accepted that theoretical design cannot be used alone to achieve this. *Dimensionless numbers or groups are a useful way of scaling up a process since certain heat, mass, and momentum transfer phenomena are independent of scale. For example, in mixing processes, it may be necessary to ensure that the power-to-volume ratio remains the same between lab-scale and full-scale to ensure homogeneous mixing characteristics. The scaled-up equipment is also assumed to have geometric similarity. The testing of a small-scale process is therefore quick, cost-effective, and reliable such that the experimental information gained can be reliably used on a larger scale.

scf An abbreviation for **s**tandard **c**ubic **f**eet, it is an Imperial unit of volume used in the measurement of gas, and pronounced as 'scuf'.

scheduled maintenance *See* MAINTENANCE.

schedule number A classification for pipes in terms of their wall thickness. There are ten schedule numbers in common usage: 10, 20, 30, 40, 60, 80, 100, 120, 140, and 160.

schematic flow diagram *See* BLOCK FLOW DIAGRAM.

Schmidt number A dimensionless number, Sc, characterizing mass transfer by convection that relates viscosity and diffusion of a gas through itself. It is the ratio of the kinematic viscosity to the molecular diffusivity:

$$Sc = \frac{\mu}{\rho D}$$

where μ is viscosity, ρ is density, and D is diffusivity. It is named after German scientist E. W. H. Schmidt (1892–1975).

scraped surface heat exchanger A type of double pipe heat exchanger used for heating or cooling highly viscous, fouling, and crystallizing materials such as margarine and ice cream. It consists of two concentric pipes with agitated scrapers in the inner product tube with the heating or cooling medium passing through the outer surrounding pipe. The scrapers consist of knives positioned in such a way that a screw effect is achieved. This ensures that the product near the wall is fully mixed with the bulk material. The product to be heated or cooled is pumped using a *positive displacement pump. Mounted either vertically or horizontally, it is relatively expensive and is only fully justified for highly viscous materials with viscosities in excess of 10 Pa s. The complex flow patterns, including *backmixing which can reduce the heat transfer rate, mean that the design is based on empirical or experimental *overall heat transfer coefficients.

screening A physical separation process used to separate solid particles of differing sizes. Also known as **sieving**, the separation uses a mesh with a fixed aperture allows only particle sizes below the aperture size to pass through. It is used for separating ores, grains, and many other solid–solid mixtures requiring grading or separating.

screw pump A type of *positive displacement pump used to transport fluids and solid materials. It consists of a helical screw that revolves within a cylinder transporting the materials along the axis of the screw. The Archimedes' screw pump is the simplest form of screw pump. More complex designs consist of multi-axis screws and the use of similarly shaped pumping casings (**stator**) trapping portions of the materials in cavities and carried along the pump. These are used for high-flow applications and noted for their suitability for handling viscous fluids, slurries, and shear-sensitive fluids. The *Mono pump is a well-known type of screw pump. Screw pumps are widely used in the *water treatment, food, paper, chemical, petrochemical, and mineral processing industries.

scrubber A tall cylindrical column used for gas adsorption. The separation process involves the removal of one component from a mixed gas stream by means of a selective solvent. The solvent with the adsorbed gas is sent to another column known as a *stripper, where the gas is separated again and the solvent recirculated to the scrubber for reuse. *See* ABSORPTION TOWER.

scrubbing The process of removing a component such as an impurity from a mixed gas stream or extract phase, by means of contacting it with a selective liquid solvent. The absorption process takes place within an *absorption tower or *scrubber. Scrubbing is used to remove sulphur dioxide from the flue gases from power stations, hydrocarbons from air, NOx from combustion gases, and hydrogen sulphide from natural gas. It is also known as **gas scrubbing**. *Compare* STRIPPING.

SD&P An abbreviation for **s**imultaneous **d**rilling & **p**roduction, it is a technique used to drill a new oil or gas well on a platform while continuing to produce oil or gas.

Seaborg, Glenn Theodore (1912–99) An American chemist noted as one of the discoverers of plutonium (plutonium-238 and plutonium-239). Gaining his doctorate in 1937 from the University of California, he was appointed professor of chemistry in 1945. He was responsible for nuclear chemical research at the Lawrence Radiation Laboratory and headed the Manhattan Project group from 1942 to 1946 that devised the chemical extraction processes used in the production of plutonium. He codiscovered nine other transuranium elements, including the element seaborgium, atomic number 106, which is named after him. He was awarded the Nobel Prize in Chemistry in 1951.

second 1. (Symbol s) A base SI unit of time taken as 9 192 631 770 cycles of the radiation from the transition between two hyperfine transition levels of the ground state of caesium-133. It is equal to $1/60^{th}$ of a minute and $1/3600^{th}$ of an hour. **2.** (Symbol ") An angle equal to $1/60^{th}$ of a minute or $1/3600^{th}$ of a degree.

second-generation biofuel A type of *biofuel that is produced from cellulose, hemi-cellulose, or lignin from plants to form cellulosic ethanol and Fischer–Tropsch fuels. They can be blended with petroleum-based fuels and combusted in modified internal combustion engines. Second-generation biofuels have been developed due to the limitations of first-generation biofuels, which have an impact on food production and biodiversity. Examples of the raw materials used include biomass from the non-food parts of crops, such as stems, leaves, and husks, as well as certain types of grass, fibre, and industry waste such as wood chips, skins, and pulp from fruit pressing.

second law of thermodynamics A law that states that it is impossible to construct a device that operates in a cycle and produces no effect other than the transfer of heat from a cooler body to a hotter body. This law sets the limit on the amount of heat energy that can be converted to useful work energy.

second-order control system A control system in which the dynamics are determined by second-order differential equations. They mainly arise in process plants as the result of the combination of two first-order systems in series, the consequence of the addition of complex control systems, or the system may itself be intrinsically second order . This is generally uncommon where only the *continuity equations are needed for process models such as mass, component, and energy balances which are inherently first order. If continuity equations for momentum are needed, the resulting balances will be inherently second order. Such balances are used for the modelling of mechanical systems such as control valves or the behaviour of the fluid dynamics of a process.

second-order differential equation A differential equation involving only first and second derivatives.

second-order reaction *See* ORDER OF REACTION.

sedimentation A separation process in which particles in suspension settle under the influence of gravity into a clear liquid and a slurry that has a higher solids content. It is commonly used in *water treatment. A thickener is a type of *continuously operated sedimentation device in which concentrated slurry is produced and the clear liquid overflows. A clarifier is another type of sedimentation device that is used to produce a more concentrated solid–liquid mixture. *See* THICKENING. The use of flocculating agents can assist with the settling process by causing the particles to form flocs or aggregates that have a larger size and enabling them to settle more quickly. For mixtures in which the particles do not influence one another, the rate of settling can be determined using *Stokes's law, which is dependent on the size of particle and relative density to the particular to the fluid. Where the difference in density is small, centrifugal force will hasten the separation process.

seed crystals Used in the process of crystallization, seed crystals are small crystals of solute used to accelerate the precipitation of crystals from supersaturated solutions by providing nuclei upon which crystal growth can continue. Nucleation is the phase transition from solute in a solution to a crystal lattice. By using small crystals, the seeding process therefore reduces the amount of time required for nucleation to occur.

segregation coefficient The ratio of the concentration of an impurity in a liquid to the concentration of an impurity in a solid. It is used as a measure of the ratio of the impurities in a liquid–solid interface in a *zone refining process used for producing high pure metals.

selective medium A *growth medium used to grow or culture selected microorganisms such as genetically modified or recombinant yeast or bacteria. The medium either contains all the necessary nutrients to sustain growth but is missing at least one component for which the necessary genes are present in the modified microorganisms or it contains genes that enable the microorganisms to grow such as provide the resistance to an antibiotic, such as ampicillin, which is supplied to the medium.

selectivity 1. The amount of a reactant converted to a product expressed as a ratio or percentage of the reactant converted to all products in a chemical reaction. **2.** The effectiveness of a solvent for separating a solution by comparing the ratio of one component to another in the solvent at equilibrium. The value must exceed unity for extraction to be possible. Selectivity is analogous to the *relative volatility used in distillation. **3.** The equilibrium constant of an ion exchange reaction. **4.** The ability of a *semi-permeable membrane to separate a component from a mixture of components.

Selexol process A process used to remove *acid gases from hydrocarbon gas streams by absorption in polyethylene glycol dimethyl ether (DMPEG). The process is used to remove or reduce the level of hydrogen sulphide, carbon dioxide and carbonyl sulphide, and other organic sulphur compounds such as mercaptans from gases such as natural gas. It involves gas absorption under pressure with the solvent in which the gas is contacted counter currently in an absorption column with the solvent. The solvent is regenerated by flashing or stripping.

self-extinguishing A substance that is incapable of sustained combustion in air after removal of external heat or flame.

self-ignition An ignition resulting from self- or spontaneous heating. *Compare* SPONTANEOUS IGNITION.

self-regulation The property of a body, process, or machine without closed-loop control to reach a new steady state after a sustained *disturbance.

semi-batch process A process that is operated on a batch basis, such as reactions taking place in a stirred tank, but where the materials can be partially added over time. This has the advantage of allowing good control of temperature and composition, while permitting efficient mixing of the materials. *See* CONTINUOUS PROCESS.

semi-continuous process A process whose overall operation is continuous but which features unit operations that are operated on a batch basis. For example, the conversion of sugar cane to alcohol as a biofuel is a continuous process in which sugar cane as the raw material is processed continuously to extract the fermentable sugars. A number of batch fermenters are used in varying states of operation to convert the sugar to alcohol whose collective output is then fed to a continuously operating distillation unit.

semi-infinite The geometry of an object in which one or two dimensions are taken to be infinite while one is finite. It is a useful way of performing calculations such as the flow of heat in a particular direction without the need to consider the influence of flow in the other directions. For example, a semi-infinite slab has a finite thickness but infinite breadth and length, while a semi-infinite cylinder has an infinite length and finite radius.

semi-permeable membrane A thin layer of material that permits certain molecules to pass through while being impervious to others. The process of *osmosis is the result of *diffusion across a semi-permeable membrane. The permeability of the membrane can permit small molecules such as oxygen, carbon dioxide, and glucose to pass, but not allow larger molecules such as proteins and sucrose through. The *selectivity is achieved by pore size within the membrane or by the presence of charged ions within the membrane.

sensible heat The thermal energy or heat which when added to a substance increases its temperature without a change in state. *Compare* LATENT HEAT.

sensor An instrument or device used to detect the condition of a *process variable. Commonly used sensors include temperature, pressure, and level transmitters, and toxic gas detectors.

SEPA An abbreviation for **S**cottish **E**nvironment **P**rotection **A**gency, it is a non-governmental public body responsible for the protection of the environment in Scotland. Its primary role is to protect and improve the environment by regulating activities that can cause harmful pollution.

 SEE WEB LINKS
• Official website of Scottish Environment Protection Agency.

separation 1. The division and parting of materials into their constituent parts. Separation is used for purification by removing contaminants or for enrichment. In some cases, separation may involve removing single components or groups of components from a mixture. Examples of physical and chemical **separation processes** include chromatography, distillation, evaporation, drying, electrolysis, desorption, and gravity. The petrochemical refining of crude oil is an industrial process used to separate the complex mixture of hydrocarbons into valuable components largely through distillation. The choice of separation is based on the chemical and physical properties of the materials such as chemical affinity with other components, size, shape, and density. **2.** The phenomenon of fluid streamlines changing direction due to changes in boundary shape as a result of fluid inertia or velocity distribution near the boundary surface.

separator A device used to separate immiscible liquids based on differences in density or for separating solids from liquids or gases. For example, in the offshore industry, oil, water, and sand, each with a different density, are separated as layers in gravitational separators. A weir is typically used to allow the top layer to overflow and separate from the layer or layers below. Centrifuge separators increase the force of the particles, increasing the rate of separation, and are used to separate cream from milk, oil, and water emulsions that are difficult to separate by gravity, and very small particles from liquids where gravitational separation would otherwise be too slow.

sequestration 1. The process of forming a complex of an ion in solution to prevent the chemical effect by removing it from solution. Some polymers can be used to sequester metal ions such as copper in water. **2.** The process of removing greenhouse gases and in particular carbon dioxide from the atmosphere. Trees and plants use photosynthesis to sequester the gases from the atmosphere to help reduce the greenhouse effect and control global warming. *See* CARBON SEQUESTRATION.

set-on tee A tee-piece in a pipe that is formed by welding one pipe over a hole made in another. It is formed by a joining weld at the junction of the tee-piece.

set point The desired value of a *process variable in a controlled system that is to be attained.

settling tank A vessel with a large capacity used to separate particles from a liquid under the influence of gravity. The design is based on a steady flow entering and leaving in which the liquid velocity is uniform at all points in the tank, allowing the particles to descend freely to the bottom. The capacity is designed such that any particle touching the base of the tank will be retained by the tank, and conversely, any other particle still in suspension will be swept out with the effluent.

settling time The time required for the output of a controlled process to a stimulus to enter and remain within a specified narrow band around the final steady-state value.

settling zone The largest section within a *sedimentation or *settling tank used to separate solid particles from a liquid. It is designed to provide a calm area in which the suspended particles are able to settle. Below is the *sludge zone in which the particles accumulate.

Seveso incident A major chemical disaster that occurred at a small chemical plant on 10 July 1976 near Milan, Italy. It involved a chemical runaway reaction in which the chemical 2,3,7,8-tetrachlorodibenzo-p-dioxin (TCDD) was released into the environment as a *side reaction caused by a kettle heater being left on overnight. The release caused the death of many animals and the evacuation of local inhabitants. Following many studies, it has led to the standardization of industrial safety regulations aimed at improving the safety of sites containing large quantities of potentially dangerous substances. The Seveso II Directive is a European Union law formed as a result of the disaster and subsequent investigation. This is known as the *COMAH regulations in the UK.

SFE *See* SUPERCRITICAL FLUID EXTRACTION.

shake flask A small glass vessel which is usually conical. It is used in preparing an inoculum of living microorganisms for a biochemical process such as fermentation. The microorganisms are first prepared by aseptically transferring colonies of the microorganism from agar plates upon which they have been stored at a chilled temperature into shake flasks containing an aqueous medium together with the necessary nutrients. The flasks are *monoseptic in that only the living microorganism is present and sealed with sterile cotton wool. They are then slowly agitated in an orbital motion and maintained at a controlled temperature desirable for growth, such as 30ºC. After a period of several hours to several days, the living cells have grown to a population sufficient to be transferred or inoculated into a bioreactor.

shale gas Natural gas trapped within sedimentary rock.

shape factor A description of the geometric shape of a particle that may not necessarily be spherical. It is defined as the ratio of one characteristic length to another and is used in determining the behaviour of particles in a fluid, such as particle settling.

shear force An applied force to a material that acts in a direction that is parallel to a plane rather than perpendicular. A material such as a solid or fluid is deformed by the application of a shear force over a surface, known as the *shear stress. The **shear strain** is the extent of the deformation defined as the ratio of the deformed distance with length. The **shear modulus** is the ratio of the shear stress to the shear strain.

shear rate (Symbol $\dot{\gamma}$) The deformation of a fluid under the influence of an applied *shear force presented as the change in velocity of the fluid perpendicular to flow:

$$\dot{\gamma} = \frac{dv}{dz}$$

It is also known as the **velocity gradient**. The SI unit is s^{-1}.

shear stress (Symbol τ) The *shear force applied to a fluid that is applied over a surface. Where the shear stress is proportional to the *shear rate, the fluid exhibits *Newtonian behaviour and the *viscosity is constant. The SI units are $N\,m^{-2}$.

shell and tube heat exchanger A device used to transfer heat from one medium to another. It consists of a shell that contains tubes. One medium is contained within the shell and the other within the tubes, and heat is transferred from one to the other across the tubes. There are many designs commonly used and the simplest is a single-pass type exchanger in which a cold liquid to be heated flows through the tubes from one side of the exchanger to the other. Steam is used as the heating medium and enters as vapour and leaves as condensate from the bottom. A *kettle reboiler is a type of shell and tube heat exchanger in which steam is admitted through the tubes. The choice of hot or cold fluid in the tubes or shell depends on the application and nature of the fluids, such as their susceptibility to fouling.

shell side The space between the outside of the tubes and the inside of the casing or shell of a *shell and tube heat exchanger. *Compare* TUBE SIDE.

Sherwood, Thomas Kilgore (1903–76) An American chemical engineer and founding member of the National Academy of Engineering. After gaining his PhD at MIT, he was briefly assistant professor at Worcester Polytechnic Institute before returning to MIT in 1930 as assistant professor, eventually rising to professor and dean of engineering. On retirement from MIT in 1969 he became professor of chemical engineering at the University of California, Berkeley. He published the first text on mass transfer in 1937 entitled *Absorption and Extraction*, which was republished in 1974 as *Mass Transfer*. The Sherwood number is named after him.

Sherwood number A dimensionless number, Sh, that represents the relationship between mass diffusivity and molecular diffusivity:

$$Sh = \frac{kL}{D_{AB}}$$

where k is the mass transfer coefficient, L is the characteristic dimension, and D is the diffusivity of the solute A in the solvent B. It corresponds to the *Nusselt number used in heat transfer. It is named after Thomas *Sherwood (1903–76).

shock wave A pressure wave of very high pressure intensity and high temperature that is formed when a fluid flows supersonically, or in which a projectile moves supersonically through a stationary fluid. It can be formed by a violent event such as a bomb blast or an explosion. A **shock-wave compression** is the non-isentropic adiabatic compression in a wave that is travelling above the speed of sound.

shutdown 1. The status of a process that is not currently in operation due to scheduled or unscheduled maintenance, cleaning, or failure. **2.** A systematic sequence of actions that is needed to stop a process safely. *See* EMERGENCY SHUTDOWN.

siccative A material that is capable of absorbing moisture. It is used as a drying agent for certain pharmaceuticals and foods.

side reaction A chemical reaction that takes place at the same time as a main reaction and produces unwanted products and therefore reduces the yield of the desired product. For example, in the high-temperature cracking reaction of propane to produce propene (propylene) $C_3H_8 \rightarrow C_3H_6 + H_2$, some of the hydrogen can react with the propane to produce methane and ethane as a side reaction $C_3H_8 \rightarrow C_3H_6 + H_2$. The conditions of the reaction must therefore be controlled to reduce this unwanted reaction.

side stream The continuous removal of a liquid or a vapour from a process such as a distillation column that is not the main process flow. For example, drawing off vapour or liquid part-way up a distillation column can have economic advantages in terms of the physical size of column and the amount of boil-up energy required.

sieve plate column A type of *distillation column which uses a stack of perforated plates to aid the distribution and intimate contact between vapour and liquid. The plates allow vapour to pass up and bubble through the liquid on the plates. The rate of flow of vapour is sufficient to prevent the liquid from draining through the sieve plates. Instead, the liquid pours over a weir and down a downcomer to the sieve plate below.

Sievert, Rolf Maximilian (1896–1966) A Swedish physicist who specialized in the study of biological effects of ionizing radiation. He was head of physics at Sweden's Radiumhemmet before heading the department of radiation physics at the Karolinska Institute. He studied and measured the effects of radiation dosage in the diagnosis and treatment of cancer. He also invented various instruments for measuring radiation doses. The derived SI unit for ionizing radiation dose equivalent, the *sievert, is named after him.

sievert (Symbol Sv) The SI derived unit for the dose equivalent of ionizing radiation. Unlike the *gray, which is the SI derived unit for the absorbed radiation, the sievert is a way of quantitatively measuring the biological effects of ionizing radiation. It measures the equivalent dose of radiation as having the same damaging effect as an equal dose of gamma rays. It is named after the Swedish physicist Rolf Maximilian *Sievert (1896–1966).

sight glass A small window located on the side of a process vessel such as a reactor or column to allow a visual observation of the contents within.

signal Transmitted information about a *process variable in a controlled system in the form of a voltage, current, pneumatic, mechanical, or digital signal. The error signal in a closed-loop system is the signal formed when subtracting a particular return signal from its corresponding input signal. The input signal is the signal applied to the system, whereas the output signal is the signal delivered by the system.

significant figures The number of digits used to express the accuracy of a figure. For example, 1.032 is taken to be accurate to four significant figures whereas 12.3 is taken to be accurate to only three significant figures as is 0.003 24.

SIL *See* SAFETY INTEGRITY LEVEL.

silo A tall, cylindrical structure used for storing particulate materials such as grain.

Simpson's rule A numerical integration method used to obtain the approximate value to a definite integral. It is applied to odd numbers of data and is given by:

$$A = \frac{S}{3}(F + L + 4E + 2O)$$

where S is the width of the intervals between data, F and L are the first and last ordinate, E is the sum of the even-numbered ordinates, and O is the sum of the remaining odd-numbered ordinates.

single phase The presence of a fluid in a system such as a vessel or pipeline as being either entirely a gas or entirely a liquid.

sintering A high-temperature process in which powdered materials below their melting point are compacted together to create a solid form. The process is based on atomic diffusion in which atoms in the powder particles diffuse across the boundaries of the particles, fusing the particles together, and creating one solid piece. Since the materials are not required to reach their melting points, the process is useful for materials with very high melting points such as tungsten (m.p. 3,422°C). Virtually all metals can be sintered, as can many non-metallic substances such as glass, ceramics, and organic polymers. Sintered bronze is used in bearings since its porosity allows lubricants to be retained within it. Sintered stainless steel and porous plastics are used as filter materials employed in the pharmaceutical and food industries. Sintered powders of silver and gold are used to make jewellery.

siphon *See* SYPHON.

SI units The system of base units of the international metric system. The numbers used express the ratio of a measured quantity to some fixed standard for which the unit is the name or symbol for the standard. The *Système International d'Unités is the name that was formally given in 1960 following the tenth meeting of the General Conference of Weights and Measures. There are three classes of units: base units, derived units, and supplementary units. The seven base units are metre (m), kilogram (kg), second (s), ampere (A), kelvin (K), candela (Cd), and mole (mol). Derived units are formed by combining base units such as newton (N), joule (j), pascal (Pa), and watt (W). Two supplementary units are the radian (rad) and steradian (sr), which are units for plane and solid angles, respectively.

Prefixes are used for the basic SI unit with the exception of weight, where the prefix is used with the unit gram (g), not the basic SI unit kilogram (kg). Prefixes are also not used for units of angular measurement (degrees, radians), time (seconds), or temperature (°C or K). The prefixes are used in a way that the numerical value of a unit lies between 0.1 and 1,000. For example, 56 kN rather than 5.6×10^4 N, 11.2 kPa rather than 11,200 Pa, and 6.2 mm rather than 0.0062 m.

(⊕) SEE WEB LINKS
• Official website of National Institute of Standards and Technology (NIST).

skid A solid platform or base upon which process equipment is attached.

slag A material that is produced during the smelting or refining of metals by reaction of a flux such as calcium oxide with impurities in the ore. The slag, which may contain calcium silicate, phosphate, and sulphide, floats on the surface of the liquid metal and can be easily separated. It can be used as a fertilizer if the phosphorus content is sufficiently high.

slip-plate *See* SPADE.

slip ratio The ratio of the superficial velocity of a gas or vapour to liquid in a two-phase flow in a horizontal pipe. The simplest approach to estimating the gas void fraction of a flowing gas–liquid mixture is to assume that the flow is homogeneous. That is, both phases flow at the same velocity.

slop oil Contaminated condensate, also known as bad oil, produced on offshore oil and gas platforms. It is held in a **slop oil tank** and returned back to the production header for recirculation.

sludge The solid or semi-solid waste layer that is deposited below a *supernatant liquid. The sludge from industrial processes may contain harmful and toxic materials and requires careful disposal. The sludge from sewage waste treatment processes that has been both aerobically and anaerobically digested, is free from harmful pathogens, offensive smells and odours, and may be safely used as an organic fertilizer.

sludge digestion A process used to stabilize concentrated wastewater sludge before disposal, usually by anaerobic biological degradation. The digestion process involves converting solids to non-cellular products in which complex fats, proteins, and polysaccharides are first hydrolyzed by facultative and anaerobic bacteria, followed by the conversion to methane and carbon dioxide by anaerobic bacteria.

sludge zone A region at the bottom of a *sedimentation tank used to separate solid particles from a supernatant liquid. The particles are allowed to settle in the *settling zone allowing them to accumulate as a *sludge at the bottom. It therefore provides a storage area for the sludge before its removal for treatment or disposal. The zone is designed to have low velocities such that the sludge accumulates and remains undisturbed, which could otherwise lead to washout. The sludge is removed by scraper or by vacuum devices that move along the bottom.

slug 1. A moving bullet-shaped gas pocket formed in multiphase fluid flow. In a vertical pipe or tube the slug is axially symmetrical and occupies most of the cross-sectional area of a pipe. In horizontal flow, the shape of the pocket has a curved nose and the shape is governed by buoyancy, ratio of gas to liquid, and their relative velocities. **2.** A unit of mass in the *f.p.s. engineering system of units that will accelerate at one foot per second per second to give a one pound (lb) force.

slug catcher A vessel located at the end of a pipeline used to entrap and separate slugs of gas entrained with flowing liquid. It is used at the end of risers from oil and gas reservoirs in which slugs of gas rise with the crude oil as well as water and sand. The slugs of gas or surges are separated in the slug catcher, which has a sufficient capacity to cope with the largest slugs that can be expected. The separated gas and liquid are then able to be processed at a controlled rate.

slug flow A type of intermittent multiphase fluid flow regime that is characterized by pockets of gas in the form of high-velocity gas bubbles. In vertical flow, the gas is in the form of axially symmetrical bullets known as slugs that occupy most of the cross-sectional area of the pipe. In horizontal flow, the gas is also in the form of pockets of gas, but they are not symmetrical although they do have a curved nose in the direction of travel. In both cases, the resulting flow alternates between high-liquid and high-gas composition. *Slug catchers at the end of pipelines are used to disengage the gas. Semi-slug flow occurs where the surges do not completely fill the pipe and is often considered to be a form of *wavy flow.

slurry 1. A general name for a viscous liquid consisting of a concentrated suspension of solid particles. **2.** A viscous liquid suspension of manure used for fertilizing fields.

smart pig *See* PIG.

smelting A high-temperature metallurgical process in which metal is separated by fusion from impurities in minerals and ores. The metals may be chemically or physically combined or mixed in the minerals or ores. The smelting process takes place in a furnace in which the ore is mixed with a reducing agent such as carbon and a fluxing agent such as limestone. The molten metal, being of a higher density than the *slag, falls to the bottom of the furnace where it is removed.

Smith–Brinkley shortcut method A quick procedure used to estimate the components in a multicomponent mixture leaving the top and bottom of a distillation column operating with continuous feed. The procedure is applicable to any stage-wise separation process. For a distillation column with a single feed and a total condenser, the fractional recovery of any component in the bottom product is calculated from details that include the reflux ratio, internal flows of liquid and vapour above and below the feed point (i.e., the rectifying and stripping sections), and the relative volatilities of the components. In the calculation, the reboiler counts as stage one.

Smith equation A relationship representing desorption isotherms for hygroscopic products of high humidities in the order of up to 95 per cent. The Smith and *BET equations complement one another in representing desorption isotherm data from low to high humidities.

smoke The dispersion in air of fine particles of carbon ranging from 0.01 to 15 μm, and other solids and liquids as the result of incomplete combustion.

smoldering A form of combustion without flame and usually incandescent with moderate smoke.

smother To extinguish a fire by blocking the oxygen supply or limiting it to a point below that required for combustion.

SNG An abbreviation for **s**ynthetic **n**atural **g**as or **s**ubstitute **n**atural **g**as, which is a gas that is produced from coal, *petroleum coke, solid waste, or biomass. Carbon-containing coal is gasified to produce *syngas that is then converted to methane. In the *steam-oxygen gasification process, coal is gasified with steam and oxygen to produce carbon monoxide, hydrogen, carbon dioxide, methane, and higher hydrocarbons such as ethane and propane. The composition of SNG is dependent on the temperature and pressures used in the gasifier conditions.

soap A substance used to remove dirt, oil, and grease from surfaces. It is a salt of a long-chain fatty acid and made by boiling fats with sodium hydroxide. Manufactured in a batch process, fat or oil is heated with a slight excess of alkali in an open kettle. Salt is then added to precipitate the soap into curds, recovered, and purified. In the more common continuous process, the fat or oil is hydrolyzed by water at high temperature and pressure in the presence of a catalyst. The fatty acids and glycerol are removed and separated by distillation and the acids neutralized with an appropriate amount of alkali to make soap. Synthetic detergents now exceed the use of ordinary soaps, since soaps give a slight alkaline solution in water due to the partial hydrolysis of sodium salts, which can be harmful to fabrics. *See* SAPONIFICATION.

Soave–Redlich–Kwong (SRK) equation of state An equation of state widely used to predict the vapour–liquid equilibria of substances. It is a development of the *Redlich–Kwong equation of state that correlated the vapour pressure of normal fluids:

$$p = \frac{RT}{V-b} - \frac{a\alpha(T)}{V(V+b)}$$

The a and b constants are obtained from critical point data. It also involves a function which was developed to fit vapour pressure data using reduced temperature, Tr:

$$\alpha = \left[1 + (0.480 + 1.574\omega - 0.176\omega^2)(1 - T_r^{0.5})\right]^2$$

where ω is an eccentricity coefficient.

soft matter A general name given to non-crystalline material and includes colloidal suspensions, surfactants, polymers, pastes, gels, and foams. They exhibit a combination of fluid and solid properties.

software package A professionally written computer program that is designed to perform a particular task. It is used to undertake complex and often repetitive computations. Software packages are used for a wide range of applications such as project management and flowsheeting, as well as for the study of complex flow of fluids and heat transfer termed as *Computational Fluid Dynamics or CFD, and for the study of stresses in process equipment such as pressure vessels, such as Finite Element Analysis software.

sol A colloidal solution in which small solid particles are dispersed in a liquid continuous phase.

solder An alloy used to join metal surfaces. Soft solders are made from tin and lead in varying amounts to adjust the melting point within the range 200–300°C. Hard solder additionally contains silver. Brazing solders are made from copper and zinc and melt at around 800°C.

solenoid valve A type of electromechanically operated valve typically used to control the flow of a fluid through a pipe. The position of the valve is controlled by an electric current through a solenoid. Solenoid valves are fast-acting and typically used to control the dosing of fluids, the release of materials, and for shut off.

solid A substance in a physical state that is resistant to physical change in size and shape. More correctly, it is a state of matter where the strength of the intermolecular and atomic forces is such that there is no translational motion within the substance. Held within a lattice framework, the molecules themselves do, however, vibrate about their average position.

solid fuel A fuel for combustion that is solid such as coke or coal rather than oil or gas.

solid solution A crystalline material in which two or more elements or compounds share a common lattice. Certain alloys such as gold and copper form solid solutions in which some of the copper atoms in the lattice are replaced by gold atoms. Isomorphous compounds can form solid solutions since they have the same crystal structure.

solidus A boundary line or curve on a *phase diagram between solid and liquid/solid at equilibrium. Below the line, the substance is solid.

solubility A measure of the ability of a solvent to dissolve a solid to form a solute in a solution at a given temperature and pressure. It can be expressed in terms of the mass of the solute per unit mass of solution, the mass of solute per unit mass of solvent, or the mass of solute per unit volume of solution or solvent.

solute The dissolved substance within a solution, the liquid part being the solvent.

solution 1. A homogenous mixture of a solute dissolved in a liquid. A solute can be a solid, liquid, or a gas. A solvent is the liquid that dissolves another substance. An aqueous solution is formed when water is the solvent. The solution is saturated when no more solute will dissolve at a particular temperature. Miscible liquids are liquids that dissolve completely in another liquid such as water and methanol. Conversely, an immiscible liquid is a liquid that does not dissolve completely in another liquid but forms a layer, such as oil on water. The separation of the components of a solution can be achieved by evaporation, crystallization, and distillation. **2.** A value that satisfies an algebraic equation.

solvation The process in which there is some chemical association between the molecules of a solvent and the molecules or ions of a solute allowing them to dissolve. For example, an aqueous solution of copper sulphate contains the complex ions of the type $[Cu(H_2O)_6]^{2+}$. **Hydration** is the process of solvation where the solvent is water. Solvation is also known as dissolution.

Solvay, Ernest Gaston Joseph (1838–1922) A Belgian industrial chemist who, in 1861, developed the ammonia-soda process or *Solvay process, used to manufacture soda ash (anhydrous sodium carbonate) from a solution of sodium chloride and limestone (calcium carbonate). This was noted as being a considerable improvement on the earlier *Leblanc process. Solvay worked at his uncle's chemical factory from an early age before founding his own company. Having made his fortune through his patents, he used his wealth for philanthropic purposes. Towards the end of his life, he was elected to the Belgium Senate and became Minister of State.

Solvay process A major industrial process used for the production of sodium carbonate known as soda ash. Also known as the **ammonia-soda process**, approximately three-quarters of all sodium carbonate is produced by this method with the remainder being mined from natural deposits. Developed by Ernest *Solvay (1838–1922) in 1861, the process is based on the fact that when excess carbon dioxide is passed into a solution of brine containing ammonia, the ammonium bicarbonate, which is first formed, interacts with the sodium chloride to give a precipitate of sodium bicarbonate as the salt is only sparingly soluble in the brine due to the common ion effect. Sodium carbonate is then readily prepared from the bicarbonate by heating at up to 230°C producing carbon dioxide that can be used again. The sodium carbonate is finally ground to a powder. The process is carried out over several stages and starts by passing concentrated brine down two towers. In the first, ammonia gas is bubbled and absorbed in the brine liquid. In the second, carbon dioxide, which is produced by the *calcination of limestone, is bubbled up through the ammoniated brine in which sodium bicarbonate precipitates out:

$$NH_4HCO_3 + NaCl \rightarrow NaHCO_3 + NH_4Cl$$

The sodium bicarbonate is the least soluble and is crystallized and filtered out from the hot ammonium chloride solution. The solution is then reacted with the quicklime (calcium oxide) remaining from the calcination of the limestone. As the ammonia is much more costly than the sodium carbonate, it is recovered by adding calcium hydroxide and recycled back to the initial brine solution:

$$2NH_4Cl + Ca(OH)_2 \rightarrow 2NH_3 + CaCl_2 + 2H_2O$$

The Solvay process is an improvement on the earlier *Leblanc process since the materials are less costly; brine is cheaper than rock salt and no sulphuric acid is required. With no evaporation involved, less energy is required, there are no by-products produced, a purer product is obtained, and the process is continuous, and around 97 per cent of the carbon dioxide in the limestone is converted into sodium carbonate.

solvent The liquid part of a solution in dissolving another substance or substances. Water is a commonly used solvent. Being polar, water is capable of dissolving ionic compounds or covalent compounds that ionize. Non-polar solvents, such as benzene, do not dissolve ionic compounds but will dissolve non-polar covalent compounds. In alloys, the solvent is taken as the major component in the *solid solution.

solvent extraction The separation of the constituents of a liquid by contact with another insoluble liquid. If the constituents distribute themselves differently between the two liquids, a certain degree of separation will result. The separation can be enhanced further by multiple contacts. An example is the separation of plutonium and uranium isotopes dissolved in nitric acid, which can be extracted to differing extents using odourless kerosene as the solvent. Since the two liquids are immiscible in one another, the two liquids are agitated in such a way as to increase the surface area contact and to promote the rate of separation. The *extract is the solvent-rich product stream while the *raffinate is the residual liquid stream from which the solute has been removed. It is also known as **liquid-liquid extraction**.

solvolysis The chemical reaction between a compound and its solvent.

sonication An *ultrasonic technique used to disrupt the cell wall and membrane of microorganisms and release the intracellular material. The ultrasonic energy causes areas of compression and rarefaction in which cavities form, resulting in violent collapse and shock waves that are believed to be the cause of the cellular damage. Being a small-scale technique, it is largely confined to use in laboratories.

Soret effect A mass transfer phenomenon used in the separation of isotopes. Its effect is small in comparison with other effects that promote mass transfer and involves applying a temperature gradient. It is named after Swiss scientist Charles Soret (1854–1904), and is also known as **thermal diffusion**.

sorption A chemical and physical process in which a substance becomes attached to another substance. Physical sorption involves the attraction of molecules to a surface by *van der Waals' forces. *Chemisorption involves the formation of a chemical bond on the surface. Sorption is an exothermic process since the resistance of motion to a mobile molecule transmits its energy in the form of heat. *See* ABSORPTION; ADSORPTION; ION EXCHANGE.

sour gas Natural gas that contains hydrogen sulphide and mercaptans at a concentration in excess of 10 ppm. It is called sour on account of its sour and foul-smelling odour. It is required to be removed on account of its corrosive properties. *Compare* SWEET GAS.

SOV An abbreviation for **s**olenoid **o**perated **v**alve. *See* SOLENOID VALVE.

SOx A general term for sulphur oxide gases that are largely based on sulphur dioxide and formed during the combustion of oil and coal that contain sulphur.

space time (Symbol τ) The holding time or mean residence time of materials in a continuous flow reactor required to process one reactor volume of feed under specified conditions to achieve a desired product composition. For a flow reactor with volume, V, and a total volumetric flow rate, Q, at the inlet:

$$\tau = \frac{V}{Q}$$

The reciprocal is the *space velocity, which is the number of reactor volumes under specified conditions that can be processed per unit time. The **space time yield** is the net yield of a product from a reactor per unit time per unit of effective reactor volume.

space velocity Applied to the processing of materials within a continuous flow chemical reactor, it is the reciprocal of the *space time. It can refer to the conditions in a specific location of the process materials in a reactor.

spade A solid plate that is inserted into a pipe to ensure isolation of material within it. It is inserted between the flanges and is made of the same material and has the same rating as the pipe. It is also known as a **slip-plate**.

Spalding number A dimensionless number, B, used in liquid droplet evaporation studies. It relates the sensible heat and latent heat of the evaporated material:

$$B = \frac{c_p \Delta t}{\lambda}$$

where c_p is the specific heat capacity, Δt is the temperature difference between the surrounding gas and the liquid, and λ is the latent heat.

span The difference between the maximum and minimum value indicated by an instrument used to measure a process variable such as temperature, pressure, and level.

sparger A perforated tubular ring positioned at the bottom of a vessel containing liquid such as a bioreactor through which air or oxygen is discharged creating a swarm of bubbles that rise up through the liquid medium promoting oxygen transfer to the liquid. The size of the bubbles and their velocity determines the rate of oxygen transfer. Their size can be controlled by the number of holes, rate of flow of air or oxygen, and the location of the sparger to a rotating impeller that can disperse the bubbles.

speciality (specialty) chemicals Chemicals produced at the high-value end of the chemicals business that are characterized by their innovative uses such as in the development or modification of existing processes or products, or in the exploitation of new or developing technologies.

specific Relating to a specified or particular thing, such as being a characteristic property especially in relation to the same property of a standard reference substance expressed per unit mass. For example, the *latent specific heat of a substance is the latent heat per unit mass. The adjective is also used in relation to other terms, such as *specific speed, in which the performance of a centrifugal pump is compared to the performance of other pumps.

specific enthalpy (Symbol h) The enthalpy of a system defined as $h = U + pV$ where U is the internal specific energy, p is the pressure and V is the specific volume of a substance. Like pressure, temperature, and volume, enthalpy is a property of a substance. The enthalpy is normally expressed with respect to some reference value. The specific enthalpy of water or steam, for example, is zero at 0.01°C and 101, 325 Pa.

For the freezing of foods, a value of 0 kJ kg^{-1} is taken at −40°C.

specific gravity The ratio of the density of a substance to the density of water at 20°C. The specific gravity of water at 20°C is therefore 1.0.

specific growth rate (Symbol μ) The rate of increase in concentration of living microbial cells per unit concentration in a growing culture such as in a bioreactor. It is a measure of the doubling time for cell division:

$$\mu = \frac{\ln\left(\dfrac{x}{x_o}\right)}{t}$$

where x and x_o are the final and initial concentration of cells in time t. It commonly has the unit of h^{-1}.

specific heat *See* SPECIFIC HEAT CAPACITY.

specific heat capacity (Symbol c_p, c_v) The amount of heat required to raise the temperature of one kilogram of a substance by a temperature of one degree K. The SI units are J kg^{-1} K^{-1}. The *molar heat capacity is based on molar mass for which the SI units are J mol^{-1} K^{-1}.

specific latent heat (Symbol L or λ) The quantity of heat absorbed or released per unit mass when a substance changes its physical phase at constant temperature and pressure. The **specific latent heat of fusion** (specific enthalpy change on fusion) of a body is the heat required to convert one kilogram of the solid at its melting point into liquid at the same temperature. The **specific latent heat of vaporization** (specific enthalpy change on vaporization) of a liquid is the heat required to convert one kilogram of the liquid at its boiling point into vapour at the same temperature. The SI units are J kg^{-1}.

specific speed A classification of centrifugal pump impellers at optimal efficiency with respect to *geometric similarity. It is useful for the scale-up and selection of centrifugal pumps and is a measure of pump pressure, head, and speed. Although the specific speed represents the numerical value for a rotational speed, and is usually expressed simply as a number, it actually has dimensions L$^{3/4}$T$^{-3/2}$. *See* SUCTION SPECIFIC SPEED.

specific surface area A measure of the total surface area of a solid per unit mass or volume. It is used in *catalysis and gas *adsorption where a chemical reaction is promoted on the surface of a *catalyst, or the rate of adsorption of a component is dependent on contact with a surface. The specific area of highly porous solids can be determined using the BET isotherm. The specific surface area of a gram of activated carbon is typically 500 m^2. The SI units are m^2 kg^{-1} or m^2 m^{-3} (or m^{-1}).

specific volume The volume occupied by a substance per unit mass. It is the reciprocal of density and has the SI units of m^3 kg^{-1}.

speed The rate of change of distance with time. It is a scalar quantity and has the SI units of m s^{-1}. Velocity is the speed of a body in a specified direction and is therefore a vector quantity.

sphere A body formed from the rotation of a circle about its diameter. It has a volume of $\frac{4}{3}\pi r^3$ and surface area of $4\pi r^2$ where r is the radius. Spherical vessels are commonly used to store process fluids under high pressure, such as liquefied petroleum gas (LPG), since there is an equal distribution of stress throughout the wall of the vessel. Cylindrical vessels and columns that operate under high internal pressure feature domed or hemispherical ends to distribute stresses. *See* HORTONSPHERE.

sphericity (Symbol Φ) The extent to which an irregularly shaped or non-spherical particle equates to being spherical. It is independent of particle size and defined as the ratio of the surface area of a particle to its actual surface area:

$$\Phi = \frac{6v_p}{d_p s_p}$$

where d_p is the equivalent or nominal particle diameter, v_p is the volume of a particle, and s_p is its surface area. For a spherical particle, $\Phi = 1$. Tables of sphericity are used for particles of various defined geometric shapes.

spigot 1. A fitting at the end of a pipe that fits into another to form a joint. **2.** A type of tap to control the flow of liquid and used in wooden casks.

splitter A device used to divide the flow of process material into two or more streams. It can be as simple as a T-piece in a pipe or as complex as a mechanically operated diverter used to control the continuous or incremental portions of flow in particular directions. In simulation software packages, a splitter is an operation used to separate a process stream into two or more streams.

spontaneous ignition, combustion The initiation of combustion of a material by heating without the use of an external ignition source such as a spark or flame. The *auto-ignition temperature is the temperature at which the material is heated by its surroundings to the point that spontaneous ignition takes place.

spray column A simple type of liquid–gas contactor in which a liquid is sprayed into a gas contained within a column. It is typically used to absorb gases into a liquid in which the liquid is sprayed as fine droplets from the top of the column and the gas to be absorbed enters at the bottom of the column and leaves at the top.

spray dryer A device used to remove the moisture from a high moisture-containing fluid that contains a solid to be dried. The solid is often heat-labile, such as milk, and is continuously atomized into small droplets within a large chamber into which is fed a continuous flow of warm drying air or gas. Evaporation of the suspended droplets is rapid and the dried product is quickly carried away with the current of air or gas and separated, usually in a cyclone. Rapid drying in this way is suitable for materials that may be heat-sensitive such as certain biological and food products.

spray dryer absorbers *See* DRY SCRUBBING.

spud A term used to start the drilling of an oil or gas well. **Spudding** is used where a large drill bit forms the hole that is first lined and sealed before the main drill bit is inserted.

sputtering A process in which atoms from an electrode are removed from its surface by the impact of high-energy ions as in a discharge tube. It can be used to clean the surface or to deposit a uniform film of metal on an object within an evacuated chamber.

SRV *See* SAFETY RELIEF VALVE.

stabilizer A substance that is added to a colloid to prevent it from coagulating.

stage 1. A tray or plate in a distillation column in which equilibrium of a vapour and a liquid is reached. **2.** A part of a continuous process in which some form of separation process takes place.
   The **stage efficiency** is the deviation from the equilibrium condition. In a distillation column, the stage efficiency is the composition of the mixture passing a tray divided by the composition if it were to be in equilibrium. *See* MURPHREE PLATE EFFICIENCY

stagnant film An assumption used in mass transfer calculations in which there is an assumed stationary film of gas or liquid surrounding an object such as a particle through which a component diffuses.

stagnation point A point in a flowing fluid where the fluid is stationary or brought to rest, such as in the mouth of a *Pitot tube.

stainless steel Alloys of iron noted for their resistance to corrosion and containing small amounts of carbon as well as chromium. Most stainless steels also contain nickel. Various other metals and non-metals are also often added to provide particular properties. They are resistant to corrosion due to a thin protective oxide coating on the surface. An example is 18:8:1 stainless steel which contains 18 per cent nickel, 8 per cent chromium, and 0.01 per cent carbon. Being of a high tensile strength and with excellent corrosion resistance, stainless steels are used extensively in the chemical and process industries for pipework and vessels.

standard atmospheric pressure The pressure of the atmosphere taken to be 101,325 Pa (i.e. 1013.25 mbar). Atmospheric pressure is not constant but variable and is influenced by meteorological conditions.

standard candle A former name for *candela, which is the unit of luminous intensity. It is no longer used due to confusion with another unit called the international candle.

standard cell A voltaic cell used to produce a constant and accurately known electromotive force (e.m.f.). It is used as a standard e.m.f. to calibrate voltage-measuring instruments.

standard deviation (Symbol σ) A statistical measure of the dispersion of a set of data from the mean and equal to the square root of the *variance. In a sample of n observations, the standard deviation is:

$$\sigma = \sqrt{\frac{\sum_{i=1}^{n}\left(x_i - \bar{x}\right)^2}{n-1}}$$

where $\bar{x}$ is the mean of the sample. The standard deviation is therefore the square root of the mean of the sum of squared differences of the data points from the mean. A small value indicates a cluster around the mean whereas a large value indicates a wider spread of data.

standard electrode An electrode used in measuring electrode potential.

standard enthalpy of combustion *See* HEAT OF COMBUSTION.

standard enthalpy of formation *See* HEAT OF FORMATION.

standard error The estimated *standard deviation of a parameter whose value is not known exactly.

standard solution A solution of known concentration used in volumetric analysis.

standard state A state of a system used as a thermodynamic reference point. Reference points are usually taken as a temperature of 298.15K, a pressure of 101.325 kPa, and concentration of 1 M. The standard state is denoted by the superscript symbol $^{\bullet}$. For example, the standard molar heat of formation of water in the reaction:

$H_{2(g)} + \frac{1}{2}O_{2(g)}$ H_2O_l is $\Delta H_f^{\bullet}$ is $-286\,kJ\,mol^{-1}$.

standard temperature and pressure *See* S.T.P.

Stanton number A dimensionless number, St, used for forced convection heat transfer and relates the rate of heat transfer to the thermal capacity of a fluid:

$$St = \frac{h}{\rho v c_p} = \frac{Nu}{Re\,Pr}$$

where h is the surface heat transfer coefficient, ρ is the density, v the velocity, and c_p is the specific heat capacity . Nu is the *Nusselt number, Re is the *Reynolds number, and Pr is the *Prandtl number. It is named after British scientist Thomas Edward Stanton (1865–1931).

Stanton–Pannell chart A chart that presents the variation of friction factors for fluids across a wide range of *Reynolds numbers. It was published by British scientists Thomas Edward Stanton (1865–1931) and J. R. Pannell in 1914.

start-up A systematic sequence of events that is required in order to operate fully a chemical plant or item of process equipment.

state function A thermodynamic quantity whose value depends only on the state of a substance. The change in value depends only on the initial and final states of the system and is independent of the route taken to reach that state. *Enthalpy is a state function since a change in enthalpy depends only on the initial and final states, and is independent of the route between these states, and thus forms the basis of *Hess's law.

state of matter One of three physical forms for matter being *gas, *liquid, and *solid.

static equilibrium *See* EQUILIBRIUM.

static head The potential energy of a liquid expressed in *head form:

$$h = \frac{p}{\rho g}$$

where p is the pressure, ρ is the density, and g is the gravitational acceleration. It is used directly in the *Bernoulli theorem for which the other two head forms are *velocity head and *pressure head.

static mixer *See* IN-LINE MIXER.

static pressure The measure of the pressure of a gas or liquid without movement. *Compare* IMPACT PRESSURE.

statics The study of mechanics in which balanced forces act on bodies resulting in the body remaining at rest. *Compare* DYNAMICS.

stationary phase The stage in the growth of a culture of microorganisms in a batch-operated bioreactor where the rate of growth ends. It occurs once all the limiting substrate has been exhausted such that no further growth is possible. The death phase then follows with the reduction of the number of viable cells.

stationary point *See* TURNING POINT.

statistical error *See* ERROR.

statistical mechanics The study of the properties of physical systems that can be predicted by the statistical behaviour of their constituent parts.

statistical process control A set of measurement techniques used to monitor a process in order to assess variability of performance and allow for predictions of when corrective action needs to be taken to prevent a problem from occurring. It uses data collected from various points within the process. Variations that may affect the quality of the end product or service are detected and corrected. The emphasis is on early detection and the prevention of problems. **Multivariable process control** is a form of statistical process control that uses a set of manipulated and control variables to control a process plant.

statistical tables Published tables of the values of cumulative distributions functions, probability functions, and probability density functions. They are used to determine whether or not a particular statistical result exceeds a required significance level. Examples of commonly used statistical tables include the normal distribution curve, chi-square distribution curve, Student's t-distribution curve, and F-distribution curve.

statistics A branch of mathematics that involves the planning of experiments, study of the classification and analysis of data using probability theories, and the application of interpretation methods such as hypothesis testing. It is used to form decisions and derive conclusions particularly where data may have a considerable degree of *error or uncertainty.

steady state A condition in which the net rate of change between the input and output to a process or system is zero and there is no dependence on time. For example, a steady-state material balance is where the total material entering a process and subsequently undergoing chemical reaction is equal to the total amount of material leaving the process. Where there is an accumulation of material or where there is a loss of material, the process or system is said to be in an *unsteady state. *Compare* EQUILIBRIUM. **Steady-state flow** is the flow of a fluid into a space such that there is no loss or accumulation, and it is therefore unvarying with respect to time.

steam The gaseous form of water formed when water boils. At atmospheric pressure, steam is produced at 100°C by boiling water. It is widely used in the chemical and process industries as a utility for heating processes such as in *kettle reboilers for distillation columns. It is also used in power generation where steam is produced or 'raised' from a thermal or nuclear process and expanded through turbines. Scottish engineer James *Watt (1736–1819) understood the value of steam and his improvements to the Newcomen steam engine were an important contribution to the Industrial Revolution. Other uses of steam include sterilization, which is used in the food and medical industries. Steam is effective at destroying harmful pathogens and is a harmless substance once cooled. Wet steam is water vapour that contains water droplets. When heated further, the water evaporates. The **dryness fraction** of steam is the ratio of the amount of water in steam to the total amount of water vapour. *Superheated steam is produced by heating the steam above the boiling point of water. The thermodynamic properties of steam are presented in published *steam tables.

steam cracking The high-temperature reduction in length or cracking of long-chain hydrocarbons in the presence of steam to produce shorter-chain products such as ethylene, propylene, and other small-chain alkenes.

steam cycle A closed thermodynamic cycle used for power generation and involves raising steam from water in a boiler, expansion through a turbine, condensation, and return to the boiler (see Fig. 55). All steam turbine systems are based on adaptations of the *Rankine cycle. Represented on a *temperature-entropy diagram, its features include superheating, reheating, and regenerative feed heating, which are used to raise the overall cycle efficiency.

steam distillation The separation of immiscible organic liquids by distillation using steam. It involves the injection of live steam into the bottom of the distillation column and into the heated mixture for separation. The steam reduces the partial pressure of the mixture and reduces the temperature required for vaporization. When distilled, the components operate independently of one another, with each being in equilibrium with its own

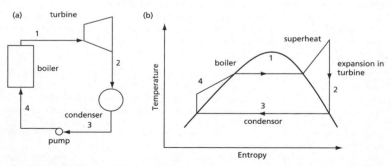

Fig. 55a and 55b

vapour. Steam distillation is used in the primary separation of crude distillation in a *fractionation column.

steam injection The use of live steam fed directly into a process to provide water and heat, and to enhance either extraction or reaction. It is commonly used as an *enhanced oil recovery method to recover oil from depleted reservoirs or from oil sands in which viscous heavy oil is recovered using steam injection to reduce the viscosity of the oil, aid transport, and recovery. Steam is also directly used in the separation of crude oil and fed to the bottom of the *fractionation column. This is the primary separation of crude oil into fractions that have differing boiling points. *Steam cracking uses steam for *thermal cracking and reforming hydrocarbons.

steam jacket *See* JACKET.

steam jet ejector A type of fixed operation pump that uses high-pressure steam passed through a constriction to create a low pressure due to the *venturi effect, and to which the equipment to be evacuated is connected such as a distillation column condenser. In spite of requiring high-pressure steam, the device has no moving parts and therefore has low maintenance costs. It can also handle corrosive vapours.

steam point The temperature that corresponds to the maximum vapour pressure of water at standard atmospheric pressure (101.325 kPa). This corresponds to a temperature of 100°C.

steam reforming The conversion of methane from natural gas into hydrogen. It is used in production of ammonia in which the methane is first produced from desulphurized and scrubbed natural gas, mixed with steam and passed over nickel catalyst packed in tubes at a high temperature of around 900°C:

$$CH_4 + H_2O \rightarrow CO + 3H_2$$
$$CH_4 + 2H_2O \rightarrow CO_2 + 4H_2$$

The reactions are endothermic.

steam tables Published tables that present thermodynamic data for enthalpy, entropy, and specific volume of steam at various temperatures and pressures. Steam is a commonly encountered material in chemical processes and its properties have been extensively tabulated. Steam tables therefore provide a quick and valuable reference point.

steam tracing An internal pipe or tube used in process vessels and pipelines carrying steam to provide sufficient heating to a fluid to keep it at a controlled temperature. The amount of steam or heat supplied is sufficient to overcome losses. Steam tracing typically is used in pipelines carrying molten bitumen and other fluids prone to solidification on cooling, to ensure that they remain in a liquid state.

steam trap A device used to automatically drain and remove condensate from steam lines to protect the steam main from condensate build-up. Various types of steam traps are used and generally consist of a valve that can be operated by a float, spring, or bellows arrangement. Discharge of the hot condensate may be either to the environment or into a collection pipe and returned to the boiler for reuse, if appropriate.

steam turbine *See* TURBINE.

steel An alloy of iron used extensively for the fabrication of process vessels, columns, pipes, heat exchangers, ancillary equipment, and supporting structures. Steels contain up to 2.1 per cent carbon and varying amounts of other elements such as manganese, nickel, chromium, molybdenum, and silicon. There are many alloy steels with varying properties and used for specific applications. Steel with a chromium content of 11 to 12 per cent is known as stainless steel. Pipes and process vessels are often made from steel due to its high tensile strength and resistance to corrosion. Steel is also used for the support structures for process plant. It is manufactured by the *basic-oxygen process, which involves a charge of molten pig iron and scrap being blown with high-pressure oxygen on the surface through a water-cooled lance. *See* STAINLESS STEEL.

Stefan–Boltzmann constant (Symbol σ) A proportionality constant representing the thermal radiation heat loss by emission from a *black body. In SI units, it has a value of 5.669 7 x 10^{-8} W m^2 K^{-4}. It is named after Austrian physicists Josef Stefan (1853–93) and Ludwig Eduard Boltzmann (1844–1906).

Stefan's law (Stefan–Boltzmann law) A law of thermal radiation in which the radiation of all wavelengths per second per square metre from a *black body at an absolute temperature T to surroundings at temperature T_o is proportional to the fourth power of the absolute temperatures: $q = \sigma(T^4 - T_o^4)$. The proportionality constant, σ, is known as the *Stefan–Boltzmann constant equal to 5.669 7 x 10^{-8} W m^{-2} K^{-4}. The law is named after Austrian physicists Josef Stefan (1853–93) and Ludwig Eduard Boltzmann (1844–1906) who theoretically derived the equation.

STEL An abbreviation for short term exposure limit. *See* WORKPLACE EXPOSURE LIMITS.

stenching The addition of a pungent-smelling substance to another to give a distinctive and strong odour. Substances such as diethylsulphide and mercaptans are added to odourless natural gas and LPG for safety purposes so that leakage can be readily detected. It is also known as **odourizing**.

step A discrete activity or stage within an overall process or chemical reaction.

step response The behaviour of a controlled process to adjust from one steady-state condition to another.

steradian (Symbol sr) A supplementary SI unit defined as the solid angle which, having its vertex in the centre of a sphere, cuts off an area of the surface of the sphere equal to that of a square with sides of length equal to the radius of the sphere.

stere A unit of volume used to measure the volume of stacked timber and equal to one cubic metre.

stereochemistry The branch of chemistry that is concerned with the study of the shape of molecules.

sterility A state of being free from microorganisms or spores. An *autoclave uses steam to kill all active microorganisms or spores in liquid media.

still A batch distillation vessel. It is used in the *whisky industry to distil fermented liquor known as wash. In the Scotch whisky industry, the still or pot still is made from copper and heated by steam. It has a characteristic swan neck and the distillate is condensed to a liquid and collected before being redistilled a second or occasionally a third time. *See* WHISKY.

still gas A mixture of gases produced in petroleum refineries that includes methane, ethane, and ethylene as by-products of upgrading heavy petroleum fractions into more valuable and lighter products. Also known as *refinery gas, it is used as a refinery fuel or petrochemical feedstock.

Stirling engine A type of heat engine that involves the cyclic compression and expansion of air or any other gas as the working fluid. As a closed thermodynamic system, it uses both a hot and cold cylinder separated by a regenerator, in which the working fluid is recycled around the engine. There is a net conversion of heat energy to mechanical work. The use of the regenerator as a form of internal heat exchanger and thermal store makes the Stirling engine distinct from other types of hot air engine. It was invented in 1816 by Scottish engineer Robert Stirling (1790–1878) and is noted for its quiet operation and high efficiency. There has been much recent interest in its redevelopment particularly for use in *combined heat and power (CHP) plants.

stirred tank *See* AGITATED VESSEL.

stochastic process A statistical way of generating a series of random values of a mathematical variable and building up a particular statistical distribution from these values. Stochastic processes are used in non-equilibrium *statistical mechanics.

Stockholm Convention on Persistent Organic Pollutants An international environmental treaty that aims to eliminate, reduce, or restrict the production and use of chemicals described as *persistent organic pollutants (POPs). These are chemical substances capable of persisting in the environment and may pose a risk to human health and to the environment. Formed by the United Nations Environment Programme, the environmental treaty was established with cooperation of many international members and signed in 2001 in Stockholm, Sweden. The Convention includes requirements that developed countries provide resources to ensure that POP production and use are eliminated, whether intentional or unintentional, and that they are disposed in environmentally responsible ways.

((⊕)) SEE WEB LINKS
• Official website of the Stockholm Convention.

stoichiometric coefficient The coefficients used in balanced chemical reactions to show the relative number of molecules reacting. For example, in the reaction $3H_2 + N_2 \rightarrow 2NH_3$ the stoichiometric coefficients are 3, 1, and 2, respectively.

stoichiometric combustion The complete combustion of a fuel with oxygen, with the balanced molecular amount of fuel to oxygen. In practice, an excess of oxygen is used in the combustion of fuels. *See* COMBUSTION.

stoichiometry The study of the quantitative relationship between reactants and products in chemical reactions based on using the relative amounts of elements. A **stoichiometric reaction** is a chemical reaction involving the exact proportion of elements. In biochemical reactions, which tend to be more complex, a mass rather than a mole-based approach is used and is related to parameters such as the carbon content and *chemical oxygen demand.

Stokes, Sir George Gabriel (1819–1903) An Irish physicist and mathematician noted for his contributions to fluid mechanics. The youngest of six sons of a Protestant minister, he moved to Bristol at the age of 16 to pursue his studies, which prepared him for studying mathematics at Cambridge. Appointed as the Lucasian professor of mathematics at Cambridge in 1849, he studied the science of hydrodynamics and established his law of viscosity describing the velocity of a small sphere descending through a viscous fluid. He is best known for his work on fluid dynamics including his contribution to the *Navier-Stokes's equations. He also formulated *Stokes's law. A unit of kinematic viscosity is also named after him. A member of the Royal Society, he was first secretary and then president from 1885 to 1890.

stokes (Symbol St) A c.g.s. unit of measure of the *kinematic viscosity of a fluid expressed as the ratio of the viscosity, μ, in poise divided by the density, ρ, in grams per cubic centimetre:

$$v = \frac{\mu}{\rho}$$

The units of centistokes, cSt, are more commonly used where 1 cSt is equal to 10^{-6} m^2 s^{-1}. The kinematic viscosity of water is exactly 1 cSt at 20.2°C.

Stokes–Einstein equation An equation used to determine the diffusion coefficient of particles through a liquid with low *Reynolds number. It was first derived by Albert *Einstein and based on a Stokes's particle undergoing Brownian motion:

$$D = \frac{k_B T}{6\pi\eta r}$$

where k_B is the Boltzmann constant, T is the absolute temperature, η is the viscosity, and r is the particle radius. The equation is of significance since it was first used to confirm molecular theory.

Stokes's law An equation based on a small particle suspended in a fluid to accelerate from stationary to achieve its terminal velocity. The equation suggests that the terminal velocity is reached after an infinite period of time but is usefully expressed as:

$$v_t = \frac{g(\rho_p - \rho)d_p^2}{18\mu}$$

where g is gravitational acceleration, $\rho_p - \rho$ is the difference in density between the particle and fluid, d_p is the diameter of the particle and μ is the viscosity of the fluid. The particle, which is usually solid although it also applies to small bubbles and droplets, has a maximum size of particle governed by a correlation of *Reynolds number based on the particle dimension and drag coefficient as:

$$Re_p = \frac{24}{C_d} \leq 0.4$$

as well as a minimum size for the fluid viscosity to operate (Brownian motion). The external forces on the particles may be gravity, centrifugal, electrostatic, and magnetic. Stokes's law is named after Sir George *Stokes (1819–1903).

stonewall The maximum stable flow and maximum head condition for a centrifugal compressor.

storage tank A large container or vessel used to hold liquid or powdered substances. Storage tanks may hold raw materials, intermediate products, final products, cooling water, solvents, waste products, etc. Storage tanks are usually located in a *tank farm.

s.t.p. An abbreviation for **s**tandard **t**emperature and **p**ressure, and is the standard conditions used as the basis for many thermodynamic calculations and tabulations involving temperature and pressure, and used for comparing the properties of gases. It is defined as 0°C (273.15 K) and one standard atmosphere (101 325 Pa).

strain The dimensionless change in a material produced by an applied stress divided by the original dimension. For a wire held vertically and stretched by a weight, the strain is the extended length expressed as a ratio of the original upstretched length.

stratified flow A two-phase flow regime that occurs in horizontal or slightly inclined pipes and channels where the liquid phase flows as a layer at the bottom of the channel with the gas phase above. This type of flow occurs at low gas velocities where separation of liquid and gas has occurred in which the liquid flows in the lower part of the pipe with the gas above it. At low gas velocities, the liquid–gas interface is smooth without ripples. At higher gas velocities, ripples and waves form on the liquid surface, eventually leading to the breakdown of this flow regime.

stream The flow of process materials to or from a process, plant unit, or within a pipeline. They are indicated on *process flow diagrams in which the direction of flow is shown by an arrow. The details of the stream are presented in a *stream table, which includes the components, their mass or molar flows, phase, temperature, and pressure. In *process integration studies, the hot and cold streams correspond to the relative temperatures and are distinguished by their function for either heating or cooling.

streamline flow An imaginary line in a flowing fluid such that the tangent to it at every point gives the direction of flow, and its velocity at any instant. It is also known as *laminar flow.

stream table Usually accompanying a *process flow sheet, a stream table presents the complete material accountancy throughout a process. It shows each process *stream, its composition, individual and total flows, as well as process conditions of temperature, pressure, and phase (solid, liquid, gas, or mixed). The flows are typically presented on a mass or molar basis per unit time such as kmol h^{-1}. The stream table data is usually given as the design *steady-state values. Maximum design values may also be indicated.

stress An applied force over a given area of a material that produces a strain. The SI units for stress are N m^{-2}.

stress corrosion cracking A cracking effect of a material held under tension due to the action of chemicals leading to its failure. The cracks formed are either intercrystalline or transcrystalline, in that they either form between and around crystals, or across and through them. Examples of chemicals capable of leading to stress corrosion cracking include chlorides on stainless steels, nitrates and hydroxides on mild steels, and ammonia on brass.

string The full length of tubing or a drill pipe that is used in an offshore drilling operation.

stripper A separation vessel used to reduce the amount of a volatile component in a liquid mixture. This is usually a column containing a packing material in which liquid material is fed at the top and cascades down, intimately contacting another fluid stream flowing up through which mass transfer takes place. For example, ammonia can be stripped out of water, where ammonia gas leaves at the top of the column and water with a low ammonia content from the bottom.

stripping A separation process in which chemicals are removed from one phase and transferred to another by absorption using a stripping agent. **Gas stripping** is the process of removing a component from a gas by contact with a liquid. For example, volatile organics and hydrogen sulphide can be removed from wastewater by contacting it with steam or air within a **stripper**. The chemicals are therefore stripped out of the wastewater and recovered from the steam or air from the top of the column. *Compare* SCRUBBING.

stripping section The section in a distillation column located below the feed point, in which the less volatile component or components in the mixture undergoing separation increase in concentration towards the bottom of the column as the more volatile component or components are stripped out. The upper section is the *rectification section.

structured packing A type of packing material used within vessels to increase the contact area between two phases to enhance a chemical reaction or separation. It consists of many arranged sheets of corrugated metal which cause the phases to make good contact with one another. They are used in distillation and absorption columns as well as some types of chemical reactor. *See* PACKED BED.

Stubs' Wire Gauge *See* BIRMINGHAM WIRE GAUGE.

subcritical A condition in a nuclear reactor in which the rate of nuclear fission is not sufficient to sustain a chain reaction. *See* CRITICAL MASS.

sublimate The solid that is formed through the process of *sublimation.

sublimation A direct change of state from the solid to the gaseous state without the appearance of the liquid state. The principle of sublimation is used in *freeze drying. **Desublimation** is sometimes used to describe the condensation of a gas to a solid and **ablimation** is used to describe the condensation of water vapour to ice.

subnatant A layer of liquid, particles, or sediment that lies beneath an upper layer of liquid, known as the *supernatant.

substance Any material that has a definite chemical composition. It may be a chemical element, a compound, or an alloy. Ores and minerals are naturally occurring substances and comprise mixtures of elements and compounds. A **substance hazardous to health** is any substance that can cause harm to human health by being toxic, irritant, corrosive, harmful, sensitizing, carcinogenic, mutagenic, or toxic to reproduction. *See* LETHAL DOSE.

substituted mechanism A reaction involving an enzyme (E) and a substrate (S_1) to produce a product (P_1) and a modified enzyme (E^*). The modified enzyme then reacts with another substrate (S_2) to form another product (P_2) with the regeneration of the enzyme:

$$E + S_1 \rightarrow E^* + P_1$$
$$E^* + S_2 \rightarrow E + P_2$$

substitute natural gas *See* SNG.

substitution The process of solving mathematical problems by replacing one variable with another. For example, for the simultaneous equations: $x - y = 1$ and $2x - y = 3$, in the first equation $y = x - 1$ can be substituted into the second as $2x - (x - 1) = 3$ and be readily solved to show $x = 2$ and so $y = 1$.

substitution reaction A chemical reaction in which an atom or a molecule is substituted for another atom or molecule. An example is the reaction of zinc with hydrochloric acid to produce zinc chloride and hydrogen:

$$Zn + 2HCl \rightarrow ZnCl_2 + H_2$$

substrate 1. A substance upon which an enzyme acts. For example, starch is the substrate of the enzyme amylase that is hydrolyzed to maltose. **2.** The medium on which microorganisms grow in a *bioreactor. **3.** The substance on which some other substance is absorbed or in which it is absorbed.

suction boot A vertical leg that extends downwards from the suction or inlet pipe to a gas compressor. The purpose is to catch and collect moisture during the ascent of the gas into the compressor such that it can be removed safely without harm to the compressor. A valve is used to periodically drain the accumulated liquid.

suction pressure The pressure entering a duct or pump at a pressure below the system pressure.

suction specific speed A dimensionless number used as a measure of centrifugal pump performance for a particular application. Evaluated at the *best efficiency point, which corresponds to the maximum efficiency of the pump, the suction specific speed refers to the suction side of the pump and is used to identify issues of cavitation and the type of pump appropriate for a particular application, such as multi- or single stage, mixed or axial flow. It is calculated from:

$$S_n = \frac{NQ^{\frac{1}{2}}}{gH^{\frac{3}{4}}}$$

where N is the rotational speed of the impeller, Q is the flow rate, g is the gravitational acceleration, and H is the delivered head. *Compare* SPECIFIC SPEED.

suffocation Interference with the entrance of air into the lungs with resultant asphyxia. It can be the result of the release of gas or vapour in a *confined space in which oxygen is depleted.

sulfate process *See* KRAFT PROCESS.

Sulfinol process A regenerative process used to remove or reduce the level of hydrogen sulphide, carbon dioxide, carbonyl sulphide, and other organic sulphur compounds such as mercaptans from natural gas. It involves gas *absorption with solvents in which the gas is contacted countercurrently in an absorption column with the solvent. The *scrubbing process uses di-isopropanolamine dissolved in a mixture of sulfolane ($C_4H_8SO_2$) and water. Regenerated solvent is introduced at the top of the absorber. The solvent in which the sulphur compounds are absorbed is heated in a heat exchanger with the regenerated solvent and is fed back to the regenerator where it is further heated to release the dissolved gases using steam. The gas is passed to a *Claus process to recover elemental sulphur. The process was developed in the 1960s.

summing point Used in process control, it is any point in which the process signal is added algebraically.

sump A vessel located at a low place with a capacity sufficient to collect or retain liquids. It is often located at the bottom of a process or a machine to collect waste liquids such as oil.

supercooling A *metastable state in which a liquid is cooled to below its normal freezing point without a change of phase. The particles of the liquid lose their energy but do not form a lattice structure of the solid crystal. By seeding with a small crystal, crystallization of the liquid then occurs and the temperature returns to the freezing point. Supercooling in the air occurs in which supercooled water droplets in the absence of freezing nuclei may exist at temperatures as low as -40°C.

supercritical The thermodynamic state of a substance that is above its *critical temperature and *critical pressure. The substance has a density greater than that of a gas but less than that of a liquid. The viscosity is also greater than that of a gas but less than that of a liquid. Supercritical fluids such as carbon dioxide are used as solvents in many extraction processes involving organic materials.

supercritical fluid extraction (SFE) An extraction process that uses pressures and temperatures above the critical point of the extracting solvent. Carbon dioxide is a popular solvent on account of its low critical point (31.4°C and 72 bar). The extraction process involves compressing the carbon dioxide and heating it. The supercritical carbon dioxide has the density of a liquid but properties of a gas, which aids diffusion and helps solubility. The solvent and dissolved extract is then transferred to a separator tank and the pressure reduced precipitating the extract. The carbon dioxide is recycled into the extractor via a condenser. A small amount of *make-up is required to allow for losses. Examples of SFE using carbon dioxide include the decaffeination of tea and coffee, flavour extraction from hops, the removal of pesticides from rice, and the dry cleaning of cloths. Other solvents used in supercritical fluid extraction include nitrous oxide, propane, and water.

superficial velocity The velocity of a fluid through a pipe that also contains another fluid of another state expressed in terms of the overall cross-section area as if no other fluid were present.

superfluid A fluid in a state being characterized by a very low viscosity and therefore possessing frictionless flow. It also has a high thermal conductivity. The only known fluid is cryogenically cooled liquid helium at a temperature close to absolute zero. **Superfluidity** is the state of being, or property of becoming, a superfluid.

superfractionation The separation of liquid mixtures by distillation that have close boiling points. Due to the relative volatility between two components being close to unity, the separation therefore requires a large number of theoretical plates and high reflux ratios.

superheat The heat added to a saturated vapour such as steam to raise the temperature above the saturation point at a given pressure. The **degrees of superheat** correspond to the temperature to which the vapour is heated above the normal boiling point.

superheated steam Steam produced in a boiler that has been heated to a temperature above that of the boiling point of water at a given pressure. It is produced by heating saturated steam in a boiler and passing it through a **superheater**, which further heats the steam. The number of degrees of superheat refers to the temperature of steam above the saturation temperature. Superheated steam is used to prevent harmful and wasteful condensation in steam turbines.

supernatant (supernate) A liquid that lies above the surface of another liquid or layer of sediment. *Subnatant means lying under, whereas **infranatant** means lying below.

supersaturation An unstable solution that contains more of a solute in solution than it can hold, at a particular temperature, in the presence of crystals of that solute. It is prepared by cooling an unsaturated solution and by isothermal evaporation of a solvent. Used in the process of *crystallization, supersaturation is more common with organic solvents than aqueous solvents.

supersonic flow The velocity of a fluid greater than the speed of sound i.e. greater than Mach 1.

supervisory control A form of automatic process control in which control loops operate independently subject to intermittent corrective action such as *set point changes arising from an external source.

supply pressure The pressure at the entry port to an item of process equipment.

surface finish The degree of roughness of a vessel or pipe surface, important for its durability. A clean surface is essential for providing maximum resistance to corrosion.

surface flux The amount of thermal radiation per unit area emitted from a flame or some other hot body. It is also known as the **surface emissive power**.

surface roughness The surface finish of pipes and vessels. There are various ways it is measured: The parameter Ra is the mean of the absolute values of height difference measured along the surface. Rt is the total roughness and describes the peak-to-valley height in a profile. Rz is the average roughness based on the successive measuring points or an individual peak and valley height. A profilometer is commonly used to measure a surface profile and uses a diamond stylus, which moves vertically in response to the roughness as it is guided over the surface. Other techniques include visual methods and scanning electron microscopy.

surface tension (Symbol σ) The force that acts on the surface of a liquid tending to minimize the surface area. The effect is caused by the attraction of molecules of the liquid to produce a film of tension over the surface. The SI unit is N m^{-1}. The c.g.s. unit of dyne per cm is still in common usage where 1 dyne cm^{-1} = 0.001 N m^{-1}. The surface tension of a liquid can be measured from the *contact angle of a drop of a liquid on a surface. Other methods include a Du Nouy tensiometer, which consists of a ring of wire placed on the surface of a liquid and measuring the force to remove it from the surface. The drop weight method involves a drop of liquid hanging from the end of a capillary tube. The downward force is supported by the surface tension just before detachment. The weight and radius of the drop can be measured, from which the surface tension can be determined.

surfactant A molecule that contains both polar and non-polar parts causing the molecule to act at the surface where different substances meet. Soaps are surfactants. Their structure allows them to detach grease and oil particles from a surface being cleaned and to emulsify them so that they can be washed away.

surge An unstable operating condition when the flow through a compressor is decreased to the point that momentary flow reversals can occur. This can lead to major damage of the compressor. **Surge control** is therefore used to prevent this and uses a **surge tank** which is a storage vessel, drum, or reservoir used to absorb unexpected rises in flow or pressure. It can also be used to provide additional fluid in the event of a drop in pressure or flow. Surge tanks are used in hydroelectric power stations to provide additional capacity and used to mitigate pressure variations due to rapid changes in the velocity of water.

suspension A mixture of particles suspended in a fluid and independent from one another. The particles may be either solid or liquid.

sustainability A process, business, or activity that is capable of being maintained at a steady level without exhausting natural resources or causing adverse ecological damage. Founded on the three elements of economics, effects on society, and the environment, it considers the wider effects and longer-term implications on the planet. The concept of sustainability has developed over the decades and can be traced back to United Nations Conference on the Human Environment held in 1972 that added the effect of the environment to the list of problems facing the existence of humankind. The UN Conference also led to the creation of the United Nations Environment Programme, which provides leadership and partnership in caring for the environment and enables nations to improve the quality of life without compromising future generations.

((()) SEE WEB LINKS
• Official website of the United Nations Environment Programme.

sustainable development The development of businesses, processes, and products that meets the needs of the present without compromising the ability of future generations to meet their own needs. It is based on the three elements of environment, economics, and society. A sustainable process requires all three. The 1992 Earth Summit held in Rio de Janeiro, Brazil, was the first global conference to address the issues of the environment being integrated into the issues of the global economy. An outcome from the conference was Agenda 21, which was a non-binding agenda that set goals and recommendations related to environmental, economic, and social issues. World leaders reaffirmed the principles of sustainable development at the 2002 World Summit held in Johannesburg, South Africa, and set an agenda for reducing world poverty and improving the lives of humans through responsible use of limited resources, while respecting the environment for future generations.

sweet gas Natural gas that contains very small amounts of hydrogen sulphide and carbon dioxide. North Sea gas is considered to be naturally sweet. *Compare* SOUR GAS.

Swiss cheese model A safety management tool used to explain how different but connected systems are related in achieving process safety. The barriers that prevent, detect, control, and mitigate accidents are depicted as slices of the cheese with each having a number of holes of differing sizes. The size and number of holes in each slice represents the imperfections in the barrier and are defined as specific performance standards. Well-managed processes have few or small holes. When two or more slices are put together, protection can be achieved and represented as the coverage of holes.

symbol A letter or a character used to present a quantity of something, a chemical reaction, a physical constant or variable, a mathematical operation, or relation.

syneresis The consolidation of a gel by the reduction in volume or contraction by forcing out interstitial water. It is used in the drying of solids.

syngas *See* SYNTHESIS GAS.

synthesis The formation of complex chemical compounds from simple compounds.

synthesis gas Also known as **syngas**, it is a mixture of carbon monoxide and hydrogen, made by steam reforming natural gas:

$$CH_4 + H_2O \rightarrow CO + 3H_2$$

A principal use of hydrogen is in the *Haber process. Before the Second World War, *water gas was previously used.

synthetic A substance that has been created artificially by chemical reaction and does not come from a natural source.

synthetic fibres Used in textiles, they are made from various raw materials. Those derived from petroleum, coal, and natural gas include polyesters, acrylics, nylon, polyethylene, polypropylene, polyvinylchloride, polyurethane, and synthetic rubbers. Fibres derived from cellulose include rayon, acetate, and triacetate. Inorganic fibres include glass and metal. Synthetic fibres such as nylon are produced by extruding the molten thermoplastic through extrusion dies called spinnerets into air that cools the fibres. The most widely used polyester fibre is PET (polyethylene terephthalate), which is used in carpeting.

synthetic hydrocarbons Hydrocarbon products that resemble petroleum products formed in high-temperature and pressure catalytic processes. The first products developed included methanol manufactured in the 1920s in Germany in a process operated at 400°C and 200 atmospheres using zinc and chromium oxide catalysts. The *Fischer–Tropsch process is used to produce saturated hydrocarbons of different molecular weight.

synthetic medium *See* DEFINED MEDIUM.

synthetic natural gas *See* SNG.

syphon The transfer of liquid from one vessel to another at a lower elevation by means of a pipe or flexible tube whose highest point is above the surface of the liquid in the upper vessel. It is a useful technique when the original (or upper liquid) has a layer of sediment, which must not be disturbed as in racking homemade wine.

system Refers to a quantity of a substance, or a group of substances, or energy under consideration contained within a space or transferred across a boundary. A system may be a mass of material or an energy contained within a boundary such as a vessel and isolated from the surroundings. Within an *isolated system the mass remains constant and the system is entirely uninfluenced by changes in its environment. In an *open system, it is possible to exchange energy and matter with its surroundings, whereas in a *closed system, energy can be transferred across the boundary but not matter, such as heating a vessel. This is an idealized system since there will be some exchange of energy and possibly material.

In a *steady-state flow system there is a transfer of energy or matter in and out such that the system remains constant. In a **cyclic system**, the final state is identical to the initial state. That is, the heat absorbed is equal to the work done by the system. In an **adiabatic process**, there is no heat exchange with the surroundings. The process is therefore thermally isolated or the process is very rapid such that heat has no time to enter or leave the system. This is an idealized process since insulation is not perfect and there will be some transfer of heat. In an **isothermal process**, there is no exchange of temperature with the surroundings. The type of system may depend on the timeframe of interest. For example, a vessel containing volatile liquid can be considered to be closed for a very short period after which the system can be considered to be open.

systematic error *See* ERROR.

Système International d'Unités Known more commonly as *SI units, it is a system comprising seven base units of the international metric system. The units are metre (m) for length, kilogram (kg) for mass, ampere (A) for electrical current, second (s) for time, kelvin (K) for temperature, candela (cd) for luminosity, and mole (mol). Derived units are the newton, joule, pascal, and watt.

S

tail end The final stage in the reprocessing of *nuclear fuel. The end products are uranyl and plutonium nitrate solutions. These can then be converted into new nuclear fuel.

tail gas The gas arising from the *Claus process that contains sulphur and sulphur dioxide. The gas is treated to remove sulphur vapour, sulphur dioxide, and traces of other sulphur compounds for release to the atmosphere. The recovered sulphur is returned to the process. The treatment involves reducing the sulphur compounds to hydrogen sulphide by passing through a bed of cobalt-molybdenum catalyst. The gases are cooled to remove excess water vapour.

tailings The largely uneconomic and non-metallic minerals separated from ores in mining processes known as *gangue.

tailrace A channel used for carrying away tailings from mining processes in water.

tails The heavy products recovered from the bottom of a fractional distillation column.

tangent 1. A straight line that has a contact point with a curve at which point it has the same slope as the curve. **2.** A trigonometric function. The tan of the angle, α, is the ratio of the side opposite to the angle to the side adjacent to the right-angled triangle.

tank A vessel used to contain liquids usually at atmospheric pressure. For closed tanks that are likely to become pressurized, such as due to pumping operations, a pressure relief system and vacuum breaker for pumping out may be fitted. A **tank farm** is a dedicated area of a chemical plant that is used for storage tanks. The tanks are used for storing liquids that may be feed, product, or reagents used in the plant. The location is generally near the process according to needs and safety issues.

tapping A point on a process vessel or pipe used to gain access to the material within. On a furnace, the tapping is used for drawing off molten metal. In a pipeline, it can be used to extract a small sample for analysis, or can be connected to a pressure-measuring device such as a manometer involving two tapping points across an *orifice plate meter.

tar A dark viscid substance obtained by the destructive distillation of organic matter such as coal, wood, or by petroleum refining. Coal tar fuels are produced by tar distillation and consist of either batch stills or pipe stills for continuous distillation. Tar oil is the product obtained from the distillation of coal tar.

tatoray process A catalytic process used for the transalkylation of toluene to a mixture of benzene and *xylene. The vapour phase process involves a fixed bed of zeolite catalyst and hydrogen. It is an abbreviation of **t**ransalkylation **a**romatics **Tor**ay developed by Toray Industries Inc. *Compare* XYLENE-PLUS PROCESS.

Taylor bubble *See* PLUG FLOW.

Taylor series An infinite power series used to determine the development of a given function. Obtained from Taylor's theorem, a function $f(x)$ can be expanded to the nth degree as:

$$f(x) = f(a) + \frac{f'(a)}{1!}(a-x) + \frac{f''(a)}{2!}(x-a)^2 + \ldots$$

Taylor's formula gives $f(x)$ with increasing accuracy, the larger the series used. The Taylor series can be used to develop a given function $f(x)$ in powers of $(x-a)$. It was developed by Brook Taylor (1685–1731).

Taylor vortices A secondary fluid flow pattern that can occur in the gap or annulus of a concentric cylinder, or cup and bob system known as a Couette. For a rotating bob and stationary cup, a shear rate may reach a critical value such that a series of rolling toroidal flow patterns occur in the annulus of the Couette. In a cup and bob-type rheometer, this gives rise to inaccurate measurements of viscosity.

TCE Formerly known as *The Chemical Engineer*, it is the official monthly magazine of the *Institution of Chemical Engineers covering news, current affairs, and events in the chemical and process engineering industry worldwide. It also includes updates on industrial developments, chemical engineering education, recruitment, and advertising.

 SEE WEB LINKS
• Official website of the magazine *TCE*.

tellerette A ring-shaped spiral used as a packing material in adsorption columns. It has a high specific surface area and is used to provide effective contact between a gas and liquid. Having a high voidage, the pressure drop across the packing is low.

Temkin isotherm An empirical adsorption isotherm that relates the quantity of gas molecules absorbed onto a surface with pressure: $\theta = k\ln(np)$ where θ is the measure of the sites occupied per unit area of surface measured as the ratio of the mass of absorbate to the mass of absorbent, p is pressure, k and n are empirical constants for a particular temperature.

temperature A measure of the intensity of heat that will flow into or out of a body or medium from another body or medium, and in which direction the heat flows. As a physical property of a body, it is proportional to the kinetic energy of the atoms or molecules. Where there is no heat flow, the body or medium is in thermodynamic equilibrium and at the same temperature as the other body or medium. Where they are not in equilibrium, the heat flows in the direction of the higher to the lower temperature body. There are various *temperature scales used to quantify the property of temperature including *kelvin, *centigrade or *Celsius, and *Fahrenheit scales.

temperature-entropy diagram A graphical representation of experimental thermodynamic properties for a substance in which absolute temperature, T, is presented on the y-axis and entropy, S, as the x-axis. The T–S diagram shows lines of constant pressure and distinguishes between saturated liquid, vapour, and superheated vapour. They are convenient to show the optimum efficiencies in various thermodynamic cycles such as *refrigeration.

temperature gradient The difference in temperature between two points of a known distance apart and is a measure of the direction of heat flow. The flow of heat is from the body or surface with the higher temperature to the lower temperature. It is the *driving force for heat flow.

temperature scales An empirical scale used to represent the physical property of temperature of a body. There are a number of scales in use and each is calibrated with fixed points to represent zero degrees, such as the freezing point of water and another temperature, such as the boiling point of water at standard atmospheric pressure, with a division between them to permit interpolation. In the case of the *Celsius this is 100 degrees. For scientific purposes, the *kelvin scale is used, which has absolute zero as a fixed point and other fixed points with divisions used to interpolate between them. They commonly are related as:

Fahrenheit to Celsius: $T_{(°C)} = 5/9(T_{(°F)} - 32)$
Celsius to Fahrenheit: $T_{(°F)} = 9/5(T_{(°C)} + 32)$
Celcius to kelvin: $T_{(K)} = T_{(°C)} + 273.16$

tempering A process used to harden alloys by heating to a particular temperature, holding for a given period of time, and then cooling at a controlled rate to room temperature. Steel is tempered to allow excess carbide to precipitate out of a supersaturated solution of *solid solution of martensite and then rapidly cooled by quenching in cold water to prevent further precipitation or grain growth.

temporary refuge A safe place on an offshore oil and gas platform where process operators and other personnel can take temporary shelter during emergency situations such as fire and gas leaks. This is usually within the accommodation block.

tensile strength A mechanical property of a material as a measure of the resistance to tensile stress. This is the force per unit area required to break the material. Steel is commonly used in chemical plants for pipework, vessels, and support structures on account of its high tensile strength and resistance to corrosion.

tensor A mathematical entity that is the general equivalent in any n-dimensional coordinate system. A *vector in a two- or three-dimensional coordinate system is a special case of a tensor. Tensors are used to describe how all the components of a quantity behave under certain transformations.

terminal velocity The velocity of a moving body that has attained a constant maximum velocity in which the forces on the body such as those due to gravity are balanced by the resistive drag forces. Particles can be separated from liquids in settling tanks such as lagoons under the influence of gravity. Where the terminal velocity is very slow and a more rapid separation is required, centrifugal separators can sometimes be used where much higher centrifugal forces on the particles increase the terminal velocity and therefore reduce the time for separation. *See* STOKES'S LAW.

ternary Composed of three parts. For example, a ternary mixture has three compounds or components; a ternary *alloy has three elements.

tesla (Symbol T) The derived SI unit for magnetic flux density and is equal to the flux of one weber in an area of one square metre. It is named after the Croatian-born US electrical engineer Nikola Tesla (1857–1943) who invented transformers, generators, and dynamos.

test separator A horizontal cylindrical vessel used on offshore platforms for the separation of gas and water from crude oil and is identical to a *production separator except that it is used to process the contents of a single well. The oil enters the separator through a manifold. The reduction of pressure causes the release of dissolved gases, which are removed from the top of the vessel. The water and oil separate by virtue of being immiscible and having different densities. The oil and condensate is separated from the water by overflowing a weir. The results from the well test separator are used to provide information on the performance of the well.

textile Fabric and other material made from combinations of fibres that are woven, knitted, braided, and tufted. The fibres are long, thread-like materials from natural sources such as animal, plant, or mineral, or chemically synthesized. They occur or are made into different forms, such as filaments, which are long continuous fibres; tow, which is a bundle of untwisted continuous fibres; and yarn, which is a bundle of twisted fibres.

TFR *See* TUBULAR FLOW REACTOR.

theorem A conclusion from a mathematical argument that has been proved based on certain assumptions. A **corollary** is a result that follows on from a theorem such that a separate theorem is not required.

theoretical air The amount of air that is required to burn completely a given amount of fuel. The amount is determined from the chemical composition of the fuel. In practice, excess air is required for complete combustion.

theoretical stage A part of a process in which two fluids are in equilibrium. Such an equilibrium stage may also be referred to as a **theoretical tray, equilibrium stage**, or **ideal stage**. In a *distillation column, a liquid and a vapour are close to being in equilibrium in the reboiler and partial condenser. On each of the trays within the column, the enrichment of vapour of the *more volatile component is less than one theoretical stage. This means that more stages or trays are required in practice to achieve a desired separation. In absorption columns, the amount of a gas absorbed is a fraction of the amount absorbed in a theoretical stage. In a liquid–liquid solvent extraction process, a *mixer-settler is close to one theoretical stage. *See* MURPHREE PLATE EFFICIENCY.

theory A description of a mathematical, physical, or chemical principle that does not fully cover all of the circumstances and has not fully achieved the incontrovertible status of a law.

therm The former non-SI unit of thermal energy equal to 10^5 British thermal units (Btu) equal to 105,505,600 joules.

thermal analysis A technique used to determine the chemical analysis of a substance by heating it. *Differential scanning calorimetry is an example in which a sample under investigation is heated or cooled under controlled conditions to allow the enthalpy change due to thermal decomposition to be studied.

thermal capacitance Another name for the *thermal mass of a material, which is the ability of a body to store thermal energy, and is expressed as the heat required to raise the temperature of a body by 1°C. It is the product of the mass of the body and the heat capacity. The SI units are J K^{-1}.

thermal conductivity (Symbol k or λ) The measure of the movement of heat through a body by kinetic molecular activity. It is used in *Fourier's law, which states that the thermal conductivity is independent of the temperature gradient but not necessarily of temperature itself. The thermal conductivity is the proportionality constant between heat flux and temperature gradient. That is, the rate of flow of heat (dQ/dt) through a surface of area A in a medium is given by:

$$\frac{dQ}{dt} = -kA\frac{dT}{dt}$$

where dT/dt is the temperature gradient measured in the direction normal to the surface. Values for the thermal conductivity vary widely for substances, with metals having the highest and finely powdered materials the lowest. The SI units are $W\,m^{-1}\,K^{-1}$.

thermal cracking A process that uses heat and pressure to break down and chemically alter heavy petroleum hydrocarbon molecules into smaller, lighter molecules. Typical pressures range from 7 bar to 70 bar and temperatures range from 450°C to 540°C. It is less frequently used than catalytic cracking since the yields of high-octane products are lower.

thermal death time (Symbol t_D) Used in the thermal sterilization process of foods that may be contaminated with harmful microorganisms, it is the time in minutes to bring about complete sterilization at a particular temperature, T. It is calculated using the *F-value and *z-value:

$$t_D = F_{121}10^{\frac{T-121}{z}}$$

thermal diffusion *See* SORET EFFECT.

thermal diffusivity (Symbol α) The rate at which thermal energy moves through a body due to a change in temperature. It is used in unsteady-state heat transfer calculations in which a body with a non-uniform temperature approaches equilibrium:

$$\alpha = \frac{k}{c_p \rho}$$

where k is the thermal conductivity, c_p is the specific heat capacity, and ρ is the density. Materials that have a high thermal diffusivity, such as metals, diffuse heat more quickly than materials with a low thermal diffusivity. When the temperature around the material changes, heat flows in or out of the material until thermal equilibrium is reached, assuming the environment around the material remains unchanged. Materials that have a high thermal diffusivity reach thermal equilibrium more quickly than materials with a low thermal diffusivity. The SI units are $m^2\,s^{-1}$.

thermal efficiency The ratio of heat output from a thermal device to the heat input expressed as a decimal or percentage. Examples include heat exchangers, furnaces, and dryers.

thermal expansion The increase in volume of a substance with temperature. Most substances increase in volume with increasing temperature. The coefficient of expansion, α, is the ratio of the change in length of a substance with temperature to the length at 0°C. The coefficient of volumetric expansion, β, is approximately three times the value of α. Water is a notable exception in which above 0°C, the volume decreases with temperature to approximately 4°C and increases in volume thereafter. The SI unit is K^{-1}.

thermal hysteresis When a body is heated and then cooled through the same temperature range, the temperature path taken is different.

thermal mass The ability of thermal materials, particularly those used for building and construction purposes, to moderate internal temperatures and to regulate heat release, and therefore to delay the time at which peak temperatures occur. It is the product of the mass and heat capacity of the material and has the SI units of $J\,K^{-1}$. A high thermal mass can store and later release large quantities of heat without a large temperature rise on the surface, whereas a low thermal mass can release its heat quickly. See THERMAL CAPACITANCE.

thermal oxidation process 1. A process used to produce a thin layer of silicon dioxide on the surface of a semi-conductor wafer of silicon. The high-temperature process causes the oxidizing agent to diffuse into the wafer. Using temperatures of up to 1,200°C, water or oxygen reacts with the silicon to form silicon dioxide as either **wet oxidation** or **dry oxidation**, respectively. **2.** A high-temperature process involving the oxidizing of combustible materials by above the auto-ignition temperature in the presence of oxygen, reducing them to carbon dioxide and water.

thermal radiation The energy emitted by all bodies above absolute zero temperature due to the excitation by molecular vibration. Energy is transmitted between bodies without heating the space in between unless the medium is capable of absorbing energy. The *Stefan–Boltzmann constant, σ, is used in thermal radiation calculations in which the rate of heat loss from a *black body is:

$$q = \varepsilon \sigma A\left(T_1^4 - T_2^4\right)$$

where ε is the emissivity, σ the Stefan–Boltzmann constant, A is the area, and T is the absolute temperature. The SI units are $J\,s^{-1}$.

thermal resistance (Symbol R) The resistance of a body to transmit thermal energy. It is the ratio of the driving force to the rate of heat transfer. A lagged pipe carrying a hot fluid has thermal resistances that include convective resistance on the inside of the pipe, resistance due to conductivity of the pipe wall material, resistance due to the lagging material, convective resistance on the outside of the pipe, and possibly the resistance to heat transfer due to fouling on the inner pipe wall surface. It has SI units $m^2\,K\,W^{-1}$.

thermal runaway See RUNAWAY REACTION.

thermite process A highly exothermic process used to produce molten iron for in-situ welding such as railway lines. The reaction involves igniting a mixing iron oxide with aluminium to form aluminium oxide and molten iron:

$$2\,Al + Fe_2O_3 \rightarrow Al_2O_3 + 2\,Fe$$

It is also known as aluminothermy.

thermochemistry A branch of chemistry involving the study of the heat energy from a process and includes chemical reactions and physical changes of state.

thermocouple A temperature-sensing instrument that consists of a pair of dissimilar metal wires joined together. One pair of wires operates as a reference junction and the other as the sensing junction. Where a temperature difference exists, an e.m.f. difference is measured and a current flows from which the temperature is determined.

thermodynamic cycles Heat engines, refrigeration cycles, and steam cycles can all be represented using ideal thermodynamic cycles. Heat engines are usually represented on pressure-volume or *temperature-entropy diagrams. Refrigeration and steam cycles are usually represented on temperature-entropy diagrams. Reciprocating machines and simple air compressors are usually shown on pressure-volume diagrams.

thermodynamic diagrams Charts used to present complex thermodynamic data for materials over a wide range of conditions. They are used to simplify thermodynamic calculations. Commonly used diagrams include enthalpy-concentration, pressure-enthalpy, temperature-entropy, enthalpy-entropy (*Mollier), and *psychrometric charts. The reference conditions are the pure compounds at some specified condition such as pressure.

thermodynamic equilibrium The condition of a system in which the quantities that specify the system, such as temperature and pressure, remain unchanged. It is often abbreviated to *equilibrium.

thermodynamics The study of the relationship between properties of matter, changes in these properties, and transfers of energy between matter and its surroundings that bring about these changes. These changes may be both physical and chemical. There are four laws of thermodynamics that define the relationships in terms of temperature, energy, and entropy. The *zeroth law of thermodynamics states that for two bodies being in thermal equilibrium with a third body, then they must all be in thermal equilibrium with each other. The *first law of thermodynamics refers to the conservation of energy in which the change in internal energy of a system is equal to the difference in the heat added to the system and the work done by the system. The *second law of thermodynamics states that it is impossible to construct a device that operates in a cycle and produces no effect other than the transfer of heat from a cooler body to a hotter body. The law sets the limit on the amount of heat energy that can be converted to useful work energy. The *third law of thermodynamics enables absolute values to be stated for entropies by stating that the entropy of a system approaches a constant value as the temperature approaches absolute zero. It therefore provides an absolute scale of values for entropy by stating that for changes involving only a pure crystalline substance at absolute zero, the change of the total entropy is zero.

thermodynamic temperature The temperature defined in terms of the laws of thermodynamics and therefore independent of the properties of the body being measured. It is usually expressed using the *kelvin scale. It is not measured directly but is usually inferred from measurements with a gas thermometer that contains a nearly ideal gas. The thermodynamic method to specify temperature was proposed by Lord *Kelvin.

thermolysis The decomposition of a substance as the result of heating. It is used in the process of thermal *cracking of hydrocarbons to produce smaller hydrocarbons of lower molecular weight.

thermometer An instrument used to measure temperature based on the thermal expansion of a gas or liquid. Commonly used liquid-in-glass thermometers consist of a bulb containing a liquid such as mercury or some other liquid such as alcohol coloured with a dye, and a long graduated capillary. As the temperature of the liquid rises, the liquid expands out of the bulb and moves along the graduated scale. The temperature is read directly. Other principles of operation include the expansion of metal and bi-metallic materials, the change in resistance to the flow of electricity and semi-conductors (thermistors). Copper, platinum, and nickel are the most commonly used metals in resistance thermometers.

thermonuclear reactor A type of reactor in which nuclear fusion takes place with the controlled release of a considerable amount of energy. While such reactors are still at the experimental stage, the main challenges are the ability to reach the very high temperatures needed that are in excess of a million degrees Celsius, and containing the reacting nuclides for a sufficient period of time to achieve the required ignition temperature. In a *tokamak, powerful magnets are used to guide the charged plasma around the toroidal-shaped reactor and to prevent collisions with the walls.

thermophilic bacteria A type of bacteria that is tolerant of heat or temperature. They can survive at temperatures in excess of most other bacteria and have been found in sulphur-rich thermal volcanic vents such as geysers and fumaroles. Their resistance has been exploited in some biotechnological processes. For example, heat-resistant **thermophilic enzymes** are used in products such as washing powders.

thermostat A type of on–off device used to automatically control the temperature of a system, a piece of equipment such as a domestic boiler in which heat is applied when the temperature falls below a desired value. The heat is applied to the point that a maximum allowable temperature is reached, at which point the application of heat is halted.

thermo-syphon reboiler A type of *reboiler heat exchanger used to boil up the liquid from the bottom of a distillation column. It consists of vertical tubes heated with condensing steam in which natural circulation of the liquid is caused by the reduced density of the heated liquid in the tubes drawing in more liquid. No pump is therefore required. They are also known as *calandrias.

thickening A separation process used to remove liquid from a suspension of particles in a solution, usually under the influence of gravity. The settling of the particles gives rise to an increase in the concentration of particles in a solution as a sludge or slurry in which clear liquid above overflows and is removed. In a thickener, which consists of wide tank with a slightly conical base, a revolving rake moves the sludge towards the centre of the base for removal. *Compare* CLARIFICATION.

Thiele–Geddes A procedure used for stage-wise calculations for multicomponent distillation problems. The procedure involves computations from both ends of the column and works towards the middle. Numerical instabilities occur when stage-wise calculations cross a feed stage. The procedure is not suitable for multiple-feed or draw-off columns. *Compare* LEWIS–MATHESON.

third law of thermodynamics The law which states that the entropy of a perfect crystal is zero at a temperature of absolute zero.

thixotropic fluids Fluids such as certain gels, paints, and lubricants that have a viscosity that decreases when a stress is applied, as in stirring, and is also dependent on the time that the stress has been applied. *Compare* ANTITHIXOTROPIC FLUIDS.

Thomson, James (1822–92) A Scottish engineer who studied civil engineering at the University of Glasgow where his younger brother William (later Lord *Kelvin) was also a student. He worked on water-power engineering and thermodynamics, and invented various water wheels and turbines. He became professor of civil engineering. He calculated the effect of pressure on the melting point of ice, which was shown by experiment to be correct by the work of Lord Kelvin.

Thomson, Sir William *See* KELVIN, LORD.

Three Mile Island accident A major nuclear accident that occurred on 28 March 1979 at the Three Mile Island Generating Station in Pennsylvania, USA, involving a *core meltdown and release of radioactive material into the environment. The cause was attributed to a *pilot-operated relief valve that had become stuck open allowing reactor coolant to the pressurized water reactor to escape. Human factors were also attributed to the accident including inadequate training and an operator manually overriding the automatic emergency cooling.

three-term control *See* PID CONTROL.

threshold limit value (TLV) The airborne concentration of a particular substance used to define conditions under which nearly all people may be repeatedly exposed for a working lifetime without causing adverse effects. The time weighted average or TWA is the time concentration of a particular substance for an eight-hour day or 40-hour working week to which nearly all workers may be exposed. The ceiling TLV is the airborne concentration that should not be exceeded at any time.

throttling A way of controlling the rate of flow of a fluid in a pipe by means of a regulating valve. For example, it is used to control the rate of vaporized fuel to an internal combustion engine. Throttling is used in the suction line to a compressor in which a pressure regulator adjusts a control valve, normally a butterfly valve.

throughput The total amount of material fed to or produced as a product from a process per unit time. It can be expressed as either a mass or a volumetric rate. In oil refineries, the throughput refers to the stream of feedstock supplied.

tie-in The joining of two sections of pipeline. Tie-ins may be used in modifications to an existing pipeline to make it join a new pipeline, or in the joining of various sections of a newly laid pipeline constructed in sections.

tie line A horizontal line used in a liquid–vapour phase diagram for two substances. The line extends from the point that a liquid can coexist with its vapour to the point where only vapour exists as the pressure is reduced. A tie line is perpendicular to an *isopleth.

tie substance An inert and unreactive substance, such as nitrogen, which passes through a system or process in a completely unchanged form. It enters in an input stream and leaves via an output stream. It therefore forms a tie between the input and output streams and as a result is useful for undertaking material balance calculations such as for combustion processes.

time constant (Symbol τ) A parameter that determines the dynamic path of a process to respond to a disturbance and approach a new steady state. It is a product of both capacity and resistance. A low resistance means that the final steady-state condition is reached quickly. 63.5 per cent of the final change corresponds to one time constant after the input change.

titration A laboratory technique used to find the exact volume of acid of unknown concentration that is needed to neutralize a certain volume of alkali of known concentration. It uses a burette to carefully administer the acid and a colour indicator, such as phenolphthalein, to indicate the end point in the technique.

TLV *See* THRESHOLD LIMIT VALUE.

TNT An abbreviation for trinitrotoluene, it is a pale yellow material used as a high explosive first prepared by German chemist Julius Wilbrand (1839–1906). The **TNT equivalent** is a convenient way of expressing the magnitude of an explosion by calculating the amount of TNT which, when detonated at a particular point, would cause the same level of *blast wave damage. The ton of TNT is a unit of energy equal to 4.184 gigajoules of energy, and is the approximate amount of energy released on the detonation of one ton of TNT.

tokamak A toroidal *thermonuclear reactor used in thermonuclear fusion experiments originally developed in Russia in the 1960s. It has a strong axial magnetic field that keeps the plasma inside the ring-shaped vacuum from touching the external walls—a hot plasma that hits the walls will rapidly cool, ending the fusion process. The name comes from the Russian for 'toroidal chamber with magnetic coils'. *See* JET.

tolerance A range of physical dimensions of an object within which the true dimensions lie. It is used in machining components such as the meshing gears where clearance must be controlled.

tomography A non-destructive imaging technique used to examine the internals of process pipes and equipment. It uses penetrating waves such as X-rays and produces a sectional image known as a tomogram. Invented by electrical engineer Sir Godfrey Newbold Hounsfield (1919–2004), it uses a computed tomographic or CT scanner. The scanner rotates around the object under investigation and takes a series of X-ray measurements. The accumulated data is then used to construct a three-dimensional image. Tomography has made a substantial contribution to medicine as well as many other areas of science and engineering.

tonne A metric unit of weight equal to 1,000 kg. The long or gross ton of 2,240 lb (1,016.05 kg) was removed from official UK measures in 1985, whereas the short ton is still used in the US and is equal to 2,000 lb (907 kg).

tons refrigeration A unit signifying the capacity of a refrigeration plant or air-conditioning system using the Imperial system of measurement in which 1 ton of refrigerant will freeze 1 ton of water at 32°F (0°C) in 24 hours. A one-ton unit is equivalent to 3.52 kW.

topping The separation of *crude oil to remove light fractions by distillation. A **topping refinery** separates crude oil into light components with no further refining being involved. The components are used as fuel.

top product The vapour or liquid drawn from the top of a distillation column. In a continuous distillation process, a portion is returned to the column for further enrichment known as *reflux. In a batch distillation process, the composition of the top product changes with time as volatile components are progressively drawn off.

topsides The oil and gas processing equipment installed in an offshore platform located above the splash zone. These include drilling, gas and liquid processing, accommodation, utilities, services, and safety equipment. On floating production systems, the topsides equipment is normally pre-assembled into skids, which are located above the deck to provide some protection against large breaking waves.

torque (Symbol τ) The ability of a mechanism to do work, usually by rotation, such as a bolt or a rotating shaft on a pump or compressor, or viscometer, which takes into account the distance through which force is applied:

$\tau = FL$

where F is the force and L is the length from the axis of the shaft. The SI units are N m.

torr A unit of pressure used in high vacuum applications and equal to one millimetre of mercury or 133.322 Pa. The unit is named after Evangelista *Torricelli (1608–47) who discovered the principle of the barometer.

Torricelli, Evangelista (1608–47) An Italian scientist and mathematician who was professor of mathematics in Pisa, and did work on the thermal expansion of liquids. For a short time he was an assistant to *Galileo and started building telescopes after the death of his teacher. He discovered the attraction between the Earth and air molecules that results in pressure and concluded by saying 'We live submerged at the bottom of an ocean of the element air'. He wrote on a number of subjects in applied mathematics, including the movement of a stream of water through a small hole in the side of a container. The **Torricelli theorem**, which he proposed in 1643, is named in honour of him: $v = \sqrt{2gh}$, where v is the velocity of the discharging jet of liquid and is proportional to the square root of the supplied head, h. The equation was derived from a balance between potential and kinetic energies for which energy losses are neglected. His most important development was the mercury barometer in 1643 with the help of his pupil Vincenzo Viviani (1622–1703) who demonstrated the existence of a vacuum. He was able to show that the pressure of the atmosphere varied with weather conditions. This important work led to the development of meteorology for weather prediction. Pascal's siphon, Samuel Morland's diagonal barometer, and Robert Hooke's quadrant barometer are all variations and improvements on Torricelli's instrument.

Torricellian vacuum The vacuum formed in a vertical glass tube filled with mercury and closed at the upper end. The level of mercury in the tube is used to measure atmospheric pressure as a barometer. The pressure of the Torricellian vacuum formed above the level of the mercury is equal to its vapour pressure, which is about 10^{-3} torr. It is named after Evangelista *Torricelli (1608–47) who discovered the principle of the barometer.

total acid number (TAN) A measure of the potential corrosivity of crude oils. It is expressed as the number of milligrams of potassium hydroxide that are required to neutralize one gram of crude oil. Values of greater than 1 mg are considered corrosive and labelled as High-TAN crudes.

total annual costs The costs incurred on a process on an annual basis. It is the sum of the *fixed costs, *variable costs, and taxes. Along with the *economic potential, it is a useful economic indicator of a process when considering the preliminary design of a process.

total condenser A type of heat exchanger used in a distillation process in which all the vapour from the top of the column is condensed and part of it is drawn off as the top product. *Compare* PARTIAL CONDENSER.

total dissolved solids (TDS) The total amount of organic and inorganic substances dissolved in water. The principal constituents in fresh water are sodium, potassium, calcium, and magnesium, sodium, and potassium cations, and carbonate and hydrogen carbonate, chloride, sulphate, and nitrate anions. It is a measure of the quality of water for drinking and for rivers. The main source of total dissolved solids is from sewers, urban sources, and agricultural surface runoff. The total dissolved solids are measured in parts per million (ppm). The World Health Organization considers drinking water to be unacceptable if the total dissolved solids content is in excess of 1,200 ppm. *Compare* TOTAL SUSPENDED SOLIDS.

((⊕)) SEE WEB LINKS
• Official website of the World Health Organization information on water sanitation.

total heat The sum of the sensible heat, latent heat, and superheat (if any) of a substance. As the temperature and pressure rise in a substance, the amount of sensible heat increases, while the latent heat decreases.

total moisture The total amount of water that is retained within the pores of a solid substance. The inherent moisture is the moisture that is stored within the pores of particles and which can only be removed by heating.

total pressure The sum of the partial pressures of all the constituents in a mixture of gases and vapours. *See* DALTON'S LAW.

total reflux A condition in which all the overhead product from a distillation column is returned back to the column as reflux. On a *McCabe–Thiele diagram, increasing the *reflux ratio moves the operating lines of both stripping and rectifying sections away from the equilibrium line reducing the number of theoretical plates required to bring about a desired separation until they follow the 45° diagonal. This corresponds to the minimum number of trays required in the column for the separation.

total solids The solids remaining in a liquid solution after evaporation and may include colloidal, soluble, and suspended solids.

total suspended solids The amount of suspended material in water. It is a measure of the quality of water for drinking and for rivers, and is determined by filtering a volume of water through a glass fibre filter and drying it. The total suspended solids is expressed as the collected dry weight per volume of water filtered. *Compare* TOTAL DISSOLVED SOLIDS.

tower Another name for a column, which is a tall, cylindrical vessel used for carrying out processes such as distillation, absorption, or extraction.

town gas A manufactured fuel gas for domestic and industrial use such as coal gas, substitute natural gas (SNG), and natural gas. It is typically composed of about 50 per cent hydrogen, 25 per cent methane gas, between 7 and 17 per cent carbon monoxide, and lesser amounts of carbon dioxide, nitrogen, and hydrocarbons. It is toxic and its characteristic unpleasant smell is due to its sulphur content.

toxic The harmful properties of a chemical substance or some physical agent on humans. The harmful effects are the result of absorption, inhalation, or digestion and result in illness, injury, or death. A **toxic hazard** is a type of atmospheric hazard that may be poisonous, toxic, or harmful if inhaled.

trace heating The heating of a fluid in a pipe by means of steam or electrical heating elements. The fluid may have a relatively low melting point such as bitumen and is required to remain in the liquid state to prevent it from solidifying and blocking the pipe. The trace heating consists of an internal pipe carrying the steam, which runs down the centre of the pipe carrying the fluid to provide sufficient heat transfer that prevents solidification.

tracer A radioactive isotope used to follow a chemical or biochemical reaction.

train A group of similar process units that are operated in series. A *distillation train is a series of distillation columns used to progressively remove or cut components from multi-component feed mixtures.

transducer A device used to convert a signal from a sensed process variable into another form of signal. Sensed variables may be temperature and pressure, and converted to pneumatic pressure, voltage, or current. These signals are then used either directly to control valves or used by computers where the signals are digitized and used with other signals to control a process or system. An **I/P transducer** is a type of transducer that converts an electrical input signal to a pneumatic output signal while a **P/I transducer** converts a pneumatic output signal to an electrical input signal. The electrical signal is in the range 4 to 20 milliamps, while the pneumatic signal is 3 to 15 psig (20–100 kPa).

transfer function Used in process control to represent the relationship between the output to input signal from and to a process in which the differential equations used to describe the signals as functions of time have been transformed using Laplace transforms. Laplace transforms are used to simplify the calculations since differential equations do not readily enable the relationship between the output to the input to be discerned. The transformation is said to be from the time domain to the s-domain. If the input is $y(t)$ and the output is $y(t)$ then the transformed function $G(s)$ is:

$$G(s) = \frac{Y(s)}{X(s)}$$

Capital letters are used for variables in the s-domain and lower-case letters for the time domain.

transformation 1. The conversion of reactants to products irrespective of the chemical, physical, or biochemical process route involved. **2.** The change in a mathematical expression or equation resulting from the substitution of one set of variables by another. **3.** The change in an atomic nucleus to a different nuclide as the result of the emission of either an *alpha particle or a *beta particle.

transient response A short, brief, or temporary change in the state of a system when there has been a disturbance. The transient effects in a well-controlled system tend to die away with time, allowing the system to settle to a steady-state condition. For example, a thermometer initially reading a temperature T_1 immersed in a hot oil of temperature T_o will read a new temperature of T_2:

$$T_2 = T_o - (T_o - T_1)e^{\frac{-t}{\tau}}$$

where τ is the *time constant. The exponential term will eventually die away with time to give the steady-state value of the thermometer reading the oil temperature.

transition 1. The change from one state to another. **2.** The period of time during which a change takes place, passing from one state to another.

transition flow A flow regime that exists between laminar or streamline flow and turbulent flow. Velocity fluctuations may be present and impossible to predict.

transition point 1. The point at which a moving fluid increases in velocity and changes from laminar flow to turbulent flow. **2.** The temperature at which a substance changes phase such as from liquid to solid. **3.** The temperature at which a crystalline substance changes into another crystalline form.

transition temperature The temperature at which an immediate change of physical properties occurs, such as a change of phase, crystalline structure, or conductivity.

transmissivity The portion of radiant energy falling on a surface that is transmitted through the body. *Compare* ABSORPTIVITY, REFLECTIVITY.

transmitter A transducer that responds to a measured *process variable by way of a sensing element and converting it to a signal that is a function only of the measurement.

transuranic element An element whose atomic number is greater than that of uranium (i.e. exceeds 92). These elements do not occur in nature. Neptunium, with an atomic number of 93, is the first.

trapezium rule A form of *numerical integration used to find the approximate area under a curve by dividing it into parts of trapezium-shaped sections that form columns of equal width with bases that lie on the horizontal axis. Also known as the trapezoidal rule, it is calculated from:

$$\int_b^a f(x) \approx (b-a)\frac{f(a)+f(b)}{2}$$

trays Perforated plates used in distillation columns that permit vapour to rise up through a layer of liquid as bubbles and designed in such a way that the liquid is unable to flow down at the same time. There are various designs such as sieve trays, which are located both above and below the feed tray. Stripping trays are located below the feed tray in which the concentration of the less volatile component in the liquid increases as the liquid flows down the column. The purity of the bottom product is increased by increasing the number of trays. Rectifying trays are located above the feed tray in which the concentration of the more volatile component is enriched as the vapour rises from tray to tray up the column. Again, the purity is increased by increasing the number of trays.

trenched piping The organized routing of pipework within a wide, open trench to transport fluids such as raw materials, products, and utilities such as water and steam to and from process equipment. The trench is usually trapezoidal in cross section. Being open and at ground level, leaks from pipes into the trench are easy to detect. In the event of leakage, liquids and heavy gases are contained in the bottom of the trench. *Compare* PIPETRACK.

trial and error A problem-solving technique used to obtain a solution to a problem by using reasoning. The method is useful when there is insufficient knowledge or information within a problem to reach a solution by analytical means. A reasoned judgement is therefore made and further adjustments are made based on the effects. *Compare* GUESS AND CHECK.

triangular diagrams A graphical presentation of the interaction of three components in a mixture known as a *ternary liquid system. The diagram (see Fig. 56) uses equilateral triangular coordinates, with each liquid being presented as either a mass fraction or percentage in terms of the other two. Any point on the side of the diagram represents a binary mixture. Where one pair of liquids is partially soluble in each other and both are fully soluble in the third at a particular temperature, the diagram is known as an isotherm. An example is benzene, water, and acetic acid with benzene dissolving completely in the other two. For example, in Fig. 56a, any *ternary liquid outside the solubility curve at P is a homogenous solution of one liquid. Any ternary liquid below at Q will form two insoluble, saturated liquid phases of equilibrium compositions R and S. The line RS is the tie line, which passes through Q. The plait point T is the last tie line and the point where A- and B-rich solubility curves merge. Where the two pairs of liquids are partially soluble, such as chlorobenzene (A) and water (B), and water and methyl ethyl ketone (C), the isotherm appears as Fig. 56b. Homogeneous ternary solutions are formed at P while two liquids phases at equilibrium appear with R and S corresponding to tie lines in the heterogeneous area.

triboelectrification The generation of static electricity caused by friction in flowing fluids and solids.

tribology The study of friction, lubrication, and wear between moving surfaces.

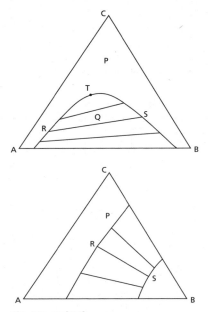

Fig. 56a and 56b

trigonometric functions The representation of relationships between the sides and angles of a right-angled triangle, such that for an angle θ:

sine θ = opposite side/hypotenuse
cosine θ = adjacent side/hypotenuse
tangent θ = opposite side/adjacent side
secant θ = hypotenuse/adjacent side
cosecant θ = hypotenuse/opposite side

trigonometric series An expression for the sine and cosine trigonometric functions as convergent power series:

$$\sin x = \frac{x}{1!} - \frac{x^3}{3!} + \frac{x^5}{5!} - \frac{x^7}{7!} + ...$$

$$\cos x = 1 - \frac{x^2}{2!} + \frac{x^4}{4!} - \frac{x^6}{6!} + ...$$

trip The fast shutdown of an item of chemical plant or process equipment such as a pump. The shutdown is the result of a process condition being exceeded such as an abnormal flow, pressure, temperature, or concentration, etc. In a nuclear power plant, a trip can lead to the fast shutdown of a nuclear reactor by rapid insertion of the neutron-absorbing *control rods.

triple point The temperature and pressure at which the gas, liquid, and solid phases (or states) of a substance are in equilibrium. The triple point of water, in which vapour, liquid, and ice phases exist in equilibrium, is 0.01°C and 611.2 Pa. The triple point of water forms the basis of the *thermodynamic temperature scale proposed by Lord *Kelvin.

triplex pump A type of reciprocating pump that has three cylinders. The pistons or plungers operate out of phase with one another such that the outflow is continuous and is the sum of the discharging cylinders. They are commonly used in offshore drilling operations and can handle a wide range of fluids including slurries and abrasive and corrosive fluids.

trouble-shooting A form of problem-solving used to identify, solve, and eliminate problems within a process that has failed or has the potential to fail. It is a logical and systematic search for the source or cause of the problem, and solutions presented to ensure that the process is restored back to its full operability. Trouble-shooting is often applied once a problem has arisen and the process stops functioning. In its simplest form, it can take the form of a systematic checklist and requires critical thinking. Computer techniques are used for more complex systems where a sequential approach is either too lengthy or not practical, or where the interaction between the elements in the system are not obvious.

Trouton's rule A method used to determine the approximate latent heat of vaporization of a substance based on its normal boiling point as:

$$\frac{\lambda}{T_{BP}} = \Delta s \approx 90 \ Jmol^{-1}K^{-1}$$

It is used where no heat of transformation data exists or can be readily found. For polar liquids, the value increases while for non-polar liquids, the value decreases. It should not be used for highly polar or non-polar molecules. It is named after Irish physicist Frederick Thomas Trouton (1863–1922).

tube bundle Pipes in a shell and tube heat exchanger that are packed into an arrangement to ensure effective heat transfer from the outer surface and good transport for fluids through the tubes. The tubes in the tube bundle are spaced and typically set with a rectangular or triangular pitch, and held and sealed with a tube plate. Baffles also provide rigidity and encourage turbulent flow of fluids through the shell side. The tubes can be a straight single-pass or hairpin double-pass arrangement. The tube bundle can be removed from the shell for periodic cleaning. Lugs are welded to the baffles for lifting purposes.

tube sheet A mounting plate used for the support and spacing of tubes in heat exchangers, boilers, coolers, and filters.

tube side A reference to the inside of the tubes of a shell and tube heat exchanger through which a heat transfer medium flows. *Compare* SHELL SIDE.

tubing A conduit with a circular cross section used for the transportation of fluids. Available in a wide variety of materials such as metals, glass, and plastic, there is no clear-cut distinction between tubing and pipes. In general, tubing is thin-walled and comes in coils with long lengths. Sizing is usually indicated by the outside diameter. The wall thickness is given by the BWG (*Birmingham Wire Gauge) number.

tubular flow reactor (TFR) A commonly used chemical reactor consisting of parallel pipes or tubes contained within a cylindrical vessel. Reactants are fed to one end and the reacted product is withdrawn from the other. A heat transfer medium can be circulated between the tubes making them useful for the conversion of raw materials to products in chemical reactions that require heat exchange.

tundish A vessel with a broad opening or funnel at the top with one or several holes at the bottom. It is used in plumbing and metal-founding.

turbidity The clarity of water as a way of determining its quality. An optical measure of the turbidity of a fermenting broth in a bioreactor can be used to determine the rate of growth of microorganisms. Within limits, there is usually a direct relationship between *cell dry weight and turbidity.

turbidostat cultivation *See* CONTINUOUS CULTIVATION.

turbine A machine used to generate electricity by the expansion of a gas or vapour at high pressure through a set of blades attached to a rotor. The blades rotate as the result of the expansion and conversion of energy. *Gas turbines and **steam turbines** are commonly used to generate electricity. A nozzle is used to direct the high-speed gas or steam over a row of turbine blades. The fluid pushes the blades forwards causing them to rotate due to the change in momentum. A row of stationary blades within the turbine redirects the fluid in the correct direction again before it passes through another set of nozzles and expands to a lower pressure. A steam turbine may have several pressure sections and operate at high pressure, medium pressure, and as the steam expands, a low-pressure section, all linked to the same shaft. The steam in the medium-pressure section may be returned to a boiler and reheated before doing further work, to prevent the formation of water in the turbine.

turbulent flame The propagation of a flame under turbulent flow conditions such as a jet engine flame.

turbulent flow A fluid flow regime characterized by the fluctuating motion and erratic paths of particles. In pipes of circular cross section, this occurs at *Reynolds numbers in excess of 4,000. Turbulent flow occurs when inertial forces predominate resulting in macroscopic mixing of the fluid. *Compare* LAMINAR FLOW.

turnaround A term used for a planned downtime of a process plant for maintenance. The **turnaround time** is the time required to prepare and restore a process or an item of equipment back into operation. It may include cooling, emptying and cleaning, charging, and reheating to bring the process or equipment back on stream.

turning point A point on a graph at which the gradient of a tangent of a mathematical function changes sign. Where a gradient changes from a positive to a negative, the point represents a maximum point; where a gradient changes from a negative to a positive, the point represents a minimum. Turning points are also known as **stationary points**. Where the mathematical function $y = f(x)$ is known, the turning point can be obtained from the first derivative:

$$\frac{dy}{dx} = 0$$

A maximum or minimum can be identified from the sign of its second derivative.

turn-key A process or system that has been designed and built by a contractor that is ready for immediate use and for a fixed fee.

TWA An abbreviation for **t**ime **w**eighted **a**verage, it is an occupational exposure limit used to protect the health of employees against exposure to harmful airborne substances. It is averaged over a defined period of time to which workers may be exposed by inhalation. *See* WORKPLACE EXPOSURE LIMITS.

12-D process *See* BOTULINUM COOK.

two-dimensional flow An approach used to determine the flow of a fluid in which all velocities are parallel to a given plane. The flow between two parallel flat plates is considered to be two-dimensional. Either rectangular (x,y) or polar (r,θ) coordinates are used to describe the flow characteristics.

two-phase multiplier A factor used in the determination of pressure drop in two-phase flow. It is determined from the superficial velocities of both the gas and liquid phases for which the respective pressure drops can be determined and then combined to determine the overall pressure drop using the two-phase multipliers for both phases. They are obtained from charts, graphs, and mathematical correlations.

two-property rule A rule used to uniquely define a system and requires specification of two independent properties such as specific internal energy, specific volume, specific enthalpy, absolute temperature, and specific entropy. All of the other properties can be found if the two independent properties are known. The properties are independent if one can be varied while the other does not change. *See* PROPERTY DIAGRAM.

Uchatius process A process used in the nineteenth century to make cheap steel that involves pouring molten cast iron into water. The iron granules are then mixed with fresh iron containing manganese and fireclay, and heated. The molten product is then poured into moulds. It was named after Austrian engineer Franz von Uchatius (1811–81), an officer in the Austrian army who invented the process in England in 1855.

UFD *See* UTILITY FLOW DIAGRAM.

Ufer process A process once used for refining light oil produced from the carbonization of coal. The oil is first washed with sulphuric acid before adding water. The resulting mixture of resins passes into the oil phase. The dilute sulphuric acid was then used to make ammonium sulphate. The process, named after its German inventor A. Ufer, was used in the 1920s and 1930s.

UFL *See* UPPER FLAMMABLE LIMIT.

UHT *See* ULTRA HIGH TEMPERATURE.

ultimate analysis A type of compositional analysis of fuels expressed in percentages for carbon, hydrogen, oxygen, nitrogen, sulphur, and ash. The mnemonic 'no cash' is an easy way to remember the elements. The *proximate analysis is the analysis of fuels in terms of moisture content, volatile matter, ash, and fixed carbon content.

ultimate-cycle tuning method An empirical procedure used to tune a controller using *PID control with optimum controller settings. Developed by J. G. Ziegler and N. B. Nichols in 1942, it assumes that open loop transfer functions can be approximated by a first order system with a time delay. The settings for the tuned controller result in an under-damped transient response with a *decay ratio of a quarter.

ultimate oxygen demand (UOD) A method used to determine the demand for oxygen in wastewater and is a measure of the conversion of all the carbon being converted to carbon dioxide and nitrogen to nitrates ions. It is less widely used than the *biochemical oxygen demand (BOD) and *chemical oxygen demand (COD) tests, which are used to measure the effects of pollution in water.

ultimate tensile strength The maximum strength that a material can withstand before fracture and failure. It is therefore the maximum stress, which is the applied load divided by the cross-sectional area of a test piece of material. The SI unit is the pascal and is more commonly expressed as MPa. For materials that don't deform under an applied load, this is the nominal stress at the point the material breaks. For materials that deform, this is the point that necking forms before breakage.

ultracentrifugation A high-speed centrifugation process used to separate very small particles such as colloids from liquids, and macromolecules such as proteins and nucleic acids from solutions. It uses speeds of up to 60,000 rpm and can generate a force on a particle of up to two million times greater than the force of gravity.

ultrafiltration The separation of very fine particles and molecules by filtration through a microporous or *semi-permeable membrane. It is used to separate molecules with a molecular weight in the range of 3,000 to 100,000. Ultrafiltration therefore can separate out macromolecules such as proteins, polysaccharides, and fat globules while allowing smaller water and lactose molecules to pass through the membrane as permeate.

ultraforming process A catalytic reforming process of naphtha used in the petroleum industry that uses platinum and rhenium catalyst in a swing reactor. The reactor can periodically be taken off-line so that the catalyst can be regenerated.

ultra high-pressure processing *See* HIGH-PRESSURE PROCESSING.

ultra high temperature (UHT) A process used in the food industry to sterilize liquid beverages such as fresh fruit drinks and milk. It involves heating the liquid to a temperature of 140°C using a *plate heat exchanger and holding the temperature for a couple of seconds. This is sufficient to destroy all potentially harmful pathogenic microorganisms and spores. The effect of the process on milk, however, is to caramelize the sugars giving a sweeter taste. It also denatures some of the proteins making processed milk unusable for making cheese. The process is also known as **ultra heat treatment**.

ultra orthoflow process A fluid catalytic cracking process used to convert petroleum distillates and heavier fractions to components of lower molecular weight.

ultrasonic cleaning The use of ultrasonic pressure waves to vibrate an object to be cleaned while the object is immersed in a cleaning fluid. The process is used to produce a very clean surface and is used for medical equipment and jewellery.

ultrasonic flow meter A type of non-intrusive flow meter used to measure the velocity of a fluid flowing in pipes and open channels. It uses the principle of ultrasound in which an ultrasonic signal is passed through the fluid, which is detected by another sensor located a short distance away. The time is measured between the pulses of emitted and detected ultrasound. The delay between sending and receiving the signal is related to the velocity of the fluid. While inexpensive to operate and with no moving parts, ultrasonic flow meters are sensitive to changes in fluid density and to distortions of flow profile.

ultrasonics The study of ultrasonic pressure waves for the purposes of testing metals for faults, flaws, and thickness, as well as for surface cleaning. It uses frequencies above audio frequencies and cannot be detected by the human ear. **Ultrasonic testing** is a non-destructive technique used to identify the thickness of metal such as vessels and pipework, and the presence of corrosion. It uses short pulses of ultrasound with frequencies typically of around 2 MHz.

ultraviolet radiation Electromagnetic radiation with a wavelength shorter than that of visible light. It lies just beyond the visible spectrum and has typical wavelengths in the order of 10^{-7} m.

unbound moisture A liquid that is held by a solid, which is in excess of the equilibrium moisture content corresponding to the saturation humidity.

unconfined vapour cloud explosion A loose term used to describe a *vapour cloud explosion, which is an explosion due to the ignition of a cloud of flammable vapour in air.

underdamped *See* DAMPING.

underflow A liquid that leaves a continuously fed mixer-settler or some other solvent extraction device. The liquid outflow is divided into two streams by a weir arrangement based on density difference. The liquid with the highest density or heavy phase leaves as the underflow. The light phase leaves as the *overflow.

underground storage The storage of gas such as methane or ethylene in vast underground reservoirs and natural rock strata instead of using above-ground gas holders.

Underwood equation A shortcut method used to estimate the minimum reflux ratio in a multicomponent distillation process. It was proposed by A. J. V. Underwood in 1948.

unicracking A *hydrocracking process used to produce hydrocarbon fuels by simultaneous hydrogenating and cracking of liquid petroleum fractions to form hydrocarbon mixtures of low molecular weight. It uses an aluminosilicate catalyst either contained within a *zeolite or in an amorphous state.

unidak process A catalytic process used in petroleum refining to extract naphthalene from reformer residues. It involves a dealkylation stage to form naphthalene. The catalyst is based on cobalt and molybdenum and the process is operated at 600°C.

UNIFAC A semi-empirical thermodynamic model used to predict the behaviour of components in complex mixtures, which uses structural groups to estimate component interactions. It is an abbreviation for **UNI**QUAC **F**unctional-group **A**ctivity **C**oefficients and is used to predict non-electrolyte activity in non-ideal mixtures. It is used to predict the activity coefficients as a function of composition and temperature. It is useful when experimental data is not available.

uniform flow A fluid flow condition in which there is no change in fluid velocity at a given time with respect to distance.

unimolecular reaction A chemical reaction involving only one molecule as the reactant. An example is the decomposition of ammonia to nitrogen and hydrogen on a metal surface. Unimolecular reactions are always first order.

union A ring-like device used to couple or link together pipes and tubes.

unionfining processes One of a number of petroleum *hydrodesulphurization and hydrodenitrogenation processes used to produce a high-quality diesel fuel.

 SEE WEB LINKS
• Website of Honeywell UOP Company.

UNIPOL process A process used for polymerizing ethylene to polyethylene and also for polymerizing propylene to polypropylene. Unlike the *Ziegler–Natta process, it uses a gas phase process at low pressure. The catalyst is continuously added to the process and the granular product withdrawn. A co-monomer is also usually used in the process.

u

(⊕) SEE WEB LINKS
• Website for Dow Company's UNIPOL process.

UNIQUAC *See* UNIVERSAL QUASI-CHEMICAL.

unit 1. The fundamental measure of a physical quantity such as length, mass, and time. Derived units include area, velocity, density, etc. SI units have replaced previous systems for scientific and engineering purposes. **2.** An item of process equipment or plant designed to carry out a specific task.

unit operation A basic step in a process or operation carried out in a chemical plant. A unit operation involves the study of physical, chemical, or biochemical changes that occur during the processing of materials. The design of equipment and systems is based on these operations and includes mixing, reaction, and separation. Unit operations may be classified as:
1 Fluid flow processes including filtration and fluidization.
2 Heat transfer including evaporation and condensation.
3 Thermodynamic processes including heat pumps.
4 Mechanical processes such as solids and particle handling.
5 Mass transfer including distillation, drying, solvent extraction, and adsorption.
Arthur D. *Little is credited with coining the concept of unit operation.

universal constants The parameters that do not change such as gravitational constant, the speed of light, the Planck constant, and the charge on an electron, etc.

universal gas constant (Symbol R) The constant or proportionality in the law of ideal gases:

$$pV = nRT$$

where p is pressure, V is volume, n is the number of moles, and T is absolute temperature. The SI units are $8.314 \, kJ \, kmol^{-1} \, K^{-1}$.

universal quasi-chemical (UNIQUAC) A thermodynamic model tested against experimental data used to obtain vapour liquid equilibria data. It is used in computer simulation software packages particularly for carrying out complex distillation calculations.

UNOX process An activated sludge sewage treatment process used for treating domestic effluents. It uses oxygen aeration in closed tanks rather than air.

unsaturated 1. A chemical compound having double or triple bonds within its structure. Unsaturated compounds can undergo addition as well as substitution reactions, such as the hydrogenation of vegetable fatty acids. **2.** A solution containing less than the maximum equilibrium amount of a solute at a given temperature. *Compare* SATURATED; SUPERSATURATED.

unscheduled maintenance *See* MAINTENANCE.

unstable 1. A chemical compound that readily decomposes. **2.** The spontaneous decomposition of a radionuclide by nuclear decay. **3.** A process or mechanical or electrical system that has the tendency of self-oscillation. Process control is used to ensure that the dynamics of a process are controlled.

unsteady state A condition in which the transport of material or energy in and out of a process is not balanced; instead there is either a loss or an accumulation over time. An **unsteady-state mass balance** involves the flow of materials into a process together balanced with the flow of materials out with any accumulation or loss. For example, the flow of a liquid into a tank, Q_{in}, with an open drain valve that has a flow out, Q_{out}, resulting in a change in capacity dV/dt, can be expressed mathematically as:

$$Q_{in} = Q_{out} + \frac{dV}{dt}$$

In an **unsteady-state energy balance** the same principle applies. The accumulation of energy within a process where all the energy forms are considered including kinetic, potential, heat flow rates, enthalpies, and stirrer works may result in an increase in the thermal energy and a rise in temperature. **Unsteady-state heat transfer** involves the transfer of heat under conditions where the temperature changes with time. For the simple case of one-dimensional conduction in a solid slab, the accumulation of heat is a product of the mass and specific heat of the material and the increase in temperature where:

$$\frac{\partial T}{\partial t} = \frac{k}{\rho c_p} \frac{\partial^2 T}{\partial x^2} = \alpha \frac{\partial^2 T}{\partial x^2}$$

where α is the thermal diffusivity of the material. General solutions of unsteady-state conduction for simple geometries are available such as for slabs, infinitely long cylinders, and spheres. For a semi-infinite slab, the integration for the heating or cooling from both sides by a medium of constant surface temperature is:

$$\frac{T_S - T_1}{T_S - T_2} = \frac{8}{\pi^2} \left(e^{-aFo} + \frac{1}{9} e^{-9aFo} + \frac{1}{25} e^{-25aFo} + \ldots \right)$$

where T_S is the average temperature of the surface, T_1 is the initial temperature, T_2 is the temperature at time t, Fo is the Fourier number, and a is $(\pi/2)^2$. **Unsteady-state heat transfer** occurs where there is a change of material within a space with time. Similar one-dimensional mathematical principles apply.

unstructured model A simple mathematical description of the rate of growth of microorganisms based on the cells being represented by a single variable such as cell concentration, X. The rate of cell growth is proportional to the cell concentration:

$$r_X = \frac{dX}{dt} = \mu X$$

A widely used expression that relates the specific growth rate μ to the amount of substrate in a fermenting medium is the *Monod equation.

unstructured packing Small objects that are randomly arranged in distillation and absorption columns to provide a high surface area thereby allowing intimate contact between a rising vapour or gas with a descending liquid to allow effective mass transfer to take place. Many types are commonly used such as Berl saddles, Raschid rings, Intalox saddles, and Pall rings amongst others, and are made from a variety of materials such as

u

plastic, metal, and ceramic, which are inert to the substances in which they are in contact. *Compare* STRUCTURED PACKING.

UOD *See* ULTIMATE OXYGEN DEMAND.

updraft The upward movement of air or a gas in a structure or through a product. It is used to designate the direction in a dryer or fluidized bed. The opposite is called downdraft.

upper flammable limit The highest concentration of a flammable vapour or gas mixed with an oxidant such as air that will propagate a flame at a specified temperature and pressure. *See* FLAMMABILITY LIMITS.

upset An unscheduled alteration to the operation of a process. *See* PROCESS UPSET.

upstream A stream of material for processing that has not yet entered the process for chemical transformation in reactors, etc. In the oil and gas industry, it includes production facilities, pipelines, and receiving terminals. *See* DOWNSTREAM.

upthrust *See* ARCHIMEDES' PRINCIPLE.

uranium enrichment The process of purifying uranium. The difference in mass between uranium-235 and uranium-238 allows the isotopes to be separated and to increase or enrich the percentage of uranium-235. All enrichment processes, either directly or indirectly, make use of this small mass difference enabling the uranium-235 to be used as a nuclear reactor fuel. *See* FLUORINATION.

uranium series *See* RADIOACTIVE SERIES.

Urbain process A process used to produce activated charcoal by pulverizing a precarbonized material such as peat or lignite, then heating with phosphoric acid, followed by washing with hydrochloric acid. The process, developed in the 1920s, is named after its inventor Edouard Urbain, who patented many gas treatment processes by carbonaceous materials.

USGPD An abbreviation for **US g**allons of oil **p**er **d**ay, it is an Imperial unit of volumetric flow used in the oil industry. *See* GALLON.

utilities The services used to support a process such as fuel, heating, cooling, steam, electricity, refrigeration, compressed air, fire water, power generation, instrument tool air, and process water. After the raw materials consumed in a process, they make up the most significant *variable costs in a process. *See* PROCESS ECONOMICS.

utility flow diagram (UFD) A type of *engineering flow diagram showing the layout of utility services that connect process equipment. These include steam and condensate lines, water supply and return, air, fuel gas, refrigeration, and flare systems as well as various flush and priming lines for pumps and instruments.

utility waste *See* WASTE.

U-tube manometer An instrument used to determine the differential pressure between two points such as across an orifice plate in a pipeline. It consists of a U-tube of glass or clear plastic with the top of each vertical leg being attached to the *tapping points across the flow meter. The U-tube contains a manometric fluid that is opaque, unreactive, has a low volatility, and has a higher density than the process fluid. The difference in levels between the two vertical legs of the manometer provides a measure of the differential pressure. It is also known as a **differential manometer**.

U-value The *overall heat transfer coefficient for a heating or cooling system, such as a heat changer, and is a measure of the thermal efficiency of the heat transfer device. It is dependent on the tube and shell side film coefficients as well as the thermal conductivity of the material of the heat exchanger. The SI units are $W\,m^{-2}\,K^{-1}$.

UV disinfection A *water-treatment process used to eradicate harmful bacteria and viruses by exposing potentially contaminated water to *ultraviolet radiation. At a certain level of intensity, ultraviolet light is fatal to all microorganisms that inhabit water. Mercury arc lamps are used to generate the ultraviolet radiation with low-pressure lamps being the most common and effective. The lamp is made of fused silica or quartz to allow transmission of the ultraviolet light. The efficiency of UV disinfection is diminished by turbidity and by the build-up of scale on the tubes.

UVOX process A process used to purify water using air and *ultraviolet radiation. The ultraviolet radiation converts the oxygen in the air to ozone, which kills the pathogenic organisms in the water as well as other parasitic dissolved matter. It is used for purifying water in swimming pools and other water systems.

((⊕)) SEE WEB LINKS
• Website of UVOX Redox Systems, with page describing UV disinfection and ozone oxidation with UVOX Redox Systems.

u

vacuum A space that is devoid of matter and in which there is a very low pressure of gas. A perfect vacuum contains no atoms or molecules. However, this is not actually obtainable as the surrounding container itself has a finite vapour pressure. A soft or low vacuum has a pressure of around 10^{-2} Pa whereas a hard vacuum is around 10^{-2} to 10^{-7} Pa. An ultra-high vacuum has a pressure below 10^{-7} Pa.

vacuum breaker A method used for pumping liquids out of sealed tanks to prevent a vacuum from forming. It involves the injection of an inert gas, such as nitrogen, to maintain a slight positive pressure.

vacuum distillation A distillation process that operates at a reduced pressure. The effect of the low pressure is to reduce the boiling point temperature of the mixture being separated. This is useful where the materials may be heat-labile, such as the separation of volatiles from fruit juices, or prevent the decomposition or cracking of materials being distilled in crude oil distillation. In the vacuum distillation of petroleum fractions, residues may be used as the feedstock for cracking other products, or may be blended to produce fuel oil or asphalt. *See* MOLECULAR DISTILLATION.

vacuum drying The use of a vacuum for drying foods that are sensitive to temperature. Using a vacuum of 1 to 70 torr (133 to 9,333 Pa) lowers the temperature for vaporization of the moisture in the food. Applied to concentrating fruit juices, it differs from *freeze-drying in that heat is applied to the food material.

vacuum filtration A separation process in which a suspension of solid particles in a liquid is separated using a filter. The filter is used to trap the particles and consists of a wire or fabric mesh or cloth. The rate of filtration is increased by using a vacuum to draw the filtrate through the filter, leaving a cake on the filter. Continuously operated vacuum filters consist of either a rotating drum covered in fabric that drips into the suspension allowing a filter cake to build up and be scraped off, or consists of a slowly rotating flat filter plate in which a vacuum is drawn from the underside. The suspension is dripped onto one location and as the filter plate is rotated, the filtrate is drawn through leaving a cake that is scraped off with a fixed blade at another location. *See* ROTARY VACUUM FILTER.

vacuum pressure safety valve (VPSV) An automatic valve system that relieves the pressure of a gas. It is used for small or negative pressure differences between a vessel and atmospheric pressure.

vacuum pump A type of pump used to reduce the pressure of a gas. Rotary pumps can readily achieve pressures of 0.1 Pa. For very low pressures down to 10^{-7} Pa, a diffusion pump is required. Ion pumps and cryogenic pumps can achieve even lower pressures of 10^{-9} Pa and 10^{-13} Pa, respectively.

valency (valence) The property of atoms or radicals equal to the number of atoms of hydrogen that an atom or group can combine with or displace in forming compounds. Hydrogen therefore has a valency of 1. In ionic compounds, it is equal to the ionic charge. For example, in $MgCl_2$, magnesium has a valency of 1 while chlorine has a valency of 2. In covalent compounds, it is equal to the number of bonds formed. For example, in SO_2, sulphur has a valency of 4 and oxygen has a valency of 2.

validation A documented form of evidence that a process is capable of producing a product to a specified standard or quality. It is used for various manufacturing processes and, in particular, the production of pharmaceutical products. *See* PROCESS VALIDATION.

valve A device used to regulate and control the rate of flow of a fluid. Used in pipes, they consist of a controlled flow area, which is activated electronically, pneumatically, or manually. There are various designs and sizes including butterfly valves, diaphragm valves, ball valves, globe valves, and gate valves. In gate valves, the flow area is adjusted by the movement of a stem to increase or decrease the available area. Globe valves have a ball with a hole that sits in the passage of flow in which the available flow area is adjusted by rotation of the ball away from the axis of the pipe. *See* CONTROL VALVE.

valve tray A perforated plate within a *distillation column that is equipped with caps that lift when the pressure of rising vapour from the tray below exceeds the hydrostatic pressure of the liquid on the tray above. When the pressure falls, the valves drop and cover the perforations, preventing drainage of the liquid.

van Arkel–de Boer process A process once used to produce pure metal, such as tungsten, titanium and zirconium, by the formation of metal iodides followed by their thermal decomposition on a hot tungsten filament. It was developed by Dutch chemists Anton Eduard van Arkel (1893–1976) and Jan Hendrik de Boer (1899–1971). It is also known as the **iodide process** or **crystal bar process**. The process has been replaced by the *Kroll process.

van der Waals, Johannes Diderik (1837–1923) A Dutch physicist and professor at Amsterdam University. He completed a doctoral dissertation in 1873 at the University of Leiden entitled *On the continuity of the liquid and gaseous states*. The significance of this work was to account for many phenomena in vapours and liquids, which had already been observed by Thomas *Andrews (1813–85). Van der Waals derived a new equation of state by postulating the existence of intramolecular forces and a finite molecular volume that predicted more accurately experimental data particularly under conditions near the critical point. He was awarded the Nobel Prize for Physics in 1910 and the weak attractions between molecules called *van der Waals' forces are named after him.

van der Waals' equation A cubic *equation of state developed by *van der Waals, which was an early attempt to describe the behaviour of real gases. The equation allows for the reality of the volume occupied by the gaseous molecules (V–b) where b is a constant appropriate for each gas, and attractive forces between the gaseous molecules, a. Allowing for these two corrections, van der Waals' equation is given by:

$$\left(p - \frac{a}{V^2}\right)(V - b) = RT$$

where p is the pressure of the gas, V is the molar volume, R is the universal gas constant, and T is the absolute temperature.

van der Waals' force The weak attractive forces between atoms or molecules. Named after Johannes Diderik *van der Waals (1837–1923), the force accounts for the a/V^2 term in the *van der Waals' equation. These weak forces are inversely proportional to the seventh power of the distance between the atoms or molecules and are responsible for the non-ideal behaviour of gases and for the lattice energy of molecular crystals. Van der Waals' forces are due to dipole–dipole interactions. That is, the electrostatic attractions between molecules.

van Laar equations The equations used for the activity coefficients of the components in a liquid as a function of the composition of the mixture. Developed in 1929 by Dutch chemist Johannes van Laar (1860–1938) for liquid solutions based on the *van der Waals' equation of state, it was based on observations that strongly non-ideal solutions are associated with large heats of mixing. The van Laar equation gives good agreement for many binary mixtures. The equations can be extended to mixtures of more than two components, but the equation becomes unwieldy, and is therefore rarely used beyond binary mixtures.

van't Hoff, Jacobus Henricus (1852–1911) A Dutch professor of chemistry in Amsterdam, and then Leipzig and Berlin, who had studied in Holland, France, and Bonn. He began his research at twenty years of age and his simple theory that the four valency bonds of a carbon atom are directed to the four corners of a tetrahedron explained asymmetrical carbon compounds and established stereochemistry in 1874. In 1886 he explained osmosis, and showed the connection between osmotic and gas pressures, which led to the *Arrhenius' theory of electrolytic dissociation. He received the first Nobel Prize in Chemistry in 1901.

van't Hoff's isochore An equation formulated by *van't Hoff for the variation of the equilibrium constant in a chemical reaction with temperature:

$$\frac{d\log_e K}{dT} = \frac{\Delta H}{RT^2}$$

where K is the equilibrium constant, R is the gas constant, T is the thermodynamic temperature, and ΔH is the enthalpy of the reaction.

vaporization The process of changing a substance into a vapour or into the gaseous state.

vapour A gaseous state of matter that can be reduced to a liquid by way of condensation. A vapour is therefore a gas below its critical temperature.

vapour-absorption cycle A thermodynamic cycle used in refrigeration systems. It has no moving parts and energy is supplied as heat either by an electric heater or a gas burner. Ammonia is usually used as the refrigerant. The absorption cooling cycle has three phases: the liquid refrigerant is evaporated to a vapour thereby extracting heat from its surroundings providing the refrigeration effect; the vapour is then absorbed into another liquid; the refrigerant-containing liquid is regenerated by heating, releasing the refrigerant. The adsorption refrigerator was invented by French engineer Ferdinand Philippe Edouard Carré (1824–1900).

vapour cloud explosion (VCE) An explosion resulting from the ignition of a large cloud of flammable vapour in air. It may involve in excess of one tonne of explosive fuel. The effect of the explosion results in significant overpressure and may be more serious than the fire since it has the potential to cause fatalities more readily due to the rapid release of energy, giving little to no time for evasive action to be taken by personnel.

vapour-compression cycle A thermodynamic cycle used in refrigeration systems. It uses a volatile liquid refrigerant, which is pumped through the cooling coils of an evaporator where the latent heat used in its evaporation cools the materials requiring refrigeration. The vapour is then compressed to a high pressure by a compressor and then condensed in a condenser back to a liquid, liberating heat. The liquid refrigerant is contained in a storage vessel before finally passing through an expansion valve to reduce its pressure and being pumped back to the evaporator and the cycle repeated. In a domestic refrigeration system, the cycle is repeated until the temperature reaches the desired level of around 4°C in the cool box and –18°C in the freezer compartment. A thermostat is used to maintain a steady state by turning the compressor on and off. *See* REFRIGERATION.

vapour density The ratio of the density of a vapour or gas to hydrogen at the same temperature and pressure. The vapour ratio is therefore equal to the ratio of the mass of the vapour or gas to the mass of hydrogen occupying the same volume for the same temperature and pressure.

vapour–liquid equilibria (VLE) The relationship between a liquid and its vapour at equilibrium. At a given temperature and pressure, the composition of the vapour will be determined by the composition of the liquid. The relationship between the two is frequently expressed using K-factor values of the components involved. For binary components A and B showing ideal behaviour, the relative volatility, α_{AB}, can be used where the mole fraction of the vapour of the more volatile component y_A is related to the liquid mole fraction x_A by:

$$y_A = \frac{\alpha_{AB} x_A}{1 + x_A(\alpha_{AB} - 1)}$$

There are various liquid activity models available including Margules, van Laar, Wilson, non-random two-liquid (NRTL), and universal quasi-chemical (UNIQUAC) models. Mixing rules are used for mixtures to combine pure component parameters.

vapour–liquid separator A vessel used to separate a vapour–liquid mixture. The vessel is designed to allow the liquid to settle under the action of gravity, accumulating at the bottom of the vessel, where it is withdrawn, while the vapour travels upwards at a design velocity that minimizes the entrainment of any liquid droplets in the vapour.

vapour pressure The pressure exerted by the vapour of a solid or a liquid in which it is in contact and at equilibrium for a specified temperature. The vapour pressures of pure substances can be obtained from published data or from empirical equations such as the *Antoine equation. The *Clausius–Clapeyron equation can also be used but is less accurate. The vapour pressure can be lowered by the addition of a solute. This is related to the decrease in freezing point and increase in boiling point.

vapour recovery unit (VRU) The part of a *fluidized catalytic cracker unit in which light hydrocarbon gases are produced by the cracking process, compressed, and then separated into different product streams.

variable A mathematical expression representing either a general point taken from a specific set of values, or as a name for an unknown point to be determined from within a specified set of possible values. **Dependent variables** have values that are influenced by changes in other values, whereas **independent variables** are not influenced.

variable area flow meter A type of flow meter used for the measurement of flow of gases and liquids by virtue of the elevation of a solid float within a vertical, tapered tube. The tube is usually made of glass while the float is made of metal, ceramic, or plastic with a float density greater than that of the fluid. The float is usually bomb-shaped having a cylindrical body with a coned bottom and short top piece, which is slightly larger in diameter and often grooved to encourage the float to spin, thereby improving stability due to a gyroscopic effect. The upwards flow of the fluid causes the float to rise and reach an equilibrium elevation. The flow is read from a calibrated scale on the tube. There is a roughly linear variation of mass flow rate with float position in the tube. In practice, a scale is marked on the tube. Calibration curves are used for a particular fluid, temperature, and float. Unlike the *venturi or *orifice type flow meters, it operates with a fixed pressure drop across the float and has a variable area around the float. The ***Rotameter** is a registered name for a type of variable area flow meter.

variable costs The costs of a process that include the cost of raw materials and all other chemicals that are consumed in the process such as catalysts, the cost of the utilities such as steam, electricity, fuel, cooling water, process water, compressed gases, etc. The variable costs also include *unscheduled maintenance costs, royalties, licence costs, transportation costs, and costs associated with quality control and other forms of monitoring. The variable costs of a process are dependent on the rate of production. The sum of the *fixed costs and variable costs is the total cost of the process. *See* PROCESS ECONOMICS.

variance A statistical measure of the dispersion of a set of data from the mean. The variance is equal to the square of the *standard deviation and is used to distinguish between probability distributions.

VCE *See* VAPOUR CLOUD EXPLOSION.

vector 1. A representation of a quantity having both magnitude and direction. Force, velocity, and acceleration are all vectors. **2.** A carrier of genetic material such as a bacteria carrying recombinant DNA fragments for another microbial species.

vector analysis A quantity with magnitude and direction. Examples include force, velocity, acceleration, and momentum. The decomposition of a vector in three dimensions may be written as:

$$v = iv_x + jv_y + kv_z$$

where i, j, and k are unit vectors in the x, y, and z direction, and v_x, v_y, and v_z are the scalar magnitudes of v's components in these directions.

velocity The speed of a body in a given direction. Velocity is thus a vector quantity, whereas speed is a scalar quantity.

velocity gradient The deformation of a fluid under the influence of a shearing force presented as the change in velocity of the fluid perpendicular to the flow.

$$\dot{\gamma} = \frac{dv}{dz}$$

It is also known as the *shear rate.

velocity head The kinetic energy of a moving fluid expressed in head form:

$$h = \frac{v^2}{2g}$$

It is therefore seen as the equivalent height of a column of the fluid if it were brought to rest. The velocity head can be used directly in the *Bernoulli theorem. The SI unit is m.

vena contracta The minimum flow area formed downstream by a fluid flowing through a sharp-edged opening. It literally means 'contracting veins'. It is seen in a freely discharging jet of liquid just beyond the opening where the flow lines converge.

vent An opening from a vessel. It usually consists of a short tube and opens to the atmosphere to allow the pressure of the vessel to attain atmospheric pressure within. It is required to be of adequate capacity and free from blockage. Appropriate safeguards are often used in vessel design and operation, such as multiple vents, flame arresters, and the addition of inert gas.

venturi effect The phenomenon in which a fluid flowing through a restriction increases in velocity with a corresponding decrease in pressure. It is named after Italian physicist Giovanni Battista Venturi (1746–1822).

venturi meter A fluid flow-measuring device that consists of a tapered tube to form a throat such that the increase in velocity results in a decrease in pressure, known as the *venturi effect. The differential pressure produced by the flowing fluid through the throat gives a measure of the rate of flow. The pressure differential can be measured using manometers or other types of pressure measurement devices attached both upstream and at the throat. In practice, the theoretical rate of flow through the device is not achieved as friction in the device is ignored. To allow for this difference, a correction factor known as a coefficient of discharge is used. For a well-designed venturi, the coefficient lies between a value of 0.95 and 0.98. It is named after Italian physicist Giovanni Battista Venturi (1746–1822).

venturi scrubber A device used for gas cleaning in which a liquid is injected at the throat of the venturi, and due to the high velocity of the gas to be cleaned, it disintegrates into droplets. These droplets then collide with smaller droplets, particles, or gases to be absorbed, which are then collected in a cyclone or demister device downstream.

Verneuil process A process used to produce synthetic gemstones that was developed in 1902 by French chemist Auguste Victor Louis Verneuil (1856–1913). It involves melting finely powdered materials using an oxy-hydrogen flame and crystallizing the droplets on a seed crystal. It is used to make ruby, sapphire, and forms of corundum.

vertical tube evaporator A device used to concentrate a solution by boiling and evaporation, and consists of vertical tubes that are heated by steam. In short tube evaporators, steam used as the heating medium is contained within the tubes, whereas in long tube evaporators, the product is contained within the tubes and external heating by steam is used.

vessel A general name for a receptacle used to contain a liquid, solid, or gas.

vibromixer A type of low shear mixer used to mix shear-sensitive biological reactions and polymer solutions. It consists of a disc with a conical perforation, perpendicularly attached to a shaft that moves up and down with a controlled frequency. The turbulence is increased by adjusting the amplitude and frequency of the translatory movement.

view factor The fraction of thermal radiation leaving a surface that is intercepted by another surface. The view factor depends only upon the geometric arrangement of the surfaces, and satisfies the reciprocity relation:

$$A_i F_{ij} = A_j F_{ji}$$

where A_i is surface i and F_{ij} refers to the fraction of radiation that leaves surface i and is directly intercepted by surface j. The view factor has a numerical value of between 0 and 1. Tables and equations are available for calculating the view factors of some simple and most commonly encountered geometries. It is also known as the configuration factor or shape factor.

virial equation of state A power-series expansion of the compressibility for a substance expressed in terms of molar density:

$$Z = 1 + B\rho + C\rho^2 + D\rho^3 + \dots$$

The equation assumes that the compressibility is unity for zero density. The coefficients B, C, and D are known as the second, third, and fourth **virial coefficients**, and are functions of temperature. It is used to fit experimental PVT data of compressed real gases. The equation was originally known as the Kammerlingh–Onnes equation.

visbreaking A *thermal cracking process that involves reducing the viscosity of petroleum residues from heavy atmospheric or vacuum still bottoms. The purpose is to produce distillate products that can reduce the viscosities of hydrocarbon distillation residues used for producing fuel oil blends. It involves a high temperature, non-catalytic process in the presence of steam.

viscoelasticity The property of certain materials to exhibit both viscous and elastic behaviour when deformed under an applied shear stress. Some complex polymer solutions exhibit viscoelasticity.

viscometer An instrument used to measure the viscosity of a liquid. Rotational type viscometers include concentric cylinders, and cone and plate instruments. They operate at constant speed, which defines the shear rate. The liquid forms a continuous film between the rotating element and the stationary element. The torque to resist the motion defines the shear rate. The Ostwald viscometer, which is used to measure the viscosity of liquids, consists of a bulb into which the liquid is filled and is allowed to flow under gravity through a capillary tube. It is the time taken for the meniscus to reach a mark on the capillary from which the viscosity can be calculated. A falling-sphere viscometer is based on *Stokes's law, which consists of timing a steel sphere at its terminal velocity under the influence of gravity through the liquid.

viscometry The study of the deformation and flow of fluids that considers only shearing stresses. *Compare* RHEOMETRY.

viscose process A process that is used in the production of regenerated cellulose fibres in a product known as rayon. Cellulose is obtained from wood or cotton and reacted with sodium hydroxide. The alkali cellulose is then dissolved in carbon disulphide to produce cellulose xanthate, which is dissolved in a solution of sodium hydroxide. The injection of this solution into an acid bath produces the regenerated cellulose product. It is known as viscose on account of its high viscosity. The process was invented in 1882.

viscosity A measure of the flow transport behaviour of a fluid. It is the phenomenon in which a fluid will withstand a slight amount of molecular tension between particles, which will cause an apparent shear resistance between two adjacent layers. The term 'viscosity' is used to describe the fact that certain fluids flow easily, such as gases, water, and mercury, while others do not, such as tar, treacle, and glycerine. These fluids are broadly classified as thin and thick fluids. Sir Isaac *Newton (1642–1727) proposed that the shear stress is proportional to the velocity gradient or shear rate. By considering a fluid sandwiched between two parallel plates set at a distance dz apart in which the upper plate moves with some small velocity dv in comparison with the lower plate, there will be a small resisting force over the plate area due to viscous frictional effects in the fluid. This force per unit area of plate (F/A) is known as the shear stress, τ. Newton's law of viscosity is therefore given as:

$$\tau = \mu \frac{dv_x}{dz} \text{ which is sometimes written as } \tau = \mu\dot{\gamma}$$

The proportionality constant, μ, is known as the coefficient of dynamic viscosity and is also known as the absolute or dynamic viscosity of the fluid; it is influenced by process conditions such as temperature. In the case of Newtonian fluids, μ is a constant. Examples of Newtonian fluids include water, ethanol, and benzene. The viscosity of a fluid may, however, not always be constant for certain fluids for different applied shear stresses. These fluids are known collectively as non-Newtonian fluids. In such cases, the term 'apparent viscosity' is conveniently used. Examples of non-Newtonian fluids include paint, polymers, most slurries, and many foodstuffs. The calculation remains the same and is the ratio of shear stress to shear rate. In SI units, it has examples of non-Newtonian fluids include paint, polymers, most slurries, and many foodstuffs. the units $kgm^{-1}s^{-1}$ or using derived SI units Nsm^{-2} or Pa.s. In the c.g.s. system it is measured in poise (P) or centipoise (cP), where 1 cP is equivalent to 10^{-3} Nsm^{-2}. The viscosities of gases are significantly less than for liquids. Oils, such as olive oil, are an order of magnitude higher than liquids such as water. In general, high-viscosity fluids can be considered in the order of 10 Nsm^{-2} and above. *See* KINEMATIC VISCOSITY.

viscous force The drag effect that occurs on a body when placed in a viscous fluid. For two parallel plates sandwiching a viscous fluid, the viscous force is proportional to the difference in velocity between the two plates.

VITOX process A process used in sewage treatment and other microbiological processes in which oxygen is added through the throat of a *venturi to form fine bubbles to aid oxygen transfer. Unlike the *UNOX process, the VITOX process uses open tanks.

vitrification A high-temperature process used to convert highly radioactive liquid waste into glass form for long-term storage purposes. The process heats the material to 1,100°C, evaporating the liquid either to the glass transition temperature to form an amorphous solid that is free from a crystalline structure, or is added to molten glass. In both processes, the molten glass is loaded into steel containers. The hardened glass material is therefore free from leakage and can be stored in safe and secure underground repositories.

VLE *See* VAPOUR–LIQUID EQUILIBRIA.

voidage The space between solid particles expressed as a percentage or fraction of the total volume. The voidage of a *packed bed is used to indicate the available space for the flow of gas or liquid for chemical reaction. For spherical particles packed in face-square arrangement in which each particle touches six others, the voidage is 0.46. *Compare* POROSITY.

volatility The tendency and measure of the ability of a liquid to go into the vapour phase expressed as the ratio of the mole fraction in the gas phase (y_A) to that in the liquid phase (x_A):

$$K_A = \frac{y_A}{x_A}$$

The *relative volatility is a comparison of the lightness of two components and can be found by taking a ratio of the vapour phases of the two components.

volt (Symbol V) The derived SI unit of electrical potential defined as the potential difference between two points on a conductor carrying a current of one ampere when the power dissipated between the points is equal to one watt. It is named after Italian physicist Count Alessandro Volta (1745–1827).

voltaic cell A device that produces an e.m.f. as the result of a chemical reaction that takes place within it. The reaction takes place at the surface of two electrodes, each of which is immersed in an electrolyte. It was devised by Italian physicist Count Alessandro Volta (1745–1827). It is also known as a *galvanic cell.

voltaic pile An early form of battery that consists of a number of dissimilar metals such as copper and zinc with each pair being joined in series by paper pads moistened with an electrolyte. It was devised by Italian physicist Count Alessandro Volta (1745–1827).

volume The three-dimensional space enclosed within or occupied by a body. For example, the volume occupied by a cylindrical vessel is equal to the cross-sectional area multiplied by the length. The volume occupied by a liquid can be determined from its mass divided by its density. The SI unit is m^3.

volumetric analysis A method of chemical analysis that involves measuring volumes of liquids or gases, such as titration.

volumetric efficiency (Symbol η) Used to quantify the efficiency of fans, blowers, compressors, and pumps. It is ratio of the volume of a fluid discharged divided by the volume displaced by the moving part during the same time.

volumetric flow rate The volume of a fluid that passes a given surface per unit time. SI units are $m^3\,s^{-1}$.

volumetric flux The volumetric flow rate of a fluid per unit area. It is also known as the *superficial velocity and used in multiphase flow systems, such as gas and liquid flow through pipelines. The SI units are $m\,s^{-1}$. The integration of the volumetric flux over a flow area gives the *volumetric flow rate.

volumetric oxygen transfer coefficient (Symbol $k_L a$) A measure of the capacity of a bioreactor, or any other aeration reactor or system, to transfer oxygen from air into a liquid phase. It is the product of the oxygen transfer coefficient k_L and the interfacial area, a, between the air bubbles and the liquid, per unit volume of the liquid.

volute chamber A spiral-shaped casing around the impeller of a centrifugal pump. It is used to convert kinetic energy of the moving fluid into pressure energy at the point of discharge.

von Kármán, Theodore (1881–1963) A Hungarian-American mathematician and physicist noted for his significant contribution to advances in aerodynamics, and in particular on supersonic and hypersonic airflow behaviour. Born Szőllőskislaki Kármán Tódor, he studied engineering in Budapest before moving to Germany to join *Prandtl at the University of Göttingen. He emigrated to the US in 1930 to become director of the Guggenheim Aeronautical Laboratory at the California Institute of Technology (GALCIT). At the age of 81, he was the recipient of the first National Medal of Science, presented by President John F. Kennedy.

vortex A rotating fluid. *See* FREE VORTEX; FORCED VORTEX.

vortex breaker A device used in tanks and vessels to prevent the formation of a vortex when discharging through an orifice. It typically consists of a cross of vertical plates or baffles on the outlet from a vessel.

vortex flow meter An instrument used to measure the volumetric rate of flow of a fluid and is based on the principle that the velocity is directly proportional to the frequency at which vortices are shed from a body in a flow stream. One design relies on a body that lies in the direction of flow to generate vortices that are detected with the aid of a capacitive transducer, which consists of two electrodes separated from two welded diaphragms by a dielectric of fluid. When the fluid flows, vortices cause asymmetric movement of the diaphragms varying the capacitances generating a current output.

vortex shedding An unsteady and oscillatory flow behaviour of a moving fluid around a body in which vortices are formed at certain velocities. The vortices are formed on the downstream side of the body and become detached from either side of the body known as a **von Kármán vortex street**. The body will tend to move towards the region of low pressure and forms the basis of a vortex-shedding flow meter. The vibrating oscillations are also responsible for the whistling sound of overhead power lines. Tall stacks and chimneys are protected from the potentially destructive oscillatory effects by the use of helical fences known as strakes or spoilers.

VPSV *See* VACUUM PRESSURE SAFETY VALVE.

VRU *See* VAPOUR RECOVERY UNIT.

vulcanization A process for hardening rubber and improving its elasticity by heating it with sulphur or sulphur compounds such as sulphur monochloride. The process involves cross-linking the polymer chains, adding strength to the rubber, and acting as a form of 'memory' that allows the polymer to recover to its original shape after stretching. The process was invented in 1839 by the American Charles Goodyear (1800–60).

v

Wacker process 1. A catalytic process used to oxidize aliphatic hydrocarbons such as ethylene to ethanol, aldehydes, and ketones using oxygen. The process uses an aqueous solution of mixed palladium and copper chlorides either in solution or on a support of activated carbon through which the ethylene is bubbled. The process was invented in 1957 and is named after the chemical company. **2.** A process used for the production of sodium salicylate through the reaction of sodium phenate and carbon dioxide.

Waelz process A process used for the extraction of zinc and lead from ores using a rotary kiln. Zinc is still largely extracted from ores although around 40 per cent of zinc is recycled from galvanized steels and other scrap metals. A high-temperature rotary Waelz kiln is used to convert zinc dust from electric arc furnaces into zinc oxide, which is then converted to zinc metal. The process of zinc volatilization was first proposed in 1881.

Walker process A catalytic oxidation process used to partially oxidize natural gas or LPG to form a mixture of methanol, formaldehyde, and acetaldehyde. The process uses oxygen in the form of air and aluminium phosphate as the catalyst. It was developed in the 1920s and is named after its inventor J. C. Walker.

wall shear stress (Symbol τ_w) The shear force applied to a flowing fluid at the point of contact with a surface. Where it is assumed that the wall shear stress is proportional to the kinetic energy of the moving fluid per unit volume, the proportionality constant is known as the *Fanning friction factor, f. That is $\tau_w = f \rho v^2 / 2$ where ρ is the density of the fluid and v is the velocity. The SI units are N m^{-2}.

washback *See* FERMENTATION.

washings The liquid waste stream leaving a process in which a liquid has been in contact with a solid for the purposes of extraction and leaching. In a multistage washing process operated with countercurrent flow in which the liquid flows in the opposite direction to the flow of process stream, the liquid becomes progressively more dilute.

Washoe process A process used to extract silver from sulphide ores. It involves heating the ores with a solution of sodium chloride in an iron vessel. The sodium chloride dissolves the silver while the iron reduces it. It was invented in 1860 and is named after the district in Nevada in the US where the ore was originally mined.

waste The material left over from a process that no longer has an economic value nor has potential as a valuable resource. Its lack of value may be the result of difficulties in recovery that may be otherwise uneconomic, expensive to transport, or possess hazardous properties. **Process waste** is waste generated from within a process such as from by-products in reactors or loss of solvents from separations and recycle systems, whereas **utility waste** is waste generated through the provision utility services to the main process such as waste from steam boiler *blowdown or heat loss from a heat exchanger network.

waste heat The generation of low grade heat from a process discharged into the environment. The excess heat is produced due to the low efficiencies of thermodynamic cycles used in nuclear and thermal power stations. The efficiency of power generation is the electrical energy generated to the heat generated. The waste heat can be partly recovered and reused such as by using *waste heat boilers.

waste heat boiler A type of boiler used to produce very hot water or to raise steam using hot by-product gases or hot liquids from a chemical process. By recovering the waste thermal energy, the thermal efficiency of a process is improved. The hot water and steam can be used for other useful purposes elsewhere, such as for heat transfer.

waste management The methods and procedures used to dispose of waste materials arising from chemical processes. All chemical processes produce waste in various forms and at the various stages of production and materials transportation. The objectives of the management of waste materials are to minimize responsibly the impact on the environment and human health, and to conserve scarce resources, as well as to maximize waste reuse, reclamation, and recycling. In the case of nuclear waste, the material is categorized in terms of its radioactivity and the waste managed accordingly. Low radioactive waste may go to landfill, medium radioactive waste can be recycled, and highly radioactive waste, which poses a serious risk to the environment and human health over very long periods of time (tens of thousands of years for some isotopes), is stored in deep caverns in geologically stable rock.

water A colourless, odourless, tasteless, and non-flammable liquid with the chemical formula H_2O that covers two-thirds of the surface of the planet. With a freezing point of $0°C$ and boiling point of $100°C$ at atmospheric pressure, it is used in huge quantities in the chemical industry. It is the most commonly used solvent and has the ability to dissolve a wide range of materials. It is also extensively used as a heat transfer medium for both heating and cooling purposes. It can be converted to ice or to steam, and in some cases a supercritical state is used as a solvent. It is also used as a transport medium in conveying other materials, and widely used in cleaning process plant equipment and process materials. Depending on the application, water quality is an important consideration and there is increasing pressure to reduce the amount of water that is used. The cost of transportation and scarcity are also increasingly important considerations.

water activity (Symbol a_w) A measure of the amount of water in a substance such as food and expressed as an equilibrium relative humidity. That is, it is the vapour pressure of water in the substance or in a solution divided by the vapour pressure of pure water at a particular temperature. Distilled water therefore has a water activity of unity. Since contaminating microorganisms in foods are dependent on the amount of available water, water activity is a useful parameter in controlling the growth of microorganisms.

water gas A mixture of carbon monoxide and hydrogen gas produced by the exothermic reaction of steam on heated carbon:

$$C + H_2O_{(steam)} \rightarrow CO + H_2$$

Water gas burns with a non-luminous bluish flame and is also known as **blue water gas**. Water gas was once produced from coke, and was in turn produced from coal. Prior to the use of steam reforming of natural gas, the main use before the Second World War was for the production of hydrogen in the *Haber process and was combined with the *water-gas shift reaction to increase the amount of hydrogen produced.

water-gas shift A catalytic process used to convert carbon monoxide gas to carbon dioxide and hydrogen using steam:

$$CO + H_2O_{(steam)} \rightarrow CO_2 + H_2$$

The reaction takes place in a fixed bed reactor known as a **water-gas shift reactor**. It uses a mixed catalyst of iron and chromium oxides. The reaction is exothermic and is completed either before or after the acid gas removal. The products are known as *synthesis gas or **syngas**, for short.

water hammer A violent and potentially damaging shock wave in a pipeline caused by the sudden change in flow rate, such as by the rapid closure of a valve. The effect is avoided by controlling the speed of valve closure, lowering the pressure of the fluid, or lowering the fluid flow rate.

water pollution The presence of harmful or objectionable material in water in sufficient quantity so as to cause a change in its chemical, physical, biological, or radiological quality that may be injurious to its existing or intended potential use. This includes use for human consumption and general domestic use for washing, cooking, etc., industrial use, aquatic organism health such as fish, and ecosystem health such as rivers, lakes, and sea.

water treatment A process used to improve the quality of water by removing or reducing the presence of pollutants so that the water is fit for purpose, such as being potable and fit for human consumption, for industrial use, or for release into the aquatic ecosystem. A water-treatment plant involves the separation of solids from wastewater by sedimentation and flocculation, and reducing the *BOD.

water vapour The gaseous state of water dispersed with air at a temperature below the boiling point of the water. The amount present in air is designated by the humidity. The *relative humidity is the amount of water vapour in a mixture of dry air. A relative humidity of 100 per cent corresponds to the partial pressure of water vapour being equal to the equilibrium vapour pressure, and depends on the temperature and pressure.

Watt, James (1736–1819) A Scottish engineer who made a significant improvement to the Newcomen steam engine. The Watt engine proved so effective that he is credited with the invention of the steam engine, since it may be argued that the Newcomen engine is essentially only a pump. He was therefore instrumental in the Industrial Revolution in Great Britain and around the world, since factories no longer needed to be tied to locations where there was a strong source of water power. Massive machines could therefore be built and housed in factories which proved to be the dawn of mass production.

watt (Symbol W) An SI unit of power defined as the rate of energy output or consumption of one joule per second. The power consumed over a period of time is measured in watt-hours. The watt is named after Scottish engineer James *Watt (1736–1819) who developed a steam engine to replace horses in coal pits. The term *horsepower was used by Watt to demonstrate the equivalent power of his new machine, where one horsepower (hp) is equal to 746 watts.

watt-hour A unit of energy expressing the power in watts used over a period of time of one hour. The consumption of electricity is often expressed in kilowatt-hours or kWh.

wavelength (Symbol λ) The distance between each crest or trough of a wave. It is related to frequency, f, by the velocity of the wave, v, as:

$$\lambda = \frac{v}{f}$$

wavy flow A type of stratified two-phase fluid flow in a horizontal pipe in which waves form at the surface between a liquid and gas or vapour. This type of fluid occurs where there is a notable difference between the superficial velocities of the vapour or gas and the liquid sufficient to cause waves to form. It is also known as the **stratified-wavy flow regime**.

wax A solid or semi-solid substance with may be either a monoester of fatty acids or a mixture of hydrocarbons of high molecular weight. As a monoester, such as beeswax, the acid and alcohol proportions both have long saturated carbon chains:

$$C_{25-27}H_{51-55}COOC_{30-32}H_{61-65}$$

Waxes are used to make polishes, cosmetics, pharmaceutical preparations, ointments, and candles. In nature, waxes coat leaves and stems in plants that grow in arid conditions and have a protective effect.

WCEC *See* WORLD CHEMICAL ENGINEERING COUNCIL.

Weber number A dimensionless number, We, used in thin film gas–liquid flows where surface tension has a major effect on the strongly curved interface between two fluids such as in the formation of droplets and bubbles. It is a measure of the relation between surface tension and gravity forces:

$$We = \frac{\rho v^2 l}{\sigma}$$

where ρ is the density of the fluid, v is the velocity, l is the characteristic dimension such as droplet diameter, and σ is the surface tension. It is named after German engineer Moritz Weber (1871–1951).

weeping A phenomenon that occurs in a distillation column in which liquid on a sieve plate passes down through the perforations intended for the vapour to pass up. Weeping occurs when the velocity of the upward vapour is too low. This may be caused by insufficient boil-up.

weight The property of a body to exert a force due to the influence of gravity $F = mg$. The SI unit is the newton (N). *Compare* MASS.

weir A vertical obstruction across a channel carrying a liquid over which the liquid discharges. In a distillation column, a weir is used to retain an amount of liquid on a sieve tray or plate. While the vapour enriched with the more volatile component rises up through the perforations on the sieve tray or plate, the liquid cascades over the weir into the *downcomer to the tray below. The **weir crest** is the top of the weir over which the liquid flows.

Weissenberg effect A phenomenon that occurs when a spinning rod is placed into a liquid polymer solution. Molecular entanglements within the polymer cause the polymer chains to be drawn towards the rod and appear to climb up the rod instead of being thrown

w

outwards and away from the rod. It is named after Austrian physicist Karl Weissenberg (1893–1976).

Weizmann process A fermentation process used to produce acetone, butanol, and ethanol using the acid-resistant bacterium *Clostridium acetobutylicum*. The bacteria derived from soil and cereals is able to convert whey, sugar, and starch. The process was developed by Russian-born chemist Chaim Weizmann (1874–1952) and was used in the UK in the First World War for the production of acetone, which was used in the production of cordite. He became a UK citizen in 1910 and then the first president of Israel in 1949. The process is also known as the **ABE fermentation**.

weld The attachment of two metal surfaces by the use of heat to melt and fuse the metals.

well A natural oil or gas reservoir that exists below a layer of sedimentary rock. *Wildcat wells are the first wells to be drilled in a particular geographic location. **Appraisal wells** are drilled after hydrocarbon has been identified. **Development wells** are drilled according to a predetermined pattern to maximize the amount of oil hydrocarbon that can be recovered. This is based on available seismic surveys and other geological data. **Injection wells** are used for the injection of fluids into the reservoir for enhanced oil hydrocarbon recovery. The **wellhead** is the top of the oil or gas well.

Wellman–Lord process A process used for flue gas desulphurization. Sulphur dioxide in the flue gas is absorbed in sodium sulphite in a wet spray scrubber column forming sodium bisulphite:

$$Na_2SO_3 + H_2O + SO_2 \rightarrow 2NaHSO_3$$

The solution is regenerated by heating the solution to release the sulphur dioxide and then either collected or used in the production of sulphuric acid. The sodium sulphite is returned to the process.

Welsh process A copper smelting process used in South Wales that originates from the early eighteenth century and was used through to the end of the nineteenth century. It used a furnace in which roasting, fusing, and refining were carried out. The process was superseded by bigger furnaces.

(((⊕))) SEE WEB LINKS
• Official website of the Copper Development Association Inc., Education pages.

West Texas Intermediate A grade of crude oil used as a benchmark in oil pricing. Also known as Texas light sweet, it has a relatively low density and low sulphur content. Other price benchmarks include Brent crude and Dubai crude.

wet and dry bulb hygrometer *See* HYGROMETER.

wet basis A method of representing the moisture content of a substance in which the amount of water is taken as a ratio of the combined amount of substance and water. The moisture content of a very wet substance on a wet basis will approach but not equal 100 per cent. *Compare* DRY BASIS.

wet bulb temperature The temperature indicated by a glass bulb thermometer whose surface is kept wet by a thin film of liquid, usually water, and which is exposed to a current of air. It is therefore the dynamic equilibrium temperature attained by a liquid surface subject to the action of a rapid stream of gas. The use of both wet and dry bulb temperature is used to determine the *humidity of a gas such as water vapour in air for a particular temperature and pressure. The difference between the wet and dry bulb temperatures is known as the **wet bulb depression**.

wet gas 1. A term used to describe light hydrocarbon gas dissolved in heavier hydrocarbons. Wet gas is an important source of LPG. **2.** Water that is present in natural gas in offshore platforms. It is necessary to remove the water from the gas for export through subsea pipelines. The pipeline is dosed with corrosion inhibitors to prevent hydrate formation.

wet process 1. A process used to remove the skin of coffee cherries before they are dried. The cherries are immersed in water to separate unripe fruit before pressing. The pulp is fermented to digest the cellulose using large amounts of water. The **dry process** uses only the juices from the cherries. **2.** A process used to produce phosphoric acid from phosphate rock using a strong acid such as sulphuric, nitric, or hydrochloric acid. Sulphuric acid is used to digest tricalcium phosphate mineral rock, apatite, to produce a solution of phosphoric acid and insoluble calcium sulphate. This is then concentrated by evaporation.

wet scrubbing A process used to remove polluting or harmful gases and particles from a gas stream. For example, carbon dioxide can be removed from a flue gas stream by solvent extraction using an amine solution such as methyl diethanolamine to form a stable salt. The rate of absorption increases with the amine concentration. The amine solution can be regenerated by heating. *Compare* DRY SCRUBBING.

wet steam A mixture of droplets of water and steam that are both at the saturated temperature. If additional heat is added at constant pressure, the temperature remains constant until the point where all the water has evaporated and converted into saturated steam. Beyond this saturation temperature, the steam becomes superheated. The *quality of the steam is the amount of water in the steam.

wetted perimeter The part of the cross section of a pipe or channel that is in contact with a liquid.

wetting agent A surface active agent that reduces the surface tension of a liquid. It has the effect that small amounts of a wetting agent cause the liquid to spread out over a flat surface.

whisky (whiskey) A distilled spirit made from a fermented mash of sugars derived from cereal. Scotch malt whisky and Irish whiskey are made from malted barley from which the starch is converted to sugars through germination. Bourbon whiskey is made from a grain mixture containing at least 51 per cent corn. Rye whiskey is made from at least 51 per cent rye grain. The malting process releases enzymes used to convert the starch to sugars. It is then roasted and mashed as a fermentation that converts the sugars into alcohol. The fermented wort liquid is distilled and the distillate collected. Maturation takes place in wooden casks. For Scotch whisky, this must be for a minimum of three years. Single malt whisky is from a single distillery. Most whiskies are blended.

Whitman two-film theory A theory used in gas adsorption to describe the mass transfer of a solute from a gas into a liquid in which either side of the interface exists a stagnant gas film and a similar stagnant liquid film (see Fig. 57). The two films are assumed to offer the only resistance to mass transfer. In the bulk gas and liquid, the solution partial pressure

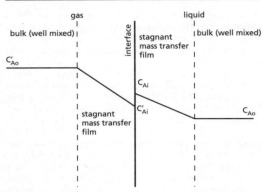

Fig. 57

and concentration are considered to be constant. At the interface, they are assumed to be in equilibrium. It was developed by W. K. Lewis and W. G. Whitman in 1924.

whole numbers The set of integers {1, 2, 3 . . . } excluding zero.

Wien's displacement law A law that states that the most strongly emitted wavelength in the continuous spectrum from a full radiator is inversely proportional to the absolute temperature of that body (see Fig. 58). $\lambda T = b$ where b is Wien's constant equal to 2.898×10^{-3} m K. The maximum temperature in the spectral distribution from a *black body is therefore displaced towards the shorter wavelengths with increasing temperature. For example, an electric bar heater emits no colour when cold and on heating initially glows red and is white (i.e. a mixture of all coloured light) when very hot. The law was stated by German physicist Wilhelm Wien (1864–1928).

wildcat well An oil and gas exploration test *well. Not all drilling operations are successful. Where oil or gas is found in the wildcat, a series of appraisal wells are drilled at locations to represent the boundaries of the field before a permanent installation is located in position.

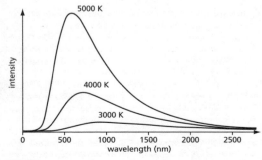

Fig. 58

Wilke–Chang correlation A semi-empirical correlation used for the estimation of the diffusion coefficient of a substance present in infinite dilution in a solvent.

Windscale nuclear accident The UK's worst nuclear accident, which occurred on 10 October 1957 at the Windscale nuclear power station at Sellafield. The first of two *atomic piles, built in 1950 and 1951 as part of Britain's atomic weapons project, caught fire releasing radioactive material into the environment. The 120-metre-high piles were solid graphite moderated and air cooled with horizontal channels within which uranium cartridges could be passed and exposed to neutron radiation to produce plutonium. After the accident, pile one was sealed and pile two was permanently shut down shortly after.

wine A fermented product made from fermented sugar derived from grapes. The process involves harvesting grapes, from which the juice is extracted using roller or rotary paddle crushers. White wine involves filtration and centrifugation of the skins and pulp whereas red wine does not. Fermentation takes place in vessels that were traditionally made of wood but large-scale processes now use stainless steel. Water and yeast are added to convert the sugar to alcohol. Sulphur dioxide is added to control potential contamination by bacteria. The fermentation is maintained at a controlled temperature and may last several days to two weeks, at the end of which yeast and any other solid material is removed by racking. The wine is clarified by filtration and fining using bentonite clay or some other absorbing material. The wine is stored at a controlled temperature for several months and bottled.

Winkler process An early coal gasification process that uses air and steam in a fluidized bed operated at atmospheric pressure. The air provides the oxygen in the process. The process is named after its inventor Fritz Winkler who developed it in the 1920s.

wiped film evaporator A type of evaporator used for concentrating heat-labile solutions. It consists of a vertical cylindrical column or cylinder surrounded by a steam jacket. The solution to be evaporated enters the top of the vessel and is allowed to descend the inner walls under the influence of gravity. A rotating assembly of blades with a close clearance to the wall ensures a short hold-up and corresponding residence time, where longer residence times at the boiling temperature would otherwise damage the solution.

wispy annular flow A two-phase flow regime in a pipe or tube characterized by a continuous gas core and wall film of liquid. The flow regime occurs at high gas velocities. At very high liquid flow rates, liquid concentrations in the gas core are sufficiently high that droplets coalescence in the gas core leading to streaks or wisps to occur instead of droplets.

Wobbe index A number used to compare the combustion energy of different fuels such as natural gas, *SNG, and *LPG, and used for comparative purposes of gas burners. Fuel gases with the same Wobbe index can be used interchangeably without a change in the air-to-fuel ratio. It is calculated from the calorific value of the gas divided by the square root of the specific gravity. The SI units are MJ m^{-3}. It is also often still expressed in Imperial units of BTU per standard cubic foot.

Wöhler's synthesis The synthesis of urea from the evaporation of a solution of ammonium isocyanate. First carried out by Friedrich Wöhler in 1828, the synthesis was significant at the time as organic substances were thought to be produced only by living organisms.

Wohlwill process An early electrolytic process used to refine gold. It used crude gold as the anode and pure gold as the cathode with a solution of gold chloride in hydrochloric acid as the electrolyte. By applying an electrical current across the electrodes, pure gold accumulates at the cathode. Silver present deposits as silver chloride. The process was developed in 1874 and is named after its German inventor Emil Wohlwill.

work (Symbol W) The forms of energy transfer that can be accounted for in terms of changes in the external macroscopic-scale physical constraints on a system. The work done is the product of a force and the distance moved by its application. Displacement work is the energy that goes into expanding the volume of a system against an external pressure such as by driving a piston out of a cylinder against an external force or driving a piston into a cylinder of a given volume and pressure. Work done can also be achieved by a paddle acting upon a fluid or by a fluid, propeller, or turbine blade. Frictional work is the work done overcoming friction. The SI unit is the joule.

work-hardening The process of increasing the hardness or strength of metals by working using a mechanical process such as compression, tension, or torsion. The process causes a permanent change in the crystal structure. Iron, copper, and aluminium can all be work-hardened, whereas metals such as zinc and lead are not capable of being work-hardened as they have the capacity to recrystallize at room temperature.

working capital The current assets of an organization minus the current liabilities. It is the part of the capital that is available for operations.

working drawing A scale drawing of a part of a process or assembly that is used as a guide for fabrication, manufacture, or construction.

working pressure The normal operating pressure of a process.

workover A major form of maintenance or remedial operation on an operating oil or gas well to either restore or increase the rate of production.

work permit An authority written by a responsible person specifying work that can be carried out, the precautions which have to be taken to ensure it is carried out safely, any particular procedures to be followed or particular equipment to be used, and the period for which the permit is valid. Work permits are usually required by law before any work is done. *See* PERMIT TO WORK.

workplace exposure limits (WEL) The UK occupational exposure limits used to protect the health of employees. The exposure limits are defined as the maximum concentration of an airborne hazardous substance averaged over a defined period of time to which workers may be exposed by inhalation and that should not cause adverse health effects. The **short-term exposure limit** (STEL) is averaged over fifteen minutes while the **time weighted average** (TWA) is averaged over eight hours. STELs are aimed to help prevent effects such as eye irritation that may occur following exposure for only a few minutes.

works A place, such as a factory, chemical plant, or refinery, where a number of people are employed.

World Chemical Engineering Council (WCEC) An association of leading professional and learned chemical engineering bodies from around the world. It was formally launched in 2001 at the closing ceremony of the Sixth World Congress of Chemical Engineering in Melbourne, Australia, with the aim of raising public awareness of the work and contribution of chemical engineering and chemical engineers. It comprises prominent chemical engineers who influence the scientific structure of the World Congress of

Chemical Engineering. The Executive Committee comprises representatives from leading societies of chemical engineering, including IChemE, AIChE, DECHEMA, and the regional federations from Europe (EFCE), Asia Pacific (APCChE), and the Americas (IACChE).

((⊕)) SEE WEB LINKS
• Official website of the World Chemical Engineering Council.

World Congress of Chemical Engineering (WCCE) A four-yearly international conference organized under the auspices of the *World Chemical Engineering Council. The conference rotates amongst the major regions of the world and is hosted by one of the professional societies for the subject. The 2009 conference was held in Montreal, Canada, the 2013 conference in Seoul, South Korea, and the 2017 conference is due to be held in Barcelona, Spain.

wrought iron An iron alloy with a very low carbon content containing fibrous inclusions, known as slag. The fibrous structure gives a graining effect. Wrought iron is noted as being tough, malleable, ductile, and easily welded. Before the development of effective methods of steelmaking, wrought iron was the most common form of malleable iron. A modest amount of wrought iron was used as a raw material for the manufacturing of steel, which was mainly to produce swords, cutlery, and other blades. Demand for wrought iron reached its peak in the 1860s with the adaptation of ironclad warships and railways, but then declined as mild steel became more available. Wrought iron rusts less readily than other forms of metallic iron and used to be used for rivets, nails, wire, chains, railway couplings, water and steam pipes, nuts, bolts, horseshoes, handrails, straps for timber roof trusses, and ornamental ironwork. Wrought iron is no longer produced on a commercial scale.

Wulff process A process used for the production of acetylene (ethyne) by the cracking of a saturated aliphatic hydrocarbon gas with superheated steam in a high-temperature furnace. The gas is first heated to around 1,300°C and then passed to a refractory brick reactor operating at 400°C. It is named after its American inventor R. G. Wulff who developed it in 1927.

w

xenobiotic Any substance foreign to living systems. Xenobiotic substances include drugs, pesticides, and carcinogens. Detoxification of such substances occurs mainly in the liver.

Xmas tree *See* CHRISTMAS TREE.

X-ray A penetrating short-wavelength electromagnetic radiation (10^{-10} m) that is produced when a stream of cathode rays (i.e. electrons) strikes a target surface. On impact, the target may emit secondary X-rays which are characteristic of the elements in it. The ability of X-rays to penetrate many types of materials makes them useful for inspecting cracks and flaws such as in *welds. *See* RADIOGRAPHY.

X-ray crystallography The use of diffraction of X-rays for the determination of the structure of crystals or molecules. It uses a beam of X-rays directed onto the target sample that is diffracted and captured on a photographic plate as a pattern of light and dark spots of differing intensities. These are analysed, and from which, the structure of the crystal or molecule is deduced. Since the wavelength of X-rays is comparable to the distance between the atoms in most crystals, the repeating structure acts like a diffraction grating for X-rays.

X-ray fluorescence The emission of X-rays from excited atoms that are produced by the impact of high-energy electrons, or other particles, or a primary beam of other X-rays. The wavelengths of the fluorescence X-rays can be measured by an X-ray spectrometer as a means of chemical analysis.

xylene A mixture of three closely related aromatic hydrocarbons of the benzene group having isomeric forms of orthoxylene (o-xylene), metaxylene (m-xylene), and paraxylene (p-xylene). They each have the formula $C_6H_4(CH_3)_2$ and have close boiling points. They can be separated by crystallization and by distillation. They are important members of the *BTX group of petrochemical feedstocks and used in small amounts in aircraft fuel. They are widely used to produce plastics, vitamins, synthetic fibres, dyes, and pharmaceutical products. The non-technical name for xylenes is xylol.

xylene-plus process A catalytic process used for the isomerization of toluene to a mixture of benzene and *xylene. The process involves a moving bed of silica-alumina catalyst. It differs from the *tatoray process in that no hydrogen is used in the process.

xylofining process A catalytic process used for the isomerization of petrochemicals that contain ethyl benzene and *xylenes. The process developed in India is operated in the vapour phase with an iron-based zeolite at a temperature of around 360°C. The reaction converts the ethyl benzene to ethylene and benzene. It is an abbreviation of **xylol** re**fining**. **Xylol** is a non-technical name for xylene.

yard An Imperial unit of length equal to three feet (36 inches). It is equal to 0.9144 metres.

yeast A fungus that forms single cells that reproduce by budding or binary fission. They are exploited commercially as biocatalysts in biotechnological processes such as in the production of alcoholic beverages wine and beer, as well as for the production of bioethanol, which is used as a motor fuel or petroleum additive. Fungi have the ability to convert sugars to alcohol through the process of fermentation. They are also used in single-cell protein production as well as in baking. They are also a dietary source of protein and vitamin B. There are many genera and important yeasts used in biotechnological processes include *Saccharomyces*, *Kluyveromyces*, and *Candida*.

yield The amount of a product that is recovered from a process or chemical reaction. It is usually expressed as a fraction or a percentage based on the raw materials used or as a ratio of the final product to the starting materials without considering any *side reactions. The **yield coefficient** is a measure of the amount of product produced to the raw materials consumed. In a biotechnological process such as fermentation, the yield coefficient is used to express the amount of microbial cells produced per amount of substrate consumed. It is expressed on a weight per weight basis such as kilograms of cells or product produced per kilogram of substrate consumed. The yield coefficient can also apply to the production of a biochemical such as ethanol to the substrate.

yield per pass The net yield of any product in a chemical reactor effluent with a recycle stream back to the reactor feed. It is expressed as a percentage of the limiting reactant in the combined reactor feed. Yields are expressed on a molal, weight, or volume basis.

yield point The stress that is required to be applied to a fluid such as a *Bingham plastic in order for it to begin to flow. Paints, gels, and toothpaste all have a yield point that must be exceeded in order for them to flow.

yield stress 1. The lowest stress that is applied to a material at which extension increases without an increase in applied load. **2.** The shear stress that is required to be applied to a fluid to make it begin to flow. *See* HERSCHEL–BULKLEY FLUID.

Young, James (1811–83) A Scottish chemist noted for his 1852 patent for the distillation of shale, and as the founder of the paraffin industry. Born in Glasgow, James (Paraffin) Young began his career working for his father's carpentry business and at a young age chose to attend evening classes on chemistry delivered by Thomas *Graham at the Andersonian Institution (now the University of Strathclyde). Graham recognized Young's talents and offered him a position as assistant in his laboratory in 1832. When Graham went to University College London in 1837 he took Young with him. There Young distinguished himself as a technical chemist not only in terms of producing chemicals but also in the difficult task of avoiding nuisance and injury, which is a precursor to health and safety. Young's career took him to St Helens College, and then on to Manchester where he became interested in oil that was found flowing from a pit in Derbyshire. This led to his work on oil production by the low-temperature distillation of shale. When the oil seepage ran out, he returned to Scotland in 1851 to set up a works near Bathgate to produce and refine oil from a rich coal seam, which was being mined nearby. Young patented his method, which earned him a considerable fortune. Today, the Grangemouth refinery nearby is a lasting legacy of oil processing and refining.

SEE WEB LINKS
• Official website of the Museum of the Scottish Shale Oil Industry, page devoted to James Young biography.

Young–Laplace equation An equation used to determine the difference in pressure between either side of a curved surface such as a bubble of radius r_1 and r_2 due to *surface tension:

$$\Delta p = \sigma \left(\frac{1}{r} + \frac{1}{r} \right)$$

where σ is the surface tension. It is named after the British scientist Thomas Young (1773–1829) and the French mathematician Pierre-Simon, Marquis de *Laplace (1749–1827).

zenith process A process used for refining vegetable oils. It involves passing droplets of oil down a column of aqueous sodium hydroxide.

zeolite A natural or synthetic material used to selectively separate substances from mixtures such as water from alcohol. They have an open three-dimensional crystal structure that naturally contains water molecules. Synthetic zeolites include hydrated aluminosilicate. The water can be released by heating and the zeolite can be used to absorb other substances of an appropriate molecular size. Literally meaning 'boiling stone', they are often called **molecular sieves**. They are also used in sorption pumps for vacuum systems, separating organic compounds such as high-octane petroleum for lead-free petrol, and in ion exchange for applications such as in water softening. Zeolites are one of the most diverse and industrially useful mineral groups.

zepto- (Symbol z) A prefix used to denote 10^{-21}.

zero energy state A term used to denote that all stored or residual energy within devices such as rotating flywheels, hydraulic and pneumatic systems, and springs has been dissipated.

zero order reaction The rate of a chemical reaction that is independent of the concentration of a particular reactant. *See* ORDER OF REACTION.

zeroth law of thermodynamics A law which states that if two bodies are in thermal equilibrium with a third body at the same time, then all three are in thermal equilibrium with each other. Essentially, it means that all the bodies are at the same temperature and therefore form the basis for the comparison of temperatures. It is known as the zeroth law since it precedes the first and second laws of thermodynamics.

zetta- (Symbol Z) A prefix used to denote 10^{21}.

Ziegler, Karl Waldemar (1898–1973) A German chemist noted for his work on the controlled polymerization of hydrocarbons through the use of organometallic catalysts. He developed a catalytic system enabling low-pressure polymerization of ethylene to linear polyethylene. He also worked on the development of plastics. He was awarded the Nobel Prize in Chemistry in 1963.

Ziegler–Natta A general name for processes used for the polymerization of olefins (alkenes) that were invented by the German chemist Karl Waldemar *Ziegler (1898–1973) and the Italian chemist Giulio *Natta (1903–79). They were jointly awarded the Nobel Prize in Chemistry in 1963.

Ziegler–Natta catalyst A type of catalyst used in the synthesis of polymers such as high-density polyethylene from olefins (alkenes). They are based on compounds of titanium such as titanium (IV) chloride (TiCl$_4$) and organoaluminium compounds such as triethyl-aluminium (Al(C$_2$H$_5$)$_3$). They are named after the German chemist Karl Waldemar *Ziegler (1898–1973) and the Italian chemist Giulio *Natta (1903–79). They were jointly awarded the Nobel Prize in Chemistry in 1963.

Ziegler–Nichols tuning A tuning method used to select optimum controller settings. It is used for the control of complex systems where little is known of the dynamics of the system or where calculation of the response is too difficult or impractical. The method was developed by American control engineers John Ziegler and Nathaniel Nichols and is based on empirical tests that gave good performance for a step response of an *open loop system. It is based on the assumption that an open loop transfer function can be approximated by a first-order system with a time delay. The recommended controller settings have been proposed for proportional, proportional and integral, and proportional, integral, and derivative action control. Applied to both open and *closed loop systems, the settings are designed to give underdamped transient responses with a *decay ratio of a quarter.

Ziegler process 1. A process used for the manufacture of high-density polyethylene using catalysts known as *Ziegler–Natta catalysts. First introduced in 1953, they improved on the original manufacturing process and were able to use a lower temperature of 60°C and pressure of around one atmosphere. They are named after German chemist Karl Waldemar *Ziegler (1898–1973) and the Italian chemist Giulio Natta (1903–79) who developed the process further in 1954 for the use with other alkenes. They were awarded the Nobel Prize in Chemistry in 1963. **2.** A process used for the production of tetraethyl lead by electrolyzing a molten complex of ethyl potassium with triethyl aluminium. It was invented by Karl Waldemar *Ziegler (1898–1973) in 1963.

Zimmerman process A thermal process used for oxidizing aqueous organic wastes and waste sewage sludge. It involves pressurizing the sludge with air and heating with steam to a temperature of 250°C. The resulting sterile product is then filtered. It is named after its inventor J. F. Zimmerman who developed it in the 1950s. This has formed the basis of a number of related processes such as the **Zimpro process**.

zincex process A process used for the extraction of zinc from pyrite cinder leachate. It uses organic solvents to remove chloride leachate before an acid treatment is used to remove the iron. It is an abbreviation of **zinc ex**traction. The **zinclor process** is a modified version of the zincex process that uses dipentyl pentylphosphate (DPPP) as the extracting solvent.

zirpro process A process used to flameproof textiles by the treatment of aqueous solutions of zirconium complexes. The textiles, such as wool, are treated with potassium hexafluoro-zirconate and citric acid, while cotton is treated with zirconium acetate and citric acid.

zone levelling A process used to distribute a solute evenly throughout a purified solid material in the form of a single crystal. For example, in the preparation of a transistor or diode semi-conductor, an ingot of germanium is first purified by *zone refining. A small amount of antimony is then added into the molten zone, which is passed evenly through the pure germanium. This technique is also used for the preparation of silicon for use in computer chips.

zone refining A process used to reduce the impurities in certain metals, alloys, and semiconductors. It relies on the solubility of an impurity being different in the liquid and solid phases. The process involves a narrow molten zone being progressively moved along the length of a specimen of the material such that the impurities are segregated at one end of the ingot and pure material at the other. Where the impurities have a lower melting point than the material being purified, the impurities travel with the moving zone, and vice versa.

Zone refining was developed by American materials scientist William Gardner Pfann (1917–82) as a method of preparing high-purity materials for the manufacture of transistors. Its early use was for purifying germanium, but it can be extended to any solute–solvent system having an appreciable concentration difference between the solid and liquid phases at equilibrium. It is also known as the **float zone process** used in semi-conductor materials processing.

z-value The temperature rise required in degrees Celsius that is able to bring about a ten-fold decrease in the thermal death time of microorganisms. It is used in the thermal sterilization process of foods that may be contaminated with harmful microorganisms. The thermal death time is the time required to reduce the number of surviving microorganisms to an acceptable level. A typical value for the thermal destruction of vegetative microorganisms is around 10°C.

Appendix 1: SI prefixes and multiplication factors

Multiplication factor		Prefix	Symbol
1 000 000 000 000 000 000 000	$= 10^{18}$	exa	E
1 000 000 000 000 000 000	$= 10^{15}$	peta	T
1 000 000 000 000	$= 10^{12}$	tera	T
1 000 000 000	$= 10^{9}$	giga	G
1 000 000	$= 10^{6}$	mega	M
1 000	$= 10^{3}$	kilo	k
100	$= 10^{2}$	hecto	h
10	$= 10^{1}$	deca	da
0.1	$= 10^{-1}$	deci	d
0.01	$= 10^{-2}$	centi	c
0.001	$= 10^{-3}$	milli	m
0.000 001	$= 10^{-6}$	micro	μ
0.000 000 001	$= 10^{-9}$	nano	n
0.000 000 000 001	$= 10^{-12}$	pico	p
0.000 000 000 000 001	$= 10^{-15}$	femto	f
0.000 000 000 000 000 001	$= 10^{-18}$	atto	a

Appendix 2: Derived units

Through common usage, fundamental units have been given names. None of them is a recognized SI unit.

Length	Area
Micron (μm)	$= 10^{-6}$ m
Angstrom (Å)	$= 10^{-10}$ m
Fermi (fm)	$= 10^{-15}$ m
Are (a)	$= 100$ m^2
Barn (b)	$= 10^{-28}$ m^2
Mass Tonne (t)	$= 10^3$ kg
Time Minute (min)	$= 60$ s
Hour (h)	$= 3,600$ s
Day (d)	$= 86,400$ s
Year (a)	$\approx 3.1557 \times 10^7$ s

Appendix 3: Derived units in SI and c.g.s.

Quantity	Dimensions	SI	c.g.s.	Ratio c.g.s./SI
Acceleration	LT^{-2}	$m\ s^{-2}$	$cm\ s^{-2}$	10^{-2}
Area	L^2	m^2	cm^2	10^{-4}
Density	ML^{-3}	$kg\ m^{-3}$	$g\ cm^{-3}$	10^{-3}
Energy	ML^2T^{-2}	joule	erg	10^{-7}
Force	MLT^{-2}	newton	dyne	10^{-5}
Length	L	metre	centimetre	10^{-2}
Mass	M	kilogram	gram	10^{-3}
Power	ML^2T^{-3}	watt	$erg\ s^{-1}$	10^{-7}
Pressure	$ML^{-1}T^{-2}$	pascal	$dyne\ cm^{-2}$	10^{-1}
Surface tension	MT^{-2}	$N\ m^{-1}$	$dyne\ cm^{-1}$	10^{-3}
Time	T	second	second	1
Velocity	LT^{-1}	$m\ s^{-1}$	$cm\ s^{-1}$	10^{-2}
Viscosity	$ML^{-1}T^{-1}$	$kg\ m^{-1}\ s^{-1}$	poise	10^{-1}
Volume	L^3	m^3	cm^{-3}	10^{-6}

Appendix 4: Abbreviations used for piping and instrumentation diagrams (P&IDs)

Analyser controller	AC
Analyser transmitter	AT
Differential pressure controller	DPC
Differential pressure transmitter	DPT
Flow indicator	FI
Flow controller	FC
Flow indicator and controller	FIC
Level controller	LC
Level indicator	LI
Level indicator and controller	LIC
Level transmitter/sensor	LT
High level alarm	LAH
Very high level alarm	LAHH
Low level alarm	LAL
Very low level alarm	LALL
pH controller	pHC
pH sensor/transmittor	pHT
Pressure indicator	PI
Pressure controller	PC
Pressure indicator and controller	PIC
Pressure transmitter	PT
Temperature controller	TC
Temperature indicator	TI
Temperature indicator and controller	TIC
Temperature sensor/transmitter	TT

Appendix 5: Dimensions and units

Quantity	Dimension	Unit
Acceleration	$[LT^{-2}]$	ms^{-2}
Area	$[L^2]$	m^2
Density	$[ML^{-3}]$	$kg\,m^{-3}$
Energy	$[ML^2T^{-2}]$	J
Force	$[MLT^{-2}]$	N
Length	$[L]$	m
Mass	$[M]$	kg
Momentum	$[MLT^{-1}]$	N s
Pressure	$[ML^{-1}T^{-2}]$	Pa
Shear stress	$[ML^{-1}T^{-2}]$	Pa
Shear rate	$[T^{-1}]$	s^{-1}
Time	$[T]$	s
Viscosity	$[ML^{-1}T^{-1}]$	Pa s
Volume	$[L^3]$	m^3
Work	$[ML^2T^{-1}]$	W

Appendix 6: Greek alphabet

Letters		Name
A	α	alpha
B	β	beta
Γ	γ	gamma
Δ	δ	delta
E	ε	epsilon
Z	ζ	zeta
H	η	eta
Θ	θ	theta
I	ι	iota
K	κ	kappa
Λ	λ	lambda
M	μ	mu
N	ν	nu
Ξ	ξ	xi
O	o	omicron
Π	π	pi
P	ρ	rho
Σ	σ	sigma
T	τ	tau
Y	υ	upsilon
Φ	φ	phi
X	χ	chi
Ψ	ψ	psi
Ω	ω	omega

Appendix 7: Periodic table

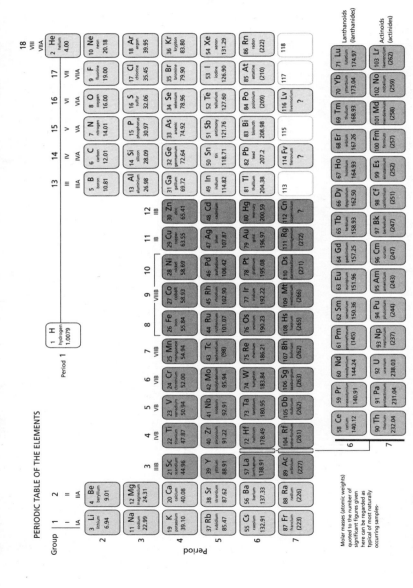

PERIODIC TABLE OF THE ELEMENTS

Molar masses (atomic weights) quoted to the number of significant figures given here can be regarded as typical of most naturally occurring samples.

© Peter Atkins 2013

Appendix 8: Fundamental constants

Constant	Units
Absolute zero	$-273.15\ °C$
Acceleration due to gravity	$9.806\ m\ s^{-2}$
Avogadro constant	$6.022\ 169\ 169 \times 10^{23}\ mol^{-1}$
Boltzmann constant	$1.380\ 622 \times 10^{-23}\ J\ K^{-1}$
Faraday constant	$9.648\ 670 \times 10^{4}\ C\ mol^{-1}$
Molar volume at s.t.p.	$2.241\ 36 \times 10^{-2}\ m^{3}\ mol^{-1}$
Planck constant	$6.626\ 196 \times 10^{-34}\ J\ s$
Speed of light	$2.999\ 925 \times 10^{8}\ m\ s^{-1}$
Stefan–Boltzmann constant	$5.670\ 400 \times 10^{-8}\ W\ m^{-2}\ K^{-4}$
Triple point of water	$273.16\ K$
University gas constant	$8.314\ 34\ J\ mol^{-1}\ K^{-1}$

Appendix 9: Recommended web links

[(🌐) SEE WEB LINKS]
This is a web-linked dictionary. To access the websites listed below, go to the dictionary's web page at www.oup.com/uk/reference/resources/chemeng, click on **Web links** in the Resources section and locate the entry in the alphabetical list, then click straight through to the relevant websites.

ACHEMA Ausstellungs-Tegung fuer Chemisches Apparatewesen
- Official website of ACHEMA.

air pollution
- Official website of Environmental Protection UK.

ALARP
- Official website of the Health and Safety Executive UK offering risk assessment advice.

American Institute of Chemical Engineers
- Official website of the American Institute of Chemical Engineers.

American National Standards Institute
- Official website of the American National Standards Institute.

American Petroleum Institute
- Official website of the American Petroleum Institute.

American Society for Testing Materials (ASTM International)
- Official website of ASTM International.

American Society of Mechanical Engineers
- Official website of the American Society of Mechanical Engineers.

Arrhenius, Svante August
- Official website of the Nobel Prize organization, with a transcript of Arrhenius' lecture of 1903.

ATEX
- Official website of the Health and Safety Executive, UK, outlining information on ATEX and explosive atmospheres.

BSI
- Offical website of BSI, with information about BSI standards.

carbon credit
- Official website of Carbon Futures, with an explanation of how carbon trading works.

carbon footprint
- Official website of the Carbon Trust.

CEFIC
- Official website of Conseil Européen des Fédérations de l'Industrie Chimique.

CERN
- Official information website of CERN, accessible to the public.

chartered chemical engineer
- Official website of the Engineering Council.

CHEMECA
- Official website for CHEMECA conference 2012.

Chemical & Engineering News
- Online access to *Chemical & Engineering News*.

Chemical Engineering
- Online access to *Chemical Engineering* magazine.

Clean Air Act
- Official website of HM Government managed by the National Archives to publish all enacted legislation in the UK.

DECHEMA
- Official website of DECHEMA.

DSEAR
- Official website of the Health and Safety Executive, DSEAR regulations.

ECHA
- Official website of the European Chemical Agency.

electrolytic separation
- Official website of the Norsk Industriarbeider Museum, Vemork (Norwegian Industrial Workers Museum).

Environment Agency (EA)
- Official website of the UK Environment Agency.

Environmental Protection Agency (EPA)
- Official website of the US Environmental Protection Agency.

European Federation of Chemical Engineering
- Official website of the European Federation of Chemical Engineering.

fire
- Official website of the National Fire Protection Association in the US.

fluoridization
- Official website of the British Fluoridation Society.

Hortonsphere
- Official website of the Chicago Bridge and Iron Company.

HSE
- Official website of the Health and Safety Executive in the UK.

Institution of Chemical Engineers
- Official website of the Institution of Chemical Engineers.

International Council of Chemical Associations
- Official website of the International Council of Chemical Associations.

JET
- Official website of EFDA JET programme

Mono pump
- Official website of Mono Company.

Nelson–Farrar cost index
- Official website of *Oil and Gas Journal* containing the Nelson–Farrar cost index.

Organization of the Petroleum Exporting Countries
- Official website of OPEC.

personal protective equipment
- Official website of the UK Health and Safety Executive.

radioactive waste
- Official website of the UK Nuclear Decommissioning Authority.

Royal Australian Chemical Institute
- Official website of the Royal Australian Chemical Institute.

Rutherford, Lord Ernest
- Official website of McGill University and its museum to Rutherford.

SEPA
- Official website of Scottish Environment Protection Agency.

SI units
- Official website of National Institute of Standards and Technology (NIST).

Stockholm Convention on Persistent Organic Pollutants
- Official website of the Stockholm Convention.

sustainability
- Official website of the United Nations Environment Programme.

TCE
- Official website of the magazine *TCE*.

total dissolved solids
- Official website of the World Health Organization information on water sanitation.

unionfining processes
- Website of Honeywell UOP Company.

UNIPOL process
- Website for Dow Company's UNIPOL process.

UVOX process
- Website of UVOX Redox Systems, with page describing UV disinfection and ozone oxidation with UVOX Redox Systems.

Welsh process
- Official website of the Copper Development Association Inc., Education pages.

World Chemical Engineering Council (WCCE)
- Official website of the World Chemical Engineering Council.

Young, James
- Official website of the Museum of the Scottish Shale Oil Industry, page devoted to James Young biography.

Bibliography

Albright, L. F. (ed.) (2009) *Albright's Chemical Engineering Handbook*, London: CRC Press.

Austin, D. G. (1974) *Chemical Engineering Drawing Symbols*, George Godwin Ltd.

Comyns, Alan E. (1993) *Dictionary of Named Processes in Chemical Technology*, Oxford: Oxford University Press.

Coulson, J. M. and J. F. Richardson (1996–) *Chemical Engineering*, vols 1 to 6, Oxford and Boston: Butterworth-Heinemann.

Gary, J. H., G. E. Handwerk, and M. J. Kaiser (eds.) (2007) *Petroleum Refining: Technology and Economics* (5th edn), London: CRC Press.

Ireland, N. O. (1962) *Index to Scientists of the World from Ancient to Modern Times*, Boston: F. W. Faxon Co.

Jordan, D. and P. Smith (2009) *Mathematical Techniques* (4th edn), Oxford: Oxford University Press.

Levenspiel, O. (1999) *Chemical Reaction Engineering* (3rd edn), New York: Wiley.

McCabe, W. L., J. C. Smith, P. Harriott (2005) *Unit Operations of Chemical Engineering* (7th edn), Boston: McGraw-Hill.

Mather, Angus (2000) *Offshore Engineering: An Introduction* (second edn.) London: Witherby & Co. Ltd.

Muir, Hazel (ed.) (1994) *Larousse Dictionary of Scientists*, Larousse.

Porter, Roy (ed.) (1994) *Hutchinson Dictionary of Scientific Biography*, Abingdon: Helicon Publishing Ltd.

Seider, W. D., Seader, J. D., and Lewin, D. R. (1999) *Process Design Principles: Synthesis, Analysis, and Evaluation*, New York: John Wiley and Co.

Smith, J. M., H. C. Van Ness, and M. M. Abbott (eds.) (2004) *Introduction to Chemical Engineering Thermodynamics*, New York and London: McGraw-Hill.

Smith, R. (1995) *Chemical Process Design*, New York and London: McGraw-Hill.

Treybal, R. E. (1980) *Mass Transfer Operations*, New York and London: McGraw-Hill.

Welty, J. R., C. E. Wicks, and R. E. Wilson (eds.) (2008) *Fundamentals of Momentum, Heat and Mass Transfer* (5th edn), New York: John Wiley.

Oxford Paperback Reference

A Dictionary of Chemistry

Over 4,700 entries covering all aspects of chemistry, including physical chemistry and biochemistry.

'It should be in every classroom and library ... the reader is drawn inevitably from one entry to the next merely to satisfy curiosity.'

School Science Review

A Dictionary of Physics

Ranging from crystal defects to the solar system, 4,000 clear and concise entries cover all commonly encountered terms and concepts of physics.

A Dictionary of Biology

The perfect guide for those studying biology — with over 5,500 entries on key terms from biology, biochemistry, medicine, and palaeontology.

'lives up to its expectations; the entries are concise, but explanatory'

Biologist

'ideally suited to students of biology, at either secondary or university level, or as a general reference source for anyone with an interest in the life sciences'

Journal of Anatomy

Oxford Paperback Reference

A Dictionary of Psychology
Andrew M. Colman

Over 9,000 authoritative entries make up the most wide-ranging dictionary of psychology available.

'impressive ... certainly to be recommended'
Times Higher Education Supplement

'probably the best single-volume dictionary of its kind.'
Library Journal

A Dictionary of Economics
John Black, Nigar Hashimzade, and Gareth Myles

Fully up-to-date and jargon-free coverage of economics. Over 3,400 terms on all aspects of economic theory and practice.

'strongly recommended as a handy work of reference.'
Times Higher Education Supplement

A Dictionary of Law

An ideal source of legal terminology for systems based on English law. Over 4,200 clear and concise entries.

'The entries are clearly drafted and succinctly written ... Precision for the professional is combined with a layman's enlightenment.'
Times Literary Supplement

A Dictionary of Education
Susan Wallace

In over 1,250 clear and concise entries, this authoritative dictionary covers all aspects of education, including organizations, qualifications, key figures, major legislation, theory, and curriculum and assessment terminology.

Oxford Paperback Reference

Concise Medical Dictionary

Over 12,000 clear entries covering all the major medical and surgical
specialities make this one of our best-selling dictionaries.

'"No home should be without one" certainly applies to this splendid
medical dictionary'

Journal of the Institute of Health Education

'An extraordinary bargain'   *New Scientist*

A Dictionary of Nursing

Comprehensive coverage of the ever-expanding vocabulary of the
nursing professions. Features over 10,000 entries written by medical
and nursing specialists.

An A-Z of Medicinal Drugs

Over 4,000 entries cover the full range of over-the-counter and
prescription medicines available today. An ideal reference source for
both the patient and the medical professional.

A Dictionary of Dentistry
Robert Ireland

Over 4,000 succinct and authoritative entries define all the important
terms used in dentistry today. This is the ideal reference for all members
of the dental team.

A Dictionary of Forensic Science
Suzanne Bell

In over 1,300 entries, this new dictionary covers the key concepts within
Forensic Science and is a must-have for students and practitioners of
forensic science.

Oxford Paperback Reference

A Dictionary of Sociology
John Scott and Gordon Marshall

The most wide-ranging and authoritative dictionary of its kind.

'Readers and especially beginning readers of sociology can scarcely
do better ... there is no better single volume compilation for an up-to-
date, readable, and authoritative source of definitions, summaries and
references in contemporary Sociology.'
A. H. Halsey, *Emeritus Professor, Nuffield College,*
University of Oxford

The Concise Oxford Dictionary of Politics
Iain McLean and Alistair McMillan

The bestselling A-Z of politics with over 1,700 detailed entries.

'A first class work of reference ... probably the most complete as well as
the best work of its type available ... Every politics student should have
one'
Political Studies Association

A Dictionary of Environment and Conservation
Chris Park

An essential guide to all aspects of the environment and conservation
containing over 8,500 entries.

'from *aa* to *zygote*, choices are sound and definitions are unspun'
New Scientist

More History titles from OUP

The Oxford Companion to Black British History
David Dabydeen, John Gilmore, and Cecily Jones

The first reference book to explore the full history of black people in the British Isles from Roman times to the present day.

'From Haiti to Kingston, to Harlem, to Tottenham, the story of the African Diaspora is seldom told. This Companion will ensure that the history of Black Britain begins to take its rightful place in mainstream British consciousness.'

David Lammy, MP, former Minister for Culture

A Dictionary of Contemporary World History: From 1900 to the present day
Jan Palmowski

Discover the facts behind the headlines with this indispensable A-Z of world history during the last century.

'Concise, current information ... highly recommended'

Choice

The Concise Oxford Dictionary of Archaeology
Timothy Darvill

The most wide-ranging, up-to-date, and authoritative dictionary of its kind.

'Comprehensive, proportionate, and limpid'

Antiquity

Oxford Paperback Reference

The Concise Oxford Companion to English Literature
Dinah Birch and Katy Hooper

Based on the best-selling *Oxford Companion to English Literature*, this is an indispensable guide to all aspects of English literature.

Review of the parent volume
'the foremost work of reference in its field'

Literary Review

A Dictionary of Shakespeare
Stanley Wells

Compiled by one of the best-known international authorities on the playwright's works, this dictionary offers up-to-date information on all aspects of Shakespeare, both in his own time and in later ages.

The Oxford Dictionary of Literary Terms
Chris Baldick

A best-selling dictionary, covering all aspects of literature, this is an essential reference work for students of literature in any language.

A Dictionary of Critical Theory
Ian Buchanan

The invaluable multidisciplinary guide to theory, covering movements, theories, and events.

'an excellent gateway into critical theory'   *Literature and Theology*

Oxford Paperback Reference

A Dictionary of Marketing
Charles Doyle

Covers traditional marketing techniques and theories alongside the latest concepts in over 2,000 clear and authoritative entries.

'Flick to any page [for] a lecture's worth of well thought through information'
Dan Germain, Head of Creative, innocent ltd

A Dictionary of Media and Communication
Daniel Chandler and Rod Munday

This volume provides over 2,200 authoritative entries on terms used in media and communication, from concepts and theories to technical terms, across subject areas that include advertising, digital culture, journalism, new media, radio studies, and telecommunications.

'a wonderful volume that is much more than a simple dictionary'
Professor Joshua Meyrowitz, University of New Hampshire

A Dictionary of Film Studies
Annette Kuhn and Guy Westwell

Features terms covering all aspects of film studies in 500 detailed entries, from theory and history to technical terms and practices.